Inverse Problems with Applications in Science and Engineering

Inverse Problems with Applications in Science and Engineering

Daniel Lesnic
Leeds University, United Kingdom

CRC Press is an imprint of the
Taylor & Francis Group, an **informa** business
A CHAPMAN & HALL BOOK

First edition published 2022
by CRC Press
6000 Broken Sound Parkway NW, Suite 300, Boca Raton, FL 33487-2742

and by CRC Press
2 Park Square, Milton Park, Abingdon, Oxon, OX14 4RN

CRC Press is an imprint of Taylor & Francis Group, LLC

Library of Congress Cataloging-in-Publication Data

Names: Lesnic, D. (Daniel), author.
Title: Inverse problems with applications in science and engineering /
Daniel Lesnic, Leeds University, United Kingdom.
Description: First edition. | Boca Raton : Chapman & Hall/CRC Press, 2022.
| Includes bibliographical references and index.
Identifiers: LCCN 2021028848 (print) | LCCN 2021028849 (ebook) | ISBN 9780367001988 (hardback) | ISBN 9781032125381 (paperback) | ISBN 9780429400629 (ebook)
Subjects: LCSH: Inverse problems (Differential equations)
Classification: LCC QA378.5 .L47 2022 (print) | LCC QA378.5 (ebook) | DDC 515/.357--dc23
LC record available at https://lccn.loc.gov/2021028848
LC ebook record available at https://lccn.loc.gov/2021028849

ISBN: 978-0-367-00198-8 (hbk)
ISBN: 978-1-032-12538-1 (pbk)
ISBN: 978-0-429-40062-9 (ebk)

DOI: 10.1201/9780429400629

Typeset in CMR10
by KnowledgeWorks Global Ltd.

To my family
and my friends.

Contents

Preface xiii

About the Author xv

1 Introduction 1

1.1 Classification of Inverse Problems 2
1.2 Systems of Linear Algebraic Equations 4
1.3 Regularization Methods 5
1.3.1 Tikhonov regularization 5
1.3.2 Truncated singular value decomposition (TSVD) . . . 6

2 Inverse Boundary-Value Problems 7

2.1 Cauchy Problem for the Laplace Equation 7
2.1.1 BEM for Laplace's equation 9
2.1.2 Iterative methods 11
2.1.3 Non-local quasi-boundary value methods 15
2.2 Cauchy Problem for the Stokes System 18
2.2.1 Landweber-Fridman method (LFM) 20
2.2.2 Parameter-free methods 21
2.2.2.1 Conjugate gradient method (CGM) 22
2.2.2.2 Minimal error method (MEM) 23
2.2.3 Numerical results and discussion 24
2.3 Cauchy-type Problems for the Biharmonic Equation 26
2.4 Cauchy Problems for the Heat Equation 32
2.4.1 Inverse heat conduction problems 32
2.4.2 Inverse modelling in elution chromatography 34
2.5 Conclusions 44

3 Inverse Initial-Value Problems 45

3.1 Quasi-Reversibility Methods 46
3.2 Logarithmic Convexity Methods 47
3.3 Non-Local Initial-Value Methods 48

4 Space-Dependent Heat Sources 53

4.1 Space-Dependent Heat Source Identification 53
4.1.1 Mathematical analysis 55
4.1.2 Variational formulation 59
4.1.3 Conjugate gradient method (CGM) 62
4.1.4 Numerical results and discussion 64
4.2 Simultaneous Identification of the Space-Dependent Heat Source and Initial Temperature 66

5 Time-Dependent Heat Sources 71

5.1 Time-Dependent Heat Source Identification 71
5.2 Non-Local Variants 73
5.2.1 The first variant 73
5.2.1.1 BEM for the first variant 76
5.2.1.2 Numerical examples and discussion 82
5.2.2 The second variant 87
5.2.2.1 BEM for the second variant 94
5.2.3 The third variant 99
5.2.3.1 BEM for the third variant 101
5.2.4 The fourth variant 104
5.2.4.1 BEM for the fourth variant 105
5.2.5 Other non-local variants 106

6 Space- and Time-Dependent Sources 109

6.1 Additive Space- and Time-Dependent Heat Sources 109
6.1.1 BEM for additive source 111
6.1.2 Truncated singular value decomposition (TSVD) . . . 115
6.1.3 Tikhonov regularization 116
6.1.4 Numerical example 117
6.2 Additive Space- and Time-Dependent Heat Sources. Integral Observations 119
6.3 Multiplicative Space- and Time-Dependent Source 125
6.3.1 BEM for multiplicative source 127
6.3.2 Solution of the inverse problem 129
6.3.3 Numerical results and discussion 129

7 Inverse Wave Force Problems 133

7.1 Determination of a Space-Dependent Force in the One-Dimensional Wave Equation from Cauchy Data 133
7.1.1 BEM for solving the direct problem 135
7.1.2 Method for solving the inverse problem 139

7.1.3 Numerical example 141
7.2 Determination of the Force Function in the Multi-Dimensional Wave Equation from Cauchy Data 144
7.2.1 Numerical results 147
7.3 Determination of a Space-Dependent Force Function from Final or Time-Averaged Displacement Data 148
7.3.1 Direct and adjoint problems 149
7.3.2 Inverse problem 151
7.3.3 Variational formulation of the inverse problem 154
7.3.4 Landweber-Fridman method (LFM) 157
7.3.5 Numerical results and discussion 159

8 Reconstruction of Interfacial Coefficients **171**

8.1 Introduction 171
8.2 Mathematical Formulation 173
8.2.1 Ill-posedness of the inverse problem 174
8.3 Conjugate Gradient Method (CGM) 175
8.3.1 Sensitivity problem 176
8.3.2 Adjoint problem 176
8.3.3 Preconditioning 178
8.3.4 Stopping criterion 179
8.3.5 Algorithm 180
8.4 Numerical Results and Discussions 180

9 Identification of Constant Parameters in Diffusion **183**

9.1 Homogeneous and Isotropic Diffusion 183
9.2 A Two-Dimensional Tracer Dispersion Problem 187
9.3 Determination of Constant Thermal Properties 190
9.3.1 Finite homogeneous conductor 191
9.3.2 Semi-infinite homogeneous conductor 194

10 Time-Dependent Conductivity **199**

10.1 Identification of the Time-Dependent Conductivity 199
10.1.1 Satisfier function and Ritz-Galerkin method 206
10.1.2 Numerical example 209
10.2 Identification of the Time-Dependent Conductivity of an Inhomogeneous Diffusive Material 210
10.2.1 Numerical example 214
10.3 Finding the Time-Dependent Diffusion Coefficient from an Integral Observation 215
10.3.1 Mathematical analysis 216
10.3.2 Numerical solution 217

10.4 Determination of a Time-Dependent Thermal Conductivity and a Free Boundary 218
10.4.1 Numerical results and discussion 220

11 Space-Dependent Conductivity **223**

11.1 Reconstruction of a Permeability Function from Core Measurements and Pressure Data 223
11.1.1 Results 226
11.2 Discontinuous Anisotropic Conductivity 228
11.2.1 Size estimation of the inclusion 230
11.2.2 Boundary integral formulation 234
11.3 Reconstruction of an Orthotropic Conductivity 235
11.3.1 Mathematical formulation and analysis 236
11.3.2 Formulation of a simplified inverse problem 238
11.3.3 Numerical solution of the direct problem 239
11.3.4 Numerical solution of the inverse problem 240
11.3.5 Numerical results and discussion 241

12 Nonlinear Conductivity **243**

12.1 Determination of Nonlinear Thermal Properties 243
12.1.1 Linearization 244
12.1.2 Inverse formulation 245
12.1.3 Inverse analysis 246
12.1.4 Results and discussion 248
12.1.4.1 The infinite slab 248
12.1.4.2 The semi-infinite slab 251
12.1.4.3 The finite slab 253
12.2 Nonlinear and Heterogeneous Conductivity 255
12.2.1 An example 257
12.2.2 Extention to transient problems 260

13 Anti-Reflection Coatings **261**

13.1 Mathematical Model and Analysis 262
13.1.1 Uniqueness of solution 264
13.2 Numerical Implementation 266
13.3 Conclusions 274

14 Flexural Rigidity of a Beam **275**

14.1 Distributed Parameters in Beam-Type Systems 275
14.1.1 Identifiability results for the constant case 278
14.1.2 Identifiability results for the space-dependent case . . 278
14.1.3 Steady-state analysis 282

14.1.3.1 Clamped-clamped beams 284
14.1.3.2 Clamped-free beams 286
14.1.4 Conclusions 288
14.2 The Comparison Model Method 288
14.3 Determination of the Flexural Rigidity of a Beam from Limited Boundary Measurements 297
14.3.1 Quasi-solutions 299
14.3.2 Combined inverse problem formulations 300
14.3.3 An engineering approach 302
14.3.3.1 Numerical examples 302
14.3.4 Finite-difference approximation 304

Bibliography **309**

Index **341**

Preface

For the past 50–60 years, the study of inverse problems has gained an increasing interest due to both its beauty of mathematical model formulation and the extending domain of applications. More clearly, inverse modelling has become the main mechanism of formulating physical problems in engineering and industry. It is nowadays a commonly used approach adopted by technological translators/facilitators to enable a two-way flow of communication between academics and industry at meetings, sandpits and workshops. At these industrial mathematics study group-type meetings, typical questions, which are nothing else than inverse problems, arise, such as: (i) what are the initial and/or boundary conditions?; (ii) is the geometry known or unknown?; (iii) are all the physical property coefficients accessible to measurement?; etc.

Driven by desire dictated by both the advancement of applied mathematics and requirements for impact case studies, this book seeks to describe, analyze and discuss a feasible classification of inverse problems with respect to the missing versus additional information. In contrast to the direct approach in which the prescription of boundary and initial conditions, coefficients and geometry lead to a single, in general well-posed, mathematical model, inverse modelling generates a wide variety of formulations with respect to what data is missing and what additional information needs to be acquired/measured. Common sense seems to indicate that the more additional data is available, the better would be the chances of a reliable reconstruction of the missing data, but care should be taken to avoid redundant information, outliers and other artefacts, as what is important is how meaningful is the data and not necessarily how big it is. Because these inverse formulations partially interchange the input and output data, they lead to non-classical mathematical problems, which are usually ill-posed by violating either the existence, uniqueness or stability of a solution. In many aspects, direct and inverse problems are opposite to each other by interchanging the cause with the effect.

The book thoroughly examines some representative classes of inverse and improperly-posed problems for partial differential equations governing a wide range of physical phenomena. The natural practical applications arise in heat transfer, electrostatics, porous media, fluids, solids and acoustics. Intended readers of the book include postgraduate students and post-doctoral fellows from qualitatively-oriented fields of mathematical sciences and engineering, as well as scientists and researchers from academia and industry, where inverse problems naturally occur.

About the Author

Over the past 30 years, Daniel Lesnic (PhD 1995, Leeds University, Professor in Applied Mathematics since 2008) has worked on a diverse range of industrial and environmental mathematical inverse problems which have involved close and lasting contact with various scientists, engineers and experimentalists both nationally and internationally. Topics include heat and mass transfer, porous media, rock mechanics, elasticity, fluid flow, bio-heat conduction, mechanics of aerosols and acoustics, with particular applications in the oil, nuclear and glass industries, medicine, corrosion engineering, river pollution, thermal barriers and anti-reflective coatings, etc. Currently, he is Associate Editor of the *Journal of Inverse and Ill-Posed Problems, Inverse Problems in Science and Engineering* and the *IMA Journal of Applied Mathematics*, and he has published over 400 papers (http://www1.maths.leeds.ac.uk/applied/staff.dir/lesnic/papers.html) in applied mathematics, edited three conference proceedings and was a Guest Editor of several journal issues on inverse problems. He is a member of the London Mathematical Society (LMS) and the Eurasian Association for Inverse Problems. He has obtained several research grants and scholarships from the LMS, the Engineering and Physical Sciences Research Council, the Royal Society, the British Council, the Leverhulme Trust, the EU and industry. He has successfully supervised 12 postdocs and 20 PhD students in inverse problems who have followed successful careers in both academia and industry. Currently, he supervises two PhD students. He has been an invited speaker at international conferences, e.g. the 3rd and 5th International Symposium on Inverse Problems, Design and Optimization, 2010, Joao Pessoa, Brazil and 2019, Tianjin, China; Inverse Problems and Applications, 2013, Linkoping, Sweden, and the 12th and 13th International Conference on "Inverse and Ill-Posed Problems: Theory and Numerics", 2020 and 2021, Novosibirsk, Russia.

Chapter 1

Introduction

Many important real-world applications give rise to an **inverse problem**, which entails to recovering unknown causes (question) based on observation of their effects (answer). This is in contrast with the corresponding **direct problem**, whose solution involves predicting effects (answer) based on a complete description of their causes (question). Clearly, inverse problems are more difficult to address than their direct problem counterpart because, in general, they are **ill-posed**; i.e. the solution either does not exist, is not unique or it does not depend continuously on the input data. For example, uniqueness and non-uniqueness of solution of direct and inverse problems, respectively, can easily be understood by considering the question "What is the capital of France?" and the answer "Paris".

Nevertheless, *uniqueness* is relevant in situations where we search for the cause (which it always exists for exact data) that generates an *observed* effect; on the other hand, if we search for a cause giving rise to a *desired* effect then the non-uniqueness does not matter but the *existence* is meaningful, as it is important to be able to search for a desired effect [112]. Concerning the third criterion of ill-posedness, often inverse problems recast as an (operator) equation $AX = Y$, where $A : \mathcal{X} \to \mathcal{Y}$ is a linear compact operator (i.e. the image of any bounded set in $\mathcal{X}$ is pre-compact in $\mathcal{Y}$) between two infinite-dimensional separable Hilbert spaces $\mathcal{X}$ and $\mathcal{Y}$, (for e.g., indirect sensing measurements lead to a Fredholm integral equation of the first kind). Then, if further the range $R(A)$ is infinite-dimensional, it follows that $R(A)$ is not closed and hence, the inverse A^{-1} of A, if it exists, is unbounded [170], and the solution $X = A^{-1}Y$ will be unstable. The way the singular values $(\sigma_n)_{n\geq 1}$ of the linear and compact operator A decay to zero indicates the *degree of ill-posedness* of the inverse problem. If $\sigma_n = O(n^{-\alpha})$, the problem is classified as *mildly* ill-posed if $\alpha \in (0, 1]$ and *moderately* ill-posed if $\alpha \in (1, \infty)$, whilst if $\sigma_n = O(e^{-\alpha n})$ with $\alpha > 0$, the problem is classified as *severely* ill-posed. For example, the numerical differentiation of a noisy function and the inversion of two-dimensional Abel integral equation are only mildly ill-posed because they have $\sigma_n = O(n^{-1})$ and $\sigma_n = O(n^{-1/2})$, respectively. On the other hand, the backward heat conduction problem is severely ill-posed because $\sigma_n = O(e^{-n^2})$. The instability can be overcome by regularization methods [224, 375].

Applications of inverse problems include medical imaging, non-destructive testing, oil and gas exploration, cryptography and forensics, tomography and process control/monitoring, to mention only a few. For instance, a typical

DOI: 10.1201/9780429400629-1

example is concerned with inferring information about the interior of a specimen subjected to a scanning process from measured data at its surface.

The background needed to formulate and solve inverse problems [158, 230] consists of mainly mathematical analysis and algebra, and it uses techniques from functional analysis, spectral theory, ordinary/partial differential equations, integral equations, etc. In recent years, tools from differential geometry and stochastic analysis are becoming important. Moreover, to realize the solution to many applied problems in a useful/impact-full way, tools of numerical analysis and scientific computing are also needed. To summarize, a complete treatment would have to include the following milestones:

1. **Mathematical Modelling** (inverse problem formulation including governing partial differential equations (PDEs), geometry, initial and boundary conditions, sources and coefficients, additional information that is available or needs to be supplied).
2. **Mathematical Analysis** (existence, uniqueness, stability).
3. **Numerical Analysis** (convergence, regularization, error estimates).
4. **Computational Analysis** (numerical implementation in Python, Matlab (including toolbox routines), Fortran (including Numerical Algorithms (NAG) routines), etc.).
5. **Experimental Analysis** (inversion of raw experimental data supplied by laboratory or industry).
6. **Design** (optimal placement of sensors and duration of measurements, suggestions for radical changes in equipment and installation).

Different expertize, skills and background are required to completely accomplish all the above tasks, making inverse modelling highly interdisciplinary. Research in inverse problems is currently very active and, at present, there exist various series of international conferences, e.g. *International Conference on Inverse Problems in Engineering: Theory and Practice*; *Applied Inverse Problems*; *Inverse Problems: Modelling and Simulation*, etc. and societies and networks, e.g., *Eurasian Association for Inverse Problems*; *Inverse Problems International Association*; *IPNet: Inverse Problems Network*, etc.

1.1 Classification of Inverse Problems

Simulating a real-world application inherently involves a process of mathematical modelling given by a governing equation (or a system):

$$\mathbb{L}(c; u) = F \quad \text{in } \Omega_T := \Omega \times (0, T) \tag{1.1}$$

and some initial/boundary/interior conditions

$$\mathbb{L}_0(c_0; u) = F_0 \quad \text{on } D_T \subset \overline{\Omega}_T, \tag{1.2}$$

where $\mathbb{L}$ is an operator (differential, integral), c and c_0 represent lists of physical property coefficients, F and F_0 contain (external) excitations, boundary and initial conditions, $0 < T \leq \infty$ is a final time of interest, $\Omega \subset \mathbb{R}^d$, $d = 1, 2, 3$, is the space solution domain (with sufficiently smooth boundary $\partial\Omega$), D_T is a subset of $\overline{\Omega}_T = \overline{\Omega} \times [0, T]$, which usually includes the lateral boundary $\partial\Omega \times (0, T)$, and u is the underlying dependent variable.

Depending on the field of aplication, the governing equation (1.1) requires specifying the operator $\mathbb{L}$. Typically, in potential flow or steady-state heat conduction the governing operator $\mathbb{L}$ is the Laplacian, in acoustics $\mathbb{L}$ is the Helmholtz operator, in unsteady-state heat conduction $\mathbb{L}$ is the heat/diffusion operator and in wave propagation phenomena $\mathbb{L}$ is the wave operator. Equation (1.1) may also include systems of PDEs, e.g. the Lamé system in elasticity, the Navier-Stokes system in viscous fluids, the Maxwell system in electromagnetism, or a combination of the above, e.g. coupled systems in thermoelasticity or thermoacoustics. The operator $\mathbb{L}$ in (1.1) often involves the material properties c at its leading part, e.g. a conductivity or a capacity, as well as lower-order terms modelling convection and reaction, whilst the sources, forces or loads acting on the system are encapsulated in the free term F. In addition to (1.1), one also needs to supply through F_0 in (1.2) appropriate boundary conditions, as well as initial conditions in case of transient phenomena, in order to make the mathematical model (1.1) and (1.2) well-posed. However, in real-world applications such specifications are not always possible in the presence of a hostile envoironment characterized by, say, high temperatures/high pressures. In such situations, one is faced to either perform an inaccurate and possibly expensive measurement of the physical quantities of interest or adopt a more efficient approach based on formulating, analysing and solving a novel, but challenging, inverse problem. In summary, inverse problems usually falls into one (or combination of more) of the following categories:

(i) *inverse boundary-value problems* (e.g. the Cauchy problems considered in Chapter 2) modelling situations in which the specimen under investigation is placed in an inaccessible or highly unfriendly ambient, e.g. re-entry vehicles into an atmosphere, where boundary data is not available;

(ii) *inverse initial-value problems* (e.g. the backward heat conduction problem considered in Chapter 3) in which already past conditions, e.g. initial temperature, displacement or velocity, are not available;

(iii) *inverse source/force problems* (considered in Chapters 4–7), concerned with identifying distributed sources of pollution, heat generation or wave excitations;

(iv) *inverse coefficient problems* (considered in Chapters 8–14), in which the physical properties c and/or c_0 of the material under testing, embedded into the structure of the operator $\mathbb{L}$ in (1.1) and/or the operator $\mathbb{L}_0$ in (1.2), are unknown;

(v) *inverse geometry problems*, not considered herein, related to obstacle identification, e.g. crack detection, boundary corrosion damage, inverse scattering and tomography.

1.2 Systems of Linear Algebraic Equations

In this section, for simplicity, the methods are explained at the practical level of computational discretized implementation. Upon discretization, a linear inverse and ill-posed problem is reduced to solving an ill-conditioned system of $M \times N$ linear algebraic equations

$$\mathbb{A}\underline{X} = \underline{Y}. \tag{1.3}$$

Then, a naive approach based on the straightforward inversion $\underline{X} = \mathbb{A}^{-1}\underline{Y}$ (in case of determined systems when $M = N$ and $\mathbb{A}^{-1}$ denotes the inverse of the matrix $\mathbb{A}$, if it exists), a least-squares solution $\underline{X} = (\mathbb{A}^{\mathrm{T}}\mathbb{A})^{-1}\mathbb{A}^{\mathrm{T}}\underline{Y}$ (in case of over-determined systems when $M > N$ and $\mathbb{A}^{\mathrm{T}}$ denotes the transpose of the matrix A), or the Moore-Penrose minimal-norm solution $\underline{X} = \mathbb{A}^{+}\underline{Y}$ (in case of under-determined systems when $M < N$ and $\mathbb{A}^{+} = \lim_{\lambda \searrow 0}(\mathbb{A}^{\mathrm{T}}\mathbb{A} + \lambda I_N)^{-1}\mathbb{A}^{\mathrm{T}}$ denotes the pseudo-inverse of the matrix $\mathbb{A}$ and I_N is the identity matrix of order N) would lead to an unstable solution due to the ill-conditioning of the matrix $\mathbb{A}$ characterized by a large *condition number* $\mathrm{cond}(\mathbb{A})$ (herein defined as the ratio between the largest and the smallest singular value of $\mathbb{A}$). This instability is manifested by unbounded and highly oscillatory solutions when the input data $\mathbb{A}$ and/or $\underline{Y}$ are contaminated with noise $\delta \geq 0$ and/or $\epsilon \geq 0$,

$$\|\mathbb{A}^{\delta} - \mathbb{A}\| \leq \delta, \quad \|\underline{Y}^{\epsilon} - \underline{Y}\| \leq \epsilon, \tag{1.4}$$

which is inherently present in any practical measurement. In this case, instead of (1.3), one would have to solve the perturbed system

$$\mathbb{A}^{\delta}\underline{X}^{\delta,\epsilon} = \underline{Y}^{\epsilon}. \tag{1.5}$$

Solving (1.3) and (1.5) yield that the relative error is estimated by [94]

$$\frac{\|\underline{X}^{\delta,\epsilon} - \underline{X}\|}{\|\underline{X}\|} \leq \mathrm{cond}(\mathbb{A}) \left(\frac{\delta}{\|\mathbb{A}\|} + \frac{\epsilon}{\|\underline{Y}\|} \right). \tag{1.6}$$

This estimate shows how the instability increases with the condition number of the matrix $\mathbb{A}$ in (1.3).

As previosly mentioned, for overdetermined systems of equations with $M \geq N$, the linear *least-squares* solution of (1.3), based on minimizing

$$\mathrm{LS}(\underline{X}) := \|\mathbb{A}\underline{X} - \underline{Y}\|^2, \tag{1.7}$$

given by

$$\underline{X} = (\mathbb{A}^{\mathrm{T}}\mathbb{A})^{-1}\mathbb{A}^{\mathrm{T}}\underline{Y}, \tag{1.8}$$

will be unstable, manifesting highly unbounded oscillations, as the dimensions of the system of equations (1.3) increase. At the same time, the (untruncated) *singular value decomposition* (SVD) solution given by

$$\underline{X} = \sum_{i=1}^{N} \frac{1}{\sigma_i}(\underline{w}_i^{\mathrm{T}} \cdot \underline{Y})\underline{v}_i, \tag{1.9}$$

where $\mathbb{A} = \mathbb{W}\mathbb{S}\mathbb{V}^{\mathrm{T}}$ with $\mathbb{W} = \mathrm{col}[\underline{w}_1, ..., \underline{w}_M] \in \mathbb{R}^{M\times M}$ and $\mathbb{V} = \mathrm{col}[\underline{v}_1, ..., \underline{v}_M] \in \mathbb{R}^{N\times N}$ orthogonal matrices, and $\mathbb{S} = \mathrm{diag}[\sigma_1, ..., \sigma_N] \in \mathbb{R}^{M\times N}$ with singular values $\sigma_1 \geq ... \geq \sigma_N > 0$, coincides with the unstable least-squares solution (1.8), [138].

1.3 Regularization Methods

Regularization is always necessary to be employed when solving ill-posed problems, especially when the data $\underline{Y}$ is contaminated with noise.

1.3.1 Tikhonov regularization

To stabilize the unrealistic solutions (1.8) (or (1.9)), one can penalize the least-squares functional (1.7) by adding a regularization term to it. This may impose extra smoothness on solution, though the level of regularity may not be known in advance. Imposing, for example, smoothness of order C^{k} with $k \in \mathbb{N}$, entails to minimizing the *Tikhonov regularization* functional [377],

$$T_\lambda(\underline{X}) := \|\mathbb{A}\underline{X} - \underline{Y}^\epsilon\|^2 + \lambda\|R_{\mathrm{k}}\underline{X}\|^2, \tag{1.10}$$

where $\lambda > 0$ is a regularization parameter to be prescribed, and R_{k} is a regularizing derivative matrix given by

$$R_0 = I_N \in \mathcal{M}_{N\times N}, \tag{1.11}$$

$$R_1 = \begin{pmatrix} 1 & -1 & 0 & 0 & ... & 0 \\ 0 & 1 & -1 & 0 & ... & 0 \\ ... & ... & ... & ... & ... & ... \\ 0 & 0 & ... & 0 & 1 & -1 \end{pmatrix} \in \mathcal{M}_{(N-1)\times N}, \tag{1.12}$$

$$R_2 = \begin{pmatrix} 1 & -2 & 1 & 0 & 0 & ... & 0 \\ 0 & 1 & -2 & 1 & 0 & ... & 0 \\ ... & ... & ... & ... & ... & ... & ... \\ 0 & 0 & ... & 0 & 1 & -2 & 1 \end{pmatrix} \in \mathcal{M}_{(N-2)\times N}, \tag{1.13}$$

etc. Minimizing (1.10) yields explicitly the solution

$$\underline{X}_\lambda = (\mathbb{A}^{\mathrm{T}}\mathbb{A} + \lambda R_k^{\mathrm{T}} R_k)^{-1}\mathbb{A}^{\mathrm{T}}\underline{Y}^\epsilon. \tag{1.14}$$

The regularization parameter λ can be chosen according to certain criteria:
• *the discrepancy principle criterion* [304, 305] selects $\lambda = \lambda(\epsilon) > 0$ for which

$$\|\mathbb{A}\underline{X}_{\lambda(\epsilon)} - \underline{Y}^\epsilon\| \approx \epsilon. \tag{1.15}$$

• *the generalized cross-validation (GSV) criterion* [126] minimizes over $\lambda > 0$, the GCV functional

$$GCV(\lambda) = \frac{\|\mathbb{A}\underline{X}_\lambda - \underline{Y}^\epsilon\|^2}{[\mathrm{Trace}(I - \mathbb{A}(\mathbb{A}^{\mathrm{T}}\mathbb{A} + \lambda R_k^{\mathrm{T}} R_k)^{-1}\mathbb{A}^{\mathrm{T}})]^2}. \tag{1.16}$$

• *the L-curve criterion* [135] is based on plotting, on a log-log graph, the residual norm $\|\mathbb{A}\underline{X}_\lambda - \underline{Y}^\epsilon\|$ versus the solution norm $\|R_k\underline{X}_\lambda\|$, for various values of $\lambda > 0$. If the resulting curve has an L-shape, then λ could be chosen at the corner of this L-curve (or more clearly at the point of its maximum curvature). Since this choice does not depend on the amount of noise, an L-curve is not always obtained.

1.3.2 Truncated singular value decomposition (TSVD)

Another way to stabilize the ill-conditioned system of linear algebraic equations (1.3) is to truncate the SVD sum in (1.9) at a suitable threshold $N_{trunc} \in \{1, ..., N\}$ given by either of the criteria mentioned above, with the truncation number playing the role of the regularization parameter.

The truncated SVD and the Tikhonov's regularization solutions can be written in the filtered form

$$\underline{X} = \sum_{i=1}^{N} \frac{f_i}{\sigma_i}(\underline{w}_i^{\mathrm{T}} \cdot \underline{Y}^\epsilon)\underline{v}_i, \tag{1.17}$$

where the filter factors $f_i \in [0, 1]$ for $i = \overline{1, N}$ are given by

$$f_i = \begin{cases} 1 & \text{for } i = \overline{1, N_{trunc}}, \\ 0 & \text{for } i = \overline{(N_{trunc}+1), N}, \end{cases} \quad \text{and} \quad f_i = \frac{\sigma_i^2}{\sigma_i^2 + \lambda}, \tag{1.18}$$

respectively. From these equations, it can be seen that the role of f_i's is to filter out the small singular values, while leaving the SVD components corresponding to large singular values almost unaffected [136].

Chapter 2

Inverse Boundary-Value Problems

This chapter considers inverse problems for partial differential equations (Laplace's, Stokes, biharmonic and heat) in which boundary conditions on a portion Γ_1 of the boundary $\partial\Omega$ of the solution domain Ω are under-prescribed. This situation occurs in physical problems where parts of the boundary of a specimen are either inaccessible to measurement, e.g. buried or hidden, or they are in contact with a hostile environment, e.g. high temperatures/high pressures. The missing boundary information on Γ_1 is compensated for by over-prescribed boundary conditions (Cauchy data) on an accessible and friendly sub-portion Γ_0 of $\partial\Omega\backslash\Gamma_1$. Three typical possibilities that usually can occur are depicted in Figure 2.1 for the Laplace equation. Because only boundary information is considered, these Cauchy inverse problems are typically associated with non-destructive testing of materials. On the other hand, nowadays, intelligent design of materials places sensors inside the specimen from the onset of its installation in a particular complex setting, such that interior/internal measurements can be made available, without the need of intrusively destroying the specimen later on. Of course, the optimal placement of such sensors is very important for experimental design and can be accomplished by performing a sensitivity analysis prior to inversion [16].

2.1 Cauchy Problem for the Laplace Equation

As someone said, 'Everybody has a Cauchy problem in his/her life'. This section addresses the simplest Cauchy problem for the Laplace equation, which is closely related to the principle of analytic continuation. Consider therefore the $d-$dimensional Laplace equation satisfied by a harmonic function u in a bounded domain $\Omega \subset \mathbb{R}^d$ given by

$$\nabla^2 u = 0 \quad \text{in } \Omega, \tag{2.1}$$

subject to the Dirichlet and Neumann boundary conditions, i.e., Cauchy data,

$$u = f, \quad \partial_n u = g \quad \text{on } \Gamma_0, \tag{2.2}$$

DOI: 10.1201/9780429400629-2

where $\partial_n u = \nabla u \cdot \underline{n}$, $\underline{n}$ is the outward unit normal to the boundary $\partial\Omega$ (assumed to be piecewise smooth), and f and g are given functions on $\Gamma_0 \subset \partial\Omega$, which have to be compatible for a solution to the inverse boundary-value problem (2.1) and (2.2) to exist. On the other hand, if the Cauchy data f and g are analytical functions and the boundary $\partial\Omega$ is sufficiently smooth then the Cauchy-Kovalevskaya theorem ensures the local existence of solution. The uniqueness readily follows from the Holmgren uniqueness theorem. Although the problem (2.1) and (2.2) may be uniquely solvable, the solution does not depend continuously on the Cauchy data (f, g) and is severely ill-posed [31].

In the next subsection, the Cauchy problem (2.1) and (2.2) is reformulated in terms of boundary integral equations.

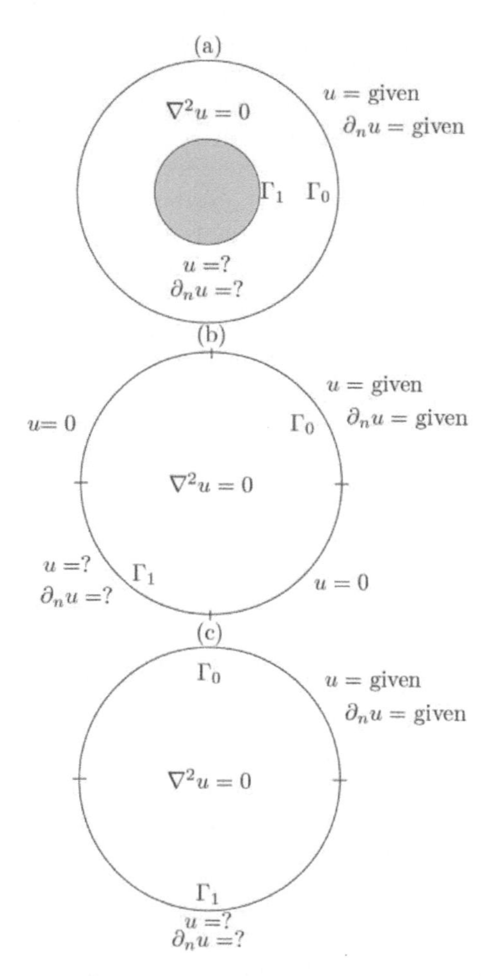

FIGURE 2.1: Sketch of Cauchy problems.

2.1.1 BEM for Laplace's equation

For the Laplace's equation (2.1), the fundamental solution is given by

$$G(\underline{x};\underline{\xi}) = \begin{cases} -\frac{1}{2\pi}\ln(r) & \text{in } d=2-\text{dimensions}, \\ \frac{1}{4\pi r} & \text{in } d=3-\text{dimensions}, \end{cases} \tag{2.3}$$

where $r := |\underline{x} - \underline{\xi}|$. Then, by employing Green's formula it follows that (2.1) is equivalent to the following integral equation, see e.g., [210],

$$\mathrm{c}(\underline{x})u(\underline{x}) = \int_{\partial\Omega}\left[G(\underline{x};\underline{\xi})\frac{\partial u}{\partial n}(\underline{\xi}) - u(\underline{\xi})\frac{\partial G}{n(\underline{\xi})}(\underline{x};\underline{\xi})\right]dS(\underline{\xi}), \quad \underline{x} \in \overline{\Omega}, \tag{2.4}$$

where $\mathrm{c}(\underline{x}) = 1$ if $\underline{x} \in \Omega$ and $\mathrm{c}(x) = \alpha_{\mathrm{t}}(x)/(2\pi)$ if $\underline{x} \in \partial\Omega$, where $\alpha_{\mathrm{t}}(\underline{x})$ is the angle between the tangents at $\partial\Omega$ on either sides of the boundary point $\underline{x} \in \partial\Omega$, (in particular $\mathrm{c}(\underline{x}) = 1/2$ if $\underline{x}$ belongs to a smooth portion of the boundary). The integral form (2.4) is more natural and convenient than the differential form (2.1) because it allows an elegant enforcement of the boundary conditions and the discretization of the boundary only, which is the essence of the boundary element method (BEM). By discretizing only the boundary $\partial\Omega$ of the solution domain instead of the full domain Ω, the BEM reduces the dimensionality of the problem by one. Consequently, in $d = 3$-dimensions, one discretizes surfaces instead of volumes, and in $d = 2$-dimensions, line contours instead of areas. In the BEM, one is solving first for the missing unspecified boundary data by collocating (2.4) at $\underline{x} \in \partial\Omega$. Once all the boundary data has been obtained, then (2.4) can be applied at any $\underline{x} \in \Omega$ to provide explicitly the solution u at any interior point inside the domain Ω, unlike domain discretization methods, e.g. the finite-difference method (FDM) and the finite element method (FEM), which require further interpolation to obtain the interior solution at other points than the mesh nodes. The BEM also deals easily with exterior unbounded domains by incorporating the infinity condition in the behavior of the fundamental solution, such that it does not require any artificial boundary be considered at large distances, as with the FDM or the FEM. The main drawback of the BEM is that it depends on whether a fundamental solution is available explicitly, which restricts its ability to mainly linear PDEs.

An immediate extension to anisotropic material is obtained by replacing the isotropic Laplace's equation (2.1) with the Laplace-Beltrami equation

$$\sum_{i,j=1}^{d} K_{ij}\frac{\partial^2 u}{\partial x_i \partial x_j} = 0, \quad \underline{x} = (x_i)_{i=\overline{1,d}} \in \Omega, \tag{2.5}$$

where $\mathbb{K} = (K_{ij})_{i,j=\overline{1,d}}$ is a symmetric and positive definite (constant) matrix. Then, the integral equation (2.4) still holds with G replaced by [72],

$$G_{\mathbb{K}}(\underline{x};\underline{\xi}) = \begin{cases} -\frac{\det(\mathbb{K}^{-1})}{2\pi}\ln\sqrt{(\underline{x}-\underline{\xi})^{\mathrm{T}}\mathbb{K}^{-1}(\underline{x}-\underline{\xi})} & \text{in } d=2-\text{dimensions}, \\ \frac{\det(\mathbb{K}^{-1})}{4\pi\sqrt{(\underline{x}-\underline{\xi})^{\mathrm{T}}\mathbb{K}^{-1}(\underline{x}-\underline{\xi})}} & \text{in } d=3-\text{dimensions}. \end{cases}$$

Other possible quasi-extensions include:

(i) nonlinear material, for which the use of Kirchhoff's transformation $\Psi(u) := \int k(u)du$, where $k(u)$ denotes the positive conductivity, changes the steady-state diffusion equation $\nabla \cdot (k(u)\nabla u) = 0$ into the Laplace equation for Ψ. Dirichlet and Neumann boundary conditions remain the same type, but linear Robin boundary conditions transform into nonlinear ones;

(ii) exponentially-graded materials governed by the equation

$$\nabla \cdot (\mathbb{K}(\underline{x})\nabla u) = 0, \quad \text{where } \mathbb{K}(\underline{x}) = \tilde{\mathbb{K}} \exp(2\underline{\beta} \cdot \underline{x}), \tag{2.6}$$

where $\underline{\beta}$ and $\tilde{\mathbb{K}}$ are constant material properties, for which the transformation $v = u \, \exp(\underline{\beta} \cdot \underline{x})$ changes (2.6) into

$$\sum_{i,j=1}^{d} \tilde{K}_{ij} \left(\frac{\partial^2 v}{\partial x_i \partial x_j} - \beta_i \beta_j v \right) = 0,$$

which possesses a fundamental solution similar to that for the anisotropic modified Helmholtz equation [33];

(iii) materials with square-root harmonic conductivity, i.e. satisfying $\nabla^2 \sqrt{\kappa} = 0$, for which the transformation $v = \sqrt{\kappa} u$ changes the steady-state equation $\nabla \cdot (\kappa(\underline{x})\nabla u) = 0$ into the Laplace's equation for v, [79].

Application of (2.4) at the boundary recasts the Cauchy problem (2.1) and (2.2) in the following system of two linear boundary integral equations:

$$\begin{aligned}
&\int_{\Gamma_1} \left[G(\underline{x};\underline{\xi})\tilde{g}(\underline{\xi}) - \tilde{f}(\underline{\xi}) \frac{\partial G}{n(\underline{\xi})}(\underline{x};\underline{\xi}) \right] d\Gamma_1(\underline{\xi}) \\
&= \mathrm{c}(\underline{x}) f(\underline{x}) - \int_{\Gamma_0} \left[G(\underline{x};\underline{\xi}) g(\underline{\xi}) - f(\underline{\xi}) \frac{\partial G}{n(\underline{\xi})}(\underline{x};\underline{\xi}) \right] d\Gamma_0(\underline{\xi}), \quad \underline{x} \in \Gamma_0, \\
&\int_{\Gamma_1} \left[G(\underline{x};\underline{\xi})\tilde{g}(\underline{\xi}) - \tilde{f}(\underline{\xi}) \frac{\partial G}{n(\underline{\xi})}(\underline{x};\underline{\xi}) \right] d\Gamma_1(\underline{\xi}) - \mathrm{c}(\underline{x})\tilde{f}(\underline{x}) \\
&= - \int_{\Gamma_0} \left[G(\underline{x};\underline{\xi}) g(\underline{\xi}) - f(\underline{\xi}) \frac{\partial G}{n(\underline{\xi})}(\underline{x};\underline{\xi}) \right] d\Gamma_0(\underline{\xi}), \quad \underline{x} \in \Gamma_1,
\end{aligned}$$

for the unknown Cauchy data $(\tilde{f}, \tilde{g})$ on $\Gamma_1 := \partial\Omega \backslash \Gamma_0$, expressed as

$$u = \tilde{f}, \quad \partial_n u = \tilde{g} \quad \text{on } \Gamma_1. \tag{2.7}$$

Discretization of the above system using the BEM with constant uniform boundary elements (M_0 on Γ_0 and M_1 on Γ_1), [193], reduces to solving a linear, but ill-conditioned, system of $(M_0 + M_1) =: M$ linear algebraic equations with $2M_1 =: N$ unknowns $(\underline{\tilde{f}}, \underline{\tilde{g}})^{\mathrm{T}}$, which can be written as in (1.3).

2.1.2 Iterative methods

The regularization methods previously outlined in section 1.3 do not preserve the structure of the operator, and additionally, the choice of the regularization matrix R_k in (1.10) requires some *a priori* information about the smoothness of solution. However, these difficulties are not encountered when using iterative regularization methods. In particular, the Landweber-Fridman or the conjugate gradient methods [113] can be applied, but in the L^2-setting they require that the boundary portions Γ_0 and Γ_1 are disjoint, in the sense that $\overline{\Gamma}_0 \cap \overline{\Gamma}_1 = \emptyset$, as in Figure 2.1(a). This restriction is not encountered in the setting of the alternating algorithm [240], which is based on solving iteratively a sequence of well-posed mixed problems constructed by alternating the Neumann and Dirichlet data $g \in H^{-1/2}(\Gamma_0)$ and $f \in H^{1/2}(\Gamma_0)$, as follows:

Step 1. Set $k = 0$ and choose an arbitrary initial guess $\tilde{f}_0 \in H^{1/2}(\Gamma_1)$ for the unknown Dirichlet data on Γ_1.

Step 2. Solve the mixed well-posed problem

$$\begin{cases} \nabla^2 u_{2k} = 0 & \text{in } \Omega, \\ \partial_n u_{2k}|_{\Gamma_0} = g, & u_{2k}|_{\Gamma_1} = \tilde{f}_k, \end{cases}$$

to determine $u_{2k} \in H^1(\Omega)$ and $\tilde{g}_k = \partial_n u_{2k}|_{\Gamma_1}$.

Step 3. Solve the mixed well-posed problem

$$\begin{cases} \nabla^2 u_{2k+1} = 0 & \text{in } \Omega, \\ u_{2k+1}|_{\Gamma_0} = f, & \partial_n u_{2k+1}|_{\Gamma_1} = \tilde{g}_k, \end{cases}$$

to determine $u_{2k+1} \in H^1(\Omega)$ and $\tilde{f}_{k+1} = u_{2k+1}|_{\Gamma_1}$.

Step 4. Set $k \to k+1$ and go to step 2 until a prescribed stopping criterion is satisfied.

Remark 2.1. (i) The sequences $(u_{2k})_{k\in\mathbb{N}}$ and $(u_{2k+1})_{k\in\mathbb{N}}$ both converge in $H^1(\Omega)$ to the solution u of the Cauchy problem (2.1) and (2.2), if it exists [240]. Furthermore, the alternating algorithm has a regularizing character, in the sense that for noisy Cauchy data it produces a stable solution, if stopped according to some appropriate discrepancy-like criterion. The same conclusions are obtained if one decides to start with an initial guess for the Neumann data on Γ_1 and modify accordingly steps 2 and 3 such that mixed problems are solved. The algorithm does not converge if, in steps 2 and 3, the mixed problems are replaced by Dirichlet or Neumann problems.

(ii) The algorithm's convergence can be accelerated through the use of a relaxation factor ζ (bounded in some suitable interval) in the marching condition $\tilde{f}_{k+1} = \zeta u_{2k+1}|_{\Gamma_1} + (1-\zeta)\tilde{f}_k$ at step 3, where ζ can be constant or variable along the curve Γ_1, [297], or variable at each iteration [222, 223].

(iii) The alternating algorithm converges for other linear, elliptic, positive-definite and self-adjoint operators such as the modified Helmholtz equation [289], the Lame system in elasticity [240, 290], or the Stokes system in slow viscous incompressible flows [23]. Recently, the algorithm was modified and made applicable to the Helmholtz equation as well [36]. Cauchy problems for the fourth-order biharmonic equation in Stokes fluids [264], or Kirchhoff plates [285], are also amenable with the alternating algorithm.

(iv) The converse is also true in the sense that the Cauchy problem (2.1) and (2.2) is solvable, i.e. the Cauchy data f and g are compatible if and only if the regularizing alternating algorithm is convergent. Below, this point connecting the convergence of a regularizing numerical algorithm to the theoretical solvability of the ill-posed Cauchy problem [99] will be illustrated.

(v) The alternating algorithm can easily be adapted to the situation when the boundary $\partial\Omega$ consists of 3 disjoint parts $\partial\Omega = \Gamma_0 \cup \Gamma_2 \cup \Gamma_1$, see Figure 2.1(b), where on Γ_2 some boundary condition (say, Dirichlet or Neumann) is prescribed, which remains unchanged throughout the iterations.

For illustration, consider $\Omega = (0,1) \times (0,1/2)$ and the boundary portions $\Gamma_0 = \{(x,0)|x \in (0,1)\}$, $\Gamma_1 = \{(x,1/2)|x \in (0,1)\}$ and $\Gamma_2 = \partial\Omega\backslash(\Gamma_0 \cup \Gamma_1)$. The Cauchy problem (2.1), (2.2) and with Dirichlet boundary condition prescribed on Γ_2 is solved iteratively (starting with the initial guess $\tilde{f}_0 = 0$ on Γ_1), as described in steps 1–4 above, using the BEM (with constant uniform boundary elements), to provide simultaneously the missing data on Γ_1, the unspecified Neumann data on Γ_2, and the solution u inside the solution domain Ω. Both consistent and inconsistent Cauchy data (2.2) are analyzed.

Example 2.1 (consistent Cauchy data)
Consider first an example for which it is known *a priori* that the solution to the Cauchy problem exists. Consider the analytical solution given by $u(x,y) = -\sin(\pi x)\sinh(\pi y)$ for $(x,y) \in \Omega$, from which the input Cauchy data (2.2) is generated. Figure 2.2 show the numerical results (obtained using the BEM with $M = 320$ uniformly distributed boundary elements) for $\tilde{f}_k$ and $\tilde{g}_k$ on Γ_1, as functions of the number of iterations, in comparison with their corresponding analytical values $\tilde{f}$ and $\tilde{g}$, respectively. It can be seen that as the number of iterations k increases the numerical solutions approach within 1% error their corresponding exact values in about 100 iterations. Although the solution remains stationary for a large number of iterations, at some stage the iteration process ought to be stopped to guarantee the stability of solution.

Example 2.2 (inconsistent Cauchy data)
Consider now the Cauchy data (2.2) given by

$$u(x,0) = f(x) = 0,$$

$$-\partial_y u(x,0) = g(x) = \begin{cases} 10 - 100|x-0.5| & \text{if } |x-0.5| \leq 0.1, \\ 0 & \text{otherwise,} \end{cases} \tag{2.8}$$

on Γ_0, and $u = 0$ on Γ_2. From the Schwartz reflection principle, the Neumann data g must be analytical function for the solution to the Cauchy problem to exist. However, as the data in (2.8) is not smooth, the respective Cauchy problem does not have a classical solution. A BEM with 240 boundary elements uniformly distributed over Γ_1 and Γ_2 is employed, whilst on the boundary Γ_0, to accurately approximate the peak of the function $g(x)$ located at $x = 0.5$, 10, 60 and 10 boundary elements uniformly distributed over the segments $[0, 9/20]$, $[9/20, 11/20]$ and $[11/20, 1]$, respectively, are used.

Figure 2.3 show the numerical solution for $u_{200}|_{\Gamma_0}$ and $\partial_n u_{201}|_{\Gamma_0}$ after $k = 100$ iterations, in comparison with the input Cauchy data (2.8). It can be seen that there is a significance discrepancy between the input and the iterate solution, pointing out that, as expected, the solution may not exist.

The conclusions obtained from the above two examples can be further justified by plotting in Figure 2.4 the reciprocity gap (see also (2.4))

$$\text{RG}(k) := \Bigg\| -0.5u_k(\underline{x}) + \int_{\partial\Omega}\left[G(\underline{x};\underline{\xi})\frac{\partial u_k}{\partial n}(\underline{\xi}) - u_k(\underline{\xi})\frac{\partial G}{n(\underline{\xi})}(\underline{x};\underline{\xi})\right] dS(\underline{\xi})\Bigg\|_{L^2(\partial\Omega)}. \tag{2.9}$$

From this figure it can be seen that for the consistent problem of Example 2.1, the reciprocity gap RG(k) decreases rapidly to zero, as $k \in [0, 100]$ increases, showing that the numerically obtained solution satisfies the Laplace equation (2.1) and the corresponding mixed boundary conditions, and thus the true solution is well-approximated. However, for the inconsistent problem of Example 2.2, as k increases, RG(k) settles to a constant value of about 1.33, which is much bigger than zero, showing the the numerically obtained

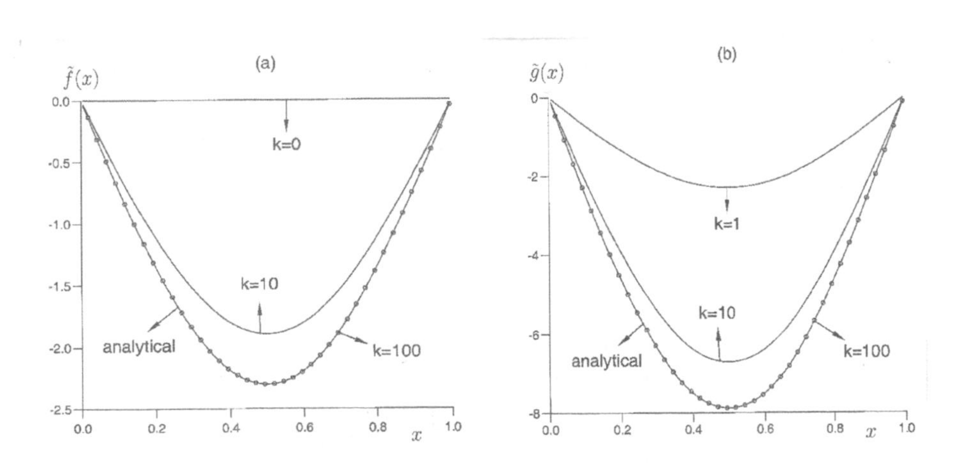

FIGURE 2.2: The numerical results for (a) $\tilde{f}_k$ and (b) $\tilde{g}_k$ on Γ_1, as functions of the number of iterations k, in comparison with their corresponding analytical values $\tilde{f}$ and $\tilde{g}$, respectively, for Example 2.1, [254].

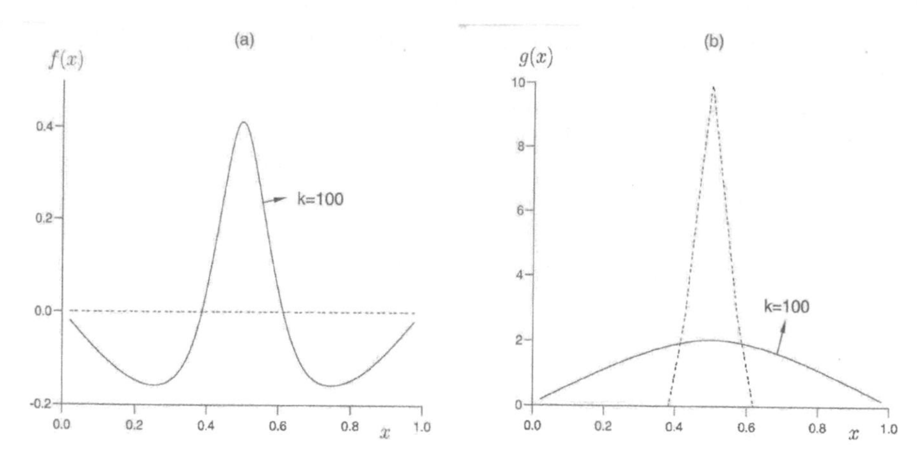

FIGURE 2.3: The numerical solution for (a) $u_{200}|_{\Gamma_0}$ and (b) $\partial_n u_{201}|_{\Gamma_0}$ after $k = 100$ iterations, in comparison with the input data (2.8) (dashed line), for Example 2.2, [254].

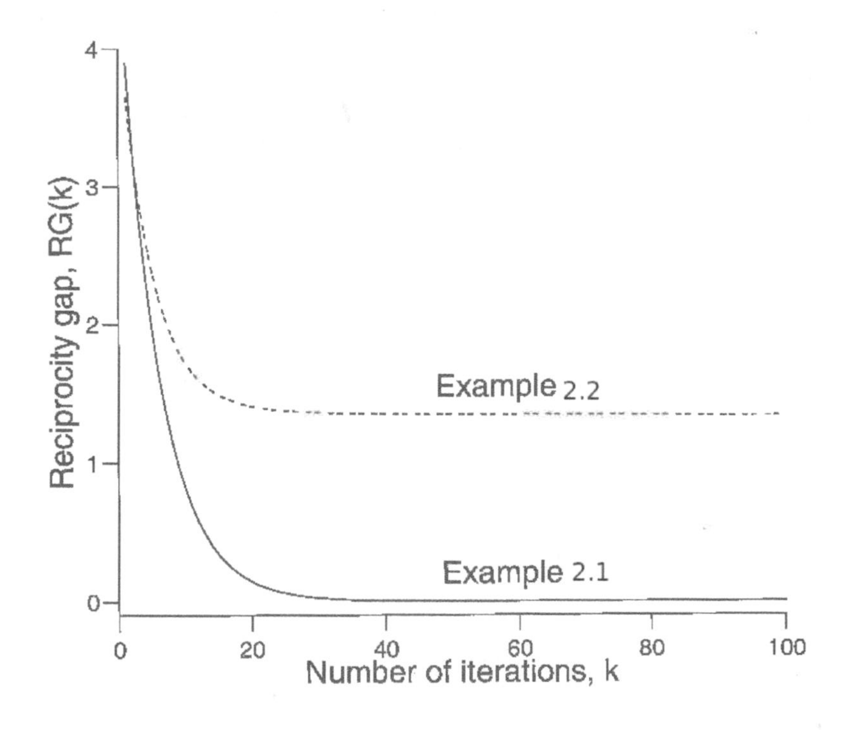

FIGURE 2.4: The reciprocity gap (2.9) versus the number of iterations, k, for the consistent (Example 2.1) and inconsistent (Example 2.2) Cauchy problems [254].

solution is artificial since it does not satisfy the Laplace equation. In fact, it proves numerically that the original Cauchy problem does not have a classical solution for the inconsistent data (2.8). From the theoretical point of view, this also shows that for the inconsistent Cauchy data problem the sequence $\{u_k\}_{k\geq 0}$ is not convergent in $H^1(\Omega)$, however this is more technical to numerically implement than the reciprocity gap introduced in equation (2.9).

To sumarize, Figures 2.2–2.4 demonstrate numerically that the iterative alternating algorithm is convergent if and only if the solution of the Cauchy problem (2.1) and (2.2) exists. This proves that the alternating algorithm is solvable in the sense of [99]. It also shows some advantage of the iterative regularizing algorithm over Tikhonov's regularization or the TSVD since the latter ones may still produce false solutions even though the original Cauchy problem does not have a solution. More advanced features of the alternating algorithm can be found in [40, 239, 293]. It would also be interesting in the future if the alternating algorithm can be extended to Cauchy problems for semilinear elliptic PDEs of the form $-\nabla^2 u = F(u)$, where the source $F(u)$ depends non-linearly on the solution u, [250].

2.1.3 Non-local quasi-boundary value methods

The method of non-local quasi-boundary value problems approximating the solutions of Cauchy problems for elliptic equations was investigated in several studies, e.g., [141, 295, 380]. In this section, the method is described in the more abstract framework considered in [141]. Thus, given T and ϵ positive numbers, $\mathcal{H}$ a Hilbert space with inner product $(\cdot,\cdot)$ and norm $\|\cdot\|$, and letting the operator $A : D(A) \subset \mathcal{H} \to \mathcal{H}$ be positive definite, self-adjoint with compact inverse, consider the Cauchy problem given by

$$\begin{cases} u_{tt} = Au, & 0 < t < T, \\ \|u(0) - f\| \leq \epsilon, & \\ u_t(0) = 0, & \end{cases} \tag{2.10}$$

where $\epsilon \geq 0$ is a given level of noise, $f \in \mathcal{H}$ is a given Dirichlet data, and the Neumann data g was taken, for simplicity, to be zero. This Cauchy problem, which requires finding the missing Cauchy data at $t = T$, is ill-posed and is regularized by the well-posed non-local quasi-boundary value problem

$$\begin{cases} (v_\alpha)_{tt} = Av_\alpha, & 0 < t < aT, \\ v_\alpha(0) + \alpha v_\alpha(aT) = f, & \\ (v_\alpha)_t(0) = 0, & \end{cases} \tag{2.11}$$

where $a > 1$ is a fixed given number and $\alpha > 0$ is the regularization parameter. The problem (2.11) is well-posed in $v_\alpha \in C^2((0, aT); \mathcal{H}) \cap C([0, aT]; \mathcal{H})$, with $v_\alpha(t) \in D(A)$ given by [296],

$$v_\alpha(t) = \sum_{n=1}^{\infty} \frac{\cosh(\sqrt{\lambda_n} t)}{1 + \cosh(a\sqrt{\lambda_n} T)} (f, \phi_n)\phi_n, \quad t \in [0, aT], \tag{2.12}$$

where $(\lambda_n^{-1}, \phi_n)_{n\geq 1}$ is an orthonormal system for A^{-1}.

The proper choice of the parameter $\alpha > 0$ yields order-optimal regularization methods, as given by the following theorem.

Theorem 2.1. ([141]) *Suppose that there exists a known upper bound $E > 0$ for $\|u(T)\| \leq E$.*
(i) Then, with the a priori choice $\alpha = \alpha(\epsilon) = (\epsilon/E)^a$,

$$\|u(t) - v_\alpha(t)\| \leq C_1(a,t,T)\epsilon^{1-t/T}E^{t/T}, \quad t \in [0,T], \tag{2.13}$$

where

$$C_1(a,t,T) := \left(\frac{(2a-2)^{1-1/a}}{a} + 1\right)^{1-t/T} \left(\frac{2^{2/a}(2a-2)^{1-1/a}}{a} + 2\right)^{t/T}.$$

For example, for $a = 2$, the choice $\alpha = (\epsilon/E)^2$ yields

$$\|u(t) - v_\alpha(t)\| \leq \left(\frac{1}{\sqrt{2}} + 1\right)\epsilon^{1-t/T}(2E)^{t/T}, \quad t \in [0,T].$$

(ii) Suppose that $0 < \epsilon < \|f\|$ and choose $\tau > 1$ such that $0 < \tau\epsilon < \|f\|$. Then, there exists a unique $\alpha = \alpha(\epsilon) > 0$ solving

$$\|v_\alpha(0) - f\| = \tau\epsilon. \tag{2.14}$$

Further, with this a posteriori choice of α,

$$\|u(t) - v_\alpha(t)\| \leq C_2(a,t,T)\epsilon^{1-t/T}E^{t/T}, \quad t \in [0,T], \tag{2.15}$$

where

$$C_2(\tau,t,T) := (1+\tau)^{1-t/T}\left(\frac{4\tau + 12}{\tau - 1}\right)^{t/T}.$$

For illustration, let us consider the rectangle $\Omega = (0,\pi) \times (0,T)$, (with $T = 1$), and the boundary portions $\Gamma_0 = \{(x,0)|x \in (0,\pi)\}$, $\Gamma_1 = \{(x,T)|x \in (0,\pi)\}$ and $\Gamma_2 = \partial\Omega\backslash(\Gamma_0 \cup \Gamma_1)$. Take $A = -\partial_{xx}$ defined on $D(A) = H_0^1(0,\pi) \subset \mathcal{H} := L^2(0,\pi)$, and consider the Cauchy problem with homogeneous Dirichlet boundary conditions prescribed on Γ_2 given by

$$\begin{cases} u_{tt} + u_{xx} = 0, \ (x,t) \in (0,\pi) \times (0,T), \\ u(0,t) = u(\pi,t) = 0, \ t \in (0,T), \\ u_t(x,0) = 0, \ u(x,0) = f(x)(1 + p\rho(x)) =: f^\epsilon(x), \ x \in (0,\pi), \end{cases} \tag{2.16}$$

where $\rho(x)$ are random variables generated (using the NAG routine G05DAF) from a uniform distribution in $[-1,1]$ and p is the percentage of noise. The solution of this ill-posed Cauchy problem is approximated by a sequence of solutions $(v_\alpha)_{\alpha>0}$ of the well-posed non-local quasi-boundary value problems

$$\begin{cases} (v_\alpha)_{tt} + (v_\alpha)_{xx} = 0, \ (x,t) \in (0,\pi) \times (0,aT), \\ v_\alpha(0,t) = v_\alpha(\pi,t) = 0, \ t \in (0,aT), \\ (v_\alpha)_t(x,0) = 0, \ v_\alpha(x,0) + \alpha v_\alpha(x,aT) = f^\epsilon(x), \ x \in (0,\pi), \end{cases} \tag{2.17}$$

where a is chosen typically to be 2 and $\alpha > 0$ is a regularization parameter to be prescribed according to the discrepancy criterion (2.14). Then, $v_\alpha(x,t)$ is taken as approximation for $u(x,t)$ in $(x,t) \in [0,\pi] \times [0,T]$.

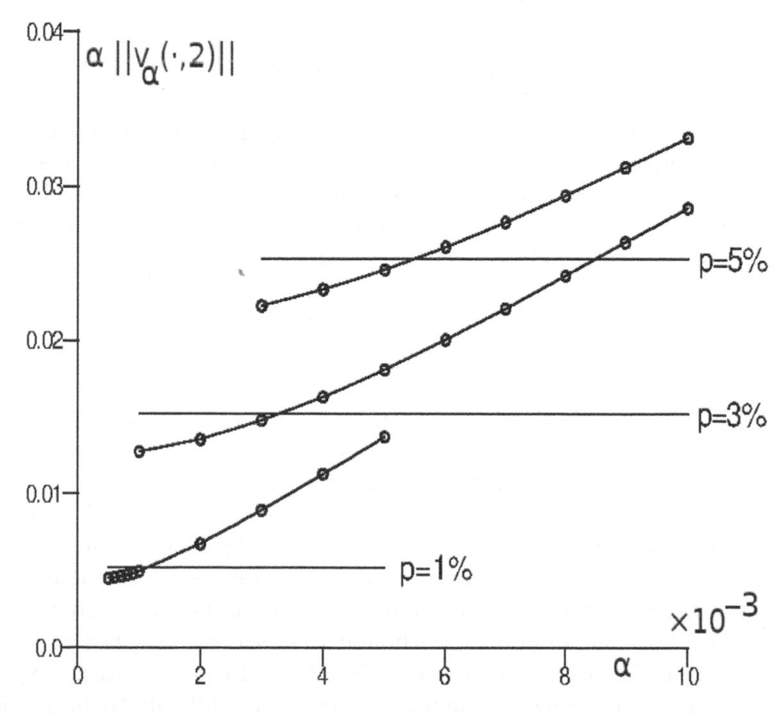

FIGURE 2.5: The graph (—o—) of the function $\alpha\|v_\alpha(\cdot, aT = 2)\|$, as a function of α, for various percentages of noise $p \in \{1,3,5\}\%$. The horizontal lines correspond to the values of $\tau\epsilon$, with $\tau = 1.2$ and ϵ given by (2.19), [141].

Application of the BEM recasts the well-posed problem (2.17) for v_α as the following system of 4 linear integral equations with the 4 unknowns $(v_\alpha)_x(0,t)$, $(v_\alpha)_x(\pi,t)$, $(v_\alpha)(x,T)$ and $(v_\alpha)_t(x,T)$:

$$\int_\Gamma \left[G(x,t;x',t') \frac{\partial v_\alpha}{\partial n}(x',t') - v_\alpha(x',t') \frac{\partial G}{\partial n(x',t')}(x,t;x',t') \right] dS(x',t')$$
$$= \begin{cases} \frac{1}{2}(f^\epsilon(x) - \alpha v_\alpha(x,aT)), & (x,t) \in \Gamma_0, \\ 0, & (x,t) \in \tilde{\Gamma}_2, \\ \frac{1}{2} v_\alpha(x,aT), & (x,t) \in \tilde{\Gamma}_1, \end{cases} \tag{2.18}$$

where $\Gamma = \Gamma_0 \cup \tilde{\Gamma}_1 \cup \tilde{\Gamma}_2$, $\Gamma_0 = (0,\pi) \times \{0\}$, $\tilde{\Gamma}_1 = (0,\pi) \times \{aT\}$ and $\tilde{\Gamma}_2 = \{0,\pi\} \times (0,aT)$ and $G(x,t;x',t') = -\frac{1}{2\pi}\ln\left(\sqrt{(x-x')^2+(t-t')^2}\right)$ is the fundamental solution in (2.3) of the two-dimensional Laplace's equation.

Take $T=1$, $a=2$, $f(x)=\sin(x)$, such that the analytical solution of the Cauchy problem (2.16) is $u(x,t)=\sin(x)\cosh(t)$. The BEM based on (2.18) with $N=320$ constant boundary elements, ($N/4=80$ of which discretizing uniformly each side of the rectangle $[0,\pi]\times[0,aT=2]$), is employed. For various percentages of noise $p \in \{1,3,5\}\%$, the values of $\epsilon = \|f^\epsilon - f\|_{L^2(0,\pi)}$ are

$$\epsilon = \begin{cases} 0.0043, & \text{if } p=1\%, \\ 0.0127, & \text{if } p=3\%, \\ 0.0211, & \text{if } p=5\%. \end{cases} \tag{2.19}$$

To choose the regularization parameter α according to the discrepancy principle (2.14), Figure 2.5 shows the graph of $\alpha\|v_\alpha(\cdot,aT=2)\|$, as a function of $\alpha>0$. According to (2.14), the regularization parameter α_ϵ is chosen as that for which the graph of this function intersects the horizontal line $y=\tau\epsilon$, where $\tau=1.2$. Any other choice of τ greater than 1 produces a stable solution. However, in the numerical computations it is recommended that the value of τ is not too large, e.g. in [134] a value of $\tau=1.1$ was suggested. From Figure 2.5, this selection process gives $\alpha(\epsilon) \in \{0,3,5\}\times 10^{-3}$ for $p\in\{1,3,5\}\%$, respectively. Figure 2.6(a) shows that the numerical solutions for $u(x,T=1)$, obtained as $v_{\alpha(\epsilon)}(x,1)$, are accurate and stable in comparison with the analytical solution $\sin(x)\cosh(1)$. Moreover, as the amount of noise p decreases, i.e. as $\epsilon \searrow 0$, the numerical solution converges to the corresponding analyical solution. Once the Dirichlet data $u(x,1)$ has been obtained accurately, the BEM provides the flux Neumann data $u_t(x,1)$ by solving the well-posed direct problem for $u(x,t)$. The numerical results presented in Figure 2.6(b) shows good agreement between the numerical Neumann data and the exact solution $u_t(x,1)=\sin(x)\sinh(1)$. Although the agreement is not as good as that observed in Figure 2.6(a) for the Dirichlet data, this is to be expected since higher-order derivatives (Neumann) are more difficult to be retrieved accurately than the lower-order ones (Dirichlet).

2.2 Cauchy Problem for the Stokes System

The slow viscous steady-state incompressible flow of fluids at low Reynolds numbers ($Re \ll 1$) with inertia neglected, in a fluid domain Ω is governed by the equations (taking, for simplicity, the dynamic viscosity equal to unity)

$$\begin{cases} \nabla^2 \underline{u} = \nabla p & \text{in } \Omega, \\ \nabla \cdot \underline{u} = 0 & \text{in } \Omega, \end{cases} \tag{2.20}$$

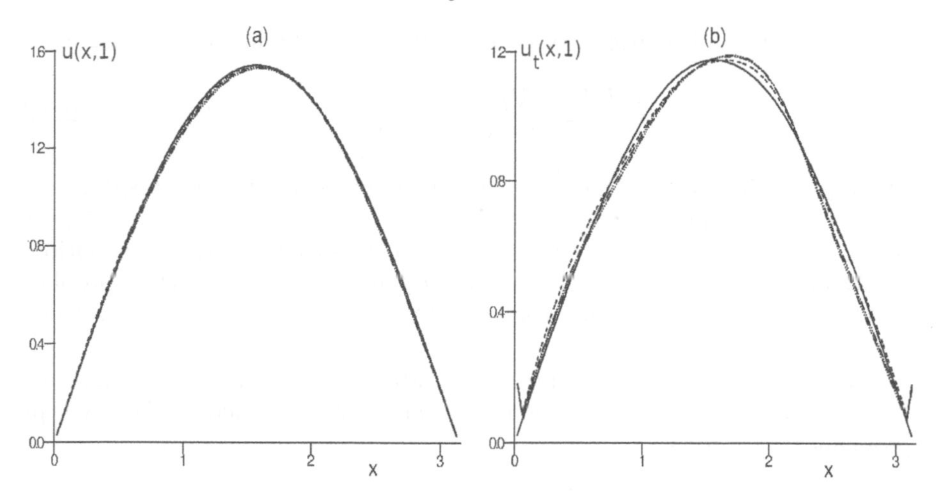

FIGURE 2.6: (a) The analytical Dirichlet data $u(x, T = 1) = \sin(x)\cosh(1)$, and (b) the analytical Neumann data $u_t(x, T = 1) = \sin(x)\sinh(1)$ for $x \in [0, \pi]$, shown with continuous line (——), in comparison with the numerical approximations for various percentages of noise: $p = 1\%$ (− − −), $p = 3\%$ (−..−) and $p = 5\%$ (−....−), [141].

for the fluid velocity $\underline{u} \in L^2(\Omega)^d$ and the pressure $p \in (H^1(\Omega))'$ (the dual of $H^1(\Omega)$). Then, if Ω is bounded domain with sufficiently smooth boundary, say of class C^2, over which we prescribe the full Dirichlet data on the velocity, results in a well-posed problem. However, if the boundary $\partial\Omega$ consists of two separate parts, one of which, Γ_1, is under-specified by prescribing nothing on it and the other, Γ_0, is over-specified by prescribing on it both the fluid velocity $\underline{f} \in L^2(\Omega)^d$ and the stress force $\underline{g} \in L^2(\Omega)^d$, i.e. the Cauchy data

$$\underline{u} = \underline{f}, \quad p\underline{n} - N\underline{u} = \underline{g} \quad \text{on } \Gamma_0, \tag{2.21}$$

where $N\underline{u} := (\nabla\underline{u} + (\nabla\underline{u})^{\mathrm{T}})\underline{n}$, a Cauchy problem arises. Assuming that the pieces Γ_0 and Γ_1 are disjoint closed curves (or surfaces), the domain Ω could occupy the annular region between two infinitely long cylinders in two dimensions (or between two spheres in three dimensions). Usually the no-slip condition on the fluid velocity is imposed but, for rotating walls, that it is equal to the wall's velocity which may be unknown. The above Cauchy problem can also be imagined in the exterior flow past an obstacle of arbitrary shape $\partial\Omega$, in which case $\Gamma_1 = \partial\Omega$ and Γ_0 would represent a curve in the fluid domain on which the fluid velocity and the stress force are measured (if Γ_0 is a closed curve, then the stress force could be obtained by solving the direct problem in the exterior complementary domain to the domain enclosed by Γ_0).

Remark 2.2. (i) The Stokes system (2.20) can also be inhomogeneous, i.e.,

$$\begin{cases} \nabla^2 \underline{u} = \nabla p + \underline{F} & \text{in } \Omega, \\ \nabla \cdot \underline{u} = G & \text{in } \Omega, \end{cases} \tag{2.22}$$

where $\underline{F}$ and G stand for a given force and compressibility terms, respectively. Then, one can split the solution of the Cauchy problem as $(\underline{u}, p) = (\overline{\underline{u}}, \overline{p}) + (\tilde{\underline{u}}, \tilde{p})$, where $(\tilde{\underline{u}}, \tilde{p})$ solves a homogeneous Cauchy problem given by equations (2.20) and (2.21), and $(\overline{\underline{u}}, \overline{p})$ solves the direct and well-posed Dirichlet problem for the inhomogeneous Stokes system (2.22) subject to no-slip fluid velocity zero boundary condition over the whole boundary $\partial\Omega$.

(ii) The prescription of the pressure gradient instead of the stress force on Γ_0 was proposed in [100], but the investigation of this different Cauchy-type problem is open work.

Next, three iterative procedures (Landweber-Fridman, conjugate gradient and minimal error methods) [215] are presented and compared for obtaining a stable solution to the above Cauchy problem given by (2.20) and (2.21).

2.2.1 Landweber-Fridman method (LFM)

Introduce the direct mixed and well-posed problem given by

$$\begin{cases} \nabla^2 \underline{u} = \nabla p & \text{in } \Omega, \\ \nabla \cdot \underline{u} = 0 & \text{in } \Omega, \\ \underline{u} = \underline{\eta} & \text{on } \Gamma_1, \\ p\underline{n} - N\underline{u} = \underline{g} & \text{on } \Gamma_0 \end{cases} \tag{2.23}$$

with its unique weak solution for the fluid velocity denoted by $\underline{u}(\underline{\eta}, \underline{g})$, and its adjoint mixed well-posed problem

$$\begin{cases} \nabla^2 \underline{v} = \nabla q & \text{in } \Omega, \\ \nabla \cdot \underline{v} = 0 & \text{in } \Omega, \\ \underline{v} = \underline{0} & \text{on } \Gamma_1, \\ q\underline{n} - N\underline{v} = \underline{\xi} & \text{on } \Gamma_0 \end{cases} \tag{2.24}$$

Then the iterative Landweber-Fridman method (LFM) runs as follows [217].

Step 1. Let $k = 1$ and choose an arbitrary initial guess $\underline{\eta}_0 \in L^2(\Gamma_1)^d$ for the fluid velocity on Γ_1. Then, with this Dirichlet data on Γ_1 solve the direct problem (2.23) to obtain the first approximation $(\underline{u}_0, p_0)$.

Step 2. Next, with the Neumann data $\underline{\xi} = \underline{u}_0 - \underline{f}$ on Γ_0 solve the adjoint problem (2.24) to obtain $(\underline{v}_0, q_0)$.

Step 3. Having constructed $(\underline{u}_{k-1}, p_{k-1})$ and $(\underline{v}_{k-1}, q_{k-1})$, obtain $(\underline{u}_k, p_k)$ by solving the direct problem (2.23) with the Dirichlet data

$$\underline{\eta}_k = \underline{\eta}_{k-1} + \gamma(q_{k-1}\underline{n} - N\underline{v}_{k-1}) \quad \text{on } \Gamma_1, \tag{2.25}$$

where γ is a prescribed constant in the interval $(0, \|K\|^{-2})$, where $K : L^2(\Gamma_1)^d \to L^2(\Gamma_0)^d$ is the linear, bounded, injective and compact operator that maps, via the solution of the well-posed problem,

$$\begin{cases} \nabla^2 \underline{u} = \nabla p & \text{in } \Omega \\ \nabla \cdot \underline{u} = 0 & \text{in } \Omega \\ \underline{u} = \underline{\eta} & \text{on } \Gamma_1, \\ p\underline{n} - N\underline{u} = \underline{0} & \text{on } \Gamma_0, \end{cases} \tag{2.26}$$

the Dirichlet input $\underline{u} = \underline{\eta}$ on Γ_1 to the Dirichlet output $\underline{u}(\underline{\eta}, \underline{0}) = K\underline{\eta}$ on Γ_0. Its adjoint $K^* : L^2(\Gamma_0)^d \to L^2(\Gamma_1)^d$ maps, via the solution of the adjoint well-posed problem (2.24), the Neumann input $\underline{\xi}$ on Γ_0 to the Neumann output $(q\underline{n} - N\underline{v}) = K^*\underline{\xi}$ on Γ_1.

Step 4. Next, with the Neumann data $\underline{\xi} = \underline{u}_k - \underline{f}$ on Γ_0, solve the adjoint problem (2.24) to obtain $(\underline{v}_k, q_k)$.

Step 5. Set $k \to k + 1$ and go back to Step 3 until the sequence $(\underline{u}_k, p_k)$ converges (for noiseless data) to the solution $(\underline{u}, p) \in L^2(\Omega)^d \times (H^1(\Omega))'$, which is assumed to exist for the given Cauchy data $(\underline{f}, \underline{g}) \in L^2(\Gamma_0)^d \times L^2(\Gamma_0)^d$. When $\underline{f} \in L^2(\Gamma_0)^d$ is noisy given as $\underline{f}^\epsilon \in L^2(\Gamma_0)^d$ satisfying

$$\|\underline{f}^\epsilon - \underline{f}\|_{L^2(\Gamma_0)^d} \leq \epsilon, \tag{2.27}$$

where $\epsilon \geq 0$ is a known upper bound for the error in measurements, the iterative procedure is stopped at the first iteration number $k = k(\epsilon)$ for which the following discrepancy principle is satisfied:

$$e_c(k) := \|\underline{u}_k^\epsilon|_{\Gamma_0} - \underline{f}^\epsilon\|_{L^2(\Gamma_0)^d} \leq \tau\epsilon, \tag{2.28}$$

where $\tau > 1$ is some safeguarding fixed constant prescribed to ensure that the solution is not under-regularized, and $\underline{u}_k^\epsilon$ is the solution for the fluid velocity obtained from the iterative procedure with the Cauchy data $(\underline{f}^\epsilon, \underline{g})$ on Γ_0.

Remark 2.3. (i) No statistical assumption is stipulated about the type of noise in (2.27), and this is useful for practical applications in which the noise does not follow any particular distribution.

(ii) The *a priori* rigorous choice of $\tau > 1$ (independent of ϵ) is not known; heuristic suggestions include $\tau = 1.01$, [134].

(iii) The choice of γ should be made in the interval $(0, \|K\|^{-2})$; however, if $\gamma > 0$ is too small the LFM may become prohibitively slow. In such a situation, parameter-free procedures can be employed, as described next.

2.2.2 Parameter-free methods

Two parameter-free iterative methods of regularization of the Cauchy problem (2.20) and (2.21) are introduced. First, remark that solving this problem is equivalent to finding $\underline{\eta} \in L^2(\Gamma_1)^d$ such that

$$K\underline{\eta} = \underline{f} - K_1\underline{g} =: \underline{y}, \tag{2.29}$$

or, its noisy version

$$K\underline{\eta} = \underline{f}^\epsilon - K_1\underline{g} =: \underline{y}^\epsilon, \tag{2.30}$$

where, from (2.27), $\|\underline{y}^\epsilon - \underline{y}\|_{L^2(\Gamma_0)^d} \leq \epsilon$, and the linear and bounded operator $K_1 : L^2(\Gamma_0)^d \to L^2(\Gamma_0)^d$ maps, via the solution of the well-posed problem (2.24), the Neumann input $\underline{\xi} = \underline{g}$ on Γ_0 to the Dirichlet output $\underline{v}\big|_{\Gamma_0} = K_1\underline{g}$.

2.2.2.1 Conjugate gradient method (CGM)

From (2.30), the CGM is based on minimizing the least-squares functional

$$J(\underline{\eta}) := \frac{1}{2}\|K\underline{\eta} - \underline{y}^\epsilon\|^2_{L^2(\Gamma_0)^d}, \tag{2.31}$$

with respect to $\underline{\eta} \in (L^2(\Gamma_1))^d$. This yields solving the normal equation

$$K^*(K\underline{\eta}) = K^*\underline{y}^\epsilon. \tag{2.32}$$

Application of the CGM algorithm 7.1 of [113] consists in the following steps:

Step 1. Set $k = 0$ and choose an arbitrary function $\underline{\eta}_0 \in L^2(\Gamma_1)^d$. Then solve with this Dirichlet data on Γ_1 the direct problem (2.23) to obtain the first approximation $(\underline{u}_k, p_k)$.

Step 2. Next, with the Neumann data $\underline{\xi} = \underline{u}_k - \underline{f}$ on Γ_0, solve the adjoint problem (2.24) to obtain $(\underline{v}_k, q_k)$ and determine the gradient

$$\underline{r}_k = (q_k\underline{n} - N\underline{v}_k)\big|_{\Gamma_1}$$

and $\underline{d}_k = -\underline{r}_k + \beta_{k-1}\underline{d}_{k-1}$ on Γ_1, with the convention that $\beta_{-1} = 0$ and

$$\beta_{k-1} = \frac{\|\underline{r}_k\|^2_{(L^2(\Gamma_1))^d}}{\|\underline{r}_{k-1}\|^2_{(L^2(\Gamma_1))^d}}, \quad k \geq 1.$$

Step 3. Solve (2.26) with the Dirichlet data $\underline{\eta} = \underline{d}_k$ on Γ_1 and set

$$\alpha_k = \frac{\|\underline{r}_k\|^2_{(L^2(\Gamma_1))^d}}{\|K\underline{d}_k\|^2_{(L^2(\Gamma_0))^d}}. \tag{2.33}$$

Finally, pass to the new iteration using the recurrence

$$\underline{\eta}_{k+1} = \underline{\eta}_k + \alpha_k\underline{d}_k \quad \text{on } \Gamma_1.$$

Step 4. Set $k \to k+1$ and go back to Step 2 until the sequence $(\underline{u}_k, p_k)$ converges to the solution $(\underline{u}, p) \in L^2(\Omega)^d \times (H^1(\Omega))'$, which is assumed to exist for the given Cauchy data $(\underline{f}, \underline{g}) \in L^2(\Gamma_0)^d \times L^2(\Gamma_0)^d$. When the Dirichlet data $\underline{f} \in L^2(\Gamma_0)^d$ is given in the form of the noisy $\underline{f}^\epsilon \in L^2(\Gamma_0)^d$ satisfying (2.27), the iterative procedure is stopped at the first iteration number $k = k(\epsilon)$ for which the discrepancy principle (2.28) is satisfied. It follows from [316] that this iterative procedure has a regularizing character, i.e., the solution obtained at this stopping iteration number converges to the true solution (if it exists), as the amount of noise ϵ tends to zero. Furthermore, (2.28) is an order optimal stopping rule if $\underline{y}^\epsilon \in \text{Range}(K)$, [331].

2.2.2.2 Minimal error method (MEM)

Rather than fitting the measured data in the least-squares sense, as given by minimizing the functional (2.31), assuming that $\underline{y}^\epsilon \in \text{Range}(K)$ the minimal error method (MEM) minimizes the iteration error [229],

$$\tilde{J}(\underline{\eta}) := \frac{1}{2}\|\underline{\eta} - K^{-1}\underline{y}^\epsilon\|^2_{L^2(\Gamma_1)^d}, \tag{2.34}$$

with respect to $\underline{\eta} \in L^2(\Gamma_1)^d$. Then, applying the MEM of [132] to the Cauchy problem (2.20) and (2.21) yields the following iterative algorithm:

Step 1. Set $k = 0$ and choose an arbitrary function $\underline{\eta}_0 \in L^2(\Gamma_1)^d$. Then solve with this Dirichlet data on Γ_1 the direct problem (2.23) to obtain the first approximation $(\underline{u}_k, p_k)$.

Step 2. Next, with the Neumann data $\underline{\xi} = \underline{u}_k - \underline{f} =: \overline{\underline{r}}_k$ on Γ_0, solve the adjoint problem (2.24) to obtain $(\underline{v}_k, q_k)$ and determine the gradient

$$\underline{r}_k = (q_k\underline{n} - N\underline{v}_k)\big|_{\Gamma_1}$$

and $\underline{\xi}_k = -\underline{r}_k + \gamma_{k-1}\underline{\xi}_{k-1}$ on Γ_1, with the convention that $\gamma_{-1} = 0$ and

$$\gamma_{k-1} = \frac{\|\overline{\underline{r}}_k\|^2_{(L^2(\Gamma_0))^d}}{\|\overline{\underline{r}}_{k-1}\|^2_{(L^2(\Gamma_0))^d}}, \qquad k \geq 1.$$

Step 3. Pass to the new iteration using the recurrence

$$\underline{\eta}_{k+1} = \underline{\eta}_k + \chi_k\underline{\xi}_k \quad \text{on } \Gamma_1,$$

where

$$\chi_k = \frac{\|\overline{\underline{r}}_k\|^2_{(L^2(\Gamma_0))^d}}{\|\underline{\xi}_k\|^2_{(L^2(\Gamma_1))^d}}. \quad k \geq 1.$$

Step 4. Set $k \to k+1$ and go back to Step 2 until the sequence $(\underline{u}_k, p_k)$ converges to the solution $(\underline{u}, p) \in L^2(\Omega)^d \times (H^1(\Omega))'$, which is assumed to exist for the given Cauchy data $(\underline{f}, \underline{g}) \in L^2(\Gamma_0)^d \times L^2(\Gamma_0)^d$. For the Dirichlet noisy data $\underline{f}^\epsilon \in L^2(\Gamma_0)^d$ satisfying (2.27), the iterative procedure is stopped at the first iteration number $k = k(\epsilon)$ for which

$$\left(\sum_{j=0}^{k} \|K\underline{\eta}_j + K_1\underline{g} - \underline{f}^\epsilon\|^{-2}_{L^2(\Gamma_0)^d}\right)^{-1/2} \leq \tau\epsilon. \tag{2.35}$$

It follows from [132] that this iterative procedure has a regularizing character, i.e., the solution obtained at the stopping iteration number given by (2.35) converges to the true solution (if it exists), as ϵ tends to zero.

Remark 2.4. In the above analysis, the boundaries Γ_0 and Γ_1 are assumed to be two closed and disjoint curves/surfaces, as in Figure 2.1(a). Otherwise, when they are open with the intersection of their closures possessing common points, as in Figure 2.1(c), one can apply the iterative alternating algorithm of Section 2.1.2, [23], valid for Cauchy data $(\underline{f}, \underline{g}) \in H^{1/2}(\Gamma_0)^d \times (H_{00}^{1/2}(\Gamma_0)^d)'$, to obtain a convergent and stable solution $(\underline{u}, p) \in H^1(\Omega)^d \times L^2(\Omega)$, or apply the iterative regularization in weighted L^2-spaces [214].

2.2.3 Numerical results and discussion

The well-posed mixed problems (2.23) and (2.24) of the iterative procedures can most efficiently be solved by recasting the Stokes equations (2.20) into the boundary integral form [242],

$$\mathrm{c}(\underline{x})\underline{u}(\underline{x}) = \int_\Gamma \left[\underline{\underline{K}}(\underline{x}, \underline{y})\underline{u}(\underline{y}) - \underline{\underline{U}}(\underline{x}, \underline{y})\underline{t}(\underline{y})\right] dS(\underline{y}), \quad \underline{x} \in \overline{\Omega}, \tag{2.36}$$

where $\mathrm{c}(\underline{x}) = 1$ if $\underline{x} \in \Omega$, and $\mathrm{c}(\underline{x}) = 0.5$ if $\underline{x} \in \partial\Omega$ (smooth), $\underline{t} = p\underline{n} - N\underline{u}$ denotes the fluid traction (stress force) and, in two-dimensions, the fundamental solution tensors $\underline{\underline{K}}$ and $\underline{\underline{U}}$ are given by

$$K_{kl}(\underline{x}, \underline{y}) = -\frac{(x_k - y_k)(x_l - y_l)}{\pi|\underline{x} - \underline{y}|^4} \sum_{m=1}^{2} (x_m - y_m)n_m, \quad k, l = \overline{1, 2},$$

$$U_{kl}(\underline{x}, \underline{y}) = \frac{1}{4\pi}\left[-\delta_{kl}\ln(|\underline{x} - \underline{y}|) + \frac{(x_k - y_k)(x_l - y_l)}{|\underline{x} - \underline{y}|^2}\right], \quad k, l = \overline{1, 2}.$$

Here δ_{kl} is the Kronecker tensor, $\underline{x} = (x_1, x_2)$, $\underline{y} = (y_1, y_2)$ and $\underline{n} = (n_1, n_2)$. The discretization of the boundary integral equation (2.36) applied at $\underline{x} \in \partial\Omega$ is performed using the BEM, [406].

Let Γ_0 and Γ_1 be two infinitely long circular cylinders of radii $0 < R_1 < R_0$ and let $\Omega = \{(x, y) \mid R_1^2 < x^2 + y^2 < R_0^2\}$ be the annular region in between filled with a viscous fluid flowing at low Reynolds numbers governed by the Stokes equations (2.20). The Cauchy problem is to determine the fluid velocity and the stress force (fluid traction) on the inner cylinder $\Gamma_1 = \{(x, y) \mid x^2 + y^2 = R_1^2\}$ by taking the Cauchy data measurements (2.21) on the outer cylinder $\Gamma_0 = \{(x, y) \mid x^2 + y^2 = R_0^2\}$. Let us consider the analytical example:

$$\underline{u}(x, y) = (4y^3 - x^2, 4x^3 + 2xy - 1), \quad p(x, y) = 24xy - 2x, \quad (x, y) \in \Omega, \tag{2.37}$$

which satisfies (2.20). This generates the Cauchy data (2.21) on Γ_0 given by

$$\underline{f}(x, y) = (4y^3 - x^2, 4x^3 + 2xy - 1),$$

$$\underline{g}(x, y) = (-12y^3 - 2y^2 + 2x^2 + 12x^2y, -12x^3 + 12xy^2 - 8xy)/R_0.$$

Take $R_1 = 1$ and $R_0 = 2$ and employ $M = 128$ constant boundary elements to discretize uniformly the boundary $\partial\Omega = \Gamma_0 \cup \Gamma_1$ in the BEM discretization of boundary integral equation (2.36). An arbitrary initial guess such as $\underline{\eta}_0 = \underline{0}$ is chosen to initiate the iterative algorithms described in Sections 2.2.1 and 2.2.2. To test the stability of the methods, multiplicative noise of the form $\underline{f}^\epsilon = \underline{f}(1 + 0.01\rho)$, where ρ are random real numbers drawn from a uniform distribution in $[-1, 1]$, is added to the input boundary velocity data $\underline{f}$. This generates the amount of noise $\epsilon = \|\underline{f}^\epsilon - \underline{f}\| \approx 0.49$.

TABLE 2.1: The stopping iteration numbers for the CGM and LFM (with $\gamma = 0.1$) using criterion (2.28) with $\tau \approx 1$, the MEM using criterion (2.35) with $\tau \approx 1.35$, and the errors e_c and e_a given by (2.38).

k_{CGM}	9
k_{LFM}	239
k_{MEM}	6
$e_c(k_{CGM})$	0.479
$e_c(k_{LFM})$	0.489
$e_c(k_{MEM})$	0.660
$e_a(k_{CGM})$	0.288
$e_a(k_{LFM})$	0.533
$e_a(k_{MEM})$	1.598

Let us introduce the convergence and accuracy errors defined by

$$e_c(k) = \|\underline{u}_k|_{\Gamma_0} - \underline{f}^\epsilon\|_{L^2(\Gamma_0)^d}, \quad e_a(k) = \|\underline{\eta}_k - \underline{u}|_{\Gamma_1}\|_{L^2(\Gamma_1)^d}, \tag{2.38}$$

where k is the iteration number, $\underline{u}|_{\Gamma_1}$ is the exact fluid velocity on Γ_1 which can be obtained from (2.37), and $\underline{\eta}_k$ is obtained from the iteration procedures with the Cauchy data $\underline{f}^\epsilon$ and $\underline{g}$. The LFM and CGM are ceased according to the discrepancy principle stopping criterion (2.28), whilst the MEM is ceased according to the stopping criterion (2.35). Using these criteria, Table 2.1 shows the stopping interaction numbers, the convergence errors $e_c(k)$ on Γ_0 and the accuracy errors $e_a(k)$ on Γ_1 given by (2.38). Whilst for the CGM and the LFM the constant τ can be chosen close to unity, for the MEM, values of τ between 1 and 1.1 produce numerical results which exhibit a slightly oscillatory unstable behavior and they become inaccurate, especially for the fluid traction $\underline{t}$. This shows that the MEM is more sensitive to the choice of τ than the other iterative methods [229]. From Table 2.1 it can be seen that the parameter free procedures CGM and MEM are much faster (about 10-20 times) than the LFM. In terms of accuracy, the CGM performs best, followed by the LFM and the MEM. Figure 2.7 shows the numerical solutions for $u_1(R_1, \theta)$ and $t_1(R_1, \theta)$ in comparison with the corresponding analytical solutions obtained from (2.37). The $L^2(\Gamma_1)^d$—errors between the numerical and analytical solutions are, respectively: (i) for $u_1(R_1, \theta)$: 0.2, 0.4 and 1.0 for the CGM, LFM and MEM, and (ii) for $t_1(R_1, \theta)$: 2.2, 3.1 and 4.0 for the CGM, LFM and

MEM. This shows that the CGM outperforms the LFM followed by the MEM in terms of accuracy. However, all the methods produce stable and reasonably accurate numerical solutions.

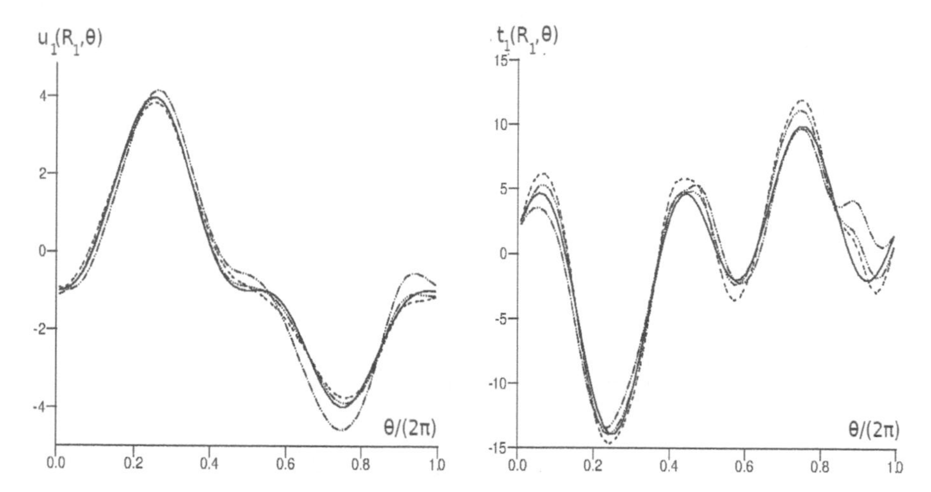

FIGURE 2.7: The numerical solution for $u_1(R_1, \theta)$ and $t_1(R_1, \theta)$ for 1% multiplicative noise, obtained using the CGM (-....-), LFM (– – –) and MEM (-..-) in comparison with the analytical solutions (—), [215].

2.3 Cauchy-type Problems for the Biharmonic Equation

Higher-order PDEs increase the number of ways in which inverse boundary-value problems can be formulated. Consider, e.g., the biharmonic equation

$$\nabla^4 \psi = 0 \quad \text{in } \Omega \tag{2.39}$$

or, equivalently

$$\nabla^2 \psi = \omega, \quad \nabla^2 \omega = 0 \quad \text{in } \Omega \tag{2.40}$$

in a bounded planar domain $\Omega \subset \mathbb{R}^2$ with sufficiently smooth boundary, governing the Kirchhoff's plate theory in elasticity or the two-dimensional Stokes flow of fluids. In these applications, the functions ψ and ω that satisfy (2.40) are called the deflection and bending moment of the plate, whilst for the Stokes slow viscous flow of fluids (related to the Stokes system of equations (2.20) and (2.21) through $\underline{u} = (\partial_y \psi, -\partial_x \psi)$ and $\omega = \nabla \times \underline{u} = \partial_x u_2 - \partial_y u_1$) they represent the stream function and vorticity, respectively. If ψ and either of $\psi' := \partial_n \psi$, ω or $\omega' := \partial_n \omega$ are prescribed on the boundary $\partial\Omega$, then the resulting boundary value direct problem is well-posed. However, in situations where it is not possible to specify two boundary conditions at all points of $\partial\Omega$, extra information needs to be supplied on a portion of the boundary or even in the interior of the domain. If a boundary portion $\Gamma_0 \subset \partial\Omega$ is over-specified by prescribing on it ψ, ψ' and ω, or ψ, ψ' and ω', or ψ, ω and ω', or ψ, ψ', ω and ω', and the remaining boundary $\Gamma_1 = \partial\Omega \backslash \Gamma_0$ is under-specified by prescribing on it only one or no boundary condition, then the resulting problems are of Cauchy-type inverse and ill-posed problems [264].

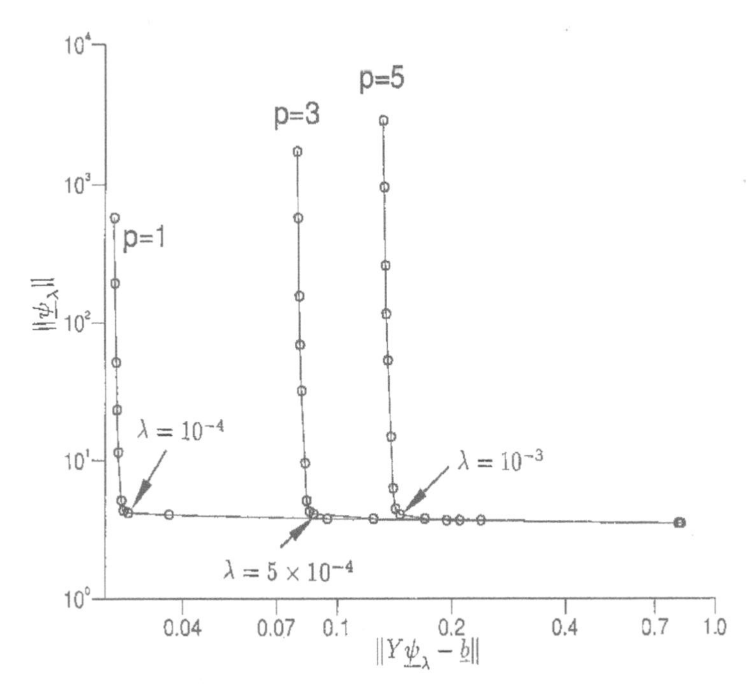

FIGURE 2.8: L-curve log-log plot of the residual $\|Y\underline{\psi}_\lambda - \underline{b}\|$ versus the solution-norm $\|\underline{\psi}_\lambda\|$ for various percentages of noise $p\% \in \{1, 3, 5\}\%$ and various values of the regularization parameter $\lambda \in [10^{-11}, 10^0]$, [271].

As Cauchy problems have been discussed before, let us consider the case when the under-specification of boundary conditions is compensated for by measurements of ψ on an interior curve $S \subset \Omega$. Therefore, let us consider the inverse problem given by (2.39), or equivalently (2.40), in a simply-connected bounded planar domain Ω, subject to the boundary condition

$$\alpha\psi(\underline{x}) + (1-\alpha)\partial_n\psi(\underline{x}) = f(\underline{x}), \quad \underline{x} \in \partial\Omega, \tag{2.41}$$

where $\alpha \in \{0, 1\}$, and the internal measurement

$$\psi = g \quad \text{on } S, \tag{2.42}$$

where S is the boundary of a simply-connected sub-domain Ω_1 compactly contained in Ω, and f and g are prescribed functions. The local existence and uniqueness of solution of the linear inverse problem given by equations (2.39), (2.41) and (2.42) hold [13]. Below, the numerical the solution based on the BEM and the Tikhonov's regularization method [271] is presented. For the sake of explanation let us consider $\alpha = 0$ in which case (2.41) becomes

$$\partial_n \psi(\underline{x}) = f(\underline{x}), \quad \underline{x} \in \partial\Omega, \tag{2.43}$$

with the mention that the case of $\alpha = 1$ was investigated elsewhere [272], using the minimization of the energy $\int_\Omega |\nabla\omega|^2 d\Omega$ of the Laplace equation satisfied by the harmonic function ω.

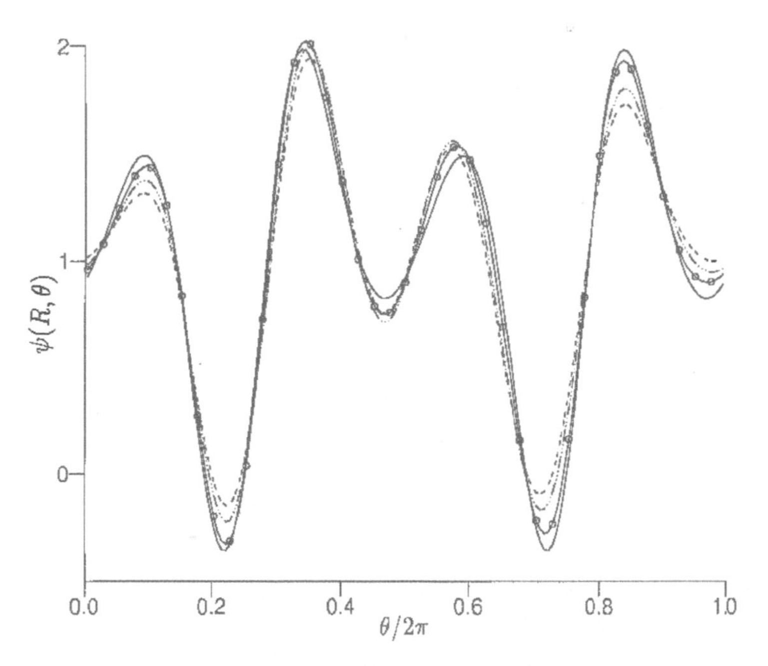

FIGURE 2.9: The numerical solution for $\psi(R,\theta)$, for various percentages of noise $p\% = 1\%$ $(-\circ-)$, $p\% = 3\%$ $(-\cdots-)$ and $p\% = 5\%$ $(- - -)$ in comparison with the analytical solution (—), [271].

Based on Green's identity, the equations (2.40), (2.42) and (2.43) can be recast into the following coupled boundary integral form:

$$\frac{1}{2}\psi(\underline{x}) = \int_{\partial\Omega} \Big(f(\underline{\xi}) G(\underline{x};\underline{\xi}) - \partial_{n(\underline{\xi})} G(\underline{x};\underline{\xi}) \psi(\underline{\xi}) \Big) \, dS(\underline{\xi})$$

$$+ \int_{\partial\Omega} \Big(\partial_n \omega(\underline{\xi}) \tilde{G}(\underline{x};\underline{\xi}) - \partial_{n(\underline{\xi})} \tilde{G}(\underline{x};\underline{\xi}) \omega(\underline{\xi}) \Big) \, dS(\underline{\xi}), \quad \underline{x} \in \partial\Omega, \tag{2.44}$$

$$\frac{1}{2}\omega(\underline{x}) = \int_{\partial\Omega} \left(\partial_n\omega(\underline{\xi})G(\underline{x};\underline{\xi}) - \partial_{n(\underline{\xi})}G(\underline{x};\underline{\xi})\omega(\underline{\xi})\right) dS(\underline{\xi}), \quad \underline{x} \in \partial\Omega \quad (2.45)$$

and

$$g(\underline{x}) = \int_{\partial\Omega} \left(f(\underline{\xi})G(\underline{x};\underline{\xi}) - \partial_{n(\underline{\xi})}G(\underline{x};\underline{\xi})\psi(\underline{\xi})\right) dS(\underline{\xi})$$
$$+ \int_{\partial\Omega} \left(\partial_n\omega(\underline{\xi})\tilde{G}(\underline{x};\underline{\xi}) - \partial_{n(\underline{\xi})}\tilde{G}(\underline{x};\underline{\xi})\omega(\underline{\xi})\right) dS(\underline{\xi}), \quad \underline{x} \in S, \quad (2.46)$$

where

$$G(\underline{x};\underline{\xi}) = -\frac{1}{2\pi}\ln|\underline{x}-\underline{\xi}|, \quad \tilde{G}(\underline{x};\underline{\xi}) = -\frac{1}{8\pi}|\underline{x}-\underline{\xi}|^2\left(\ln|\underline{x}-\underline{\xi}|-1\right)$$

are the fundamental solutions of two-dimensional Laplace's and biharmonic equations, respectively. Using a BEM discretization with M uniform boundary elements and $M_1 \geq M$ internal points uniformly distributed around the contour S and eliminating ω and $\partial_n\omega$ recast (2.44)–(2.46) as a system of M_1 linear algebraic equations in M unknowns, generically written as [271],

$$Y\underline{\psi} = \underline{b}, \quad (2.47)$$

where Y contains BEM-matrices whose elements can be evaluated analytically [192], $\underline{\psi}$ represents the discretized version of the unknown boundary function $\psi|_{\partial\Omega}$ and $\underline{b}$ is a right-hand side vector of knowns containing the input data f and g. Since the inverse boundary value problem under study is ill-posed the resulting system of discretized equations (2.47) is ill-conditioned, and, to stabilize the solution, the (zeroth-order) Tikhonov's regularization method, based on minimizing the functional

$$T_\lambda(\underline{\psi}) := \|Y\underline{\psi} - \underline{b}\|^2 + \lambda\|\underline{\psi}\|^2, \quad (2.48)$$

is applied, yielding the regularized solution

$$\underline{\psi}_\lambda = (Y^{\mathrm{T}}Y + \lambda I_M)^{-1}Y^{\mathrm{T}}\underline{b}. \quad (2.49)$$

Numerical results are illustrated for a typical example,

$$\psi(x,y) = x(\sin(x)\cosh(y) + \cos(x)\sinh(y))/2 \quad (2.50)$$

in a circle Ω of radius $R = 2$ centred at the origin. The internal curve S is taken as the boundary of circular domain Ω_1 of radius $R_1 < R = 2$ centred at the origin. For $M = M_1 = 40$, it was found that the determinant of the matrix Y in equation (2.47) decreases drastically from $O(10^{-1})$ to $O(10^{-65})$ and $O(10^{-183})$ as the ratio $\mu := R_1/R$ decreases from 0.75 to 0.5 and 0.25, respectively. This is as expected from the principle of unique continuation, with the problem becoming more ill-posed as the distance between $\partial\Omega = \partial B(\underline{0};R)$ and $S = \partial\Omega_1 = \partial B(\underline{0};R_1)$ increases. Furthermore, as $M = M_1$ increases to 80 and

160, the value of $\det(Y)$ becomes too small to be output by the NAG routine F03AAF used. Therefore, numerical results are presented only for the most ill-conditioned situation, namely for $\mu = 0.25$ and $M = M_1 = 160$. The measured internal data g in (2.42) is further perturbed by Gaussian additive noise with mean zero and standard deviation $\sigma = p\% \times \max_{\theta \in [0,2\pi)} |\psi(R_1, \theta)|$, generated using the NAG routine G05DDF.

The dependence of the regularization parameter λ on the percentage of noise $p\%$ is illustrated by the L-curves in Figure 2.8. These curves possess an under-smoothing portion for small values of λ where the numerical solution is oscillatory and the solution norm $\|\underline{\psi}_\lambda\|$ becomes unbounded, as in the least-squares method. The curves also possess an over-smoothing portion corresponding to large values of λ where the constraint on the residual $\|Y\underline{\psi}_\lambda - \underline{b}\|$ plays virtually no role and the minimization of (2.48) will eventually produce only the trivial zero function. Finally, the curves possess a flat portion over which the solution norm $\|\underline{\psi}_\lambda\|$ is constant and any value of λ in this region may be chosen to produce a stable approximate solution. Moreover, the length of this flat portion decreases as the amount of noise p increases and the input data becomes noisier. The values of λ at the corners of the obtained L-curves where the under- and over-smoothing portions meet are considered as a suitable compromise to choose for the regularization parameter [48, 135]. Stronger regularization, i.e. larger value of λ, is needed as the percentage of noise p increases. In cases when an L-curve is not obtained [381], the more rigorous discrepancy principle [305], based on the *a priori* knowledge of an upper bound for the amount of noise could be used.

Figure 2.9 illustrates the stable numerical solution for $\psi|_{\partial\Omega}$ for various percentages of noise $p \in \{1, 3, 5\}$ obtained using the L-corner values for the regularization parameter given from Figure 2.8 by $\lambda \in \{10^{-4}, 5 \times 10^{-4}, 10^{-3}\}$, respectively. However, the boundary values of ω and $\omega' := \partial_n \omega$ obtained by an inversion of the system of linear equations, generically written as

$$Z(\underline{\omega}, \underline{\omega}')^{\mathrm{T}} = \underline{c}, \tag{2.51}$$

resulted from the previous elimination of $\underline{\omega}$ and $\underline{\omega}'$ from (2.44)–(2.46), were found highly oscillatory and unbounded. This is expected since retrieving higher-order derivatives from lower-order noisy derivatives is in itself an ill-posed procedure [265]. To deal with this instability, the ill-conditioned system of linear algebraic equations (2.51) is solved using the TSVD, where the truncation level j is is chosen according to the L-curve criterion depicted in Figure 2.10. Using the corner values as the truncation levels in the SVD, the stable numerical results shown in Figure 2.11 are obtained.

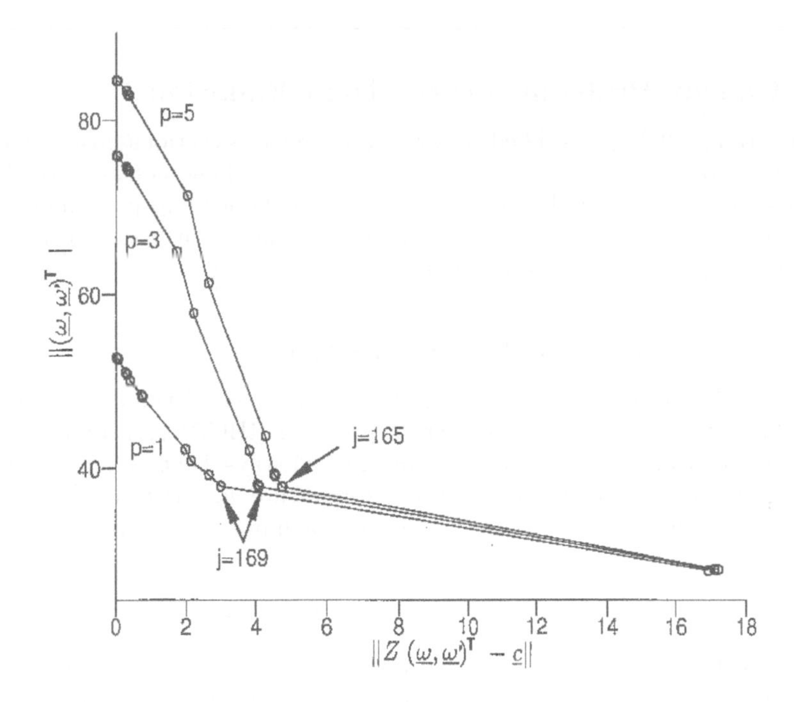

FIGURE 2.10: L-curve plot of the residual $\|Z(\underline{\omega}, \underline{\omega}')^{\mathrm{T}} - \underline{c}\|$ versus the solution-norm $\|(\underline{\omega}, \underline{\omega}')^{\mathrm{T}}\|$, for various percentages of noise $p\% \in \{1, 3, 5\}\%$ and various values of the truncation number j of the SVD, [271].

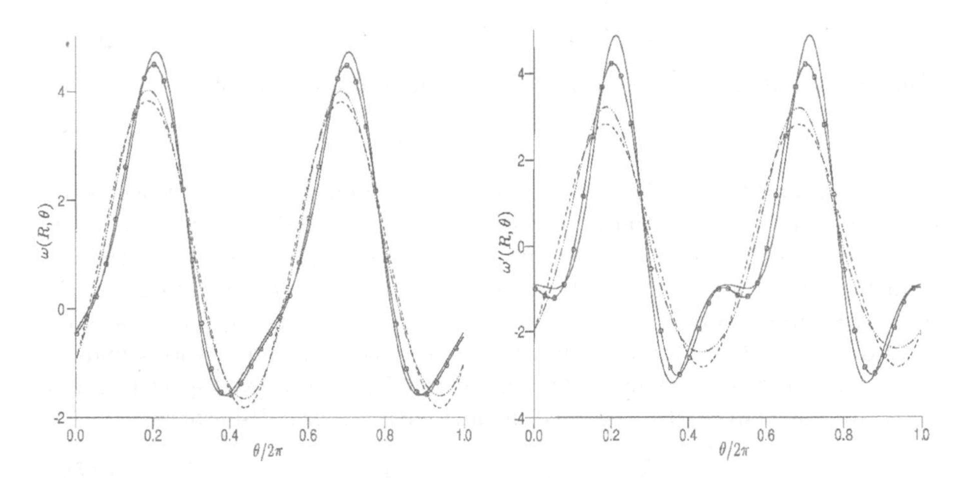

FIGURE 2.11: The numerical solutions for $\omega(R, \theta)$ and the normal derivative $\omega'(R, \theta)$ for $p\% = 1\%$ $(- \circ -)$, $p\% = 3\%$ (-...-) and $p\% = 5\%$ $(- - -)$ noise, in comparison with the analytical solution (—), [271].

2.4 Cauchy Problems for the Heat Equation

The governing PDEs considered so far in the previous sections have been at steady-state of elliptic type. In this section, unsteady phenomena governed by the non-stationary parabolic heat equation, with the leading part modelling diffusion supplemented by lower-order parts modelling convection/advection and reaction processes, are considered.

2.4.1 Inverse heat conduction problems

One classical inverse boundary-value problem, also termed the Cauchy problem [50], or the inverse heat conduction problem (IHCP) [27], or the non-characteristic Cauchy problem [144], or the sideways heat equation [109], is given by the heat equation for the temperature u, (written for a one-dimensional finite slab of length $l > 0$ over a finite time period $(0, T)$),

$$u_t = Du_{xx} \quad \text{for } (x,t) \in (0,l) \times (0,T) =: Q_T, \tag{2.52}$$

where $D > 0$ is the diffusion coefficient defined as the ratio between the thermal conductivity $\kappa > 0$ and the heat capacity ρc, where $\rho > 0$ and $c > 0$ are the density and specific heat, respectively, subject to the Cauchy boundary data at the end $x = l$ given by the specification or measurement of the boundary temperature and heat flux, i.e. the Cauchy data,

$$u(l,t) = \mu_l(t), \quad t \in (0,T), \tag{2.53}$$

$$\kappa u_x(l,t) = q_l(t), \quad t \in (0,T). \tag{2.54}$$

Usually, an initial condition at $t = 0$ is specified in the form of

$$u(x,0) = u_0(x), \quad x \in [0,l]. \tag{2.55}$$

The initial condition (2.55) may not be available (and, in fact, is not needed to be prescribed [140]) in situations when the heat conducting device is already in service and the initial temperature is an extra unknown to be determined from temperature measurements at the final time $t = T$ (see more on this backward unique continuation problem later on in Chapter 3). It can also be mentioned here the alternative practical case when instead of the boundary data (2.53) or (2.54), the internal temperature history at a fixed space location X_0 inside the slab $(0, l)$ is recorded in time with a thermal sensor, as

$$u(X_0,t) = \chi(t), \quad t \in (0,T). \tag{2.56}$$

In this case, the initial temperature needs to be supplied at least in the space interval (X_0, l), and by solving the resulting direct problem for the heat equation (2.52) in the restricted domain $(X_0, l) \times (0, T)$ subject to the Dirichlet (or mixed) boundary conditions (2.56) and (2.53) or (2.54) result in the heat flux $\kappa u_x(X_0, t)$ becoming available at $x = X_0$. This new piece of information together with the temperature data (2.56) recasts then as a Cauchy problem for the sideways heat equation in the region $(0, X_0) \times (0, T)$. So, in the above sense, the Cauchy problem given by equations (2.52)–(2.55) and the IHCP given by equations (2.52), (2.53) (or (2.54)), (2.55) and (2.56) are somewhat equivalent, but it is not obvious which of the two formulations is less ill-posed. For both problems uniqueness holds from the Holmgren unique continuation property for the sideways heat equation [248], but the existence of solution of the former Cauchy problem requires $t \mapsto q_l(t) + \frac{1}{\sqrt{\pi}} \int_0^t \frac{\mu_l'(\tau)}{\sqrt{t-\tau}} d\tau$ be a function of class two in the sense of [123] (a function $z \in C^\infty(a, b)$ is of Holmgren class two if there exist positive constants c and M such that $\|z^{(n)}\|_{L^\infty(a,b)} \leq cM^{-n}(2n)!$ for all $n \in \mathbb{N}$, [140]) and some compatibility between the resulting series expansion of the solution and the initial condition (2.55) at $t = 0$, [44]. In terms of continuous dependence on the data, both problems violate the stability of solution with respect to small errors in the input data. In a first instance, the latter problem seems less ill-posed because $X_0 < l$ and the Fourier analysis reveals this; however, numerically, the latter formulation requires solving an extra direct problem, which is well-posed in retrieving the solution $u(x, t)$ in the region $[X_0, l] \times [0, T]$ but is ill-posed in the recovery of the normal derivative heat flux at $x = X_0$ from the noisy measurement of the temperature data (2.56), [37]. Practically, the boundary measurements (2.53) and (2.54) are less destructive than the internal measurement (2.56) which, by being intrusive, may damage the material.

A physical situation for which the above IHCP models apply is that given by re-entry vehicles in atmosphere [27], where the temperature at the nozzle of a rocket, say at $x = 0$, is so high that any thermocouple that may be attached to it would be destroyed. Instead one could measure both the temperature (2.53) and heat flux (2.54) at $x = l$ inside the capsule. Other applications include [376]: (i) the safety analysis of nuclear reactors performed by measuring the temperature at the adiabatic inner wall of a hollow cylinder to determine the unavailable temperature and heat flux at the outer wall that is abruptly cooled [118]; (ii) determination of the temperature 'spike' inside a cannon at firing [358]; (iii) determination of a particle board's surface temperature and heat flux over which a lacquer coating layer is applied [34].

For simplicity, it was assumed that the heat source in (2.52) is absent since otherwise, due to the linearity of the problem one can superimpose the solution of the direct problem with homogeneous initial and boundary conditions but containing a heat source $F(x, t)$ in the right hand-side of (2.52). For the same reason, it can be assumed that the initial temperature u_0 in (2.55) is homogeneous, i.e.

$$u(x, 0) = 0, \quad x \in [0, l]. \tag{2.57}$$

Assuming further that the Cauchy data $\mu_l(t)$ and $q_l(t)$ defined in (2.53) and (2.54), respectively, belong to $L^2(\mathbb{R})$ (after extending them with zero-values for negative times), a simple Fourier transform analysis on the IHCP given by equations (2.52)–(2.54), (2.57), yields that [311],

$$\begin{pmatrix} \hat{\mu}_0(\tau) \\ \hat{q}_0(\tau) \end{pmatrix} = \begin{pmatrix} \cosh(\beta) & -\frac{\sinh(\beta)}{\beta} \\ -\beta\sinh(\beta) & \cosh(\beta) \end{pmatrix} \begin{pmatrix} \hat{\mu}_l(\tau) \\ \hat{q}_l(\tau) \end{pmatrix}, \tag{2.58}$$

where $\beta = \sqrt{|\tau|/2}(1 + i\ \mathrm{sgn}(\tau))$. Although the exact Cauchy data $(\mu_l(t), q_l(t)) \in L^2(\mathbb{R})^2$ may be such that $(\hat{\mu}_l(\tau), \hat{q}_l(\tau))$ decay faster than $\exp(-\sqrt{|\tau|/2})$ for large $|\tau|$, for the desired output $(\mu_0(t), q_0(t))$ at $x = 0$ to be in $L^2(\mathbb{R})^2$, the noisy data $(\mu_l^\epsilon(t), q_l^\epsilon(t))$ will still be in $L^2(\mathbb{R})^2$ satisfying

$$\|\mu_l - \mu_l^\epsilon\|_{L^2(\mathbb{R})} \leq \epsilon, \quad \|q_l - q_l^\epsilon\|_{L^2(\mathbb{R})} \leq \epsilon, \tag{2.59}$$

but its Fourier transform $(\hat{\mu}_l^\epsilon(\tau), \hat{q}_l^\epsilon(\tau))$ will not necessarily decay so fast as its noiseless counterpart. Therefore, the product between the matrix (containing the amplification factor $\exp(\sqrt{|\tau|/2})$) and the perturbed noisy vector $(\hat{\mu}_l^\epsilon(\tau), \hat{q}_l^\epsilon(\tau))^{\mathrm{T}}$ will not necessarily be in $L^2(\mathbb{R})^2$ anymore, and consequently, this shows that the IHCP is severely ill-posed in the high-frequency Fourier components [311], as any small noise $\epsilon > 0$ in (2.59) will drastically be amplified by the exponential factor $\exp(\sqrt{|\tau|/2})$ for large $|\tau|$. Another important remark is that one cannot obtain the temperature $u(0, T)$ at $t = T$ from the Cauchy data on the interval $[0, T]$ only [108], and for that one needs to consider the Cauchy data (2.53) and (2.54) on an extended 'future temperature' interval $[0, T']$ with $T' > T$, [263, 344].

As with the Cauchy problem for Laplace's equation tackled in section 2.1, the IHCP was thoroughly investigated in the literature using various numerical methods, e.g., the function specification method [26], the conjugate gradient method [7, 344], the 'optimal filtering' method [358], the mollification method [309], the wavelet, wavelet-Galerkin and spectral regularization methods [109, 343], the sequential windowing of the data [34], the Fourier method [394], the Tikhonov regularization method [300], the Landweber-Fridman and minimal error methods [24], the energy method [19], the hyperbolic perturbation [383], etc.

In the next subsection, an inverse boundary-value model is described for an application that requires determining the upstream flow of a diffusion column from measurements of the flow at the bottom of column [275].

2.4.2 Inverse modelling in elution chromatography

An inverse boundary-value problem for the convection-diffusion parabolic equation arises when modelling the release of hormones from secretory cells in response to a stimulus in a medium flowing past them and through a diffusion column [369]. It is assumed that the one-dimensional layer of diffusion of length $l > 0$ is accompanied by forced convection so that the concentration u of hormones satisfies the convection-diffusion equation

$$u_t + vu_x = Du_{xx}, \quad (x, t) \in (0, l) \times (0, \infty), \tag{2.60}$$

where x and v are the distance and flow velocity down the column and $D > 0$ is the diffusivity. A similar situation arises in the dispersion of pollutants in rivers [74], or in the forced convection cooling of flat electronic substrates [93].

At the initial time, the concentration is uniform and can be taken to be zero as in (2.57). The bottom of the column is insulated so a zero flux adiabatic boundary condition applies at $x = l$, namely,

$$u_x(l, t) = 0, \quad t \in (0, \infty). \tag{2.61}$$

By balancing the concentration flow across the inlet, at $x = 0$, [367],

$$-\frac{D}{v}u_x(0, t) + u(0, t) = h(t), \quad t \in (0, \infty), \tag{2.62}$$

where $h(t)$ represents the concentration of the hormones secreted at the top of the column $x = 0$. Then, in elution chromatography [368, 369], the question arises on the possibility of determining h from concentration measurements at the bottom of the column, i.e.,

$$u(l, t) = \mu_l(t) = \text{measured}, \quad t \in (0, \infty). \tag{2.63}$$

In particular, it is required to determine the concentration and flux at the top of the column given by

$$u(0, t) = \mu_0(t) =?, \quad -\kappa u_x(0, t) = q_0(t) =?, \quad t \in (0, \infty), \tag{2.64}$$

where $\kappa > 0$ is the conductivity.

In summary, the elution chromatography has been modelled as a Cauchy problem for the convection-diffusion equation (2.60) subject to the initial condition (2.57) that requires determining the upstream flow (2.64) at the top of column from the downstream quantities (2.61) and (2.63) at the bottom of column. Imposing all three conditions (2.57), (2.61) and (2.63) is an over-specification that requires consistency between this data being satisfied.

The case $v \neq$ constant

When v is not a constant, the solution of the non-characteristic Cauchy problem (2.60), (2.61) and (2.63) can be obtained in the decomposition form [256]:

$$u(x, t) = \mathrm{u}_0(x, t) + \sum_{n=1}^{\infty} u_n(x, t), \tag{2.65}$$

where

$$\mathrm{u}_0 = \mu_l(t), \quad u_{n+1} = \frac{1}{D}L_{xx}^{-1}[L_t u_n + v(x, t)L_t u_n], \quad n \in \mathbb{N}, \tag{2.66}$$

where $L_t = \partial/\partial t$, $L_x = \partial/\partial x$, $L_{xx} = \partial^2/\partial x^2$ and $L_{xx}^{-1} = \int_l^x dx' \int_l^{x'} dx''$, [2].

Example 2.1. ([256]) Take $D = l = 1$, $v(x,t) = 2(x-1) + \frac{1-t}{x-1}$ and the exact solution $u(x,t) = e^{(x-1)^2+t^2}$. Then, $\mu_1(t) = e^{t^2}$ and the recurrence relation (2.66) gives

$$u_0 = e^{t^2}, \quad u_1 = (x-1)^2 t e^{t^2},$$
$$u_2 = (x-1)^2 \left[\frac{(x-1)^2}{12}(1+4t+2t^2) + t(1-t)\right] e^{t^2},$$

etc. and, in general, for any $n \geq 1$,

$$u_n(x,t) = e^{t^2} \left(\sum_{k=0}^{n} a_k^n(t)(x-1)^{2k}\right), \tag{2.67}$$

where $a_k^n(t)$ are functions of t to be determined from the recurrence relations

$$a_0^0(t) = 1, \quad a_0^n(t) = 0, \quad a_1^n(t) = t(1-t)^{n-1}, \quad n \geq 1,$$
$$a_{k+1}^{n+1}(t) = \frac{(2t+4k)a_k^n + (2k+2)(1-t)a_{k+1}^n(t) + (a_k^n)'(t)}{(2n+1)(2n+2)},$$
$$n \geq 2, \quad k = \overline{1,(n-1)},$$
$$a_{n+1}^{n+1}(t) = \frac{(2t+4n)a_n^n(t) + (a_n^n)'(t)}{(2n+1)(2n+2)}, \quad n \geq 1. \tag{2.68}$$

From this, infer that $a_k^n(t)$ is a polynomial of degree n in t, thus seek

$$a_k^n(t) = \sum_{m=0}^{n} \alpha(n,k,m)t^m, \quad k = \overline{0,n}, \tag{2.69}$$

where $\alpha(n,k,m)$ are unknown coefficients to be determined. From equations (2.68) and (2.69), the first few terms are given by

$$\alpha(0,0,0) = 1, \quad \alpha(n,0,l) = 0, \quad n \geq 0, \quad l = \overline{0,n},$$
$$\alpha(n,1,0) = 0, \quad \alpha(n,1,l) = (-1)^{l-1} C_{n-1}^{l-1}, \quad n \geq 1, \quad l = \overline{1,n},$$
$$\alpha(2,0,0) = \alpha(2,0,1) = \alpha(2,0,2) = \alpha(2,1,0) = 0, \quad \alpha(2,1,1) = 1,$$
$$\alpha(2,1,2) = -1, \quad \alpha(2,2,0) = 1/12, \quad \alpha(2,2,1) = 1/3, \quad \alpha(2,2,2) = 1/6$$

and, in general, for $n \geq 2$ and $k, m = \overline{1, (n-1)}$,

$$\alpha(n+1,k+1,0) = \frac{4k\alpha(n,k,0) + (2k+2)\alpha(n,k+1,0) + \alpha(n,k,1)}{(2k+1)(2k+2)},$$

$$\alpha(n+1,k+1,m) = \frac{2\alpha(n,k,m-1) + 4k\alpha(n,k,m) + (m+1)\alpha(n,k,m+1)}{(2k+1)(2k+2)} + \frac{\alpha(n,k+1,m) - \alpha(n,k+1,m-1)}{2k+1},$$

$$\alpha(n+1,k+1,n) = \frac{2\alpha(n,k,n-1) + 4k\alpha(k,n,n) + \alpha(n,k+1,n)}{(2k+1)(2k+2)} - \frac{\alpha(n,k+1,n-1)}{2k+1},$$

$$\alpha(n+1,k+1,n+1) = \frac{2\alpha(n,k,n) - (2k+2)\alpha(n,k+1,n)}{(2k+1)(2k+2)},$$

$$\alpha(n+1,n+1,0) = \frac{4n\alpha(n,n,0) + \alpha(n,n,1)}{(2n+1)(2n+2)},$$

$$\alpha(n+1,n+1,m) = \frac{4n\alpha(n,n,m) + 2\alpha(n,n,m-1) + (m+1)\alpha(n,n,m+1)}{(2n+1)(2n+2)},$$

$$\alpha(n+1,n+1,n) = \frac{4n\alpha(n,n,n) + 2\alpha(n,n,n-1)}{(2n+1)(2n+2)},$$

$$\alpha(n+1,n+1,n+1) = \frac{2\alpha(n,n,n)}{(2n+1)(2n+2)}.$$

Evaluating numerically these recurrence relations for $\alpha(n,k,l)$, the general term of the decomposition (2.65) is obtained as

$$u_n(x,t) = e^{t^2}\left(\sum_{k=0}^{n}(x-1)^{2k}\left\{\sum_{m=0}^{n}\alpha(n,k,m)t^m\right\}\right).$$

Convergence of the sequence of partial sums of series (2.65) divided by e^{t^2},

$$S_N(x,t) := 1 + e^{-t^2}\sum_{n=1}^{N}u_n(x,t), \tag{2.70}$$

to the exact solution $e^{-t^2}u(x,t) = e^{(x-1)^2}$ is illustrated in Figure 2.12, which presents the relative errors between the terms $S_3(x,1)$, $S_5(x,1)$ and $S_7(x,1)$ and the exact $e^{(x-1)^2}$, as functions of $x \in [0,1]$. It can be seen that an excellent relative error of less than 0.1% can be achieved by taking only $N = 5$ to 7 terms in (2.70). However, the sequence $\{S_N\}_{N\geq 0}$ is not convergent for all values of t, and, in fact the time coefficients $a_1^n(t) = t(1-t)^{n-1}$ for the power $(x-1)^2$ in equation (2.68) suggest that the series is convergent only for $t \in [0,2)$. This is numerically illustrated in Figure 2.13 which presents the terms $S_3(0,t)$, $S_7(0,t)$ and $S_{11}(0,t)$, as a function of $t \in [0,2]$, in comparison with the analytical value for $e^{-t^2}u(0,t) = e$. Results are presented only at the boundary $x = 0$ since at this location the problem is the most ill-posed. From this figure it can be seen that as N increases the sequence $S_N(0,t)$ converges to the exact limit e for $t \in [0,2)$, but for $t \geq 2$ it starts to diverge. To summarize, it was found that the sequence (2.70), resulting from the decomposition (2.65) and (2.66), converges rapidly to the exact solution for all $(x,t) \in [0,1] \times [0,2)$.

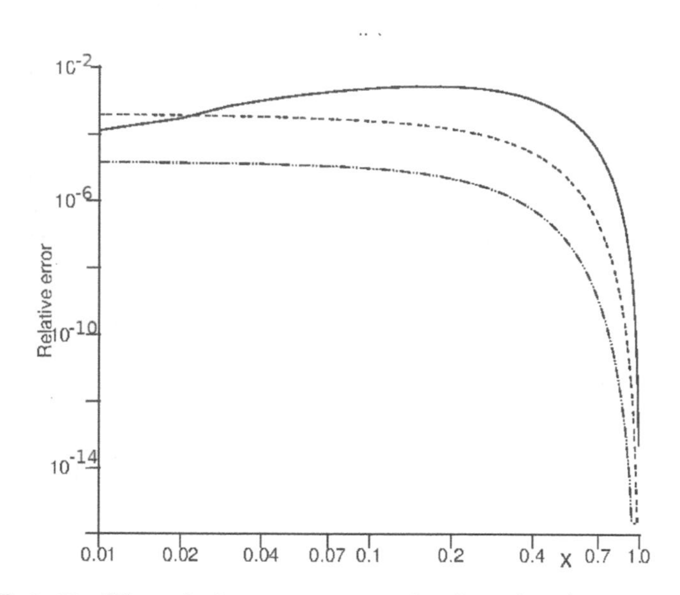

FIGURE 2.12: The relative errors, on a log-log plot, between the terms $S_3(x,1)$ (—), $S_5(x,1)$ (– – –), $S_7(x,1)$ (–..–) and the exact value $e^{(x-1)^2}$.

The case v = constant

In case v = constant, the transformation

$$u(x,t) = U(x,t)\exp\left(\frac{vx}{2D} - \frac{v^2t}{4D}\right) \tag{2.71}$$

changes the reaction-diffusion equation (2.60) into the diffusion equation

$$U_t = DU_{xx} \quad \text{for } (x,t) \in (0,l) \times (0,\infty), \tag{2.72}$$

while equations (2.57), (2.61)–(2.64) are changed into

$$U(x,0) = 0, \quad x \in [0,l], \tag{2.73}$$

$$\kappa U_x(l,t) + \frac{v\kappa}{2D}U(l,t) = 0, \quad t \in (0,\infty). \tag{2.74}$$

$$-\frac{D}{v}U_x(0,t) + \frac{1}{2}U(0,t) = h(t)\exp\left(\frac{v^2t}{4D}\right), \quad t \in (0,\infty), \tag{2.75}$$

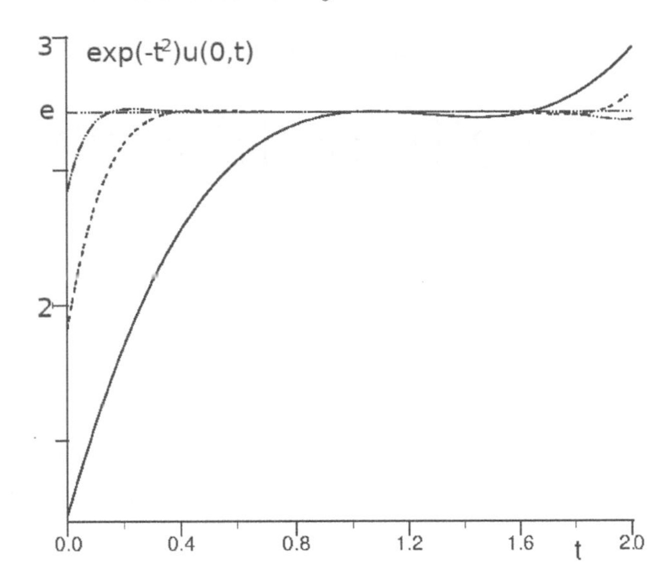

FIGURE 2.13: The terms $S_3(0,t)$ (—), $S_7(0,t)$ (− − −) and $S_{11}(0,t)$ (−..−), in comparison with the analytical $e^{-t^2}u(0,t) = e$ (−...−).

$$U(l,t) = \mu_l(t) \exp\left(-\frac{Pe}{2} + \frac{v^2 t}{4D}\right) = \text{measured}, \quad t \in (0,\infty). \qquad (2.76)$$

$$U(0,t) \exp\left(-\frac{v^2 t}{4D}\right) = \mu_0(t) =?,$$

$$-\kappa\left(U_x(0,t) + \frac{v}{2D}U(0,t)\right) \exp\left(-\frac{v^2 t}{4D}\right) = q_0(t) =?, \quad t \in (0,\infty), \qquad (2.77)$$

where $Pe = vl/D$ is the longitudinal Peclet number. If the initial condition (2.55) is not imposed, then, assuming that the function

$$t \mapsto -\frac{v\kappa}{2D}\mu_l(t) \exp\left(\frac{v^2 t}{4D}\right)$$

$$+\frac{1}{\sqrt{\pi}} \int_0^t (t-\tau)^{-1/2} \left(\mu_l'(\tau) + \frac{v^2}{4D}\mu_l(\tau)\right) \exp\left(\frac{v^2 \tau}{4D}\right) d\tau, \quad t \in (0,\infty)$$

is of Holmgren class two, it follows that the non-characteristic Cauchy problem (2.60), (2.61), and (2.63) has the unique solution [368],

$$u(x,t) = \sum_{n=0}^{\infty} \frac{(l-x)^{2n}}{D^n (2n)!}\, {}_1F_1(n; 2n+1; -v(l-x)/D)\mu_l^{(n)}(t), \qquad (2.78)$$

where ${}_1F_1(a;b;X)$ is the confluent hyper-geometric function given by [282],

$$ {}_1F_1(a;b;X) = 1 + \sum_{k=0}^{\infty} \left(\prod_{j=0}^{k} \left(\frac{a+j}{b+j} \right) \right) \frac{X^{k+1}}{(k+1)!}, \quad b \notin \mathbb{Z}_-. $$

When $v = 0$, i.e. no convection, then equation (2.60) becomes the classical diffusion equation (2.52), and (2.78) simplifies to [44],

$$ u(x,t)\Big|_{v=0} = \sum_{n=0}^{\infty} \frac{(l-x)^{2n}}{D^n (2n)!} \mu_l^{(n)}(t). \tag{2.79} $$

Similar analytical expressions are available for the 1-D diffusion of heat perpendicular to surfaces of co-axial cylinders and concentric spheres [246]. Applying the Fourier transform with respect to the time t to (2.60) yields

$$ D\hat{u}_{xx}(x,w) = iw\hat{u}(x,w) + v\hat{u}_x(x,w), \quad (x,w) \in (0,l) \times (-\infty,\infty). $$

Solving this second-order ODE gives

$$ \hat{u}(x,w) = \exp(vx/(2D)) \Big[A(w) \exp((\alpha_+ + i\alpha_-)x) + $$
$$ B(w) \exp(-(\alpha_+ + i\alpha_-)x) \Big], \quad (x,w) \in (0,l) \times (-\infty,\infty), $$

where

$$ \alpha_{\pm}(v,w) = \frac{1}{2D} \left[\frac{(v^4 + 16D^2w^2)^{1/2} \pm v^2}{2} \right]^{1/2} \tag{2.80} $$

and the coefficients $A(w)$ and $B(w)$ are determined from the Fourier transformed boundary conditions (2.61) and (2.63), where it was assumed that μ_l was extended to the whole real axis and that its Fourier transform exists and is finite. It can be observed that obtaining $\mu_0(t)$, $q_0(t)$ from $\mu_l(t)$ amplifies the error in the high-frequency components by the factor $e^{\alpha_+(v,w)}$. Since $\alpha_+(u,w) \geq \sqrt{|w|/(2D)}$, it follows that the inverse problem is ill-posed. From equation (2.80) it can be seen that the instability of the solution increases as $|v|$ increases, i.e., as $|Pe|$ increases. Also, positive v, i.e. advection, makes the problem more ill-posed than negative v, i.e. convection, [340].

If $\mu_l \in W^\infty(\mathbb{R}) = \{\mu \in S'(\mathbb{R}) \mid \text{supp}(\hat{\mu}) \text{ is a compact set of } \mathbb{R}\}$, where $\hat{\mu}$ is the modified Fourier transform of μ, namely,

$$ \hat{\mu}(w) = \frac{e^{Pe/2}}{2\pi} \int_{-\infty}^{\infty} \mu(t) e^{-t(iw+v^2/(4D))} dt $$

then the problem (2.60), (2.61) and (2.63) is well-posed in $C^2([0,l]; W^\infty(\mathbb{R}))$,

[139]. Furthermore, for a fixed $x \in [0, l]$, the series in (2.78) converges in the $W^{\infty}(\mathbb{R})$-topology [144].

In practice, only the noisy measurement μ_{lm} of μ_l is available. Then, one could first filter the data, for example, using mollifiers [309], as follows. Let us define the space of functions

$$\mathcal{S} := \{\phi \in C(\mathbb{R});$$

$$\phi(t) = \left\{ \begin{array}{cc} \phi(0) \exp\left(\frac{t^2}{t^2-(3\delta)^2}\right), & t \in (-3\delta, 0] \\ \phi(T) \exp\left(\frac{(t-T)^2}{(t-T)^2-(3\delta)^2}\right), & t \in [T, T+3\delta) \\ 0, & t \in (-\infty, -3\delta] \cup [T+3\delta, \infty) \end{array} \right\}$$

where $\delta > 0$ is the radius of mollification, which plays the role of the regularization parameter. From this definition it follows that the maximum norm on $\mathcal{S}$ is well-defined since $\|\phi\| = \max_{t\in\mathbb{R}}|\phi(t)| = \max_{t\in[0,T]}|\phi(t)| < \infty$ for all $\phi \in \mathcal{S}$. For any function $\phi \in \mathcal{S}$, define its mollification by [288],

$$J_\delta \phi(t) = (\rho_\delta * \phi)(t) = \int_{-\infty}^{\infty} \rho_\delta(t-s)\phi(s)ds, \tag{2.81}$$

where

$$\rho_\delta(t) = \frac{1}{\delta\sqrt{\pi}} \exp\left(-\frac{t^2}{\delta^2}\right) \tag{2.82}$$

is the Gaussian mollifier. Remark that $\int_{-\infty}^{\infty} \rho_\delta(t)dt = 1$. Since $J_\delta\phi \in C^{\infty}(\mathbb{R})$, the difficulty in applying formula (2.78), when μ_l is noisily measured as μ_{lm}, is overcome. The mollification method [309] then stabilizes the ill-posed problem by reconstructing the $\delta-$mollifications $J_\delta u(0,t) := J_\delta \mu_0$ and $J_\delta u_x(0,t) := J_\delta q_0$ of the functions $\mu_0(t)$ and $q_0(t)$, respectively. Mollifying (2.60), (2.61), and (2.63) with respect to t, the following associated problem is obtained:

$$\left\{ \begin{array}{ll} (J_\delta u)_t(x,t) + v(J_\delta u_x)(x,t) = D(J_\delta u_{xx})(x,t), & (x,t) \in (0,l) \times (0,\infty), \\ J_\delta u(l,t) = J_\delta \mu_{lm}(t), \quad D J_\delta u_x(l,t) = 0, & t \in [0,\infty), \end{array} \right. \tag{2.83}$$

whose solution is given by, see also (2.78),

$$J_\delta u(x,t) = \sum_{n=0}^{\infty} \frac{(l-x)^{2n}}{D^n (2n)!} \, {}_1F_1\left(n; 2n+1; -\frac{v(l-x)}{D}\right) (J_\delta \mu_{lm})^{(n)}(t). \tag{2.84}$$

The analysis in Theorems 2.2–2.4 below follows [275].

Theorem 2.2. (Existence). *The solution given by* (2.84) *exists, i.e., the series involved converges uniformly and absolutely for* $(x,t) \in [0,l] \times [0,\infty)$.

Proof. First, it is shown that for any δ and for any $\epsilon_1 > 0$ there is a constant $c_1 = c_1(\delta, \epsilon_1) > 0$ such that

$$|(J_\delta \phi)^{(n)}(t)| \leq \frac{c_1 (2n)!}{(1+\epsilon_1)^{2n}}, \quad t \in \mathbb{R}, \quad n \geq 0, \tag{2.85}$$

where $\phi \in \mathcal{S}$, [260]. Clearly, since

$$|(J_\delta \phi)(t)| \leq \|\phi\| \int_{-\infty}^{\infty} \rho_\delta(t-s)ds = \|\phi\| \leq \infty,$$

it follows that (2.85) holds for $n=0$ by taking $c_1 = \|\phi\|$. For $n > 0$,

$$|(J_\delta \phi)^{(n)}(t)| = \left| \int_{-\infty}^{\infty} \frac{d^n}{dt^n}(\rho_\delta(t-s))\phi(s)ds \right| \leq \int_{-\infty}^{\infty} |\rho_\delta^{(n)}(s)||\phi(t-s)|ds.$$

Based on (2.82), [115],

$$\rho_\delta^{(n)}(s) = \frac{(-1)^n H_n(s/\delta)\exp(-s^2/\delta^2)}{\delta^{n+1}\sqrt{\pi}}, \tag{2.86}$$

where H_n is the Hermite polynomial of order n satisfying

$$|H_n(\xi)| < A\exp(\xi^2/2)\sqrt{2^n n!}, \quad \text{where } A \approx 1.08. \tag{2.87}$$

Now, from the above it can be obtained that for any δ and for any $\epsilon_1 > 0$ there is a constant $c_1 = c_1(\delta, \epsilon_1) > 0$ such that

$$|(J_\delta \phi)^{(n)}(t)| \leq \frac{A\sqrt{2^n n!}}{\delta^{n+1}\sqrt{\pi}} \int_{\infty}^{\infty} |\phi(s)|ds \leq \frac{c_1(2n)!}{(1+\epsilon_1)^{2n}}, \quad t \in \mathbb{R},\ n \geq 0. \tag{2.88}$$

Since $\mu_{lm} \in \mathcal{S}$, then $J_\delta \mu_{lm}$ will satisfy (2.88), and using that

$$\left| {}_1F_1\left(n; 2n+1, -\frac{v(l-x)}{D}\right) \right| \leq \frac{1}{2}\left(1 + \exp\left(\left|\frac{v(x-l)}{D}\right|\right)\right),$$

the series (2.84) is majorized by the power series

$$\frac{c_1}{2}\left(1+\exp\left(\left|\frac{v(x-l)}{D}\right|\right)\right)\sum_{n=0}^{\infty}\left(\frac{x-l}{\sqrt{D}(1+\epsilon_1)}\right)^{2n}.$$

Choosing $\epsilon_1 > \max\{0, l/\sqrt{D} - 1\}$, it follows that this series is absolutely convergent for any $x \in [0,1]$.

By similar arguments $(J_\delta u)_t$, $J_\delta u_x$ and $J_\delta u_{xx}$ may be shown to exist and be the same as the result of term-by-term differentiation in (2.84). Consequently, the exact formula (2.84) holds, proving the existence of the solution $J_\delta u$. By applying (2.84) at $x=0$, the mollified boundary concentration $J_\delta \mu_0$, $J_\delta \mu_{0m}$ and the mollified flux $J_\delta q_0$, $J_\delta q_{0m}$ can be determined.

The uniqueness of the solution follows from the analyticity of the solution (2.84) of problem (2.83) in the space variable x. Next, it is shown that the solution (2.84) depends continuously on the data $\mu_{lm}(t)$.

Theorem 2.3. (Stability). *The problem (2.83) is stable with respect to perturbations in the overspecified data.*

Proof. Let us assume that the exact μ_l and the measured μ_{lm} data functions belonging to $\mathcal{S}$ satisfy $\|\mu_l - \mu_{lm}\| \leq \epsilon$. Then,

$$\left|J_\delta(\mu_{lm} - \mu_l)^{(n)}(t)\right| = \left|\int_{\infty}^{\infty} (\mu_{1m}(s) - \mu_1(s)) \frac{d^n \rho_\delta}{dt^n}(t-s)ds\right|$$

$$\leq \epsilon \int_{-\infty}^{\infty} |\rho_\delta^{(n)}(s)|ds, \qquad (2.89)$$

and, based on (2.86) and (2.87),

$$\|(J_\delta \mu_{lm})^{(n)} - (J_\delta \mu_l)^{(n)}\| \leq \frac{\epsilon A(2^{n+1}n!)^{1/2}}{\delta^n}. \qquad (2.90)$$

Based on (2.84) and (2.90)

$$\|J_\delta \mu_{0m} - J_\delta \mu_0\|$$

$$\leq \sum_{n=0}^{\infty} \left| {}_1F_1(n; 2n+1; -Pe)\right| \frac{\epsilon A(2^{n+1}n!)^{1/2} l^{2n}}{D^n \delta^n (2n)!}$$

$$\leq 2^{-1/2} \epsilon A(1 + e^{|Pe|}) \sum_{n=0}^{\infty} \frac{(2^n n!)^{1/2} l^{2n}}{(D\delta)^n (2n)!}. \qquad (2.91)$$

Now since $\frac{(2^n n!)^{1/2}}{(2n)!} \leq \frac{1}{n!}$ for all $n \in \mathbb{N}$,

$$\|J_\delta \mu_{0m} - J_\delta \mu_0\| \leq c_2 \epsilon \exp\left(\frac{l^2}{D\delta}\right), \qquad (2.92)$$

where $c_2 = A\left(1 + e^{|Pe|}\right)/\sqrt{2}$, and a similar estimate holds for the flux. Inequality (2.92) indicates that for a fixed $\delta > 0$, the error tends to zero as $\epsilon \to 0$. This shows that the mollified inverse problem (2.83) is stable with respect to perturbations in the input data.

Theorem 2.4. (Error estimate). *If the desired boundary data μ_0 belongs to $\mathcal{S} \cap C^1(\mathbb{R})$ and max$\{\|\mu_0'\|\} \leq M$, then*

$$\|\mu_0 - J_\delta \mu_{0m}\| \leq \frac{M\delta}{\sqrt{2}} + c_2 \epsilon \exp\left(\frac{l^2}{D\delta}\right). \qquad (2.93)$$

Proof. Using that μ_0 is a M−Lipschitz function

$$|\mu_0(t) - J_\delta \mu_0(t)| \leq M \int_{-\infty}^{\infty} \frac{|t-s|}{\delta\sqrt{\pi}} \exp\left(-(t-s)^2/\delta^2\right) ds = \frac{M\delta}{\sqrt{\pi}}.$$

Then, using (2.90) results in

$$\|\mu_0 - J_\delta \mu_{0m}\| \leq \|\mu_0 - J_\delta \mu_0\| + \|J_\delta \mu_0 - J_\delta \mu_{0m}\| \leq \frac{M\delta}{\sqrt{\pi}} + c_2 \epsilon \exp\left(\frac{l^2}{D\delta}\right),$$

as required. Now choosing $\delta = \overline{\delta} = -\frac{l^2}{D\ln(\sqrt{\epsilon})}$ in (2.93) it yields the error estimate

$$\|\mu_0 - J_{\overline{\delta}} \mu_{0m}\| \leq -\frac{Ml^2}{D\sqrt{\pi}\ln(\sqrt{\epsilon})} + c_2\sqrt{\epsilon},$$

whose right-hand side tends to zero as $\epsilon \to 0$.

2.5 Conclusions

In this chapter, inverse boundary-value problems for various linear operators have been considered. Uniqueness of solution holds from the unique continuation principle but the global existence and stability are in general violated. In case of elliptic stationary problems, the operators were the Laplace's, biharmonic and Stokes operators, whilst for parabolic non-stationary problems the heat or convection-diffusion operators have been considered. In these latter cases, the boundary Cauchy data has been supplied on a fixed non-characteristic curve but problems involving moving boundaries, such as inverse boundary-value Stefan problems can also be studied, see [62, 213, 342] and the references therein. Inverse boundary-value problems for the Helmholtz equation [307], wave equation [233], and in elasticity [240, 290] and thermoelasticity [241, 291] are also of interest.

Chapter 3

Inverse Initial-Value Problems

Imagine you wake up in the morning and pour water in the kettle to make tea but, because you either do not have time or simply have forgotten, you do not measure the temperature of the water at the instant of pouring. Then, after about $T = 5$ minutes the water has reached its boiling point (of 100°C), and you wonder what was the initial temperature $u(x,0) = u_0(x) =?$. This gives rise to the backward heat conduction problem (BHCP) given by

$$u_t = u_{xx}, \quad (x,t) \in (0, l = \pi) \times (0,T), \tag{3.1}$$

$$u(0,t) = u(l,t) = 0, \quad t \in (0,T), \tag{3.2}$$

$$u(x,T) = u_T(x), \quad x \in [0, l = \pi], \tag{3.3}$$

where u_T represents the temperature measurement at the final time $t = T$ (or on the 'upper base' of the cylinder $(0,l) \times (0,T)$). Since the direct problem operator $u(\cdot,0) \mapsto u_T(\cdot)$ has strong smoothing (diffusing) properties, no solution to the above BHCP exists if the data u_T is not analytic. The Fourier method of separating variables (taking $l = \pi$) yields formally

$$u(x,t) = \frac{2}{\pi}\sum_{n=1}^{\infty}\left(\int_0^{\pi} u_0(y)\sin(ny)dy\right)\sin(nx)e^{-n^2 t}. \tag{3.4}$$

Taking $t = T$ in (3.4), the Fredholm integral equation of the first kind

$$\int_0^{\pi} k(x,y)u_0(y)dy = u_T(x), \quad x \in [0,\pi], \tag{3.5}$$

is obtained, where

$$k(x,y) = \frac{2}{\pi}\sum_{n=1}^{\infty}\sin(nx)\sin(ny)e^{-n^2 T}. \tag{3.6}$$

The linear integral equation (3.5) is ill-posed because the inverse operator associated to it is not continuous. On the other hand, if the sine Fourier coefficients $u_T^n := \sqrt{\frac{2}{\pi}}\int_0^{\pi} u_T(y)\sin(ny)dy$ of u_T decay sufficiently fast such that $\sum_{n=1}^{\infty}(u_T^n)^2 e^{2n^2} < \infty$ then, the initial temperature is given by

$$u_0(x) = \sqrt{\frac{2}{\pi}}\sum_{n=1}^{\infty} e^{n^2 T} u_T^n \sin(nx), \quad x \in [0,\pi]. \tag{3.7}$$

DOI: 10.1201/9780429400629-3

This formula gives the initial temperature $u(x,0) = u_0(x)$ backwards in time. However, it is useless in practice, since it is unstable with respect to small errors in u_T being amplified by a factor of e^{n^2T}. Hence, the BHCP is severely ill-posed and, in order to obtain a stable solution, regularization needs to be enforced, leading to various methods, as described in the next sections.

3.1 Quasi-Reversibility Methods

The most well-known quasi-reversibility method considers a perturbed fourth-order problem [247],

$$\begin{cases} u_t^\alpha = u_{xx}^\alpha + \alpha u_{xxxx}^\alpha, & (x,t) \in (0,\pi) \times (0,T), \\ u^\alpha(0,t) = u^\alpha(\pi,t) = 0, & t \in (0,T), \\ u_{xx}^\alpha(0,t) = u_{xx}^\alpha(\pi,t) = 0, & t \in (0,T), \\ u^\alpha(x,T) = u_T(x), & x \in [0,\pi], \end{cases} \tag{3.8}$$

where $\alpha > 0$ is the regularization parameter. It turns out that this perturbed problem is well-posed and its solution is

$$u^\alpha(x,t) = \frac{2}{\pi} \sum_{n=1}^{\infty} \left(\int_0^\pi u_T(y) \sin(ny) dy \right) \sin(nx) \times \exp\left[n^2(1 - \alpha n^2)(T-t)\right]. \tag{3.9}$$

It can be seen that the exponentially growing term $\exp(n^2(T-t))$, present when $\alpha = 0$, is damped out by the stabilizing decaying term $\exp(n^2(1 - \alpha n^2)(T-t))$. Let us then define the linear and continuous operator

$$R_\alpha : L_2[0,\pi] \to L_2[0,\pi], \quad R_\alpha u_T := u^\alpha(\cdot, 0)$$

and let $\|u_T - u_T^\epsilon\|_{L_2[0,\pi]} \leq \epsilon$. Then, if $\alpha = \alpha(\epsilon) \in (0,1)$ is chosen such that

$$\lim_{\epsilon \to 0} \alpha(\epsilon) = \lim_{\epsilon \to 0} \epsilon \, e^{1/(4\alpha(\epsilon))} = 0,$$

$R_{\alpha(\epsilon)} u_T^\epsilon$ converges in $L_2[0,\pi]$ to $u(\cdot, 0)$, provided that the latter exists [96]. This shows that $R_\alpha u_T^\epsilon$ may be considered as an approximate solution of $u(\cdot, 0)$.

An alternative well-posed perturbation called "pseudo-parabolic", which appears in models of fluid flow in fissured material, mixed heat conduction or consolidation of clay, is given by [326],

$$\begin{cases} u_t^\alpha = u_{xx}^\alpha + \alpha u_{xxt}^\alpha, & (x,t) \in (0,\pi) \times (0,T), \\ u^\alpha(0,t) = u^\alpha(\pi,t) = 0, & t \in (0,T), \\ u^\alpha(x,T) = u_T(x), & x \in [0,\pi], \end{cases} \tag{3.10}$$

and has the solution

$$u^\alpha(x,t) = \frac{2}{\pi}\sum_{n=1}^{\infty}\left(\int_0^\pi u_T(y)\sin(ny)dy\right)\sin(nx) \times \exp\left[\left(\frac{n^2}{\alpha n^2+1}\right)(T-t)\right]. \tag{3.11}$$

One can compare the approximations given by the above two quasi-reversibility methods by solving the well-posed forward problem

$$\begin{cases} U_t^\alpha = U_{xx}^\alpha, & (x,t)\in(0,\pi)\times(0,T),\\ U^\alpha(0,t) = U^\alpha(\pi,t) = 0, & t\in(0,T),\\ U^\alpha(x,0) = u^\alpha(x,0), & x\in[0,\pi], \end{cases} \tag{3.12}$$

where $u^\alpha(x,0)$ is given by (3.9) or (3.11) applied at $t=0$, and calculate [326],

$$\|U^\alpha(\cdot,T)-u_T(\cdot)\|^2_{L_2[0,\pi]} = \begin{cases} \sum_{n=1}^\infty (u_T^n)^2(1-\exp(-n^4\alpha T))^2 & \text{for (3.9)}\\ \sum_{n=1}^\infty (u_T^n)^2\left(1-\exp\left(-\frac{\alpha n^4 T}{1+\alpha n^2}\right)\right)^2 & \text{for (3.11)}, \end{cases}$$

where $u_T^n := \int_0^\pi u_T(y)\sin(ny)dy$. The right-hand side of this equation tends to zero, as $\alpha \searrow 0$, and is smaller (thus better) in case of approximation (3.10) than is in approximation (3.8).

A well-posed hyperbolic perturbation

$$\begin{cases} \alpha u_{tt} + u_t^\alpha = u_{xx}^\alpha, & (x,t)\in(0,\pi)\times(0,T),\\ u^\alpha(0,t) = u^\alpha(\pi,t) = 0, & t\in(0,T),\\ u^\alpha(x,1) = u_T(x),\ u_t^\alpha(x,1) = u_T''(x), & x\in[0,\pi], \end{cases}$$

leads to more complicated expressions. Nevertheless, there are many quasi-reversibility ways to perturb the original problem to obtain a stable solution via PDE-regularization [301, 326].

Finally, a Cauchy-type elliptic perturbation is given by [266, 374],

$$\begin{cases} u_t^\alpha = u_{xx}^\alpha + \alpha u_{tt}^\alpha, & (x,t)\in(0,\pi)\times(0,T),\\ u^\alpha(0,t) = u^\alpha(\pi,t) = 0, & t\in(0,T),\\ u^\alpha(x,T) = u_T(x),\ u_t^\alpha(x,T) = v_T(x), & x\in[0,\pi], \end{cases}$$

assuming that it is possible to solve first the direct and well-posed heat conduction problem upward from the time $t=T$, as an initial boundary value problem, and determine the time-derivative $u_t(x,T) =: v_T(x)$.

3.2 Logarithmic Convexity Methods

Let us define the energy of the heat conducting system by

$$F(t) = \frac{1}{2}\int_0^\pi u^2(x,t)dx = \|u(\cdot,t)\|^2_{L_2[0,\pi]}, \quad t\in[0,T],$$

then $\log(F(t))$ is a convex function of t. Indeed,

$$F'(t) = \int_0^\pi u\frac{\partial u}{\partial t}dx = \int_0^\pi u\frac{\partial^2 u}{\partial x^2}dx = -\int_0^\pi \left(\frac{\partial u}{\partial x}\right)^2 dx,$$

$$F''(t) = -2\int_0^\pi \frac{\partial u}{\partial x}\frac{\partial^2 u}{\partial x \partial t}dx = 2\int_0^\pi \frac{\partial^2 u}{\partial x^2}\frac{\partial u}{\partial t}dx = 2\int_0^\pi \left(\frac{\partial u}{\partial t}\right)^2 dx.$$

It follows then that

$$\frac{d^2(\log(F(t)))}{dt^2} = \frac{d}{dt}\left(\frac{F'(t)}{F(t)}\right) = \frac{F(t)F''(t) - (F'(t))^2}{F^2(t)}$$

$$= \frac{1}{F^2(t)}\left\{\int_0^\pi u^2(x,t)dx \int_0^\pi \left(\frac{\partial u}{\partial t}\right)^2 dx - \left(\int_0^\pi u\frac{\partial u}{\partial t}dx\right)^2\right\} \geq 0.$$

The fact that $\log(F(t))$ is a convex function of t leads directly to the inequality

$$F(t) \leq [F(1)]^t[F(0)]^{T-t}, \quad t \in [0,T].$$

Then, under the additional assumption that $\|u(\cdot,0)\|_{L_2[0,\pi]} \leq M$, the following conditional stability estimate:

$$\|u(\cdot,t)\|_{L_2[0,\pi]} \leq M^{T-t}\|u_T\|^t_{L_2[0,\pi]}, \quad t \in (0,T]$$

is obtained [326]. If this result is to be of practical interest, one must be able to compute the upper bound M from the physics of the problem.

3.3 Non-Local Initial-Value Methods

The method of non-local initial-value problems approximating the solutions of parabolic equations backward in time was investigated in several studies, e.g., [41, 142, 143]. In this section, the method is described in the more abstract framework of [142]. Thus, given $T > 0$ and $\epsilon > 0$, $\mathcal{H}$ a Hilbert space with inner product $(\cdot,\cdot)$ and norm $\|\cdot\|$, and letting the operator $A : D(A) \subset \mathcal{H} \to \mathcal{H}$ be a positive definite, self-adjoint with compact inverse, consider the BHCP

$$\begin{cases} u_t = -Au, & t \in (0,T), \\ \|u(T) - u_T\| \leq \epsilon, \end{cases} \tag{3.13}$$

which requires finding the missing initial data at $t = 0$. This ill-posed problem is regularized by the well-posed non-local quasi-initial value problem

$$\begin{cases} (v_\alpha)_t = -Au, & t \in (0,aT), \\ \alpha v_\alpha(0) + v_\alpha(aT) = u_T, \end{cases} \tag{3.14}$$

where $a > 1$ (for case $a = 1$, see [8, 379]) is a fixed given number, and $\alpha > 0$ is a regularization parameter. This problem is well-posed in $v_\alpha \in C^2((0, aT); \mathcal{H}) \cap C([0, aT]; \mathcal{H})$ with $v_\alpha(t) \in D(A)$ and has the representation [296],

$$v_\alpha(t) = \sum_{n=1}^{\infty} \frac{\exp(-\lambda_n t)}{a + \exp(-a\lambda_n T)} (u_T, \phi_n)\phi_n, \quad t \in [0, aT], \tag{3.15}$$

where $(\lambda_n^{-1}, \phi_n)_{n\geq 1}$ is an orthonormal system for A^{-1}. Then, $v_\alpha(t + (a-1)T)$ is taken as an approximation to $u(t)$ for $t \in [0, T]$. The proper choice of $\alpha > 0$ yields order-optimal regularization methods, as follows.

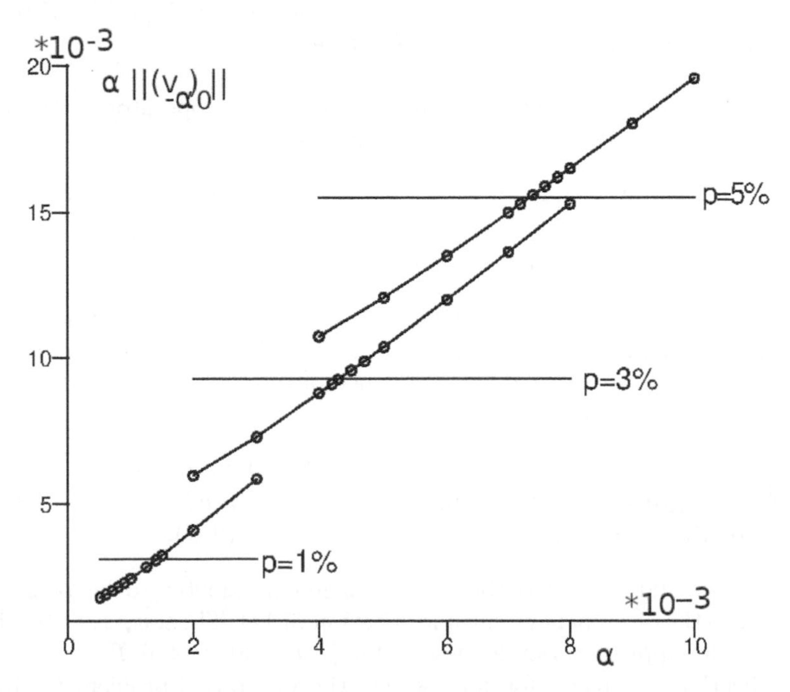

FIGURE 3.1: The graph (—o—) of the function $\alpha\|(\underline{v}_\alpha)_0\|$, as a function of α, for various percentages of noise $p \in \{1, 3, 5\}\%$. The horizontal lines correspond to the values of $\tau\epsilon$, with $\tau = 2$ and ϵ given by (3.25), [142].

Theorem 3.1. ([142]) *Suppose that there exists a known upper bound $E > 0$ for $\|u(0)\| \leq E$. Then:*
(i) with the a priori choice $\alpha = \alpha(\epsilon) = (\epsilon/E)^a$,

$$\|u(t) - v_\alpha(t + (a-1)T)\| \leq 2\epsilon^{t/T} E^{1-t/T}, \quad t \in [0, T], \tag{3.16}$$

(ii) (supposing that $\epsilon < \|u_T\|$ and letting $\tau > 1$ satisfy $\tau\epsilon < \|u_T\|$) with the a posteriori choice of $\alpha > 0$ such that

$$\alpha\|v_\alpha(0)\| = \|v_\alpha(aT) - f\| = \tau\epsilon, \tag{3.17}$$

$$\|u(t) - v_\alpha(t + (a-1)T)\| \leq (1+\tau)^{t/T} \left(\frac{2\tau - 1}{\tau - 1}\right)^{1-t/T} \epsilon^{t/T} E^{1-t/T},$$
$$t \in [0, T]. \quad (3.18)$$

Remark 3.1. The above methods are of optimal order. The continuous dependence of solution at the initial time $t = 0$ on the final-time data u_T cannot be established since the assumption $\|u(0)\| \leq E$ is too weak. However, imposing stronger assumptions lead to stability being achieved at $t = 0$ as well, as follows [142]:
(A_1) If there exist positive constants β and E_1 such that $\sum_{n=1}^{\infty} \lambda_n^{2\beta}(u(0), \phi_n) \leq E_1^2$, then one can get a logarithmic-type stability at $t = 0$.
(A_2) If there exist positive constants γ and E_2 such that $\sum_{n=1}^{\infty} e^{2\gamma\lambda_n}(u(0), \phi_n) \leq E_1^2$, then one can get a Hölder-type stability at $t = 0$.

For illustration, let us consider the BHCP (for the operator $A := -\partial_{xx}^2 : D(A) = H_0^1(0, \pi) \subset \mathcal{H} := L^2(0, \pi) \to \mathcal{H}$) given by

$$\begin{cases} u_t - u_{xx} = 0, & (x,t) \in (0,\pi) \times (0,T), \\ u(0,t) = u(\pi,t) = 0, & t \in (0,T), \\ u(x,T) = u_T(x)(1 + p\rho(x)) =: u_T^p(x), & x \in (0,\pi), \end{cases} \quad (3.19)$$

where $\rho(x)$ generates random variables from a uniform distribution in the interval $[-1, 1]$ and p is the percentage of noise. The solution of the ill-posed BHCP (3.19) is approximated by a sequence of solutions $(v_\alpha)_{\alpha>0}$ of the well-posed non-local quasi-initial value problems

$$\begin{cases} (v_\alpha)_t - (v_\alpha)_{xx} = 0, & (x,t) \in (0,\pi) \times (0,aT), \\ v_\alpha(0,t) = v_\alpha(\pi,t) = 0, & t \in (0,aT), \\ \alpha v_\alpha(x,0) + v_\alpha(x,aT) = u_T^p(x), & x \in (0,\pi), \end{cases} \quad (3.20)$$

where $a > 1$ and $\alpha > 0$ is the regularization parameter to be prescribed according to the discrepancy principle criterion (3.17). Then, $v_\alpha(x, t+(a-1)T)$ is taken as an approximation for $u(x,t)$ for $(x,t) \in [0,\pi] \times [0,T]$.

Application of Green's formula recasts the well-posed problem (3.20) for v_α as the following system of 3 linear integral equations with the 3 unknowns $(v_\alpha)_x(0,t)$, $(v_\alpha)_x(\pi,t)$ and $(v_\alpha)(x,0)$:

$$0 = \int_0^t G(x,t;\xi,\tau)\, \partial_{n(\xi)} v_\alpha(\xi,\tau)\Big|_{\xi \in \{0,\pi\}} d\tau$$
$$+ \int_0^\pi G(x,t;y,0) v_\alpha(y,0) dy, \quad t \in (0,at),\ x \in \{0,\pi\}, \quad (3.21)$$

$$u_T^p(x) - \alpha v_\alpha(x,0) = \int_0^{aT} G(x,aT;\xi,\tau)\, \partial_\xi v_\alpha(\xi,\tau)\Big|_{\xi=0}^{\xi=\pi} d\tau$$
$$+ \int_0^\pi G(x,aT;y,0) v_\alpha(y,0) dy, \quad x \in (0,\pi), \quad (3.22)$$

where $n(\xi)$ denotes the outward unit normal to the boundary $\xi \in \{0, \pi\}$ and

$$G(x, t; \xi, \tau) = \frac{\mathrm{H}(t-\tau)}{\sqrt{4\pi(t-\tau)}} \exp\left(-\frac{(x-\xi)^2}{4(t-\tau)}\right) \tag{3.23}$$

is the fundamental solution of the one-dimensional time-dependent heat equation (here H denotes the Heaviside function). The above boundary integral equations are discretized using a BEM with N boundary elements, (t_{j-1}, t_j), where $t_j = jaT/N$ for $j = \overline{0, N}$, dividing uniformly each of the boundaries $\{0\} \times (0, aT)$ and $\{\pi\} \times (0, aT)$, and N_0 cells, $[x_{k-1}, x_k]$, where $x_k = k\pi/N_0$ for $k = \overline{1, N_0}$, dividing uniformly the space domain $[0, \pi]$. Over each small boundary element and cell, the values of $v_\alpha|_{\xi \in \{0,\pi\}}$, $\partial_{n(\xi)} v_\alpha|_{\xi \in \{0,\pi\}}$ and $v_\alpha|_{\tau=0}$ are assumed constant and take the values at the midpoint (node) $\tilde{t}_j = (t_{j-1} + t_j)/2$ for $j = \overline{1, N}$, and $\tilde{x}_k = (x_{k-1} + x_k)/2$ for $k = \overline{1, N_0}$, of each boundary element and cell, respectively. Denoting by $\underline{v}_\alpha$, $\underline{v}'_\alpha$ and $(\underline{v}_\alpha)_0$ the vectors containing these values at the boundary nodes and midpoint cells, respectively, (3.21) and (3.22) recast as the following system of linear algebraic equations:

$$\begin{cases} \underline{0} = \mathcal{A}(x, \tilde{t}_i)\underline{v}'_\alpha + \mathcal{E}(x, \tilde{t}_i)(\underline{v}_\alpha)_0, & x \in \{0, \pi\},\ i = \overline{1, N}, \\ u_T^p(\tilde{x}_k) = \mathcal{A}(\tilde{x}_k, aT)\underline{v}'_\alpha + (\mathcal{E}(\tilde{x}_k, aT) + \alpha \mathbb{I}\underline{e}_k)(\underline{v}_\alpha)_0, & k = \overline{1, N_0}, \end{cases}$$

where $\mathbb{I}$ denotes the identity matrix of order N_0, $\underline{e}_k$ is the unit vector in the $k-$direction in $\mathbb{R}^{N_0}$, and the expressions

$$\mathcal{A}(x, t) = \left(\int_{t_{j-1}}^{t_j} G(x, t; \xi, \tau) d\tau\right)_{j=\overline{1,N}}, \quad \xi \in \{0, \pi\},$$

$$\mathcal{E}(x, t) = \left(\int_{x_{l-1}}^{x_l} G(x, t; y, 0) dy\right)_{l=\overline{1,N_0}},$$

can be evaluated analytically [117].

As an example, let us take $T = 1$, $a = 2$, $u_T(x) = e^{-1}\sin(x)$, such that the analytical solution of the BHCP (3.19) with $p = 0$ is

$$u(x, t) = e^{-t}\sin(x), \quad (x, t) \in [0, \pi] \times [0, 1]. \tag{3.24}$$

The above BEM is employed with $N = N_0 = 80$ boundary elements and cells. The values of $\epsilon := \|u_T - u_T^p\|_{L^2(0,\pi)}$ are

$$\epsilon = \begin{cases} 1.5 \times 10^{-3} & \text{if } p = 1\%, \\ 4.6 \times 10^{-3} & \text{if } p = 3\%, \\ 7.7 \times 10^{-3} & \text{if } p = 5\%. \end{cases} \tag{3.25}$$

To choose the regularization parameter α according to the discrepancy principle (3.17), Figure 3.1 shows the graph of $\alpha\|(\underline{v}_\alpha)_0\|$, as a function of $\alpha > 0$, intersecting the horizontal line $y = \tau\epsilon$. For $\tau = 2$, this selection process gives

$$\alpha(\epsilon) = \begin{cases} 1.4 \times 10^{-3} & \text{if } p = 1\%, \\ 4.3 \times 10^{-3} & \text{if } p = 3\%, \\ 7.4 \times 10^{-3} & \text{if } p = 5\%. \end{cases} \tag{3.26}$$

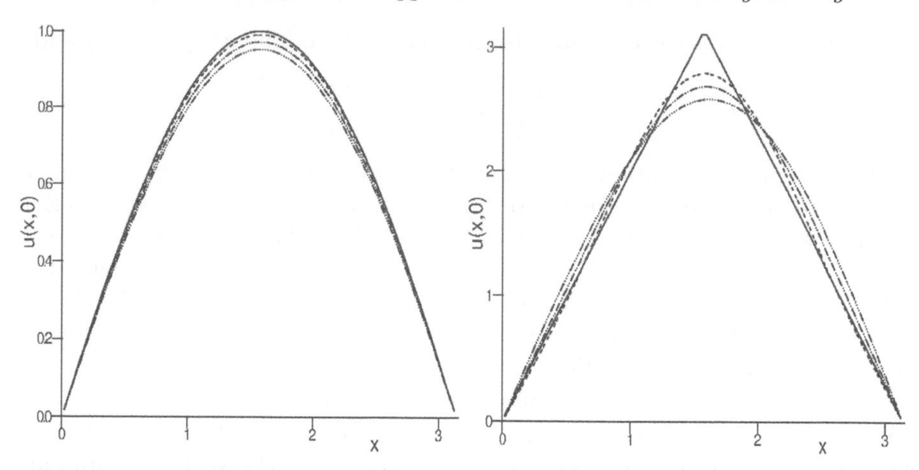

FIGURE 3.2: The analytical initial temperature $u(x,0)$ for $x \in [0,\pi]$, shown with continuous line (—), in comparison with the numerical approximations $v_{\alpha_\epsilon}(x,1)$, for various percentages of noise: $p = 1\%$ (− − −), $p = 3\%$ (-..-) and $p = 5\%$ (-....-). Example (3.24) on left and Example (3.27) on right [142].

Figure 3.2 (graph on the left) shows the analytical initial temperature $u(x,0) = \sin(x)$ in comparison with the numerical solution $v_{\alpha(\epsilon)}(x,1)$. It can be seen that the numerical solution is accurate and stable, and as the amount of noise p decreases, i.e. as $\epsilon \searrow 0$, the approximation $v_{\alpha(\epsilon)}(x,1)$ converges towards the analytical solution. The same conclusions have also been obtained [142] for a more complicated example with the analytical solution given by

$$u(x,t) = \frac{8}{\pi}\sum_{k=0}^{\infty}\frac{1}{(2k+1)^2}\cos\left[\frac{(2k+1)(2x-\pi)}{2}\right]e^{-(2k+1)^2 t},$$
$$(x,t) \in [0,\pi] \times [0,1], \qquad (3.27)$$

as shown in Figure 3.2 (graph on the right). In this case, the initial temperature profile that has to be retrieved is a non-smooth triangular function given by

$$u(x,0) = \begin{cases} 2x, & x \in [0,\pi/2], \\ 2(\pi - x), & x \in (\pi/2,\pi]. \end{cases}$$

Chapter 4

Space-Dependent Heat Sources

Inverse source problems, which are typical models for the identification of point sources of environmental pollution in a river or lake, or of the distributed heat sources in a conductor, are considered. Other applications arise in dislocation problems [1], photo-acoustic and fluorescence tomography [10, 83].

Unlike the previous chapters in which the unknowns were unspecified boundary or initial values of the solution, in this chapter the unknown source appears as a free term in the governing PDE. Consequently, this increase in the dimension of the space of unknowns requires possibly more interior domain measurement. This can easily be understood by considering the steady-state heat conduction in a bounded domain Ω given by the Poisson equation

$$\nabla^2 u(\underline{x}) = F(\underline{x}), \quad \underline{x} \in \Omega. \tag{4.1}$$

Clearly, the source $F(\underline{x})$ for $\underline{x} \in \Omega$ cannot be identified from information at the boundary $\partial\Omega$ only since one can add any function with compact support $\overline{u} \in C_0^\infty(\Omega)$ to u to generate another source solution $F(\underline{x}) + \nabla^2\overline{u}(\underline{x})$. Therefore, $u(\underline{x})$ must be measured at all points inside the domain Ω in order to uniquely retrieve the source F in (4.1) by calculating its Laplacian. Even with uniqueness ensured, this inverse problem requiring to numerically calculate the Laplacian of the noisy measured function $u(\underline{x})$+noise is still ill-posed and requires regularization [315] in order to retrieve a stable source from (4.1). Alternatively, one can assume extra information about the source itself, e.g. being a static or moving point source [11, 107], having compact support [106], poly-harmonic, additively or multiplicatively separable [355], being independent of one of the variables [155], or other restrictions [46].

In the following, the space-dependent inverse heat source identification is considered. The time-dependent case will be considered in Chapter 5, and the more general additively and multiplicatively separable cases in Chapter 6.

4.1 Space-Dependent Heat Source Identification

This section is concerned with the reconstruction of a space-dependent source from final "upper-base" or time-average temperature measurements. Although

DOI: 10.1201/9780429400629-4

applicable to general second-order linear PDEs with time-independent coefficients using the semi-group theory, for simplicity, the analysis is mainly presented for the classical one-dimensional transient heat equation.

The inverse problem of determining the temperature $u(x,t)$ and the heat source $f(x)$ in the parabolic heat equation

$$u_t(x,t) - u_{xx}(x,t) = f(x)h(x,t) + g(x,t),$$
$$(x,t) \in \Omega_T := \Omega \times (0,T) = (0,l) \times (0,T), \tag{4.2}$$

where h and g are given functions, and l and T are given positive constants, subject to the initial condition

$$u(x,0) = u_0(x), \quad x \in [0,l], \tag{4.3}$$

the Dirichlet boundary conditions

$$u(0,t) = \mu_0(t), \quad u(l,t) = \mu_l(t), \quad t \in (0,T), \tag{4.4}$$

and the additional "upper-base" final temperature measurement

$$u(x,T) = u_T(x), \quad x \in [0,l], \tag{4.5}$$

or the time-average temperature measurement

$$\int_0^T u(x,t)dt = U_T(x), \quad x \in [0,l], \tag{4.6}$$

is studied. Heat flux overspecification, e.g. $-u_x(0,t) = q_0(t)$ for $t \in [0,T]$, instead of (4.5) or (4.6), can also be considered [54].

Taking into account the superposition principle for linear problems, it can be assumed that the source g in (4.2), and the Dirichlet data μ_0 and μ_1 in (4.4) are zero. Then, (4.2) and (4.4) recast as

$$u_t(x,t) - u_{xx}(x,t) = h(x,t)f(x), \quad (x,t) \in \Omega_T, \tag{4.7}$$

$$u(0,t) = u(l,t) = 0, \quad t \in (0,T). \tag{4.8}$$

The above problems are uniquely solvable in various spaces [339, 350, 371].

Remark 4.1. (The case when h is independent of x)
When h is independent of x, i.e. $h = h(t)$, equation (4.2) becomes

$$u_t(x,t) - u_{xx}(x,t) = h(t)f(x), \quad (x,t) \in \Omega_T. \tag{4.9}$$

If $h \in L^2(0,T)$ with $h(t) \geq h_0 > 0$ for all $t \in [0,T]$, for some positive known constant h_0, and u_0, $u_T \in L^2(0,l)$, then the solution $(u(x,t), f(x))$ of the inverse problem (4.3), (4.5) or (4.6), (4.8) and (4.9) has the Fourier sine series representation [116, 155]

$$u(x,t) = \sum_{n=1}^{\infty} u_n(t)\sin(\nu_n x), \quad f(x) = \sum_{n=1}^{\infty} f_n \sin(\nu_n x),$$

where $\nu_n = \sqrt{\lambda_n}$, $\lambda_n = (\pi n/l)^2$,

$$u_n(t) = u_{0,n}e^{-\lambda_n t} + f_n \int_0^t e^{-\lambda_n(t-\tau)}h(\tau)d\tau, \quad u_{0,n} = \frac{2}{l}\int_0^l u_0(x)\sin(\nu_n x)dx,$$

$$u_{T,n} = \frac{2}{l}\int_0^l u_T(x)\sin(\mu_n x)dx, \quad f_n = \frac{u_{T,n} - u_{0,n}e^{-\lambda_n T}}{\int_0^T e^{-\lambda_n(T-\tau)}h(\tau)d\tau}, \tag{4.10}$$

in case of the "upper-base" temperature measurement (4.5), or

$$U_{T,n} = \frac{2}{l}\int_0^l U_T(x)\sin(\nu_n x)dx, \quad f_n = \frac{U_{T,n} - u_{0,n}(1-e^{-\lambda_n T})/\lambda_n}{\int_0^T \left(\int_0^t e^{-\lambda_n(t-\tau)}h(\tau)d\tau\right)dt}, \tag{4.11}$$

in case of the average time-integral temperature measurement (4.6).

Particularising even more h to be independent of t, thus a constant taken, for simplicity, to be unity, then (4.9)–(4.11) simplify further as,

$$u_t(x,t) - u_{xx}(x,t) = f(x), \quad (x,t) \in \Omega_T, \tag{4.12}$$

$$f_n = \frac{\lambda_n(u_{T,n} - u_{0,n}e^{-\lambda_n T})}{1 - e^{-\lambda_n T}}, \quad f_n = \frac{\lambda_n(\lambda_n U_{T,n} - u_{0,n}(1-e^{-\lambda_n T}))}{e^{-\lambda_n T} + \lambda_n T - 1}.$$

4.1.1 Mathematical analysis

Strong solutions in Hölder spaces

Let $\varepsilon \in (0,1)$ be fixed and assume that the data belongs to Hölder spaces (i.e. spaces of functions whose derivatives are Hölder-continuous [244]), namely, h, $g \in H^{\varepsilon,\varepsilon/2}(\overline{\Omega}_T)$, μ_0, $\mu_l \in H^{1+\varepsilon/2}([0,T])$, $u_0 \in H^{2+\varepsilon}([0,l])$, and that the compatibility conditions of order zero, namely, $\mu_0(0) = u_0(0)$, $\mu_l(0) = u_0(l)$, are satisfied. Then, there exists a unique solution $u \in H^{2+\varepsilon,1+\varepsilon/2}(\overline{\Omega}_T)$ of the direct problem (4.2)–(4.4) for any given source function $f \in H^{\varepsilon}([0,l])$ such that the compatibility conditions

$$\begin{cases} \mu_0'(0) - u_0''(0) = f(0)h(0,0) + g(0,0), \\ \mu_l'(0) - u_0''(l) = f(l)h(l,0) + g(l,0), \end{cases} \tag{4.13}$$

are satisfied [244]. For the inverse problem (4.2)–(4.5), the following solvability result holds [338].

Theorem 4.1. *Suppose that* h, h_t, $g \in H^{\varepsilon,\varepsilon/2}(\overline{\Omega}_T)$, u_0, $u_T \in H^{2+\varepsilon}([0,l])$, μ_0, $\mu_l \in H^{1+\varepsilon/2}([0,T])$,

$$h(x,t) \geq 0, \quad h_t(x,t) \geq 0, \quad |h(x,T)| \geq h_T > 0 \quad \forall (x,t) \in \overline{\Omega}_T,$$

and that the following compatibility conditions are satisfied:

$$u_0(0) = \mu_0(0), \quad u_0(l) = \mu_l(0), \quad u_T(0) = \mu_0(T), \quad u_T(l) = \mu_l(T),$$
$$\mu_0'(0) - u_0''(0) = g(0,0) + \frac{h(0,0)(\mu_0'(T) - u_T''(0) - g(0,T))}{h(0,T)},$$
$$\mu_l'(0) - u_0''(l) = g(l,0) + \frac{h(l,0)(\mu_l'(T) - u_T''(l) - g(l,T))}{h(l,T)}.$$

Then, the inverse source problem (4.2)-(4.5) *has a unique solution* $(u,f) \in H^{2+\varepsilon,1+\varepsilon/2}(\overline{\Omega}_T) \times H^{\varepsilon}([0,l])$. *Further,* f *can be obtained in* $H^{2+\varepsilon}([0,l])$ *if one assumes that* h, $g \in H^{2+\varepsilon,1+\varepsilon/2}(\overline{\Omega}_T)$ *and* u_0, $u_T \in H^{4+\varepsilon}([0,l])$.

Existence and uniqueness results are also available for the inverse source problem (4.2)–(4.4), (4.6), [337].

Weak solutions
Let $H^k(0,l)$ be the Hilbert space of functions whose generalized derivatives of order $\leq k$ belong to $L^2(0,l)$, and by $H_0^k(0,l)$ denote the closure in the H^k-norm of $C_0^\infty(0,l)$, the space of infinitely differentiable functions with compact support in $(0,l)$. In particular, $H_0^1(0,l)$ consists of functions in $H^1(0,l)$ satisfying homogeneous Dirichlet boundary conditions at the endpoints $x=0$ and l. Denote by $H^{1,0}(\Omega_T) := \{u \in L^2(\Omega_T) \mid u_x \in L^2(\Omega_T)\}$ with norm

$$\|u\|_{H^{1,0}(\Omega_T)} = \left(\int_0^T \|u(\cdot,t)\|_{H^1(0,l)}^2 dt \right)^{1/2}.$$

Similarly, for $k \geq 0$, denote by $H^{k,1}(\Omega_T)$ the Hilbert space of functions from $L^2(\Omega_T)$ having generalized space derivatives up to order k and first-order time derivative in $L^2(\Omega_T)$. The norm in this space is given by

$$\|u\|_{H^{k,1}(\Omega_T)} = \left(\int_0^T \left\{ \|u(\cdot,t)\|_{H^k(0,l)}^2 + \|u_t(\cdot,t)\|_{L^2(0,l)}^2 \right\} dt \right)^{1/2}.$$

Also define $H_0^{1,1}(\Omega_T)$ as the set of functions from $H^{1,1}(\Omega_T)$ satisfying homogeneous Dirichlet boundary conditions at the endpoints $x=0$ and l. Let $L^2(0,T;H_0^1(0,l))$ denote the space of measurable functions $u(\cdot,t) : (0,T) \to H_0^1(0,l)$ such that the norm $\|u\|_{L^2(0,T;H_0^1(0,l))} := \left(\int_0^T \|u(\cdot,t)\|_{H_0^1(0,l)}^2 dt \right)^{1/2}$ is finite, and by $C([0,T];L^2(0,l))$ the set of functions u such that the mapping

$u(\cdot,t) : [0,T] \to L^2(0,l)$ is continuous (in the usual norms).

The closure of $H^{1,1}(\Omega_T)$ in the following norm:

$$|u|_{\Omega_T} = \left(\|u_x\|^2_{L^2(\Omega_T)} + \text{supp ess}_{t\in[0,T]} \|u(\cdot,t)\|^2_{L^2(0,l)} \right)^{1/2}$$

is a Banach space denoted by $V^{1,0}(\Omega_T)$, which coincides with the space $L^2(0,T;H^1(0,l)) \cap C([0,T];L^2(0,l))$. Finally, denote by $\mathring{V}^{1,0}(\Omega_T) := L^?(0,T,II_0^1(0,l)) \cap C([0,T],L^?(0,l))$.

Let us now give two definitions for the weak solution of the direct problem (4.3), (4.7), (4.8), [140].

Definition 4.1. *The weak solution of the direct problem* (4.3), (4.7), (4.8) *in* $V^{1,0}(\Omega_T)$ *is a function* $u \in \mathring{V}^{1,0}(\Omega_T)$ *which satisfies*

$$-\int_0^l u_0(x)v(x,0)dx + \int\int_{\Omega_T} (u_x v_x - u v_t)dxdt$$
$$= \int\int_{\Omega_T} f(x)h(x,t)v(x,t)dxdt, \quad \forall v \in H_0^{1,1}(\Omega_T) \quad \text{with } v(\cdot,T) = 0. \tag{4.14}$$

Definition 4.2. *The weak solution of the direct problem* (4.3), (4.7), (4.8) *in* $H^{1,1}(\Omega_T) \subset V^{1,0}(\Omega_T)$ *is a function* $u \in H_0^{1,1}(\Omega_T)$ *which satisfies*

$$\int\int_{\Omega_t} (u_x v_x + u_t v)dxdt$$
$$= \int\int_{\Omega_t} f(x)h(x,t)v(x,t)dxdt, \quad \forall t \in (0,T], \quad \forall v \in H_0^{1,0}(\Omega_T)$$
$$\text{and} \quad u(x,0) = u_0(x), \quad \forall x \in (0,l). \tag{4.15}$$

Then, the following lemma [140] gives the well-posedness of the direct problem (4.3), (4.7), (4.8) in the weak sense of Definitions 4.1 and 4.2.

Lemma 4.1. *Assume that the product* $h\cdot f \in L^2(\Omega_T)$. *Then, the direct problem* (4.3), (4.7), (4.8) *has a unique weak solution:*
(i) $u \in \mathring{V}^{1,0}(\Omega_T)$ *in the sense of Definition 4.1 if* $u_0 \in L^2(0,l)$. *Moreover,*

$$|u|^2_{\Omega_T} \le C\left(\|u_0\|^2_{L^2(0,l)} + \|h\cdot f\|^2_{L^2(\Omega_T)} \right). \tag{4.16}$$

(ii) $u \in H^{1,1}(\Omega_T)$ *in the sense of Definition 4.2 if* $u_0 \in H_0^1(0,l)$. *Moreover,*

$$\|u\|_{H^{1,1}(\Omega_T)} \le C\left(\|u_0\|_{H^1(0,l)} + \|h\cdot f\|_{L^2(\Omega_T)} \right). \tag{4.17}$$

Remark 4.2. (The case $h(x,t) \equiv 1$.)
In the simpler case when h is constant, say $h(x,t) \equiv 1$, equation (4.2) becomes

$$u_t(x,t) - u_{xx}(x,t) = f(x) + g(x,t), \quad (x,t) \in \Omega_T. \tag{4.18}$$

Basically, one can formally differentiate (4.18) with respect to t to eliminate the source $f(x)$, [393], and solve for $u(x,t)$ independently the resulting third-order PDE. More generally, the multi-dimensional inverse source problem:

$$\begin{cases} u_t(x,t) - \nabla^2 u(x,t) = f(x), & (x,t) \in \Omega_T := \Omega \times (0,T), \\ u(x,t) = 0, & (x,t) \in \partial\Omega \times (0,T), \\ u(x,0) = u_0(x), & x \in \Omega, \\ u(x,T) = u_T(x), & x \in \Omega, \end{cases} \tag{4.19}$$

where $\Omega \subset \mathbb{R}^d$, $d \geq 1$, is a bounded domain with sufficiently smooth boundary $\partial\Omega$, has a unique solution $(u(x,t), f(x)) \in H^{2,1}(\Omega_T) \times L^2(\Omega)$, if the input data u_0 and u_T are in $H_0^1(\Omega) \cap H^2(\Omega)$, [350]. Moreover, the solution is stable with respect to errors in these data. However, the solution becomes unstable for less regular data, e.g. if u_0 and u_T are only in $H_0^1(\Omega)$, as illustrated in the following example [218]. Let $\Omega = (0,\pi)$, $u_{0m}(x) = \sin(mx)/m^{3/2}$ and $u_{Tm}(x) = (2 - e^{-m^2 T})\sin(mx)/m^{3/2}$. Then, the solution of (4.19) is

$$\begin{cases} u(x,t) = (2 - e^{-m^2 t})\sin(mx)/m^{3/2}, & (x,t) \in [0,\pi] \times (0,T), \\ f_m(x) = 2m^{1/2}\sin(mx), & x \in [0,\pi], \end{cases}$$

and it can be seen that f_m is unbounded, while u_{0m} and u_{Tm} tend to zero in $H^1(0,\pi)$, as $m \to \infty$.

The existence and uniqueness of the solution hold also true for the following inverse source problem:

$$\begin{cases} u_t(x,t) - \nabla^2 u(x,t) = f(x), & (x,t) \in \Omega_T := \Omega \times (0,T), \\ u(x,t) = 0, & (x,t) \in \partial\Omega \times (0,T), \\ u(x,0) = u_0(x), & x \in \Omega, \\ \int_0^T u(x,t)dt = U_T(x), & x \in \Omega, \end{cases} \tag{4.20}$$

when (4.5) is replaced by the average integral condition (4.6) This can be proved for instance, using a semi-group approach [350], which recasts this problem into that of determining the pair solution $(u(t), f)$ satisfying

$$\begin{cases} \frac{du}{dt} + \mathcal{A}u = f, \\ u(0) = u_0, \\ \int_0^T u(t)dt = U_T, \end{cases} \tag{4.21}$$

where $\mathcal{A} := -\nabla^2 : H_0^1(\Omega) \cap H^2(\Omega) \to L^2(\Omega)$. From the first two equations in this system, the component $u(t)$ can be expressed in terms of f as

$$u(t) = e^{-t\mathcal{A}}u_0 + \int_0^t e^{-(t-\tau)\mathcal{A}} f d\tau = e^{-t\mathcal{A}}u_0 + (I - e^{-t\mathcal{A}})A^{-1}f, \tag{4.22}$$

where I denotes the identity operator. Integrating (4.22) from 0 to T results in

$$U_T = (I - e^{-T\mathcal{A}})\mathcal{A}^{-1}u_0 + [TI + (e^{-T\mathcal{A}} - I)\mathcal{A}^{-1}]\mathcal{A}^{-1}f.$$

Denoting $\mathcal{S}(T) := \frac{1}{T}(I - e^{-T\mathcal{A}})\mathcal{A}^{-1}$ and observing that $\|\mathcal{S}(T)\|_{L^2(0,l)} < 1$, the unique solution for the source is explicitly given by

$$f = \frac{1}{T}(I - \mathcal{S}(T))^{-1}[\mathcal{A}U_T + (e^{-T\mathcal{A}} - I)u_0].$$

For the case when $h = h(t)$, uniqueness and counterexamples are provided in [366]. Below, the most general case when $h = h(x,t)$ is considered, as in [116]. The weak solution in the sense of Definition 4.1 is used.

4.1.2 Variational formulation

Assume that f, u_0, $u_T \in L^2(0,l)$ and $h \in L^\infty(\Omega_T)$. Based on Lemma 4.1, define the operator $A : L^2(0,l) \to L^2(0,l)$ by $Af := u(\cdot, T; f)$, and consider

$$(Af)(x) = u_T(x), \quad x \in (0,l). \tag{4.23}$$

for the inverse problem (4.3), (4.5), (4.7), (4.8), and

$$(\tilde{A}f)(x) = \int_0^T u(x,t)dt = U_T(x), \quad x \in (0,l). \tag{4.24}$$

for the inverse problem (4.3), (4.6)–(4.8), Let $u(x,t;f) \in \mathring{V}^{1,0}(\Omega_T)$ denote the unique weak solution of the direct problem (4.3), (4.7), (4.8) for a given source term $f \in L^2(0,l)$. The operator A is well-defined, because, on one hand, the solution of the direct problem (4.3), (4.7), (4.8) is unique and, on the other hand, the mapping $u(\cdot,t) : [0,l] \to L^2(0,l)$ is continuous for all $t \in [0,T]$. The operators A and $\tilde{A}$ are bounded and affine. They are linear only if $u_0 = 0$. Since the additional condition (4.5) or (4.6) is satisfied only approximately (due to the inherent practical measurement errors), instead of solving the equation (4.23) or (4.24), the least-squares functional $J : L^2(0,l) \to \mathbb{R}_+$ defined by

$$J(f) := \frac{1}{2}\|Af - u_T\|^2_{L^2(0,l)}, \tag{4.25}$$

or, the functional $\tilde{J} : L^2(0,l) \to \mathbb{R}_+$ defined by

$$\tilde{J}(f) := \frac{1}{2}\|\tilde{A}f - U_T\|^2_{L^2(0,l)}, \tag{4.26}$$

is minimized. The functional (4.25) is weakly continuous on closed and convex subsets of $L^2(0,l)$, [152], and weakly lower semi-continuous on bounded and convex subsets of $L^2(0,l)$, [140].

In Theorem 4.2 below it is proved that the functionals J and $\tilde{J}$ are twice Frechet differentiable and formulae for their gradients are presented. Let us first introduce the following adjoint problem:

$$\begin{cases} v_t(x,t) + v_{xx}(x,t) = h(x,t)q(x), & (x,t) \in \Omega_T, \\ v(0,t) = v(l,t) = 0, & t \in (0,T), \\ v(x,T) = \zeta(x), & x \in [0,l], \end{cases} \tag{4.27}$$

where q and $\zeta \in L^2(0,l)$. Note that the above parabolic PDE is backward in time, and due to the "final condition" at $t = T$ it is a well-posed initial boundary value problem under a reversal of time. Indeed, if t is replaced by the new variable $\tau := T - t$ in (4.27), then a boundary value problem of the same type as (4.3), (4.7), (4.8) is obtained. Therefore, there is a unique weak solution of the problem (4.27) according to Lemma 4.1. Furthermore, the following Green's formula holds [116, 218].

Lemma 4.2. *Let $u_0, \zeta \in L^2(0,l)$ and $h \cdot f, h \cdot q \in L^2(\Omega_T)$. Let $u, v \in \mathring{V}^{1,0}(\Omega_T)$ be the weak solutions of (4.3), (4.7), (4.8) and (4.27), respectively. Then,*

$$\int\int_{\Omega_T} h(x,t)[f(x)v(x,t) + q(x)u(x,t)]dxdt$$
$$= \int_0^l [u(x,T)\zeta(x) - v(x,0)u_0(x)]dx. \tag{4.28}$$

Theorem 4.2. ([116, 218]) *The functionals J and $\tilde{J}$ defined by (4.25) and (4.26), respectively, are twice Frechet differentiable and convex. Their gradients are Lipschitz continuous and given by*

$$J'(f) = \int_0^T h(x,t)v(x,t)dt, \quad \tilde{J}'(f) = \int_0^T \tilde{v}(x,t)\int_0^t h(x,s)dsdt, \tag{4.29}$$

where v and $\tilde{v}$ are the solutions of the adjoint problem (4.27) with $q = 0$, and $\zeta(x) = u_f(x,T) - u_T(x)$ and $\zeta(x) = \int_0^T u_f(x,T)dt - U_T(x)$, respectively, and u_f solves the direct problem (4.3), (4.7), (4.8) for given $f \in L^2(0,l)$. Furthermore, if $h \equiv 1$, then J is strictly convex and its gradient simplifies as

$$J'(f) = -w(x,0), \tag{4.30}$$

where w is the solution of (4.27) with $\zeta = 0$ and $q = u_f(\cdot,T) - u_T$.

Proof: For simplicity, only the properties of the functional J are proved. Let $\delta \in L^2(0,l)$ be an increment of $f \in L^2(0,l)$. Let u_δ^0 solve the direct problem (4.3), (4.7), (4.8) with $f = \delta$ and $u_0 = 0$. If $u_{f+\delta}$ also solves (4.3), (4.7), (4.8), then $u_\delta^0 = u_{f+\delta} - u_f \in \mathring{V}^{1,0}(\Omega_T)$ and

$$J(f+\delta) - J(f) = \frac{1}{2}\|u_f(\cdot,T) + u_\delta^0(\cdot,T) - u_T(\cdot)\|^2_{L^2(0,l)}$$
$$-\frac{1}{2}\|u_f(\cdot,T) - u_T(\cdot)\|^2_{L^2(0,l)} = \int_0^l u_\delta^0(x,T)(u_f(x,T) - u_T(x))dx$$
$$+\frac{1}{2}\int_0^l |u_\delta^0(x,T)|^2 dx. \tag{4.31}$$

Let v be the solution of the adjoint problem (4.27) with $q = 0$ and $\zeta(x) =$

$u_f(x,T) - u_T(x)$. From formula (4.28) applied to the functions u_δ^0 and v,

$$\int\int_{\Omega_T} h(x,t)\delta(x)v(x,t)dxdt = \int_0^l u_\delta^0(x,T)(u_f(x,T) - u_T(x))dx. \quad (4.32)$$

Then, (4.31) and (4.32) yield

$$J(f+\delta) - J(f) = \int\int_{\Omega_T} h(x,t)\delta(x)v(x,t)dxdt + \frac{1}{2}\int_0^l |u_\delta^0(x,T)|^2 dx. \quad (4.33)$$

Let us now prove the following estimate:

$$\|u_\delta^0(\cdot,T)\|_{L^2(0,l)} \le C\|\delta\|_{L^2(0,l)}, \quad \text{where } C = \frac{l\|h\|_\infty\sqrt{T}}{\pi\sqrt{2}}, \quad (4.34)$$

which, when used in (4.33), yields the formula in (4.29) for $J'(f)$. Since $u_\delta^0 \in \mathring{V}^{1,0}(\Omega_T)$ satisfies the direct problem (4.3), (4.7), (4.8) with $f = \delta$ and $u_0 = 0$, then (dropping for the time being the superscript ^{0} from u_δ^0),

$$\frac{1}{2}\int_0^l u_\delta^2(x,T)dx + \int\int_{\Omega_T} u_{\delta x}^2(x,t)dxdt = \int\int_{\Omega_T} h(x,t)\delta(x)u_\delta(x,t)dxdt. \quad (4.35)$$

Using the inequality

$$\alpha\beta \le s\frac{\alpha^2}{2} + \frac{\beta^2}{2s}, \quad \forall \alpha,\beta \in \mathbb{R}, \quad \forall s > 0$$

in the right-hand side of (4.35) and Wirtinger's inequality yield

$$\int\int_{\Omega_T} h(x,t)\delta(x)u_\delta(x,t)dxdt \le \frac{s}{2}\int\int_{\Omega_T} u_\delta^2(x,t)dxdt + \frac{1}{2s}\int\int_{\Omega_T} (\delta(x)h(x,t))^2 dxdt \le \frac{sl^2}{2\pi^2}\int\int_{\Omega_T} u_{\delta x}^2(x,t)dxdt + \frac{1}{2s}\int\int_{\Omega_T} (\delta(x)h(x,t))^2 dxdt.$$

Substituting this into (4.35) results in

$$\frac{1}{2}\int_0^l u_\delta^2(x,T)dx + \left(1 - \frac{sl^2}{2\pi^2}\right)\int\int_{\Omega_T} u_{\delta x}^2(x,t)dxdt \le \frac{1}{2s}\int\int_{\Omega_T} (\delta(x)h(x,t))^2 dxdt \le \frac{\|h\|_\infty^2 T}{2s}\int_0^l \delta^2(x)dx.$$

Taking $s = 2\pi^2/l^2$ and putting back the superscript ^{0} on u_δ^0 yield (4.34).

To prove that the functional J is in fact twice Frechet differentiable, let $\tilde{v}$ be the solution of the adjoint problem (4.27) with $q = 0$ and $\zeta(x) = u_\delta^0(x, T)$. From Green's formula (4.28) applied to the functions u_δ^0 and $\tilde{v}$,

$$\int_0^l [u_\delta^0(x,T)]^2 dx = \int\int_{\Omega_T} h(x,t)\delta(x)\tilde{v}(x,t)dxdt. \tag{4.36}$$

Following a similar argument as above, it can be obtained that

$$J''(f)\delta = \int_0^T h(x,t)\tilde{v}(x,t)dt. \tag{4.37}$$

Then, from (4.36) and (4.37),

$$(J''(f)\delta, \delta)_{L^2(0,l)} = \|u_\delta^0(\cdot, T)\|^2_{L^2(0,l)} \geq 0, \tag{4.38}$$

which implies that J is convex. Also, (4.34) and (4.38) imply that J' is Lipschitz continuous with the Lipschitz constant C explicitly given in (4.34).

Let us finally prove that in case $h \equiv 1$, J is strictly convex and that J' is given by (4.30). First, if $(J''(f)\delta, \delta)_{L^2(0,l)} = 0$, then $u_\delta^0(\cdot, T) \equiv 0$. Since also $u_\delta^0(\cdot, 0) \equiv 0$ from the uniqueness of solution of problem (4.19) it follows that $u_\delta^0 \equiv 0$ and $f = \delta \equiv 0$. Hence, J is strictly convex. Second, it is known [280], that the Laplacian operator defined over $\{u \in H_0^1(\Omega); \nabla^2 u \in L^2(\Omega)\}$ generates a C_0-continuous contraction semigroup $\{\mathrm{G}(t)\}_{t\in\mathbb{R}_+}$ in $L^2(\Omega)$. Then, the solution $v(x,t)$ of (4.27) with $q(x) = 0$ and $\zeta(x) = u_f(x,T) - u_T(x)$ can be represented as $v(x,t) = \mathrm{G}(T-t)(u_f(x,T) - u_T(x))$. Moreover, the solution $w \in L^2(0,T; H_0^1(\Omega))$ of (4.27) with $q(x) = u_f(x,T) - u_T(x)$ and $\zeta(x) = 0$ satisfies $w(x,0) = -(u_f(x,T) - u_T(x))\int_0^T \mathrm{G}(T-t)dt$. Thus, $-w(x,0) = \int_0^T v(x,t)dt$, as requested.

The main distinguished feature of Theorem 4.2 is that the gradient of the functional $J(f)$ (or $\tilde{J}$), whose minimum defines a quasi-solution of the inverse source problem, can be calculated via the well-posed adjoint problem (4.27), whose input data $q(x)$ or $\zeta(x)$ contains the measured data $u_T(x)$ (or $U_T(x)$). This and the gradient formula (4.29) suggest using gradient-type iterative methods for the minimization of the objective functional (4.25) (or (4.26)).

4.1.3 Conjugate gradient method (CGM)

The objective is to solve the inverse source problem (4.3), (4.5), (4.7), (4.8), or (4.3), (4.6)–(4.8). Decompose $u = u_{u_0} + u_f^0$, where u_{u_0} and u_f^0 solve (4.3), (4.7), (4.8) with $f = 0$ and $u_0 = 0$, respectively. Then, define the linear and bounded operator $A_0 : L^2(0,l) \to L^2(0,l)$ by

$$A_0 f := \begin{cases} u_f^0(\cdot, T) & \text{for problem } (4.3), (4.5), (4.7), (4.8), \\ \int_0^T u_f^0(\cdot, t)dt & \text{for problem } (4.3), (4.6) - (4.8). \end{cases} \tag{4.39}$$

Then, on imposing (4.5) or (4.6) results in the need to solve

$$A_0 f = \begin{cases} u_T - u_{u_0}(\cdot, T) & \text{for problem (4.3), (4.5), (4.7), (4.8),} \\ U_T - \int_0^T u_{u_0}(\cdot, t)dt & \text{for problem (4.3), (4.6) – (4.8),} \end{cases} \tag{4.40}$$

in a minimization sense, because instead of the exact data u_T and U_T, we only have available a measured approximation within some error level $\epsilon \geq 0$, say u_T^ϵ and U_T^ϵ in $L^2(0, l)$ satisfying

$$\|u_T - u_T^\epsilon\|_{L^2(0,l)} \leq \epsilon, \quad \|U_T - U_T^\epsilon\|_{L^2(0,l)} \leq \epsilon. \tag{4.41}$$

In doing this, the adjoint A_0^* of the operator A_0 needs to be found by solving the adjoint problem, and then calculate $A_0^*(A_0 f - (u_T^\epsilon - u_{u_0}(\cdot, T)))$ or $A_0^*(A_0 f - (U_T^\epsilon - \int_0^T u_{u_0}(\cdot, t)))$ for the problems (4.3), (4.5), (4.7), (4.8), or (4.3), (4.6)-(4.8), respectively. However, one can observe that these are precisely the gradients in (4.29) of the functionals J and $\tilde{J}$, respectively. Then, the GGM runs as follows [113, 116, 131, 218]:

Step 1. Choose an arbitrary initial guess $f_0 \in L^2(0, l)$ and solve the direct problem (4.3), (4.7), (4.8) with $f = f_0$ to obtain u^0.

Step 2. For $n \geq 0$, assume that f_n and u^n have been constructed. Let v_n solve the adjoint problem (4.27) with $q = 0$ and $\zeta = \zeta_n$ given by

$$\zeta_n(x) = \begin{cases} u^n(x, T) - u_T^\epsilon(x) \text{ for problem (4.3), (4.5), (4.7), (4.8),} \\ \int_0^T u^n(x, t)dt - U_T^\epsilon(x) \text{ for problem (4.3), (4.6) – (4.8),} \end{cases} \tag{4.42}$$

to determine the gradient

$$r_n(x) = \begin{cases} \int_0^T h(x, t)v_n(x, t)dt \text{ for problem (4.3), (4.5), (4.7), (4.8),} \\ \int_0^T v_n(x, t) \int_0^t h(x, s)dsdt \text{ for problem (4.3), (4.6) – (4.8).} \end{cases} \tag{4.43}$$

Step 3. Calculate $d_n = -r_n + \beta_{n-1} d_{n-1}$, where

$$\beta_{-1} = 0, \quad \beta_{n-1} = \frac{\|r_n\|_{L^2(0,l)}^2}{\|r_{n-1}\|_{L^2(0,l)}^2}, \quad n \geq 1. \tag{4.44}$$

Step 4. Solve the direct problem (4.3), (4.7), (4.8) with $u_0 = 0$ and $f = d_n$, set the direction search

$$\alpha_n = \frac{\|r_n\|_{L^2(0,l)}^2}{\|A_0 d_n\|_{L^2(0,l)}^2}, \quad n \geq 0, \tag{4.45}$$

and let $f_{n+1} = f_n + \alpha_n d_n$.

Step 5. Let u^{n+1} solve the direct problem (4.3), (4.7), (4.8) with $f = f_{n+1}$.

Step 6. Repeat step 2 until

$$\|\zeta_n\| \leq \tau\epsilon, \tag{4.46}$$

where τ is some constant greater than unity.

The above iterative CGM has a regularizing character and it converges to the solution of the inverse source problem, as ϵ tends to zero [316].

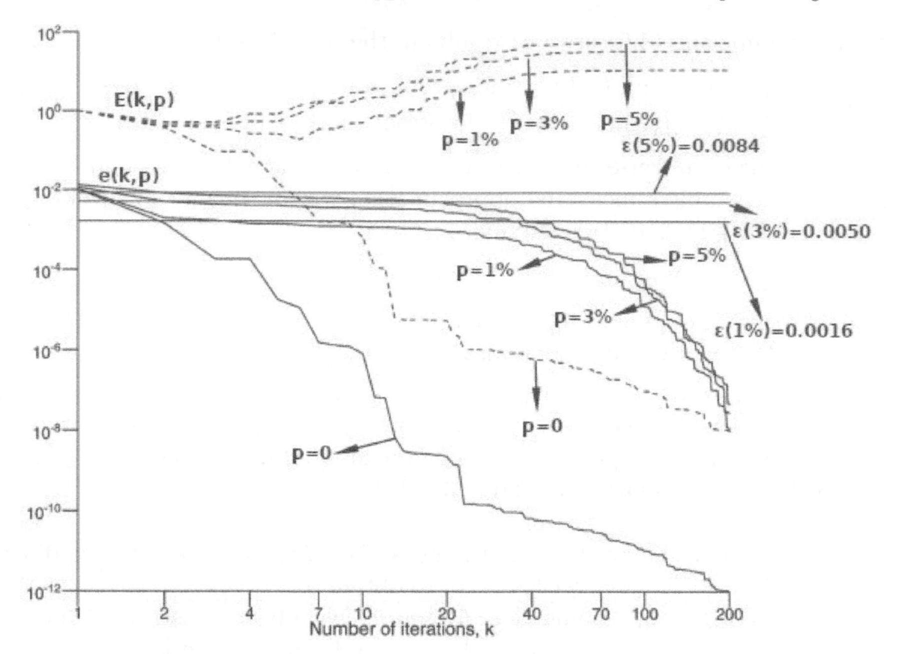

FIGURE 4.1: The errors $e(k,p)$ (—) and $E(k,p)$ (− − −), as functions of the number of iterations k, for various percentages of noise $p \in \{0, 1, 3, 5\}\%$. The values of $\epsilon(p)$ given by (4.50) are also included [218].

4.1.4 Numerical results and discussion

Numerical results of the CGM, as given in [218], are presented for the inverse problem (4.3), (4.5), (4.7), (4.8) based on the final time temperature measurement (4.5). More examples can be found in [116], where it was also indicated that the CGM results were obtained more accurate and stable than the numerical results produced by the truncated singular decomposition [399], or the Tikhonov's regularization method [397]. The well-posed direct and adjoint problems (4.3), (4.7), (4.8) and (4.27), respectively, are discretized numerically using the BEM based on the boundary integral equation

$$\begin{aligned} &\eta(x)u(x,t) \\ &= \int_0^t \left[G(x,t;\xi,\tau)\frac{\partial u}{\partial n}(\xi,\tau) - u(\xi,\tau)\frac{\partial G}{\partial n(\xi)}(x,t;\xi,\tau) \right]_{\xi\in\{0,l\}} d\tau \\ &+ \int_0^l G(x,t;y,0)u_0(y)dy + \int_0^l f(y)\int_0^t G(x,t;y,\tau)h(y,\tau)d\tau dy, \\ &\qquad (x,t) \in [0,l]\times(0,T], \qquad (4.47) \end{aligned}$$

where $\eta(x) = 1$ for $x \in (0,l)$ and $\eta(x) = 0.5$ for $x \in \{0,l\}$, and G is the fundamental solution given by (3.23).

Let $l = T = 1$ and discretize into $N_0 = 40$ uniform cells the space intervals $[0, l = 1] \times \{t = 0\}$ and $[0, 1] \times \{t = T = 1\}$, and into $N = 40$ constant-time boundary elements each of the lateral boundaries $\{0\} \times [0, T = 1]$ and $\{l = 1\} \times [0, T = 1]$. Take $h = 1$, the initial condition (4.3) specified as

$$u(x, 0) = u_0(x) = \sin(\pi x), \quad x \in [0, 1], \tag{4.48}$$

and attempt to reconstruct a source given by the Gaussian normal function

$$f(x) = \frac{1}{0.1\sqrt{2\pi}} \exp\left(-\frac{(x - 0.5)^2}{0.02}\right), \quad x \in (0, 1). \tag{4.49}$$

Since the direct problem (4.3), (4.7), (4.8) with u_0 and f given by (4.48) and (4.49), respectively, does not have a closed-form analytical solution available, the temperature data at the final time $t = T = 1$ is numerically simulated by solving this direct problem using the BEM with a sufficiently fine mesh size (and also different to that used for solving the inverse problem in order to avoid an 'inverse crime'). Moreover, in (4.5), the measured data is contaminated by additive random noise generated by the NAG routine G05DDF from a normal distribution with zero mean and standard deviation taken to be some percentage $p \in \{1, 3, 5\}\%$ from the maximum absolute value (approximatively 0.21) of the numerically simulated data. This yields

$$\epsilon(p) := \|u_T - u_T^\epsilon\|_{L^2(0,1)} = \begin{cases} 0.0016 & \text{for } p = 1\%, \\ 0.0050 & \text{for } p = 3\%, \\ 0.0084 & \text{for } p = 5\%. \end{cases} \tag{4.50}$$

Figure 4.1 shows the accuracy errors $E(k, p) := \|f - f_k\|_{L^2(0,1)}$ and the convergence errors $e(k, p) := \|u_T^\epsilon - u_k(\cdot, T)\|_{L^2(0,1)}$, as functions of the number of iterations k, for various percentages of noise $p \in \{0, 1, 3, 5\}\%$. Of course, since the inverse source problem is ill-posed, letting CGM run without any stopping will eventually produce an unstable solution. Thus, to restore stability the iterative process needs to be stopped, as illustrated in Figure 4.1, at a threshold given by the discrepancy principle (4.46). With $\tau \approx 1$, this yields the stopping iteration numbers $k(p) \in \{4, 3, 2\}$ for $p \in \{1, 3, 5\}\%$, respectively, based on which stable and accurate numerical solutions are obtained, as shown in Figure 4.2. Moreover, when compared with the Landweber-Fridman method [216], the results presented in Figures 4.1 and 4.2 are more accurate, with the number of iterations reduced by a factor of about 50-100 times.

The analysis performed in this section can also be extended to identifying space-dependent loads in vibrating beams from various measurements [153, 156]. In the next section, combined features of the inverse source problem are studied in case when additionally to the heat source the initial temperature (see the BHCP of Chapter 3) is also unknown [219, 281, 385, 402].

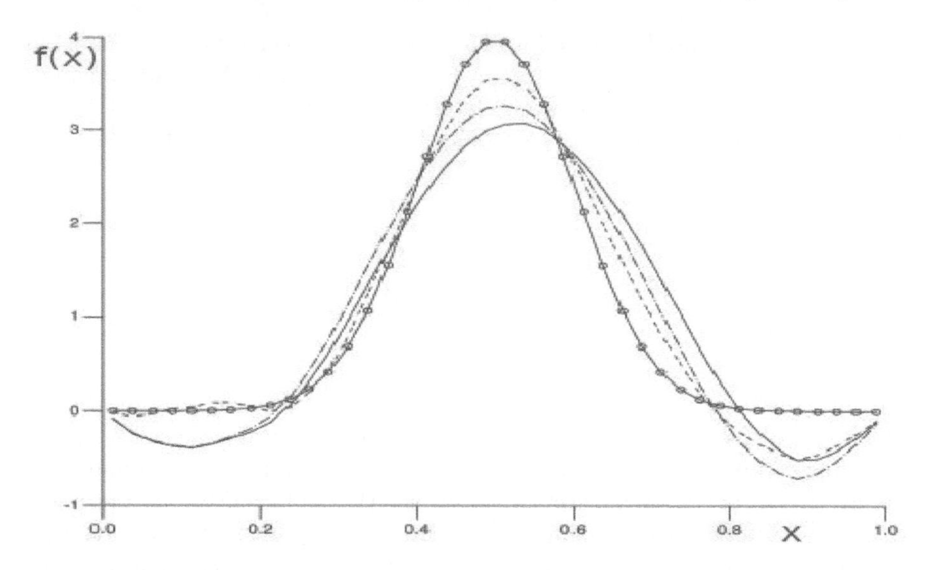

FIGURE 4.2: The analytical solution (4.49) (—) and the numerical solutions for the heat source $f(x)$, for various percentages of noise: $p = 0$ $(- \circ -)$ and $k = 100$ iterations, $p = 1\%$ $(- - -)$ and $k = 4$ iterations, $p = 3\%$ (-.-) and $k = 3$ iterations, and $p = 5\%$ (-....-) and $k = 2$ iterations [218].

4.2 Simultaneous Identification of the Space-Dependent Heat Source and Initial Temperature

Often in melting and freezing processes a possible way of designing final states is by controlling the initial temperature and heat sources. Therefore, there is interest in reconstructing these space-dependent quantities from temperature measurements at two different instants $0 < T_1 < T_2 \leq T$. The mathematical model for this combined inverse problem is then given by

$$\begin{cases} \partial_t u(x,t) - \nabla^2(x,t) = f(x), & (x,t) \in \Omega \times (0,T) =: Q_T, \\ u(x,t) = 0, & (x,t) \in (x,t) \in \partial\Omega \times (0,T), \\ u(x,0) = u_0(x), & x \in \Omega, \end{cases} \tag{4.51}$$

where the heat source $f(x)$ and the initial temperature $u_0(x)$ are to be determined from the temperature measurements

$$u(x,T_1) = u_{T_1}(x), \quad u(x,T_2) = u_{T_2}(x), \quad x \in \Omega, \tag{4.52}$$

where $\Omega \subset \mathbb{R}^d$ is a bounded domain occupying the region of a finite heat conductor with a smooth boundary $\partial\Omega$. By decoupling the problem into two time-layers, the uniqueness of solution can be established; first, for the source

$f(x)$ in the layer $\Omega \times (T_1, T_2)$, after which, for the initial temperature $u_0(x)$ in the layer $\Omega \times (0, T_1)$. More specifically, given the data $u_{T_1}, u_{T_2} \in H_0^1(\Omega) \cap H^2(\Omega)$, then the inverse problem (4.51) and (4.52) cannot have more than one solution $(u(x,t), f(x), u_0(x)) \in H^{2,1}(Q_T) \times L^2(\Omega) \times (H_0^1(\Omega) \cap H^2(\Omega))$, [219]. Although uniqueness of solution holds, the inverse problem is still ill-posed with respect to noise in the data (4.52) given by $u^\epsilon_{T_1}$ and $u^\epsilon_{T_2}$ satisfying

$$\sum_{i=1}^{2} \|u_{T_i} - u^\epsilon_{T_i}\|^2_{L^2(\Omega)} \leq \epsilon^2, \tag{4.53}$$

where $\epsilon \geq 0$ represents the level of noise. Therefore, regularization methods are necessary to deal with this instability. Below, the Landweber-Fridman method (LFM) is described for the simultaneous reconstruction of the heat source $f(x)$ and the initial temperature $u_0(x)$. For this, the adjoint problem

$$\begin{cases} \partial_t v(x,t) + \nabla^2(x,t) = q(x), & (x,t) \in \Omega \times (0,T) =: Q_T, \\ v(x,t) = 0, & (x,t) \in (x,t) \in \partial\Omega \times (0,T), \\ v(x,0) = \zeta(x), & x \in \Omega, \end{cases} \tag{4.54}$$

for given $q \in L^2(\Omega)$ and $\zeta \in H_0^1(\Omega) \cap H^2(\Omega)$, is introduced. Then, the iterative procedure runs, as follows [219]:

Step 1. Choose arbitrary initial guesses f_0 and u_0^0 in $L^2(\Omega)$ and solve the well-posed direct problem (4.51) with this data to obtain the unique solution $u^0 \in L^2(0,T; H_0^1(\Omega)) \cap C([0,T]; H_0^1(\Omega))$.

Step 2. Assume that u^k, f_k and u_0^k have been constructed. Let $v_k^{(i)}$ solve the adjoint problem (4.54) with $q = 0$ and $\zeta = u^k(\cdot, T_i) - u^\epsilon_{T_i}$ for $i = 1, 2$, and $w_k^{(i)}$ solve (4.54) with $\zeta = 0$ and $q = u^k(\cdot, T_i) - u^\epsilon_{T_i}$ for $i = 1, 2$.

Step 3. The temperature u^{k+1} is obtained by solving direct problem (4.51) with the heat source and initial temperature

$$\begin{aligned} f_{k+1}(\cdot) &= f_k(\cdot) - \gamma_1(w_k^{(1)}(\cdot, T_1) + w_k^{(2)}(\cdot, T_2)), \\ u_0^{k+1}(\cdot) &= u_0^k(\cdot) - \gamma_2(v_k^{(1)}(\cdot, T_1) + v_k^{(2)}(\cdot, T_2)), \end{aligned} \tag{4.55}$$

where γ_1 and γ_2 are positive parameters to be prescribed to ensure (neither too large) and accelerate (nor too small) the convergence of the algorithm.

Step 4. Repeat steps 2 and 3 until

$$\sum_{i=1}^{2} \|u^k(\cdot, T_i) - u^\delta_{T_i}\|^2_{L^2(\Omega)} \leq \epsilon^2. \tag{4.56}$$

To illustrate the performance of the above LFM, in the one-dimensional domain $\Omega = (0, l)$ with $l = 1$, and time interval $(0, T)$ with $T = 1$, consider [219] the inverse problem of finding the temperature

$$u(x,t) = \left(2 - e^{-\pi^2 t}\right) \sin(\pi x), \quad (x,t) \in (0,1) \times (0,1), \tag{4.57}$$

TABLE 4.1: The stopping iteration numbers $k = k(p; \gamma_1, \gamma_2)$ given by the discrepancy principle (4.56), and the corresponding accuracy errors $E_1(k;p)$ and $E_2(k;p)$ defined in (4.63), [219].

	$p = 1\%$	$p = 3\%$	$p = 5\%$
$\gamma_1 = 1, \gamma_2 = 1$	$k = 1459$ $E_1 = 0.046$ $E_2 = 0.033$	$k = 1131$ $E_1 = 0.132$ $E_2 = 0.108$	$k = 975$ $E_1 = 0.215$ $E_2 = 0.183$
$\gamma_1 = 1, \gamma_2 = 10$	$k = 668$ $E_1 = 0.043$ $E_2 = 0.020$	$k = 551$ $E_1 = 0.131$ $E_2 = 0.074$	$k = 496$ $E_1 = 0.218$ $E_2 = 0.127$
$\gamma_1 = 1, \gamma_2 = 50$	$k = 611$ $E_1 = 0.043$ $E_2 = 0.017$	$k = 507$ $E_1 = 0.131$ $E_2 = 0.063$	$k = 457$ $E_1 = 0.219$ $E_2 = 0.111$
$\gamma_1 = 10, \gamma_2 = 1$	$k = 195$ $E_1 = 0.039$ $E_2 = 0.066$	$k = 27$ $E_1 = 0.057$ $E_2 = 0.114$	$k = 24$ $E_1 = 0.132$ $E_2 = 0.120$
$\gamma_1 = 10, \gamma_2 = 10$	$k = 144$ $E_1 = 0.046$ $E_2 = 0.032$	$k = 112$ $E_1 = 0.131$ $E_2 = 0.106$	$k = 97$ $E_1 = 0.212$ $E_2 = 0.176$
$\gamma_1 = 10, \gamma_2 = 50$	$k = 72$ $E_1 = 0.043$ $E_2 = 0.024$	$k = 60$ $E_1 = 0.126$ $E_2 = 0.076$	$k = 53$ $E_1 = 0.217$ $E_2 = 0.144$
$\gamma_1 = 50, \gamma_2 = 1$	$k = 556$ $E_1 = 0.375$ $E_2 = 0.087$	$k = 296$ $E_1 = 0.668$ $E_2 = 0.213$	$k = 178$ $E_1 = 0.719$ $E_2 = 0.323$
$\gamma_1 = 50, \gamma_2 = 10$	$k = 60$ $E_1 = 0.066$ $E_2 = 0.035$	$k = 33$ $E_1 = 0.144$ $E_2 = 0.113$	$k = 21$ $E_1 = 0.207$ $E_2 = 0.186$
$\gamma_1 = 50, \gamma_2 = 50$	$k = 27$ $E_1 = 0.045$ $E_2 = 0.031$	$k = 21$ $E_1 = 0.129$ $E_2 = 0.104$	$k = 18$ $E_1 = 0.211$ $E_2 = 0.181$

the initial temperature

$$u(x,0) = u_0(x) = \sin(\pi x), \quad x \in [0,1] \tag{4.58}$$

and the spacewise-dependent heat source

$$f(x) = 2\pi^2 \sin(\pi x), \quad x \in (0,1), \tag{4.59}$$

which satisfy the heat equation in (4.51), the Dirichlet boundary conditions

$$u(0,t) = u(1,t) = 0, \quad t \in (0,1) \tag{4.60}$$

and the additional temperature measurements (4.52) at $T_1 = T/2 = 0.5$ and

$T_2 = T = 1$ given by

$$u(x, T_1) = u_{T_1}(x) = \left(2 - e^{-\pi^2/2}\right)\sin(\pi x),$$
$$u(x, T_2) = u_{T_2}(x) = \left(2 - e^{-\pi^2}\right)\sin(\pi x), \quad x \in [0, 1]. \tag{4.61}$$

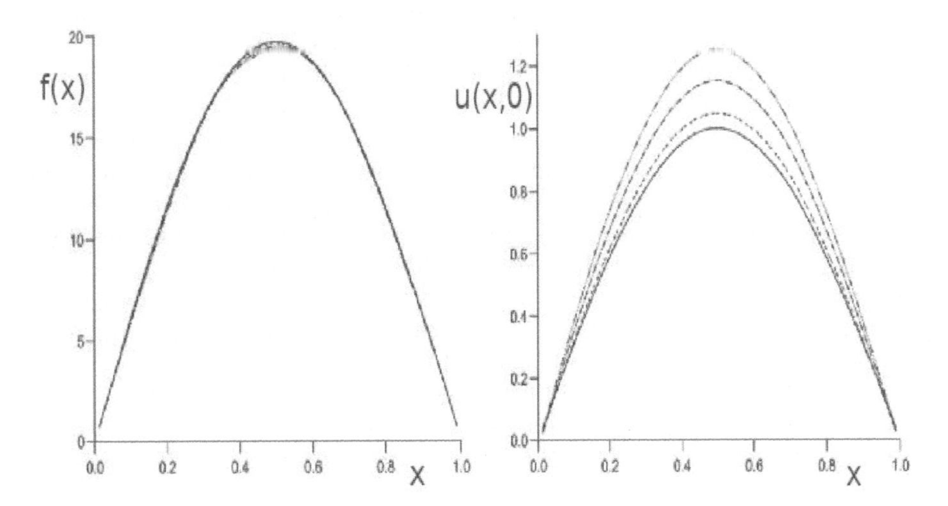

FIGURE 4.3: The analytical (——) solutions (4.59) and (4.58) for the heat source $f(x)$ and the initial temperature $u(x, 0)$, respectively, and the numerical solutions, for various percentages of noise $p = 1\%$ (– – –), $p = 3\%$ (-.-) and $p = 5\%$ (-....-), [219].

The BEM with $N = N_0 = 40$ constant boundary elements is employed for discretizing the direct and adjoint problems of the iterative algorithm, starting from the initial guess $f_0 = u_0^0 = 0$. Noise generated from a Gaussian normal distribution with mean zero and standard deviations $\sigma_1 = p\left(2 - e^{-\pi^2/2}\right)$ and $\sigma_2 = p\left(2 - e^{-\pi^2}\right)$, where p is the percentage of noise, is added to the exact data (4.61), yielding $\epsilon = \epsilon(p) \in \{0.013, 0.038, 0.064\}$ for $p \in \{1, 3, 5\}\%$, respectively. At each iteration number k, the convergence error

$$e(k; p) := \sqrt{\sum_{i=1}^{2} \|u^k(\cdot, T_i) - u_{T_i}^{\delta}\|_{L^2(0,1)}^2} \tag{4.62}$$

and the accuracy errors

$$E_1(k; p) := \|f - f_k\|_{L^2(0,1)}, \quad E_2(k; p) := \|u_0 - u_0^k\|_{L^2(0,1)}, \tag{4.63}$$

are calculated, where u_0 and f are the analytical solutions given by (4.58) and (4.59), respectively. The iterations were stopped according to the discrepancy principle (4.56). Various values of the relaxation parameters γ_1 and γ_2 up to 60 were tested, as shown in Table 4.1, with the mention that above 60 the algorithm was found divergent. From this table it can be seen that as γ_1, γ_2 or p increases, the attainability of the stopping criterion (4.56) becomes faster. Also, the smallest error E_1 in predicting the source is obtained for $\gamma_1 = 10$, $\gamma_2 = 1$, but the smallest E_2 in predicting the initial temperature is obtained for $\gamma_1 = 1$, $\gamma_2 = 50$. However, at present there is no general rule for choosing simultaneously the optimal values of γ_1 and γ_2 other than by trial and error.

Figure 4.3 shows the numerical solutions for the heat source and the initial temperature at the discrepancy principle iteration numbers given in Table 4.1 for $\gamma = \gamma_1 = \gamma_2 = 10$ for various percentages of noise p in comparison with the analytical solutions (4.59) and (4.58), respectively. From Figure 4.3 it can be seen that as p decreases the numerical solutions approximate better the analytical solutions. The errors in predicting the initial temperature are larger because of the severe ill-posedness of the BHCP in the region $0 < t < T_1 = 0.5$, [298], with more accurate results obtainable if T_1 is smaller [247]. Of course, faster iteration procedures, e.g. based on the CGM, which removes the need for the *a priori* choice of γ_1 and γ_2, could be developed.

In closure, it is mentioned that more advanced combined inverse formulations consisting in the simultaneous identification of several spacewise-dependent thermal properties, e.g. heat source, initial temperature, thermal conductivity and heat transfer (radiative, perfusion, reaction) coefficients, can be found in [67, 68, 69, 82, 396].

Chapter 5

Time-Dependent Heat Sources

This chapter is concerned with the reconstruction of time-dependent sources from various transient measurements. Since the unknown is a time-dependent quantity the rule of thumb is to seek measuring a quantity having the same time-dependent meaning, e.g. a temperature measurement at a fixed space point, a boundary heat flux or the mass/energy of the thermal system.

5.1 Time-Dependent Heat Source Identification

The objective is to solve the inverse problem of determining the temperature $u(x,t)$ and the heat source $r(t)$ in the parabolic heat equation

$$\begin{aligned} &u_t(x,t) - u_{xx}(x,t) = r(t)h(x,t) + g(x,t), \\ &(x,t) \in \Omega_T := \Omega \times (0,T) = (0,l) \times (0,T), \end{aligned} \tag{5.1}$$

where h and g are given functions, subject to the initial condition

$$u(x,0) = u_0(x), \quad x \in [0,l], \tag{5.2}$$

the Dirichlet boundary conditions

$$u(0,t) = \mu_0(t), \quad u(l,t) = \mu_l(t), \quad t \in (0,T), \tag{5.3}$$

and the additional interior temperature measurement

$$u(X_0,t) = \chi(t), \quad t \in [0,T], \tag{5.4}$$

at a fixed point $X_0 \in (0,l)$, or the heat flux measurement

$$-u_x(0,t) = q_0(t), \quad t \in [0,T], \tag{5.5}$$

or the mass/energy measurement

$$\int_0^l \omega(x)u(x,t)dx = E(t), \quad t \in [0,T], \tag{5.6}$$

DOI: 10.1201/9780429400629-5

where ω is a given weight function. The unique solvability of the inverse problem (5.1)–(5.4) in Hölder spaces is given by the following theorem, see Theorem 3 of [337].

Theorem 5.1. *Let $\varepsilon \in (0,1)$. Assume that:*
(i) μ_0, μ_l and $\psi \in H^{1+\varepsilon/2}[0,T]$;
(ii) $u_0 \in H^{2+\varepsilon}[0,l]$, g, h and $h_x \in H^{\varepsilon,\varepsilon/2}(\overline{Q}_T)$, and there exists a positive constant h_0 such that $|h(X_0,t)| \geq h_0 > 0$ for all $t \in [0,T]$;
(iii) the compatibility conditions up to the first-order

$$u_0(0) = \mu_0(0), \quad u_0(l) = \mu_l(0), \quad u_0(X_0) = \psi(0),$$
$$\mu_0'(0) = u_0''(0) + g(0,0) + \frac{h(0,0)(\chi'(0) - u_0(X_0) - g(X_0,0))}{h(x_0,0)},$$
$$\mu_l'(0) = u_0''(l) + g(l,0) + \frac{h(l,0)(\chi'(0) - u_0(X_0) - g(X_0,0))}{h(X_0,0)}.$$

are satisfied. Then, the inverse problem (5.1)–(5.4) *has a unique solution $(u,r) \in H^{2+\varepsilon,1+\varepsilon/2}(\overline{Q}_T) \times H^{\varepsilon/2}[0,T]$.*

The unique solvability of the inverse problem (5.1)–(5.3) with the heat flux measurement (5.5) is given by the following theorem [350].

Theorem 5.2. *Let $\mu_0 = \mu_l = g \equiv 0$, which can be assumed, without loss of generality, due to the linearity of the problem. Then, for $h \equiv 1$, $u_0 \in H_0^1(0,l)$ and $q_0 \in C^1(0,T)$, the inverse problem* (5.1)–(5.3) *and* (5.5) *has a unique solution $(u,f) \in (H^2(Q_T) \cap H_0^1([0,l] \times (0,T))) \times C(0,T)$.*

The numerical solution of the inverse problem (5.1)–(5.3) and (5.5) was obtained using the CGM in [157]. Below, the Tikhonov regularized solution of the inverse problem (5.1)–(5.4) is presented [117]. On applying the BEM with N constant boundary elements discretizing each of the boundaries $\{0\} \times [0,T]$ and $\{l\} \times [0,T]$, and N_0 cells discretizing uniformly the space domain $[0,l]$, the inverse problem (5.1)–(5.4) reduces to solving an ill-conditioned system of N linear algebraic equations with N unknowns $\underline{r} = (r_i)_{i=\overline{1,N}} = (r(\tilde{t}_i))_{i=\overline{1,N}}$, where $\tilde{t}_i = (2i-1)T/N$ for $i = \overline{1,N}$ denotes the boundary element node, which in a generic form can be written as

$$X\underline{r} = \underline{y}^\epsilon, \tag{5.7}$$

where the right-hand side may also contain some noise $\epsilon \approx \|\underline{y}^\epsilon - \underline{y}\|$. Since the inverse source problem is ill-posed, the straightforward inversion $X^{-1}\underline{y}^\epsilon$ of the ill-conditioned system of equations (5.7) would produce an unstable solution and in order to restore stability the Tikhonov regularization is applied. This is based on minimizing the functional

$$T_{\lambda,k}(\underline{r}) := \|X\underline{r} - \underline{y}^\epsilon\|^2 + \lambda\|R_k\underline{r}\|^2, \tag{5.8}$$

where R_k is a regularization matrix imposing various orders of C^k–smoothness, see also (1.11)–(1.13),

$$\|R_k \underline{r}\|^2 = \begin{cases} \sum_{i=1}^{N} r_i^2, & \text{if } k = 0, \\ \sum_{i=2}^{N} (r_i - r_{i-1})^2, & \text{if } k = 1, \\ \sum_{i=3}^{N} (r_i - 2r_{i-1} + r_{i-2})^2, & \text{if } k = 2, \end{cases} \tag{5.9}$$

and $\lambda \geq 0$ is a regularization parameter to be prescribed according to the L-curve criterion, by plotting the residual $\|X\underline{r}_\lambda - \underline{y}^\epsilon\|$ versus the solution norm $\|\underline{r}_\lambda\|$ for various values of λ [137] or, if an L-curve is not obtained, by the discrepancy principle based on choosing λ such that $\|X\underline{r}_\lambda - \underline{y}^\epsilon\| \approx \epsilon$, [304]. The minimization of (5.8) yields the explicit solution

$$\underline{r}_{\lambda,k} = (X^{\mathrm{T}}X + \lambda R_k^{\mathrm{T}} R_k)^{-1} X^{\mathrm{T}} \underline{y}^\epsilon. \tag{5.10}$$

The numerical results presented elsewhere [117] reveal that the numerical solution (5.10) obtained using the Tikhonov regularization method converges to the true solution, as the amount of noise ϵ decreases. Moreover, some inaccuracies recorded near the end points of the time interval $[0, T]$ are reducing, as the order k of the regularization method increases.

5.2 Non-Local Variants

Rather recently, nonlocal boundary and overdetermination conditions have become a center of interest in the mathematical formulation and numerical solution of several inverse and improperly posed problems in transient heat conduction. They opened a new area of applied numerical and mathematical modelling research. Practical applications of nonlocal boundary value problems are encountered in chemical diffusion, heat conduction and biological processes, see e.g. [65, 353]. For example, in multiphase flows involving fluids, solids and gases, the heat flux is often taken to be proportional to the difference in boundary temperature between the various phases, and the quantities γ_{ij}, $i = \overline{1,3}$, $j = \overline{1,4}$, present in the nonlocal boundary conditions (5.13) below represent those proportionality factors.

5.2.1 The first variant

Consider the inverse problem of finding the temperature $u(x,t) \in C^{2,1}(\Omega_T) \cap C^{1,0}(\overline{\Omega}_T)$ and the time-dependent heat source $r(t) \in C[0,T]$ which satisfy

$$u_t = u_{xx} + r(t)h(x,t) + g(x,t), \quad (x,t) \in \Omega_T := (0,l) \times (0,T), \tag{5.11}$$

subject to the initial condition

$$u(x,0) = u_0(x), \quad x \in [0,l], \tag{5.12}$$

and the following general boundary and overdetermination conditions:

$$\left.\begin{aligned}\gamma_{11}(t)u(0,t)+\gamma_{12}(t)u(l,t)+\gamma_{13}(t)u_x(0,t)+\gamma_{14}(t)u_x(l,t)=k_1(t)\\ \gamma_{21}(t)u(0,t)+\gamma_{22}(t)u(l,t)+\gamma_{23}(t)u_x(0,t)+\gamma_{24}(t)u_x(l,t)=k_2(t)\\ \gamma_{31}(t)u(0,t)+\gamma_{32}(t)u(l,t)+\gamma_{33}(t)u_x(0,t)+\gamma_{34}(t)u_x(l,t)=k_3(t)\end{aligned}\right\} \quad (5.13)$$

where

$$h, g \in C^{1,0}(\overline{\Omega}_T), \quad u_0 \in C^2[0,l], \quad k_i \in C^1[0,T], \quad i=1,2,3, \qquad (5.14)$$

and the matrix $\gamma(t) = \gamma_{ij} \in C^1[0,T]$ for $i = \overline{1,3}$, $j = \overline{1,4}$, has rank 3 for all $t \in [0,T]$. Moreover, by assuming, without any loss of generality, that the same third-order minor of the matrix $\gamma(t)$ is non-zero, one can express three of the four boundary data $u(0,t)$, $u(l,t)$, $u_x(0,t)$, and $u_x(l,t)$ in terms of the fourth one and distinguish the following six cases:

Case 1.

$$u_x(0,t) = \nu_1(t), \quad u_x(l,t) = \nu_2(t), \qquad (5.15)$$

$$v_1(t)u(0,t) + v_2(t)u(l,t) = \kappa(t); \qquad (5.16)$$

Case 2.

$$u(0,t) = \nu_1(t), \quad u_x(l,t) = \nu_2(t), \qquad (5.17)$$

$$v_1(t)u_x(0,t) + v_2(t)u(l,t) = \kappa(t); \qquad (5.18)$$

Case 3.

$$u(0,t) = \nu_1(t), \quad u(l,t) = \nu_2(t), \qquad (5.19)$$

$$v_1(t)u_x(0,t) + v_2(t)u_x(l,t) = \kappa(t); \qquad (5.20)$$

Case 4.

$$u(0,t) = \nu_1(t), \quad u_x(l,t) + v_1(t)u(l,t) = \nu_2(t), \qquad (5.21)$$

$$u_x(0,t) + v_2(t)u_x(l,t) = \kappa(t); \qquad (5.22)$$

Case 5.

$$u_x(0,t) = \nu_1(t), \quad u_x(l,t) + v_1(t)u(l,t) = \nu_2(t), \qquad (5.23)$$

$$u(0,t) + v_2(t)u(l,t) = \kappa(t); \qquad (5.24)$$

Case 6.

$$u_x(0,t) - v_1(t)u(0,t) = \mu_1(t), \quad u_x(l,t) + v_2(t)u(l,t) = \mu_2(t), \qquad (5.25)$$

$$v_3(t)u(0,t) + v_4(t)u(l,t) = \kappa(t), \qquad (5.26)$$

for $t \in [0,T]$. The other possible mixed boundary conditions cases corresponding to (5.17), (5.21), and (5.23) can be reduced to these ones by the change of variable $x \mapsto l-x$. The following theorems [204] give the unique solvability of the inverse problem (5.11)–(5.13) in all the above six cases.

Theorem 5.3. *Assume that the regularity conditions* (5.14) *are satisfied and:*
(i) $v_1,\ v_2 \in C^1[0,T],\ v_1^2(t)+v_2^2(t)>0,\ t\in[0,T]$;
(ii) $v_1(t)h(0,t)+v_2(t)h(l,t)\neq 0,\ t\in[0,T]$;
(iii) $\nu_1(0)=u_0'(0),\ \nu_2(0)=u_0'(l),\ v_1(0)u_0(0)+v_2(0)u_0(l)=\kappa(0)$.
Then, the inverse problem (5.11), (5.12), (5.15), (5.16) *representing Case 1 is uniquely solvable.*

Theorem 5.4. *Assume that, in addition to conditions* (5.14) *and (i) of Theorem 5.3, the following conditions are satisfied:*
$v_1(t)h(0,t)\neq 0,\ t\in[0,T]$;
$\nu_1(0)=u_0(0),\ \nu_2(0)=u_0'(l),\ v_1(0)u_0'(0)+v_2(0)u_0(L)=\kappa(0)$.
Then, the inverse problem (5.11), (5.12), (5.17), (5.18) *representing Case 2 is uniquely solvable.*

Theorem 5.5. *Assume that, in addition to conditions* (5.14), *(i) and (ii) of Theorem 5.3, the following conditions are satisfied:*
$\nu_1(0)=u_0(0),\ \nu_2(0)=u_0(l),\ v_1(0)u_0'(0)+v_2(0)u_0'(l)=\kappa(0)$.
Then, the inverse problem (5.11), (5.12), (5.19), (5.20) *representing Case 3 is uniquely solvable.*

Theorem 5.6. *Assume that conditions* (5.14) *are satisfied and that:*
$0<v_1\in C[0,T],\ v_2,\ \nu_1,\ \nu_2\in C^1[0,T]$;
and, in Case 4:
$h(0,t)-v_2(t)h(l,t)\neq 0,\ t\in[0,T]$,
$\nu_1(0)=u_0(0),\ \nu_2(0)=u_0'(l)+v_1(0)u_0(l),\ \kappa(0)=u_0'(0)+v_2(0)u_0'(l)$;
or, in Case 5:
$h(0,t)+v_2(t)h(l,t)\neq 0,\ t\in[0,T]$,
$\nu_1(0)=u_0'(0),\ \nu_2(0)=u_0'(l)+v_1(0)u_0(l),\ k(0)=u_0(0)+v_2(0)u_0(l)$.
Then, the inverse problem (5.11), (5.12), (5.21), (5.22) *representing Case 4, or* (5.11), (5.12), (5.23), (5.24) *representing Case 5 is uniquely solvable.*

Theorem 5.7. *Assume that conditions* (5.14) *are satisfied and that:*
$0<v_1\in C[0,T],\ 0<v_2\in C[0,T],\ v_3,\ v_4,\ \nu_1,\ \nu_2\in C^1[0,T]$;
$v_1(t)h(0,t)+v_2(t)h(l,t)\neq 0,\ v_3^2(t)+v_4^2(t)>0,\ t\in[0,T]$;
$\nu_1(0)=u_0'(0)-v_1(0)u_0(0),\ \nu_2(0)=u_0'(l)+v_2(0)u_0(l),\ k(0)=v_3(0)u_0(0)+v_4(0)u_0(l)$.
Then, the inverse problem (5.11), (5.12), (5.25), (5.26) *representing Case 6 is uniquely solvable.*

Although the problems of Cases 1-6 are uniquely solvable, they could still be ill-posed if small errors in the input data $\kappa(t)$ would lead to large errors in the output source solution $r(t)$. In the next section, it is described how the BEM discretizing numerically the heat equation (5.11) can be used in conjunction with the Tikhonov regularization to obtain stable solutions [162].

5.2.1.1 BEM for the first variant

Apply the Green's formula to the heat equation (5.11) to obtain the boundary integral equation

$$\eta(x)u(x,t) = \int_0^t \left[G(x,t;\xi,\tau)\frac{\partial u}{\partial n(\xi)}(\xi,\tau) - u(\xi,\tau)\frac{\partial G}{\partial n(\xi)}(x,t;\xi,\tau)\right]_{\xi\in\{0,l\}} d\tau$$
$$+ \int_0^l G(x,t;y,0)u(y,0)\,dy + \int_0^l \int_0^t G(x,t;y,\tau)r(\tau)h(y,\tau)d\tau dy$$
$$+ \int_0^l \int_0^t G(x,t;y,\tau)g(y,\tau)d\tau dy, \quad (x,t)\in[0,l]\times(0,T), \tag{5.27}$$

where $\underline{n}$, η and G have been defined before below equation (4.47). The boundaries $\{0\}\times[0,T]$ and $\{l\}\times[0,T]$ of the integral equation (5.27) are discretized into a series of N small boundary elements $[t_{j-1},t_j]$ for $j=\overline{1,N}$, where $t_j = jT/N$ for $j=\overline{0,N}$, whilst the space domain $[0,l]\times\{0\}$ is divided into N_0 small cells $[x_{k-1},x_k]$ for $k=\overline{1,N_0}$, where $x_k = k/N_0$ for $k=\overline{0,N_0}$. The boundary temperature u and the heat flux $\frac{\partial u}{\partial n}$ are assumed to be constant over each boundary element $(t_{j-1},t_j]$ and take their values at the midpoint $\tilde{t}_j = \frac{t_{j-1}+t_j}{2}$, i.e.

$$u(0,t) = u(0,\tilde{t}_j) =: u_{0j}, \quad u(l,t) = u(l,\tilde{t}_j) =: u_{lj}, \quad t\in(t_{j-1},t_j],$$
$$\frac{\partial u}{\partial n}(0,t) = \frac{\partial u}{\partial n}(0,\tilde{t}_j) =: q_{0j}, \quad \frac{\partial u}{\partial n}(l,t) = \frac{\partial u}{\partial n}(l,\tilde{t}_j) =: q_{lj}, \quad t\in(t_{j-1},t_j].$$

Also assume that the initial temperature is constant over each cell $[x_{k-1},x_k]$ and takes its value at the midpoint $\tilde{x}_k = \frac{x_{k-1}+x_k}{2}$, i.e.

$$u(x,0) = u(\tilde{x}_k,0) =: u_{0,k}, \quad x\in[x_{k-1},x_k].$$

Of course, higher-order BEM based on piecewise linear or quadratic interpolants will produce even more accurate numerical results. This is particularly useful for higher-dimensional problems. With the above piecewise constant approximations, the integral equation (5.27) is discretized as

$$\eta(x)u(x,t) = \sum_{j=1}^{N} \left[A_{0j}(x,t)q_{0j} + A_{lj}(x,t)q_{lj} - B_{0j}(x,t)u_{0j} - B_{lj}(x,t)u_{lj}\right]$$
$$+ \sum_{k=1}^{N_0} C_k(x,t)u_{0,k} + d_0(x,t) + d_1(x,t), \quad (x,t)\in[0,l]\times(0,T), \tag{5.28}$$

where

$$A_{\xi j}(x,t) = \int_{t_{j-1}}^{t_j} G(x,t;\xi,\tau)d\tau,$$

$$B_{\xi j}(x,t) = \int_{t_{j-1}}^{t_j} \frac{\partial G}{\partial n(\xi)}(x,t;\xi,\tau)d\tau \quad \text{for } \xi = \{0,l\},$$

$$C_k(x,t) = \int_{x_{k-1}}^{x_k} G(x,t;y,0)dy, \tag{5.29}$$

$$d_0(x,t) = \int_0^l \int_0^t G(x,t;y,\tau)r(\tau)h(y,\tau)d\tau dy,$$

$$d_1(x,t) = \int_0^l \int_0^t G(x,t;y,\tau)g(y,\tau)d\tau dy. \tag{5.30}$$

The integrals in (5.29) can be evaluated analytically and their explicit expressions are given by [162],

$$A_{\xi j}(x,t) = \begin{cases} 0, & t \le t_{j-1}, \\ \sqrt{\frac{t-t_{j-1}}{\pi}}, & t_{j-1} < t \le t_j,\ x = \xi, \\ \frac{|x-\xi|}{2\sqrt{\pi}}\left(\frac{e^{-z_0^2}}{z_0} - \sqrt{\pi}\mathrm{erfc}(z_0)\right), & t_{j-1} < t \le t_j,\ x \ne \xi, \\ \sqrt{\frac{t-t_{j-1}}{\pi}} - \sqrt{\frac{t-t_j}{\pi}}, & t > t_j,\ x = \xi, \\ \frac{|x-\xi|}{2\sqrt{\pi}}\left(\frac{e^{-z_0^2}}{z_0} - \frac{e^{-z_1^2}}{z_1} + \sqrt{\pi}\left(\mathrm{erf}(z_0) - \mathrm{erf}(z_1)\right)\right), & t > t_j,\ x \ne \xi, \end{cases}$$

$$B_{\xi j}(x,t) = \begin{cases} 0, & t \le t_{j-1}, \\ 0, & t_{j-1} < t \le t_j,\ x = \xi, \\ -\frac{\mathrm{erfc}(z_0)}{2}, & t_{j-1} < t \le t_j,\ x \ne \xi, \\ \frac{\mathrm{erf}(z_0) - \mathrm{erf}(z_1)}{2}, & t > t_j, \end{cases}$$

$$C_k(x,t) = \frac{1}{2}\left[\mathrm{erf}\left(\frac{x - x_{k-1}}{2\sqrt{t}}\right) - \mathrm{erf}\left(\frac{x - x_k}{2\sqrt{t}}\right)\right],$$

where $j = \overline{1,N}$, $k = \overline{1,N_0}$, $\xi \in \{0,1\}$, $z_0 = \frac{|x-\xi|}{2\sqrt{t-t_{j-1}}}$, $z_1 = \frac{|x-\xi|}{2\sqrt{t-t_j}}$ and erf, erfc are the error functions defined by $\mathrm{erf}(x) = \frac{2}{\sqrt{\pi}}\int_0^x e^{-\sigma^2}d\sigma$, $\mathrm{erfc}(x) = 1 - \mathrm{erf}(x)$, respectively. Assuming piecewise constant approximations for the functions $h(x,t)$, $g(x,t)$ and $r(t)$, namely,

$$h(x,t) = h(x,\tilde{t}_j), \quad g(x,t) = g(x,\tilde{t}_j), \quad r(t) = r(\tilde{t}_j) =: r_j, \quad t \in (t_{j-1}, t_j],\ j = \overline{1,N},$$

the integrals in (5.30) can be approximated as follows:

$$d_0(x,t) = \int_0^t r(\tau) \int_0^l G(x,t;y,\tau)h(y,\tau)dyd\tau = \sum_{j=1}^{N} D_{0,j}(x,t)r_j, \qquad (5.31)$$

$$d_1(x,t) = \int_0^t \int_0^l G(x,t;y,\tau)g(y,\tau)dyd\tau = \sum_{j=1}^{N} D_{1,j}(x,t), \qquad (5.32)$$

where

$$\begin{aligned} D_{0,j}(x,t) &= \int_{t_{j-1}}^{t_j} \int_0^l G(x,t;y,\tau)h(y,\tilde{t}_j)dyd\tau \\ &= \int_0^l h(y,\tilde{t}_j)A_{yj}(x,t)dy, \qquad (5.33) \\ D_{1,j}(x,t) &= \int_{t_{j-1}}^{t_j} \int_0^l G(x,t;y,\tau)g(y,\tilde{t}_j)dyd\tau \\ &= \int_0^l g(y,\tilde{t}_j)A_{yj}(x,t)dy, \qquad (5.34) \end{aligned}$$

are evaluated numerically using the midpoint rule. With the above approximations, the integral equation (5.28) can be rewritten as

$$\begin{aligned} &\eta(x)u(x,t) \\ &= \sum_{j=1}^{N} \left[A_{0j}(x,t)q_{0j} + A_{lj}(x,t)q_{lj} - B_{0j}(x,t)u_{0j} - B_{lj}(x,t)u_{lj}\right] \\ &\quad + \sum_{k=1}^{N_0} C_k(x,t)u_{0,k} + \sum_{j=1}^{N} D_{0,j}(x,t)r_j + \sum_{j=1}^{N} D_{1,j}(x,t), \\ &\qquad (x,t) \in [0,l] \times (0,T). \qquad (5.35) \end{aligned}$$

Collocating (5.35) at the boundary nodes $(0,\tilde{t}_i)$ and $(l,\tilde{t}_i)$ for $i = \overline{1,N}$, the following system of $2N$ equations

$$A\underline{q} - B\underline{u} + C\underline{u}^0 + D_0\underline{r} + \underline{D}_1 = \underline{0}, \qquad (5.36)$$

is obtained, where

$$A = \begin{bmatrix} A_{0j}(0,\tilde{t}_i) & A_{lj}(0,\tilde{t}_i) \\ A_{0j}(L,\tilde{t}_i) & A_{lj}(L,\tilde{t}_i) \end{bmatrix}_{2N\times 2N},$$

$$B = \begin{bmatrix} B_{0j}(0,\tilde{t}_i) + \frac{1}{2}\delta_{ij} & B_{lj}(0,\tilde{t}_i) \\ B_{0j}(l,\tilde{t}_i) & B_{lj}(L,\tilde{t}_i) + \frac{1}{2}\delta_{ij} \end{bmatrix}_{2N\times 2N},$$

$$C = \begin{bmatrix} C_k(0,\tilde{t}_i) \\ C_k(l,t_i) \end{bmatrix}_{2N\times N_0}, \quad D_0 = \begin{bmatrix} D_{0,j}(0,\tilde{t}_i) \\ D_{0,j}(l,t_i) \end{bmatrix}_{2N\times N},$$

$$\underline{D}_1 = \begin{bmatrix} \sum_{j=1}^{N} D_{1,j}(0,\tilde{t}_i) \\ \sum_{j=1}^{N} D_{1,j}(l,\tilde{t}_i) \end{bmatrix}_{2N},$$

$$\underline{q} = \begin{bmatrix} q_{0j} \\ q_{lj} \end{bmatrix}_{2N}, \quad \underline{u} = \begin{bmatrix} u_{0j} \\ u_{lj} \end{bmatrix}_{2N}, \quad \underline{u}^0 = \begin{bmatrix} u_{0,k} \end{bmatrix}_{N_0}, \quad \underline{r} = \begin{bmatrix} r_j \end{bmatrix}_N. \tag{5.37}$$

Applying the boundary and the overdetermination conditions for the 6 cases considered yield the following development.

Case 1. The Neumann heat flux boundary conditions (5.15) give

$$\underline{q} = \begin{bmatrix} q_{0j} \\ q_{lj} \end{bmatrix}_{2N} = \begin{bmatrix} -\nu_1(\tilde{t}_j) \\ \nu_2(\tilde{t}_j) \end{bmatrix}_{2N}. \tag{5.38}$$

Also, from (5.36),

$$\underline{u} = B^{-1}\left(A\underline{q} + C\underline{u}^0 + D_0\underline{f} + \underline{D}_1\right). \tag{5.39}$$

The overdetermination condition (5.16) can be rewritten as a matrix equation

$$\begin{bmatrix} V_1 & V_2 \end{bmatrix} \underline{u} = \underline{\kappa}, \tag{5.40}$$

where V_1, V_2 are $N \times N$ diagonal matrices of $v_1(\tilde{t_1}), \ldots, v_1(\tilde{t_N})$ and $v_2(\tilde{t_1}), \ldots, v_2(\tilde{t_N})$ components, respectively, and $\underline{\kappa}$ is an N column vector of the piecewise constant approximation of $\kappa(t)$, namely

$$\kappa(t) = \kappa(\tilde{t}_j) =: \kappa_j, \quad t \in (t_{j-1}, t_j],\ j = \overline{1,N}.$$

Substituting (5.39) into (5.40) yields

$$\begin{bmatrix} V_1 & V_2 \end{bmatrix} B^{-1}\left(A\underline{q} + C\underline{u}^0 + D_0\underline{f} + \underline{D}_1\right) = \underline{\kappa}.$$

Rearranging this expression, the inverse source problem in Case 1 reduces to solving the $N \times N$ linear system of equations

$$X_1\underline{r} = \underline{y}_1, \tag{5.41}$$

where $X_1 = \begin{bmatrix} V_1 & V_2 \end{bmatrix} B^{-1}D_0$, $\underline{y}_1 = \underline{\kappa} - \begin{bmatrix} V_1 & V_2 \end{bmatrix} B^{-1}\left(A\underline{q} + C\underline{u}^0 + \underline{D}_1\right)$ and

$\underline{q}$ is specified as in (5.38).

Case 2. The matrix equation (5.36) can be rearranged as follows:

$$\begin{bmatrix} A_0 & A_1 \end{bmatrix} \begin{bmatrix} q_{0j} \\ q_{lj} \end{bmatrix} - \begin{bmatrix} B_0 & B_1 \end{bmatrix} \begin{bmatrix} u_{0j} \\ u_{lj} \end{bmatrix} + C\underline{u}^0 + D_0\underline{r} + \underline{D}_1 = \underline{0},$$

where

$$A_0 = \begin{bmatrix} A_{0j}(0, \tilde{t}_j) \\ A_{0j}(l, \tilde{t}_j) \end{bmatrix}, \quad A_1 = \begin{bmatrix} A_{lj}(0, \tilde{t}_j) \\ A_{lj}(l, \tilde{t}_j) \end{bmatrix},$$
$$B_0 = \begin{bmatrix} B_{0j}(0, \tilde{t}_j) + \frac{1}{2}\delta_{ij} \\ B_{0j}(l, \tilde{t}_j) \end{bmatrix}, \quad B_1 = \begin{bmatrix} B_{lj}(0, \tilde{t}_j) \\ B_{lj}(l, \tilde{t}_j) + \frac{1}{2}\delta_{ij} \end{bmatrix}$$

are $2N \times N$ matrices. Next, the boundary conditions (5.17) are applied such that this system becomes

$$-A_0\underline{q}_0 + A_1\underline{\nu}_2 - B_0\underline{\nu}_1 - B_1\underline{u}_l + C\underline{u}^0 + D_0\underline{r} + \underline{D}_1 = \underline{0}, \tag{5.42}$$

where $\underline{\nu}_1 = \left[\nu_1(\tilde{t}_j)\right]_N$, $\underline{\nu}_2 = \left[\nu_2(\tilde{t}_j)\right]_N$, $\underline{q}_0 = \left[q_{0j}\right]_N$ and $\underline{u}_l = \left[u_{lj}\right]_N$, or

$$\begin{bmatrix} -\underline{q}_0 \\ \underline{u}_l \end{bmatrix} = \begin{bmatrix} A_0 & B_1 \end{bmatrix}^{-1} \left(A_1\underline{\nu}_2 - B_0\underline{\nu}_1 + C\underline{u}^0 + D_0\underline{r} + \underline{D}_1 \right). \tag{5.43}$$

The overdetermination condition (5.18) can be written as a matrix equation

$$\begin{bmatrix} V_1 & V_2 \end{bmatrix} \begin{bmatrix} -\underline{q}_0 \\ \underline{u}_l \end{bmatrix} = \underline{\kappa}. \tag{5.44}$$

Substituting (5.43) into (5.44) yields

$$\begin{bmatrix} V_1 & V_2 \end{bmatrix} \begin{bmatrix} A_0 & B_1 \end{bmatrix}^{-1} \left(A_1\underline{\nu}_2 - B_0\underline{\nu}_1 + C\underline{u}^0 + D_0\underline{f} + \underline{D}_1 \right) = \underline{\kappa}.$$

Rearranging this expression, the inverse source problem in Case 2 reduces to solving the $N \times N$ linear system of equations

$$X_2\underline{r} = \underline{y}_2, \tag{5.45}$$

where $X_2 = \begin{bmatrix} V_1 & V_2 \end{bmatrix} \begin{bmatrix} A_0 & B_1 \end{bmatrix}^{-1} D_0$ and
$\underline{y}_2 = \underline{\kappa} - \begin{bmatrix} V_1 & V_2 \end{bmatrix} \begin{bmatrix} A_0 & B_1 \end{bmatrix}^{-1} \left(A_1\underline{\nu}_2 - B_0\underline{\nu}_1 + C\underline{u}^0 + \underline{D}_1 \right)$.

Case 3. The Dirichlet boundary temperature conditions (5.19) give

$$\underline{u} := \begin{bmatrix} u_{0j} \\ u_{lj} \end{bmatrix}_{2N} = \begin{bmatrix} \nu_1(\tilde{t}_j) \\ \nu_2(\tilde{t}_j) \end{bmatrix}_{2N}. \tag{5.46}$$

Also, from (5.36)

$$\underline{q} = A^{-1} \left(B\underline{u} - C\underline{u}^0 - D_0\underline{r} - \underline{D}_1 \right). \tag{5.47}$$

The overdetermination condition (5.20) can be written as

$$\begin{bmatrix} -V_1 & V_2 \end{bmatrix} \underline{q} = \underline{\kappa}. \tag{5.48}$$

Substituting (5.47) into (5.48) gives the $N \times N$ linear system of equations

$$X_3 \underline{r} = \underline{y}_3, \tag{5.49}$$

where $X_3 = \begin{bmatrix} -V_1 & V_2 \end{bmatrix} A^{-1} D_0$, $y_3 = -\underline{\kappa} + \begin{bmatrix} -V_1 & V_2 \end{bmatrix} A^{-1} \left(B\underline{u} - C\underline{u}^0 - \underline{D}_1 \right)$ and $\underline{u}$ is specified as in (5.46).

Case 4. The matrix equation (5.36) can be rearranged as follows:

$$\begin{bmatrix} -\underline{q}_0 \\ \underline{u}_l \end{bmatrix} = \begin{bmatrix} A_0 & (A_1 V_1 + B_1) \end{bmatrix}^{-1} \left(A_1 \underline{\nu}_2 - B_0 \underline{\nu}_1 + C\underline{u}^0 + D_0 \underline{r} + \underline{D}_1 \right). \tag{5.50}$$

The overdetermination condition (5.22) gives

$$\begin{bmatrix} I & -V_2 V_1 \end{bmatrix} \begin{bmatrix} -\underline{q}_0 \\ \underline{u}_l \end{bmatrix} = \underline{\kappa} - V_2 \underline{\nu}_2, \tag{5.51}$$

where I is the $N \times N$ identity matrix. Substitute (5.50) into (5.51) to obtain the $N \times N$ linear system of equations

$$X_4 \underline{r} = \underline{y}_4, \tag{5.52}$$

where $X_4 = \begin{bmatrix} I & -V_2 V_1 \end{bmatrix} \begin{bmatrix} A_0 & A_1 V_1 + B_1 \end{bmatrix}^{-1} D_0$ and $\underline{y}_4 = \underline{\kappa} - V_2 \underline{\nu}_2 - \begin{bmatrix} I & -V_2 V_1 \end{bmatrix} \begin{bmatrix} A_0 & A_1 V_1 + B_1 \end{bmatrix}^{-1} \left(A_1 \underline{\nu}_2 - B_0 \underline{\nu}_1 + C\underline{u}^0 + \underline{D}_1 \right)$.

Case 5. The matrix equation (5.36) can be rearranged as follows:

$$\underline{u} = \begin{bmatrix} B_0 & (A_1 V_1 + B_1) \end{bmatrix}^{-1} \left(-A_0 \underline{\nu}_1 + A_1 \underline{\nu}_2 + C\underline{u}^0 + D_0 \underline{r} + \underline{D}_1 \right). \tag{5.53}$$

The overdetermination condition (5.24) gives

$$\begin{bmatrix} I & V_2 \end{bmatrix} \underline{u} = \underline{\kappa}. \tag{5.54}$$

Substitute (5.53) into (5.54) to obtain the $N \times N$ linear system of equations

$$X_5 \underline{r} = \underline{y}_5, \tag{5.55}$$

where $X_5 = \begin{bmatrix} I & V_2 \end{bmatrix} \begin{bmatrix} B_0 & A_1 V_1 + B_1 \end{bmatrix}^{-1} D_0$ and
$\underline{y}_5 = \underline{\kappa} - \begin{bmatrix} I & V_2 \end{bmatrix} \begin{bmatrix} B_0 & A_1 V_1 + B_1 \end{bmatrix}^{-1} \left(-A_0 \underline{\mu}_1 + A_1 \underline{\mu}_2 + C\underline{u}^0 + D_0 \underline{f} + \underline{D}_1 \right)$.

Case 6. From (5.25),

$$\underline{q} = \begin{bmatrix} q_{0j} \\ q_{lj} \end{bmatrix}_{2N} = \begin{bmatrix} -\underline{\nu}_1 - V_1 u_{0j} \\ \underline{\nu}_2 - V_2 u_{lj} \end{bmatrix}_{2N}.$$

Substituting this into the matrix equation (5.36) to yield

$$\underline{u} = \left[(A_0V_1 + B_0) \quad (A_1V_2 + B_1)\right]^{-1}\big(- A_0\underline{\nu}_1 + A_1\underline{\nu}_2 + C\underline{u}^0 + D_0\underline{r} + \underline{D}_1\big). \tag{5.56}$$

The overdetermination condition (5.26) gives

$$\begin{bmatrix} V_3 & V_4 \end{bmatrix} \underline{u} = \underline{\kappa}. \tag{5.57}$$

Substitute (5.56) into (5.57) to obtain the $N \times N$ linear system of equations

$$X_6\underline{r} = \underline{y}_6, \tag{5.58}$$

where $X_6 = \begin{bmatrix} V_3 & V_4 \end{bmatrix}\begin{bmatrix} A_0V_1 + B_0 & A_1V_2 + B_1 \end{bmatrix}^{-1} D_0$ and $\underline{y}_6 = \underline{\kappa} - \begin{bmatrix} V_3 & V_4 \end{bmatrix}\begin{bmatrix} A_0V_1 + B_0 & A_1V_2 + B_1 \end{bmatrix}^{-1}\big(-A_0\underline{\nu}_1 + A_1\underline{\nu}_2 + C\underline{u}^0 + \underline{D}_1\big)$.

From the above assembly one can see that the solution of the inverse heat source problem (5.11)–(5.13) separated into the 6 cases has been reduced to solving the $N \times N$ linear system of equations, generally written as

$$X\underline{r} = \underline{y}, \tag{5.59}$$

where X and $\underline{y}$ are the matrix and the right-hand side vector, respectively, corresponding to the linear systems of equations (5.41), (5.45), (5.49), (5.52), (5.55) and (5.58) for Cases 1–6, respectively. The system of equations (5.59) is ill-conditioned whenever the inverse problem is ill-posed. Therefore, a straightforward inversion of (5.59) such as the Gaussian elimination or the SVD will result in an unstable solution, especially when the right-hand side vector $\underline{y}$ is contaminated with noise as $\underline{y}^\epsilon = \underline{y} +$ noise. To ensure stability, the Tikhonov regularization method for (5.59) can be employed giving explicitly the solution (5.10).

5.2.1.2 Numerical examples and discussion

Let us introduce the root mean square error (RMSE) defined as

$$\text{RMSE}(r) = \sqrt{\frac{1}{N}\sum_{i=1}^{N}\big(r_{analytical}(\tilde{t}_i) - r_{numerical}(\tilde{t}_i)\big)^2}. \tag{5.60}$$

Consider a test example with the input data $T = l = 1$,

$$u(x,0) = u_0(x) = 1 + x - x^2, \quad h(x,t) = (1-x^2)e^{-t}, \quad g(x,t) = (2+x)e^t. \tag{5.61}$$

In addition, the boundary and overdetermination conditions are as follows:

TABLE 5.1: The condition numbers of the matrix X in equation (5.59), for $N = N_0 \in \{20, 40, 80\}$ for Cases 1–6, [162].

Case	$N = N_0 = 20$	$N = N_0 = 40$	$N = N_0 = 80$
1	6.1E+3	1.2E+5	7.4E+6
2	87	248	705
3	22	54	125
4	38	107	298
5	1.5E+4	7.3E+5	1.7E+8
6	1.6E+4	2.5E+6	5.8E+9

Case 1.

$$u_x(0,t) = e^t, \quad u_x(1,t) = -e^t, \quad u(0,t) + u(1,t) = 2e^t. \tag{5.62}$$

Case 2.

$$u(0,t) = e^t, \quad u_x(1,t) = -e^t, \quad u_x(0,t) + u(1,t) = 2e^t. \tag{5.63}$$

Case 3.

$$u(0,t) = e^t, \quad u(1,t) = e^t, \quad e^t u_x(0,t) + t u_x(1,t) = (e^t - t)e^t, \tag{5.64}$$

where $v_1(t) = e^t$ and $v_2(t) = t$.
Case 4.

$$\begin{aligned} u(0,t) = e^t, \quad u_x(1,t) + (1+t)u(1,t) &= te^t, \\ u_x(0,t) + e^{-t}u_x(1,t) &= e^t - 1, \end{aligned} \tag{5.65}$$

where $v_1(t) = 1 + t$ and $v_2(t) = e^{-t}$.
Case 5.

$$\begin{aligned} u_x(0,t) = e^t, \quad u_x(1,t) + e^{-t}u(1,t) &= 1 - e^t, \\ u(0,t) + (1+t)u(1,t) &= (2+t)e^t, \end{aligned} \tag{5.66}$$

where $v_1(t) = e^{-t}$ and $v_2(t) = 1 + t$.
Case 6.

$$\begin{aligned} u_x(0,t) - e^t u(0,t) = e^t - e^{2t}, \quad u_x(1,t) + (1+t)u(1,t) &= te^t, \\ tu(0,t) + (1-t)u(1,t) &= e^t, \end{aligned} \tag{5.67}$$

where $v_1(t) = e^t$, $v_2(t) = 1 + t$, $v_3(t) = t$ and $v_4(t) = 1 - t$.

In these examples, the exact solution is given by

$$u(x,t) = (1 + x - x^2)e^t, \quad r(t) = e^{2t}, \quad (x,t) \in (0,1) \times (0,1). \tag{5.68}$$

The input data in expressions (5.61)–(5.67) satisfy the conditions of Theorems 5.3–5.7 for the existence and uniqueness of the solution of the inverse problems of Cases 1–6 under investigation.

Table 5.1 shows the condition numbers of matrix X in (5.59) for $N_0 = N \in \{20, 40, 80\}$ for Cases 1–6. It can be seen that the condition numbers of the matrix X for Cases 1, 5, and, 6 are high and this ill-conditioning will need to be dealt with by the Tikhonov regularization. On the other hand, the matrices for Cases 2–4 do not have a large condition number and the system of equations (5.59) can be solved directly using, e.g., the Gauss elimination method or the SVD. To investigate the stability of the numerical solution, noise is added in the overspecified data as

$$\underline{\kappa}^p = \underline{\kappa} + random('Normal', 0, \sigma, 1, N), \tag{5.69}$$

where $random('Normal', 0, \sigma, 1, N)$ is a MATLAB command which generates random variables from a normal distribution with zero mean and standard deviation $\sigma = p \times \max_{t\in[0,1]} |\kappa(t)|$, where p is the percentage of noise. Next, numerical results are presented for $N = N_0 = 80$.

Case 1. The ill-posedness of the inverse problem and the instability of the unregularized numerical solution in Case 1 are enhanced by the presence of noise in the measured data, as illustrated in Table 5.2 by the huge RMSE(r) = 490 obtained for $p = 1\%$ noise with $\lambda = 0$. To remove the highly unbounded oscillations, the Tikhonov regularization is employed with the choice of the regularization parameter λ in (5.10) given by the discrepancy principle.

Figures 5.1(a) and 5.1(c) present the discrepancy principle curves obtained by the Tikhonov regularization of order zero and one, respectively, for $p = 1\%$ and 3% noisy data. Generated as in (5.69), this results in the amounts of noise $\epsilon = 0.32$ and 1.01 for $p = 1\%$ and 3%, respectively. The intersection between the horizontal line $y \approx \epsilon$ and the discrepancy (residual) curve yields the regularization parameter denoted by λ_d and tabulated in Table 5.2. With these values of λ_d, Figures 5.1(b) and 5.1(d) present the numerical results for $r(t)$ obtained using the zeroth- and first-order Tikhonov regularization, respectively. It can be seen that the first-order regularization produces more accurate and stable results than the zeroth-order regularization since it imposes a higher-order smoothness constraint onto the solution.

Cases 2-4. In Cases 2–4, Table 5.1 shows that the condition numbers of the matrices X_2, X_3 and X_4 are much smaller than those of the matrices for the rest of the cases. Therefore, for exact data, i.e. $p = 0$, accurate and stable numerical results of the inversion can be obtained even when no regularization is employed, i.e. $\lambda = 0$. This is consistent with the fact that the inverse problems of Cases 2–4 are less ill-posed than the inverse problem of Case 1 (and Cases 5 and 6).

Case 5. In Case 5 (and also Case 6) even higher ill-conditioning is expected to occur in the systems of equations (5.55) (and also (5.58)) by comparing the

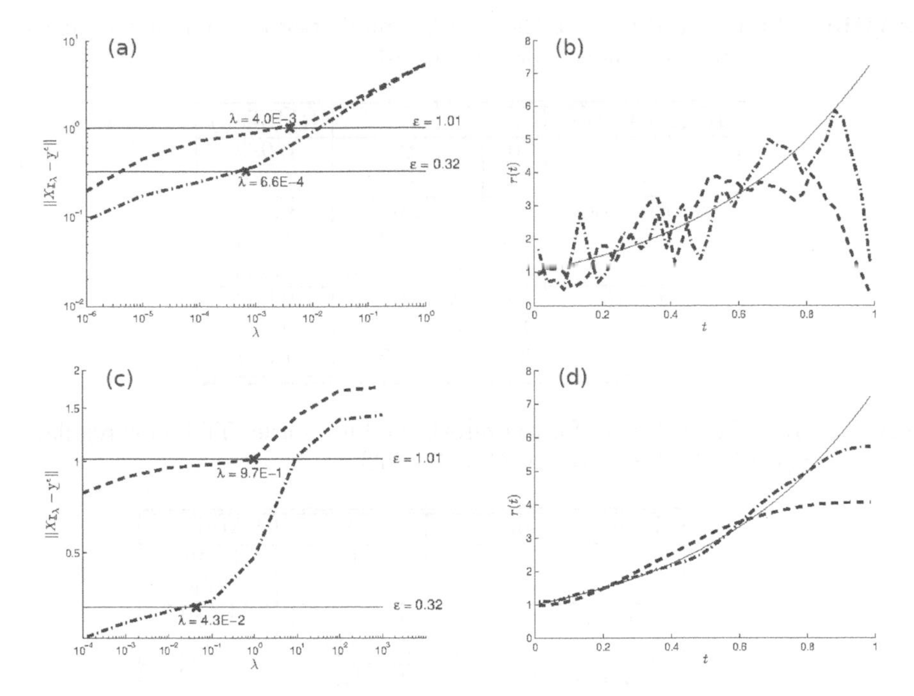

FIGURE 5.1: (a,c) Discrepancy principle curves and the analytical (—) and numerical results of $r(t)$, obtained using the (b) zeroth-order and the (d) first-order Tikhonov regularization, for $p = 1\%$ $(- \cdot -)$ and $p = 3\%$ $(- - -)$ noise with the regularization parameters λ_d given in Table 5.2, for Case 1, [162].

condition numbers in Table 5.1. Numerical results obtained using the zeroth- and first-order regularization are presented in Figures 5.2 and Table 5.3.

Case 6. In Case 6, the retrieval of a non-smooth heat source given by

$$r(t) = \left| t - \frac{1}{2} \right| + 1, \quad t \in (0, T = 1), \tag{5.70}$$

is considered, with the input data

$$\begin{aligned} u(x,0) &= u_0(x) = 1 + x - x^2, \quad h(x,t) = (1 - x^2)e^{-t}, \\ g(x,t) &= (3 + x - x^2)e^t - (1 - x^2)\left(\left| t - \frac{1}{2} \right| + 1 \right) e^{-t}, \end{aligned} \tag{5.71}$$

which are generated from the analytical temperature solution

$$u(x,t) = (1 + x - x^2)e^t. \tag{5.72}$$

TABLE 5.2: The RMSE(r) for the zeroth- and first-order Tikhonov regularization for $p = \{0, 1, 3\}\%$ noise, for Case 1, [162].

Regularization	p	λ	RMSE(r)
-	0	0	0.983
-	1%	0	490
zeroth-order	0	λ=1.0E-5	0.499
	1%	λ_d=6.6E-4	1.264
	3%	λ_d=4.0E-3	1.982
first-order	0	λ=1.0E-7	0.009
	1%	λ_d=4.3E-2	0.364
	3%	λ_d=9.7E-1	1.007

TABLE 5.3: The RMSE(r) for the zeroth- and first-order Tikhonov regularization for $p = \{0, 1, 3\}\%$ noise, for Case 5, [162].

Regularization	p	λ	RMSE(r)
-	0	0	5.598
-	1%	0	2900
zeroth-order	1%	λ_d=2.2E-3	1.584
	3%	λ_d=9.5E-3	2.033
first-order	1%	λ_d=1.7E-1	0.673
	3%	λ_d=3.4	1.334

Results are illustrated for the first-order Tikhonov regularization only, as for the zeroth-order regularization the results were obtained oscillatory. Figure 5.3 shows the numerical results obtained for various amounts of noise $p \in \{0, 0.1, 0.5, 1\}\%$ for the regularization parameters λ_d given in Table 5.4. From this figure it can be seen that the Tikhonov regularization can invert accurately up to about 0.1% noisy data. For higher levels of noise the reconstruction of the non-smooth heat source (5.70) starts to loose significant accuracy while remaining stable.

TABLE 5.4: The RMSE(r) for the first-order Tikhonov regularization for $p \in \{0, 0.1, 0.5, 1\}\%$ noise, for Case 6, [162].

p	λ	RMSE(r)
0	λ=1.0E−5	0.002
0.1 %	λ_d=2.7E−2	0.026
0.5 %	λ_d=3.1E−1	0.102
1 %	λ_d=1.1	0.128

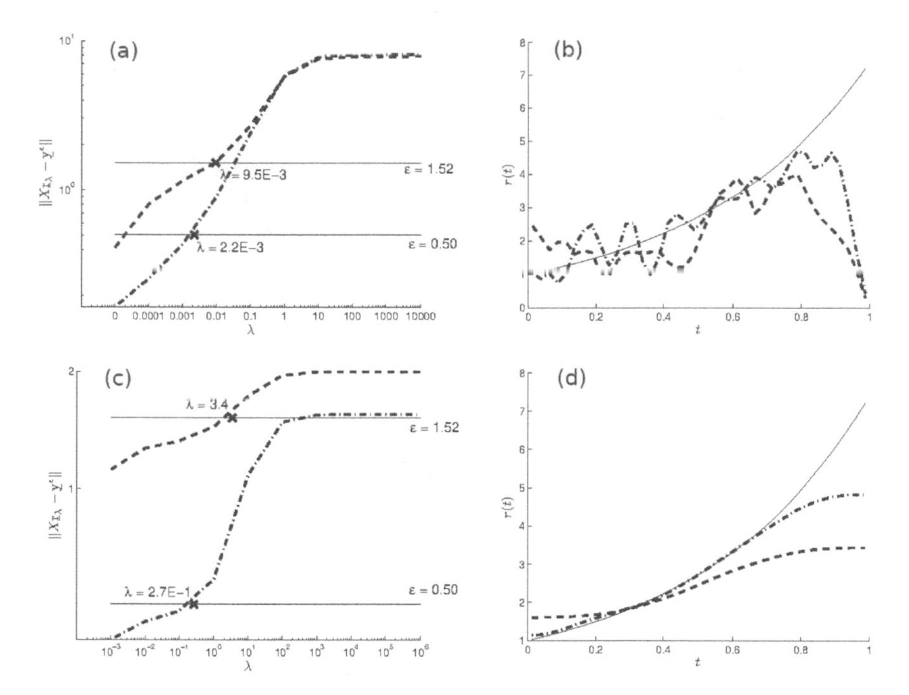

FIGURE 5.2: (a,c) Discrepancy principle curves and the analytical (—) and numerical results of $r(t)$, obtained using the (b) zeroth-order and (d) first-order Tikhonov regularization, for $p = 1\%$ $(- \cdot -)$ and $p = 3\%$ $(- - -)$ noise with the regularization parameters λ_d given in Table 5.3, for Case 5, [162].

5.2.2 The second variant

Consider the inverse time-dependent heat source problem of determining $(u(x,t), r(t))$ satisfying the one-dimensional heat equation

$$u_t - u_{xx} = r(t)h(x,t), \quad (x,t) \in (0,1) \times (0,T) =: \Omega_T, \tag{5.73}$$

subject to the initial condition

$$u(x,0) = u_0(x), \quad x \in [0,1], \tag{5.74}$$

the Robin boundary condition at $x = 1$, i.e.

$$u_x(0,t) + \alpha u(0,t) = 0, \quad t \in (0,T), \tag{5.75}$$

modelling convection, where α is a given constant representing the heat transfer coefficient, the periodic (non-local) boundary condition

$$u(0,t) = u(1,t), \quad t \in (0,T), \tag{5.76}$$

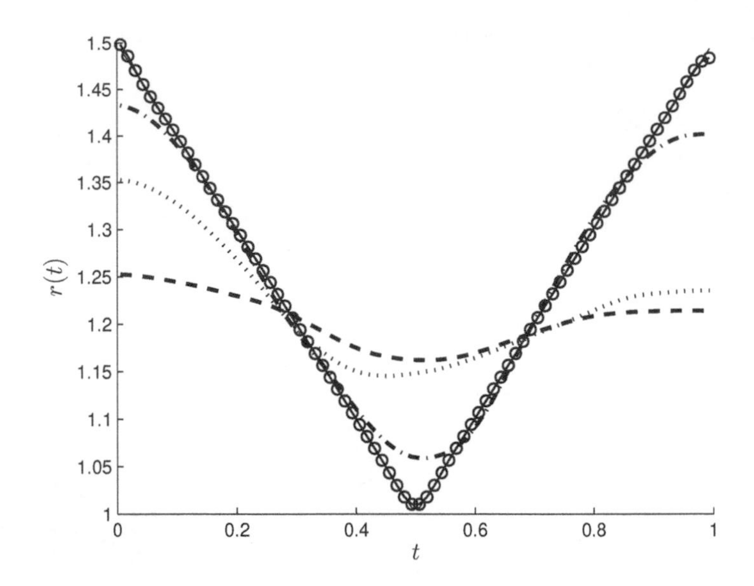

FIGURE 5.3: The analytical (—) and numerical results of $r(t)$, obtained using the first-order Tikhonov regularization, for $p = 0$ ($\circ\circ\circ$), $p = 0.1\%$ ($-\cdot-$), $p = 0.5\%$ ($\cdots$) and $p = 1\%$ ($---$) noise, with the regularization parameters λ_d given in Table 5.4, for Case 6, [162].

and the integral overdetermination measurement

$$\int_0^1 u(x,t)dx = E(t), \quad t \in [0,T]. \tag{5.77}$$

The condition (5.77) is known as the specification of energy [49] or mass [63], and has been considered in the study of particle diffusion in turbulent plasma and also in the heat propagation in a thin rod [194]. The initial condition (5.74) can also be nonlocal, e.g. of the form

$$u(x,0) + \beta u(x,T) + \int_0^T \omega(t)u(x,t)dt = \tilde{u}_0(x), \quad x \in [0,1],$$

where $\tilde{u}_0$, $\omega \geq 0$ and $\beta \geq 0$ are given quantities.

Because $r(t)$ is space-independent, α is constant, and the boundary conditions (5.75) and (5.76) are linear and homogeneous, the method of separation of variables, based on the Fourier expansion of solution u in terms of the eigenvalues and eigenfunctions of the auxiliary Sturm-Liouville spectral problem

$$\begin{cases} X''(x) + \Lambda X(x) = 0, & x \in (0,1), \\ X(0) = X(1), & X'(0) + \alpha X(0) = 0, \end{cases} \tag{5.78}$$

is a possible approach. There is a major distinction between the case $\alpha \neq 0$ and the case $\alpha = 0$ when the Robin convective boundary condition degenerate into the adiabatic zero-Neumann boundary condition

$$u_x(0,t) = 0, \quad t \in [0,T], \tag{5.79}$$

because in the case $\alpha = 0$ the eigenvalues of (5.78) have double multiplicity, whilst in the case $\alpha \neq 0$ they are simple.

Case $\alpha = 0$

The case $\alpha = 0$ follows in parts the analyzes of [4, 200]. The auxiliary Sturm-Liouville spectral problem corresponding to (5.78) given by

$$\begin{cases} X''(x) + \Lambda X(x) = 0, & x \in [0,1], \\ X(0) = X(1), & X'(0) = 0, \end{cases} \tag{5.80}$$

is not self-adjoint and the usual set of eigenfunctions $\{\cos(2k\pi x)\}_{k \in \mathbb{N}}$ is not complete in $L^2(0,1)$. To complete the basis [191], the associated eigenfunction $\tilde{X}_k$ for the eigenvalue $\Lambda_k = (2k\pi)^2$ corresponding to the eigenfunction $\cos(2k\pi x)$, defined as the solution of the problem

$$\begin{cases} \tilde{X}''(x) + \Lambda_k \tilde{X}(x) + \cos(2k\pi x) = 0, & x \in (0,1), \\ \tilde{X}(0) = \tilde{X}(1), & \tilde{X}'(0) = 0, \end{cases} \tag{5.81}$$

is considered. For $k = 0$, $\lambda_0 = 0$ and the above problem has no solution, but

$$\tilde{X}_k(x) = -\frac{x\sin(2k\pi x)}{4k\pi} \quad \text{for } k \in \mathbb{N}^*. \tag{5.82}$$

Then, the set $\left\{1, \cos(2k\pi x), -\frac{x\sin(2k\pi x)}{4k\pi} | k \in \mathbb{N}^*\right\}$ is complete and forms a Riesz basis in $L^2(0,1)$. Similarly, the adjoint problem to (5.80), namely,

$$\begin{cases} Y''(x) + \Lambda Y(x) = 0, & x \in (0,1), \\ Y'(0) = Y'(1), & Y(1) = 0, \end{cases} \tag{5.83}$$

with the eigenfunctions $\{1 - x, -\sin(2k\pi x) | k \in \mathbb{N}^*\}$ needs to be completed with the associated functions

$$\tilde{Y}_k(x) = \frac{(1-x)\cos(2k\pi x)}{4k\pi}, \quad k \in \mathbb{N}^*. \tag{5.84}$$

corresponding to the solution of the problem

$$\begin{cases} \tilde{Y}''(x) + \Lambda_k \tilde{Y}(x) - \sin(2k\pi x) = 0, & x \in (0,1), \\ \tilde{Y}'(0) = \tilde{Y}'(1), & \tilde{Y}(1) = 0. \end{cases} \tag{5.85}$$

Upon normalization, the bases $\mathcal{S} := \{X_0(x) = 2, X_{2k-1}(x) = 4\cos(2k\pi x), X_{2k}(x) = -4x\sin(2k\pi x) | k \in \mathbb{N}^*\}$ and $\mathcal{S}^* := \{Y_0(x) = 1 - x, Y_{2k-1}(x) = (1-x)\cos(2k\pi x), Y_{2k}(x) = -\sin(2k\pi x) | k \in \mathbb{N}^*\}$ form a bi-orthogonal system for $L^2(0,1)$, i.e. $(X_i, Y_j) = \int_0^1 X_i(x) Y_j(x) dx = \delta_{ij}$.

Using the Fourier method of separating variables, the solution to (5.73), (5.76), (5.77), and (5.79) is sought in the form

$$u(x,t) = u^0(t)X_0(x) + \sum_{k=1}^{\infty} u_1^k(t)X_{2k-1}(x) + \sum_{k=1}^{\infty} u_2^k(t)X_{2k}(x), \qquad (5.86)$$

where the time-dependent coefficients are to be determined. Let us define the Fourier coefficients of $h(x,t)$ and $u_0(x)$ in the basis $\mathcal{S}^*$ by

$$h_k(t) = \int_0^1 Y_k(x)h(x,t)dx, \quad u_0^{(k)} = \int_0^1 Y_k(x)u_0(x)dx, \quad k \in \mathbb{N}. \qquad (5.87)$$

Then, using the properties of the bi-orthogonal system S^*, one has $u^0(t) = \langle u(\cdot,t), Y_0\rangle$, where $\langle\cdot,\cdot\rangle$ denotes the inner product in $L^2(0,1)$. Differentiating this identity and using the boundary conditions (5.76) and (5.79) give

$$\frac{du^0}{dt}(t) = \langle u_t(\cdot,t), Y_0\rangle = \langle u_{xx}(x,t), 1-x\rangle + \langle r(t)h(x,t), Y_0(x)\rangle = r(t)h_0(t).$$

Integrating gives

$$u^0(t) = u_0^{(0)} + \int_0^t r(\tau)h_0(\tau)d\tau. \qquad (5.88)$$

Similarly, for the even terms, $u_2^k(t) = \langle u(\cdot,t), Y_{2k}\rangle$. Differentiating this identity and using the boundary conditions (5.76) and (5.79) give

$$\begin{aligned}\frac{du_2^k}{dt}(t) &= \langle u_t(\cdot,t), Y_{2k}\rangle = \langle u_{xx}(x,t), -\sin(2k\pi x)\rangle + \langle r(t)h(x,t), Y_{2k}(x)\rangle \\ &= -(2k\pi)^2\langle u(\cdot,t), Y_{2k}\rangle + r(t)h_{2k}(t) = -(2k\pi)^2 u_2^k(t) + r(t)h_{2k}(t).\end{aligned}$$

Integrating this linear ODE yields

$$u_2^k(t) = u_0^{(2k)}e^{-\Lambda_k t} + \int_0^t r(\tau)h_{2k}(\tau)e^{-\Lambda_k(t-\tau)}d\tau. \qquad (5.89)$$

For the odd terms, one gets $u_1^k(t) = \langle u(\cdot,t), Y_{2k-1}\rangle$. Differentiating this identity and using the boundary conditions (5.76) and (5.79) give

$$\begin{aligned}&\frac{du_1^k}{dt}(t) = \langle u_t(\cdot,t), Y_{2k-1}\rangle = \langle u_{xx}(x,t), (1-x)\cos(2k\pi x)\rangle \\ &+\langle f(t)h(x,t), Y_{2k-1}(x)\rangle = -(2k\pi)^2\langle u(\cdot,t), Y_{2k-1}\rangle - 4k\pi\langle u(\cdot,t), Y_{2k}\rangle \\ &+r(t)h_{2k-1}(t) = -(2k\pi)^2 u_1^k(t) - 4k\pi u_2^k(t) + r(t)h_{2k-1}(t).\end{aligned}$$

Using (5.87) and integrating this linear ODE yield

$$\begin{aligned}u_1^k(t) &= (u_0^{(2k-1)} - 4k\pi u_0^{(2k)}t)e^{-\Lambda_k t} \\ &+ \int_0^t [h_{2k-1}(\tau) - 4k\pi(t-\tau)h_{2k}(\tau)]\, r(\tau)e^{-\Lambda_k(t-\tau)}d\tau. \qquad (5.90)\end{aligned}$$

The existence and uniqueness of the solution of inverse problem (5.73), (5.74), (5.76), (5.77), (5.79) holds, as stated in the following theorem [200].

Theorem 5.8. *Suppose that the following conditions hold:*
(A_1) $E \in C^1[0,T]$;
(A_2) $u_0 \in C^4[0,1]$, $u_0(0) = u_0(1)$, $u_0'(0) = 0$, $u_0''(0) = u_0''(1)$ *and* $E(0) = \int_0^1 u_0(x)dx$;
(A_3) $h \in C(\overline{Q}_T)$, $h(\cdot,t) \in C^4[0,1]$, $h(0,t) = h(1,t)$, $h_x(0,t) = 0$, $h_{xx}(0,t) = h_{xx}(1,t)$ *and* $\int_0^1 h(x,t)dx \neq 0$ *for all* $t \in [0,T]$.
Then, the inverse problem (5.73), (5.74), (5.76), (5.77), (5.79) *has a unique solution* $(u(x,t), r(t)) \in (C^{2,1}(Q_T) \cap C^{1,0}(\overline{Q}_T)) \times C[0,T]$.

The proof of this theorem is based on recasting the problem as a Volterra integral equation of the second kind

$$r(t) = F(t) + \int_0^t K(t,\tau)r(\tau)d\tau, \quad t \in [0,T], \tag{5.91}$$

where

$$F(t) = \frac{E'(t) + 8\pi \sum_{k=1}^{\infty} k u_0^{(2k)} e^{-\Lambda_k t}}{2h_0(t) + \frac{2}{\pi}\sum_{k=1}^{\infty} \frac{1}{k} h_{2k}(t)},$$
$$K(t,\tau) = \frac{4\pi \sum_{k=1}^{\infty} k u_0^{(2k)} e^{-\Lambda_k (t-\tau)}}{h_0(t) + \frac{1}{\pi}\sum_{k=1}^{\infty} \frac{1}{k} h_{2k}(t)}. \tag{5.92}$$

Indeed, first differentiating equation (5.77) yields

$$E'(t) = \int_0^1 u_t(x,t)dx = \int_0^1 (u_{xx}(x,t) + r(t)h(x,t))dx$$
$$= u_x(1,t) - u_x(0,t) + r(t)\int_0^1 h(x,t)dx. \tag{5.93}$$

Then, using (5.79) and employing (5.86) differentiated with respect to x and evaluated at $x = 1$, namely,

$$u_x(1,t) = -8\pi \sum_{k=1}^{\infty} k \left[u_0^{(2k)} e^{-\Lambda_k t} + \int_0^t r(\tau)h(x,\tau)e^{-\Lambda_k (t-\tau)} d\tau \right], \tag{5.94}$$

as well as that

$$\int_0^1 h(x,t)dx = \sum_{k=1}^{\infty} h_k(t) \int_0^1 X_k(x)dx = 2h_0(t)$$
$$-4\sum_{k=1}^{\infty} h_{2k}(t) \int_0^1 x\sin(2k\pi x)dx = 2h_0(t) + \frac{2}{\pi}\sum_{k=1}^{\infty} \frac{1}{k} h_{2k}(t),$$

result in (5.91) with F and K defined by (5.92). The series in (5.92) are convergent because of the assumptions of the theorem, while the unique solvability of the inverse problem follows from the theory of Volterra linear integral equations of the second kind [200].

The solution of the inverse problem also depends continuously on the input data in the sense given by the following theorem.

Theorem 5.9. *Let $\mathcal{T}$ be the class of triplets of the form $\Phi = \{h, u_0, E\}$, which satisfy the assumptions $(A_1) - (A_3)$ of Theorem 5.8 and*

$$\|h\|_{C^{3,0}(\overline{\Omega}_T)} \leq N_0, \quad \|u_0\|_{C^3[0,1]} \leq N_1, \quad \|E\|_{C^1[0,T]} \leq N_2,$$
$$0 < N_3 \leq min_{t\in[0,T]} \left| \int_0^1 h(x,t)dx \right|, \tag{5.95}$$

for some positive constants N_i, $i = \overline{0,3}$. Then, the solution $(u(x,t), r(t)) \in (C^{2,1}(Q_T) \cap C^{1,0}(\overline{Q}_T)) \times C[0,T]$ of the inverse problem (5.73), (5.74), (5.76), (5.77), (5.79) depends continuously upon the input data (h, u_0, E) in $\mathcal{T}$.

Proof. First let us evaluate

$$\begin{aligned}\left|8\pi \sum_{k=1}^{\infty} k u_0^{(2k)} e^{-\Lambda_k t}\right| &\leq 8\pi \sum_{k=1}^{\infty} k \left| \int_0^1 u_0(x) Y_{2k}(x) dx \right| \\ &= 8\pi \sum_{k=1}^{\infty} k \left| -\int_0^1 u_0(x) \sin(2k\pi x) dx \right| \\ &= \frac{1}{\pi^2} \sum_{k=1}^{\infty} \frac{1}{k^2} \left| \int_0^1 u_0'''(x) \cos(2k\pi x) dx \right| \leq \frac{N_1}{6},\end{aligned}$$

where integration by parts three times and the assumption (A_2) were used. Using a similar procedure and the assumption (A_3) on the input data h give

$$\left|8\pi \sum_{k=1}^{\infty} k h_{2k} e^{-\Lambda_k (t-\tau)}\right| \leq \frac{N_0}{6}. \tag{5.96}$$

Then, based on (5.92), the following estimates are obtained:

$$\|F\|_{C[0,T]} \leq \frac{1}{N_3}\left(N_2 + \frac{N_1}{6}\right) =: N_4, \quad \|K\|_{C[0,T]^2} \leq \frac{N_0}{6N_3} =: N_5. \tag{5.97}$$

To obtain an estimate for $\|r\|_{C[0,T]}$, observe that the solution of the Volterra integral equation (5.91) is given by [200]

$$r(t) = \sum_{n=0}^{\infty} (\mathbb{K}F)^n(t), \tag{5.98}$$

where $(\mathbb{K}F)(t) := \int_0^t K(t,\tau)F(\tau)d\tau$ satisfies that

$$|(\mathbb{K}F)(t)| \leq \|F\|_{C[0,T]} \frac{(t\|K\|_{C[0,T]^2})^n}{n!}, \quad t \in [0,T],\ n \in \mathbb{N}. \tag{5.99}$$

Thus,

$$\|r\|_{C[0,T]} \leq F\|_{C[0,T]} \exp(T\|K\|_{C[0,T]^2}) \leq N_4 \exp(TN_5) =: N_6. \tag{5.100}$$

Now, let us consider two sets of input data (h, u_0, E) and $(\tilde{h}, \tilde{u}_0, \tilde{E})$ in $\mathcal{T}$ generating the unique solutions $(u(x,t), r(t))$ and $(\tilde{u}(x,t), \tilde{r}(t))$ in $(C^{2,1}(Q_T) \cap C^{1,0}(\overline{Q}_T)) \times C[0,T]$. Then, these solutions will satisfy the Volterra integral equations (5.91), which upon subtraction yield

$$\begin{aligned} r(t) - \tilde{r}(t) = F(t) - \tilde{F}(t) + \int_0^t (K(t,\tau) - \tilde{K}(t,\tau))r(\tau)d\tau \\ + \int_0^t \tilde{K}(t,\tau)(r(\tau) - \tilde{r}(\tau))d\tau. \end{aligned} \tag{5.101}$$

Denoting $C_1 := \|F - \tilde{F}\|_{C[0,T]} + TN_6\|K - \tilde{K}\|_{C([0,T]^2)}$ and $R(t) := |r(t) - \tilde{r}(t)|$ from (5.93) it is obtained that

$$0 \leq R(t) \leq C_1 + \int_0^t |\tilde{K}(t,\tau)|R(\tau)d\tau, \quad t \in [0,T]. \tag{5.102}$$

Then, from Gronwall's inequality, see Theorem 16 of [101],

$$R(t) \leq C_1 \exp\left(\int_0^t \sup_{\sigma \in [\tau,t]} |\tilde{K}(\sigma,\tau)|d\tau\right) \leq C_1 e^{TN_5}, \quad t \in [0,T]. \tag{5.103}$$

Similarly, as before

$$\begin{aligned} |F(t) - \tilde{F}(t)| \leq M_1 \left(\|E - \tilde{E}\|_{C^1[0,T]} + \|u_0 - \tilde{u}_0\|_{C^3[0,1]} + \|h - \tilde{h}\|_{C^{3,0}(\overline{D}_T)}\right), \\ |K(t,\tau) - \tilde{K}(t,\tau)| \leq M_2\|h - \tilde{h}\|_{C^{3,0}(\overline{D}_T)} \end{aligned}$$

from some positive constants M_1 and M_2, which, when combined with (5.103), yield the stability estimate

$$\|r - \tilde{r}\|_{C[0,T]} \leq M_3 \left(\|E - \tilde{E}\|_{C^1[0,T]} + \|u_0 - \tilde{u}_0\|_{C^3[0,1]} + \|h - \tilde{h}\|_{C^{3,0}(\overline{\Omega}_T)}\right),$$

for some positive constant M_3.

Case $\alpha \neq 0$
In case $\alpha \neq 0$, the well-posedness of the inverse problem (5.73)–(5.77) holds under the sufficient conditions stated in the following theorem [161].

Theorem 5.10. *Let $\alpha \neq 0$. Suppose that the following conditions hold:*
(A_4) $E \in C^1[0,T]$;
(A_5) $u_0 \in C^2[0,1]$, $u_0(0) = u_0(1)$, $u_0'(0) + \alpha u_0(0) = 0$ *and* $E(0) = \int_0^1 u_0(x)dx$;
(A_6) $h \in C(\overline{Q}_T)$, $h(\cdot,t) \in C^2[0,1]$, $h(0,t) = h(1,t)$, $h_x(0,t) + \alpha h(0,t) = 0$ *and* $\int_0^1 h(x,t)dx \neq 0$ *for all* $t \in [0,T]$.
Then, the inverse problem (5.73)–(5.77) *has a unique solution* $(u(x,t), f(t)) \in (C^{2,1}(\Omega_T) \cap C^{1,0}(\overline{\Omega}_T)) \times C[0,T]$. *Moreover, the solution depends continuously upon the input data* (E, u_0, h) *satisfying the assumptions* $(A_4) - (A_6)$.

The continuous dependence result in the above theorem was obtained, see Theorem 2 of [161], under assumptions that require high regularity of the input data, e.g.

$$\|h\|_{C^{2,0}(\overline{\Omega}_T)} \leq N_0, \quad \|u_0\|_{C^2[0,1]} \leq N_1, \quad \|E\|_{C^1[0,T]} \leq N_2,$$
$$0 < N_3 \leq \min_{t \in [0,T]} \left| \int_0^1 h(x,t)dx \right|,$$

for some positive constants N_i for $i = \overline{0,3}$. However, the smoothness condition $\|E\|_{C^1[0,T]} \leq N_2$ cannot be satisfied in practice since the data is contaminated with random noise. Therefore, the inverse problem is still practically ill-posed due to the differentiation of the noisy input energy function (5.77). In the next section, the BEM combined with the Tikhonov regularization are applied for obtaining a stable solution of the inverse and ill-posed source problem (5.73)–(5.77).

5.2.2.1 BEM for the second variant

By means of the fundamental solution (3.23) and the Green's formula, the heat equation (5.73) is transformed into the boundary integral form

$$\eta(x)u(x,t) = \int_0^t \left[G(x,t;\xi,\tau) \frac{\partial u}{\partial n(\xi)}(\xi,\tau) - u(\xi,\tau) \frac{\partial G}{\partial n(\xi)}(x,t;\xi,\tau) \right]_{\xi \in \{0,1\}} d\tau + \int_0^1 G(x,t;y,0)u(y,0)dy + \int_0^1 \int_0^t G(x,t;y,\tau) r(\tau) h(y,\tau) d\tau dy$$
$$(x,t) \in [0,1] \times (0,T). \quad (5.104)$$

The discretization of this integral equation, as described in subsection 5.2.1.1, leads to (see also (5.28) with $g = 0$),

$$\eta(x)u(x,t)$$

$$= \sum_{j=1}^{N} [A_{0j}(x,t)q_{0j} + A_{1j}(x,t)q_{1j} - B_{0j}(x,t)u_{0j} - B_{1j}(x,t)u_{1j}]$$

$$+ \sum_{k=1}^{N_0} C_k(x,t)u_{0,k} + d_0(x,t), \quad (x,t) \in [0,1] \times (0,T), \quad (5.105)$$

where $A_{\xi j}$, $B_{\xi j}$ for $\xi \in \{0,1\}$, $j = \overline{1,N}$, C_k for $k = \overline{1,N_0}$, and d_0 are defined in (5.29) and (5.30). Collocating (5.105) at the boundary nodes $(0,\tilde{t}_i)$ and $(1,\tilde{t}_i)$ for $i = \overline{1,N}$, results in the system of $2N$ equations (see also (5.36) with $g = 0$)

$$A\underline{q} - B\underline{u} + C\underline{u}^0 + D_0\underline{r} = \underline{0}. \quad (5.106)$$

From the boundary conditions (5.75) and (5.76), the boundary temperature $\underline{u}$ can be expressed as

$$\underline{u} := \begin{bmatrix} u_{0j} \\ u_{1j} \end{bmatrix} = \frac{1}{\alpha} \begin{bmatrix} q_{0j} \\ q_{0j} \end{bmatrix}.$$

Then, the system of equations (5.106) can be rewritten as

$$\left(A - \frac{1}{\alpha}(B + B^*) \right) \underline{q} + C\underline{u}^0 + D_0\underline{r} = \underline{0},$$

where

$$B^* = \begin{bmatrix} B_{1j}(0,\tilde{t}_i) & -B_{1j}(0,\tilde{t}_i) \\ B_{1j}(1,\tilde{t}_i) + \frac{1}{2}\delta_{ij} & -B_{1j}(1,\tilde{t}_i) - \frac{1}{2}\delta_{ij} \end{bmatrix}_{2N \times 2N}.$$

Then, the flux $\underline{q}$ can be expressed as

$$\underline{q} = -\left(A - \frac{1}{\alpha}(B + B^*) \right)^{-1} \left(C\underline{u}^0 + D_0\underline{r} \right). \quad (5.107)$$

Now, discretize the energy (5.77) as

$$\int_0^1 u(x,t)dx = \frac{1}{N_0} \sum_{k=1}^{N_0} u(\tilde{x}_k, t). \quad (5.108)$$

Applying this at $t = \tilde{t}_i$ for $i = \overline{1,N}$, yields

$$\frac{1}{N_0} \sum_{k=1}^{N_0} u(\tilde{x}_k, \tilde{t}_i) = \int_0^1 u(x,\tilde{t}_i)dx = E(\tilde{t}_i) =: E_i, \quad i = \overline{1,N}. \quad (5.109)$$

Using (5.104), as before, (5.109) results in the system of N equations

$$\frac{1}{N_0}\sum_{k=1}^{N_0}\left[\left(A_k^{(1)}-\frac{1}{\alpha}(B_k^{(1)}+B_k^{(1)^*})\right)\underline{q}+C_k^{(1)}\underline{u}^0+D_k^{(1)}\underline{r}\right]=\underline{E}, \tag{5.110}$$

where

$$A_k^{(1)}=\left[A_{0j}(\tilde{x}_k,\tilde{t}_i)\quad A_{1j}(\tilde{x}_k,\tilde{t}_i)\right]_{N\times 2N},$$
$$B_k^{(1)}=\left[B_{0j}(\tilde{x}_k,\tilde{t}_i)\quad B_{1j}(\tilde{x}_k,\tilde{t}_i)\right]_{N\times 2N},$$
$$B_k^{(1)^*}=\left[B_{1j}(\tilde{x}_k,\tilde{t}_i)\quad -B_{1j}(\tilde{x}_k,\tilde{t}_i)\right]_{N\times 2N},\quad C_k^{(1)}=\left[C_l(\tilde{x}_k,\tilde{t}_i)\right]_{N\times N_0},$$
$$D_k^{(1)}=\left[D_{0,j}(\tilde{x}_k,\tilde{t}_i)\right]_{N\times N},\quad \underline{E}=\left[E_i\right]_N,\quad k,l=\overline{1,N_0},\ i,j=\overline{1,N}.$$

Substituting (5.107) into (5.110) yields a system of $N\times N$ linear equations

$$X\underline{r}=\underline{y}, \tag{5.111}$$

where

$$X=\frac{1}{N_0}\sum_{k=1}^{N_0}\left[-\left(A_k^{(1)}-\frac{1}{\alpha}(B_k^{(1)}+B_k^{(1)^*})\right)\left(A-\frac{1}{\alpha}(B+B^*)\right)^{-1}D_0\right.$$
$$\left.+D_k^{(1)}\right],$$
$$\underline{y}=\frac{1}{N_0}\sum_{k=1}^{N_0}\left[\left(A_k^{(1)}-\frac{1}{\alpha}(B_k^{(1)}+B_k^{(1)^*})\right)\left(A-\frac{1}{\alpha}(B+B^*)\right)^{-1}C\right.$$
$$\left.-C_k^{(1)}\right]\underline{u}^0+\underline{E}.$$

Consider the inverse source problem (5.73)–(5.77) with $T=1$,

$$u(x,0)=\varphi(x)=1+x-x^2,\quad h(x,t)=(3+x-x^2)e^{-t},$$
$$\int_0^1 u(x,t)dx=E(t)=7e^t/6, \tag{5.112}$$

and the boundary condition (5.75) with $\alpha=-1$. The exact solution is

$$u(x,t)=(1+x-x^2)e^t,\quad r(t)=e^{2t}. \tag{5.113}$$

The condition numbers of matrix X in (5.111) are 248, 780, and 2393 for $N_0=N\in\{20,40,80\}$, respectively. These values indicate that the system of equations (5.111) is mildly ill-conditioned [328].

In what follows, numerical results obtained with $N_0=N=40$ are presented. Consider first the case of exact data, i.e., there is no noise in the energy

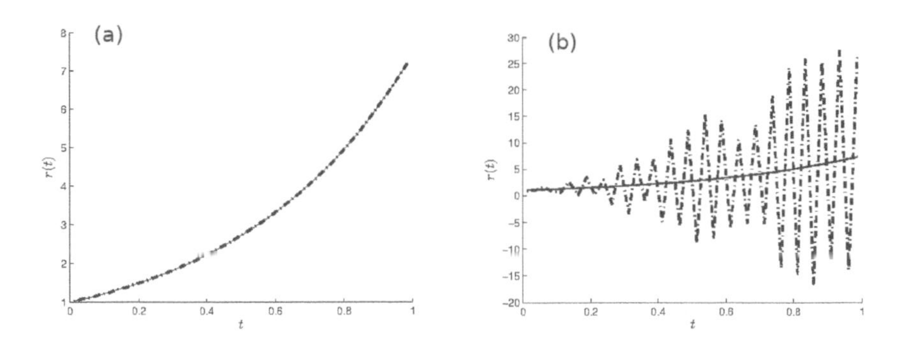

FIGURE 5.4: The analytical (——) and numerical ($- \cdot -$) results of $r(t)$ for (a) $p = 0$ exact data and (b) $p = 1\%$ noisy data, without regularization [161].

data (5.77). Figure 5.4(a) shows the analytical and numerical results of $r(t)$ from which it can be seen that the numerical solution is very accurate eventhough no regularization has been applied. Next, the stability of the numerical solution with respect to noise in the energy data, defined as

$$\underline{E}^p = \underline{E} + random('Normal', 0, \sigma, 1, N), \tag{5.114}$$

is investigated, where $\sigma = p \times \max_{t\in[0,1]} |E(t)| = 7ep/6$. Figure 5.4(b) shows the numerical solution for $r(t)$ obtained with no regularization, i.e. $\lambda = 0$, when the energy data (5.77) is contaminated by $p = 1\%$ noise. It can be clearly observed that the numerical solution for $r(t)$ is unstable if no regularization is employed. This is to be expected since the inverse problem under investigation is ill-posed. In order to stabilize the solution, the Tikhonov regularization procedure, as given by equation (5.10), is employed. Here the L-curve method and the discrepancy principle were investigated to choose the regularization parameter λ. Initially, the plot of the residual $\|X\underline{r}_\lambda - \underline{y}^\epsilon\|$ versus the norm of the solution $\|\underline{r}_\lambda\|$ was constructed, but it was found that an L-corner could not be clearly identified. This could happen because the L-curve is a heuristic method, which, because it does not assume the knowledge of the amount of noise ϵ, is not always convergent [381]. Alternatively, the more rigorous discrepancy principle which uses the knowledge of the level of noise $\epsilon = \|\underline{y}^\epsilon - \underline{y}\| = \|\underline{E}^p - \underline{E}\|$ is recommended.

Let us denote by λ_d the regularization parameter which is determined by the discrepancy principle, i.e. the largest λ for which the residual $\|X\underline{r}_\lambda - \underline{y}^\epsilon\|$ becomes less than ϵ. This is illustrated in Figures 5.5(a,c,e), where the intersection of the residual curve with the horizontal level noisy line ϵ yields the value of λ_d. The output solutions of $r(t)$ obtained with these values of λ_d are illustrated in Figures 5.5(b,d,f). Although Figure 5.5(b) represents a significant improvement over Figure 5.4(b) the numerical results are still oscillatory. However, the accuracy and stability of the numerical results can be improved by using higher-order Tikhonov regularization of order one and

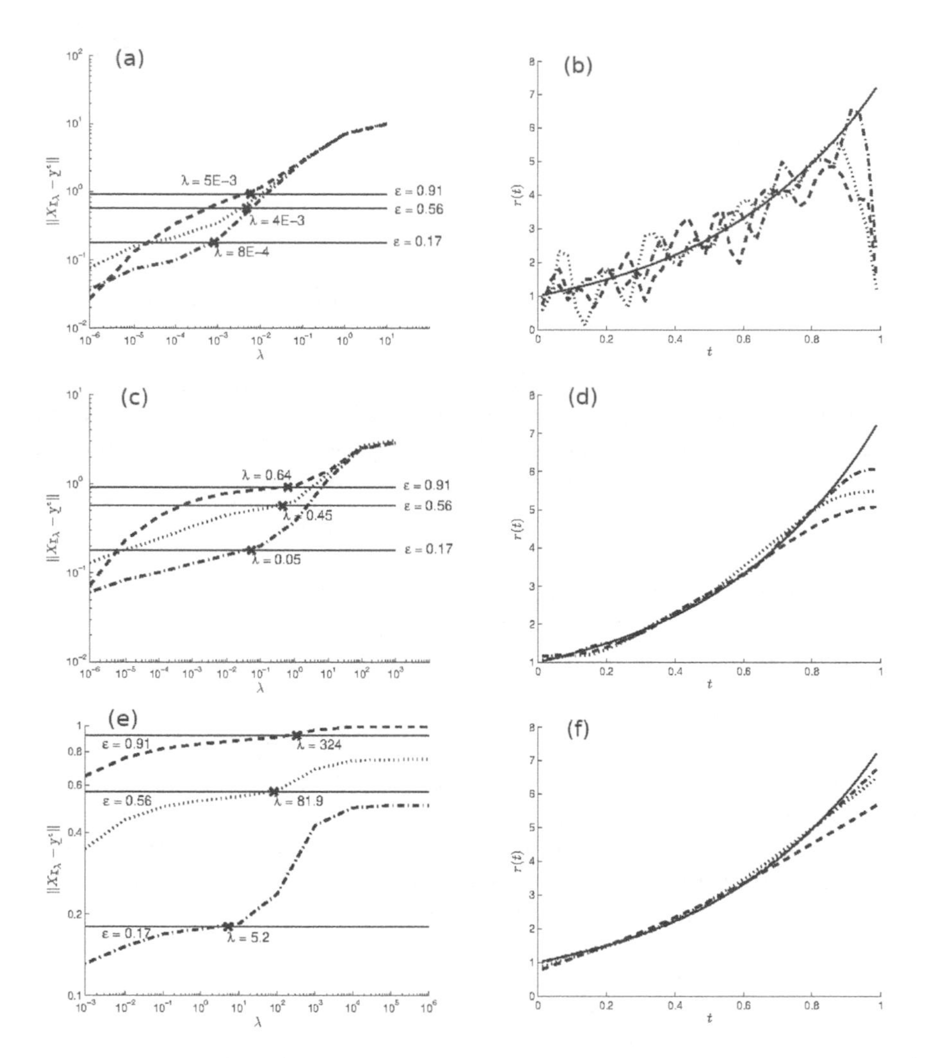

FIGURE 5.5: (a,c,e) Discrepancy principle curves and the analytical (——) and numerical results of $r(t)$, obtained using the (b) zeroth-order, (d) first-order, and (f) second-order Tikhonov regularization, for $p = 1\%$ $(- \cdot -)$, $p = 3\%$ $(\cdots)$ and $p = 5\%$ $(- - -)$ noise [161].

two, as illustrated in Figures 5.5(d) and (f), respectively. This is to be expected since the analytical solution (5.113) to be retrieved is infinitely differentiable. However, if less smooth sources are attempted to be retrieved, the use of higher-order regularization do not necessarily improve the accuracy of the numerical results [161].

An immediate extension to higher dimensions, e.g. two dimensions, is available when one considers finding the temperature $u(x, y, t)$ and the time-dependent heat source $r(t)$ which satisfy the two-dimensional heat equation

$$u_t = u_{xx} + u_{yy} + r(t)h(x, y, t), \quad (x, y, t) \in (0, 1) \times (0, 1) \times (0, T),$$

subject to the boundary conditions

$$u(0, y, t) = u(1, y, t), \; u_x(0, y, t) + \alpha u(0, y, t) = 0, \; (y, t) \in (0, 1) \times (0, T),$$

$$u(x, 0, t) = u(x, 1, t) = 0, \quad (x, t) \in (0, 1) \times (0, T),$$

the initial condition

$$u(x, y, 0) = u_0(x, y), \quad (x, y) \in [0, 1] \times [0, 1],$$

and the energy condition

$$\int_0^1 \int_0^1 u(x, y, t)dxdy = E(t), \quad t \in [0, T].$$

Then, if $h(x, y, t) = h_1(x, t)\sin(\pi y)$ and $u_0(x, y) = \phi(x)\sin(\pi y)$, the solution sought as $u(x, y, t) = v(x, t)\sin(\pi y)e^{-\pi^2 t}$ recasts the above two-dimensional problem for $u(x, y, t)$ as the following one-dimensional problem for $v(x, t)$:

$$v_t = v_{xx} + r(t)h_1(x, t)e^{\pi^2 t}, \quad (x, t) \in (0, 1) \times (0, T),$$

$$v(0, t) = v(1, t), \quad v_x(0, t) + \alpha v(0, t) = 0, \quad t \in (0, T),$$

$$v(x, 0) = \phi(x), \quad x \in [0, 1],$$

$$\int_0^1 v(x, t)dx = \frac{\pi E(t)e^{\pi^2 t}}{2}, \quad t \in [0, T].$$

This problem is of the type (5.73)–(5.77) and the techniques of the one-dimensional case apply.

5.2.3 The third variant

Consider a more general boundary condition than the boundary condition (5.75) of the form

$$au_{xx}(1, t) + u_x(1, t) + bu(1, t) = 0, \quad t \in (0, T), \tag{5.115}$$

where $a > 0$ and b are given constants, which is known as a generalized impedance boundary condition [47] in corrosion or acoustic scattering. Assuming that (5.73) holds at $x = 1$, then (5.115) yields

$$au_t(1,t) + u_x(1,t) + bu(1,t) = ar(t)h(1,t), \quad t \in (0,T). \tag{5.116}$$

The new boundary condition (5.116) is observed in the process of cooling of a thin solid bar, one end of which is placed in contact with a fluid [245], in the boundary reaction of a diffusive chemical [55], or in the transient flow pump test in a porous medium [270].

Consider a zero Dirichlet boundary condition at $x = 0$ given by

$$u(0,t) = 0, \quad t \in (0,T). \tag{5.117}$$

Because the source $r(t)$ is space-independent, a and b are constants, and the boundary conditions (5.116) and (5.117) are linear and homogeneous, the method of separation of variables in terms of the eigenvalues and eigenfunctions of the auxiliary Sturm-Liouville spectral problem

$$\begin{cases} y''(x) + \Lambda y(x) = 0, & x \in [0,1], \\ y(0) = 0, & (a\Lambda - b)y(1) = y'(1), \end{cases} \tag{5.118}$$

is a possible approach. From [38, 228], it is known that for $a > 0$, the spectral problem (5.118) with eigenvalue-dependent boundary condition at $x = 1$, has the following following properties:

(i) The eigenvalues $(\mu_n)_{n\in\mathbb{N}}$ are real and simple, and they form a strictly increasing unbounded sequence. The first eigenvalue μ_0 is negative, equal to 0 or positive if $b < -1$, $b = -1$ or $b > -1$, respectively.

(ii) For each $n \in \mathbb{N}$, the eigenfunction $y_n(x)$ has exactly n simple zeros in $(0,1)$.

(iii) For large n, the eigenvalues and eigenfunctions have the assymptotic behaviours

$$\sqrt{\Lambda_n} = \pi n + O(n^{-1}), \quad y_n(x) = \sin(n\pi x) + O(n^{-1}).$$

(iv) For any fixed $n_0 \in \mathbb{N}$, the set of eigenfunctions $(y_n(x))_{n\in\mathbb{N}\setminus\{n_0\}}$ forms a Riesz basis for $L^2(0,1)$, and the set $(u_n(x))_{n\in\mathbb{N}\setminus\{n_0\}}$, which is bi-orthogonal on it, has the form

$$u_n(x) = \frac{y_n(x) - \frac{y_n(1)}{y_{n_0}(1)}y_{n_0}(x)}{\|y_n\|^2_{L^2[0,1]} + ay_n^2(1)}, \quad n \in \mathbb{N}\setminus\{n_0\}.$$

Let us denote

$$\Phi^4_{n_0}[0,1] := \{\phi \in C^4[0,1] | \phi(0) = \phi''(0) = 0,$$
$$\phi(1) = \phi'(1) = \phi''(1) = \phi'''(1) = 0, \int_0^1 \phi(x)y_{n_0}(x)dx = 0\}.$$

Then, the well-posedness of the inverse problem (5.73), (5.74), (5.77), (5.116) and (5.117) holds, as stated in the following theorem [165].

Theorem 5.11. *Suppose that the following conditions hold:*
(A_1) $E \in C^1[0,T]$;
(A_2) $u_0 \in \Phi^4_{n0}[0,1]$ *and* $E(0) = \int_0^1 u_0(x)dx$;
(A_3) $h \in C(\overline{Q}_T)$, $h(\cdot,t) \in \Phi^4_{n0}[0,1]$ *and* $\int_0^1 h(x,t)dx \neq 0$ *for all* $t \in [0,T]$.
Then, the inverse problem (5.73), (5.74), (5.77), (5.116) and (5.117) has a unique solution $(u(x,t), r(t)) \in (C^{2,1}(\Omega_T) \cap C^{1,0}(\overline{\Omega}_T)) \times C[0,T]$. *Moreover, the solution depends continuously upon the input data* (E, u_0, h) *satisfying the assumptions* $(A_1) - (A_3)$.

The continuous dependence result in the above theorem was obtained under assumptions that require high regularity of the input data, e.g.

$$\|h\|_{C^{4,0}(\overline{\Omega}_T)} \leq N_0, \quad \|u_0\|_{C^4[0,1]} \leq N_1, \quad \|E\|_{C^1[0,T]} \leq N_2,$$

$$0 < N_3 \leq \min_{t\in[0,T]}\left|\int_0^1 h(x,t)dx\right|,$$

for some positive constants N_i for $i = \overline{0,3}$. In the next section, a BEM and Tikhonov regularization, similar to that described in subsection 5.2.2.1, is employed for obtaining a stable solution of the inverse and ill-posed source problem (5.73), (5.74), (5.77), (5.116) and (5.117).

5.2.3.1 BEM for the third variant

The BEM is applied as described in subsection 5.2.2.1. Using the homogeneous boundary condition (5.117), the boundary integral equation (5.105) becomes

$$\eta(x)u(x,t) = \sum_{j=1}^{N} [A_{0j}(x,t)q_{0j} + A_{1j}(x,t)q_{1j} - B_{1j}(x,t)u_{1j}] + \sum_{k=1}^{N_0} C_k(x,t)u_{0,k} + d_0(x,t), \quad (5.119)$$

for $(x,t) \in [0,1] \times (0,T)$. Collocating this at the boundary nodes $(0,\tilde{t}_i)$ and $(1,\tilde{t}_i)$ for $i = \overline{1,N}$ gives the system of $2N$ equations

$$A_0\underline{q}_0 + A_1\underline{q}_1 - B_1\underline{u}_1 + D_0\underline{r} + C\underline{u}^0 = \underline{0}, \quad (5.120)$$

where

$$A_0 = \begin{bmatrix} A_{0j}(0,\tilde{t}_i) \\ A_{0j}(1,\tilde{t}_i) \end{bmatrix}_{2N\times N}, \quad A_1 = \begin{bmatrix} A_{1j}(0,\tilde{t}_i) \\ A_{1j}(1,\tilde{t}_i) \end{bmatrix}_{2N\times N},$$

$$B_1 = \begin{bmatrix} B_{1j}(0,\tilde{t}_i) \\ B_{1j}(1,\tilde{t}_i) + \frac{1}{2}\delta_{ij} \end{bmatrix}_{2N\times N},$$

$$\underline{q}_0 = [q_{0j}]_N, \ \underline{q}_1 = [q_{1j}]_N, \ \underline{u}_1 = [u_{1j}]_N, \ \underline{u}^0 = [u_{0,k}]_{N_0}, \ \underline{r} = [r_j]_N.$$

To apply the boundary condition (5.116), the time-derivative $u_t(1,t)$ is approximated using the finite differences

$$u_t(1,\tilde{t}_1) = \frac{u_{1,2}/3 + u_{1,1} - 4u_0(1)/3}{\mathrm{h}}, \quad u_t(1,\tilde{t}_2) = \frac{5u_{1,2}/3 - 3u_{1,1} + 4u_0(1)/3}{\mathrm{h}},$$
$$u_t(1,\tilde{t}_i) = \frac{3u_{1,i}/2 - 2u_{1,(i-1)} + u_{1,(i-2)}/2}{\mathrm{h}}, \qquad i = \overline{3,N},$$

where $\mathrm{h} = T/N$. Applying these expressions in the boundary condition (5.116) yields the following system of N linear equations:

$$S\underline{u}_1 = F\underline{r} + \underline{\tilde{u}}_0 - \underline{q}_1, \tag{5.121}$$

where $F = \mathrm{diag}(h(1,\tilde{t}_1), \dots, h(1,\tilde{t}_N))$,

$$S = \begin{bmatrix} a/\mathrm{h}+b & a/3\mathrm{h} & 0 & . \\ -3a/\mathrm{h} & 5a/3\mathrm{h}+b & 0 & . \\ a/2\mathrm{h} & -2a/\mathrm{h} & 3a/2\mathrm{h}+b & . \\ . & . & . & . \\ 0 & a/2\mathrm{h} & -2a/\mathrm{h} & 3a/2\mathrm{h}+b \end{bmatrix}_{N\times N},$$

$$\underline{\tilde{u}}_0 = \begin{bmatrix} 4au_0(1)/3\mathrm{h} \\ -4au_0(1)/3\mathrm{h} \\ 0 \\ . \\ 0 \end{bmatrix}_N .$$

Next, collocate the over-determination condition (5.77), by using the midpoint integration rule, at the discrete time nodes $\tilde{t}_i$ for $i = \overline{1,N}$, to obtain (5.109), which when collocated at $(\tilde{x}_k, \tilde{t}_i)$, results in, see also (5.110),

$$\frac{1}{N_0}\sum_{k=1}^{N_0}\left[A_{0,k}^{(1)}\underline{q}_0 + A_{1,k}^{(1)}\underline{q}_1 - B_{1,k}^{(1)}\underline{u}_1 + C_k^{(1)}\underline{u}^0 + D_k^{(1)}\underline{r}\right] = \underline{E}, \tag{5.122}$$

where

$$A_{0,k}^{(1)} = [A_{0j}(\tilde{x}_k,\tilde{t}_i)]_{N\times N},\ A_{1,k}^{(1)} = [A_{1j}(\tilde{x}_k,\tilde{t}_i)]_{N\times N},\ B_{1,k}^{(1)} = [B_{1j}(\tilde{x}_k,\tilde{t}_i)]_{N\times N}.$$

for $k = \overline{1,N_0}$ and $i,j = \overline{1,N}$. Finally, eliminating $\underline{q}_0$, $\underline{q}_1$ and $\underline{u}_1$ between (5.120)–(5.122), the unknown discretized source $\underline{r}$ can be found by solving the generic $N \times N$ linear system of equations (5.111), where

$$X = \frac{1}{N_0}\sum_{k=1}^{N_0}\left\{ \left[(A_{1,k}^{(1)}S + B_{1,k}^{(1)}\Big| - A_{0,k}^{(1)} \right] \left[(A_1 S + B_1)\Big| - A_0 \right]^{-1} \right.$$
$$\left. \times (A_1 F + D_0) - (A_{1,k}^{(1)}F + D_k^{(1)}) \right\},$$

$$\underline{y} = \frac{1}{N_0}\sum_{k=1}^{N_0}\left\{ C_k^{(1)}\underline{u}^0 + A_{1,k}^{(1)}\underline{\tilde{u}}_0 \right.$$
$$\left. - \left[(A_{1,k}^{(1)}S + B_{1,k}^{(1)})\Big| - A_{0,k}^{(1)} \right] \left[(A_1 S + B_1)\Big| - A_0 \right]^{-1} (C\underline{u}^0 + A_1\underline{\tilde{u}}_0) \right\} - \underline{E}.$$

Consider an example for which the conditions of existence and uniqueness of solution given in Theorem 5.11 are satisfied. First, choose $T = 1$, $u_0(x) \equiv 0$, $a = 1$ and $b = 0$. For $a = 1$ and $b = 0$ the Sturm-Liouville problem (5.118) has the eigenvalues $\Lambda_n = \xi_n^2$, where ξ_n are the positive roots of the transcendental equation $\xi\sin(\xi) = \cos(\xi)$. The corresponding eigenfunctions are $y_n(x) = \sin(\xi_n x)$. Observing that $\xi_0 = \sqrt{\Lambda_0} = 0.860333$ and choosing $h(x,t) = x^3(1-x)^4(\beta_1 x + \beta_2)$, the constants β_1 and β_2 can be determined such that $h \in \Phi_0^4[0,1]$ (choosing $n_0 = 0$ for simplicity), as required by condition (A_3) of Theorem 5.11. This imposes

$$0 = \int_0^1 h(x,t)\sin(\xi_0 x)dx = \int_0^1 x^3(1-x)^4(\beta_1 x + \beta_2)\sin(\xi_0 x)dx.$$

Choosing $\beta_2 = -1$, the above condition yields $\beta_2 \approx 2.011$. Then, $\int_0^1 h(x,t)dx = -0.00037 \neq 0$, as required by condition (A_3).

Consider the retrieval of the non-smooth source function

$$r(t) = \left| t - \frac{1}{2} \right|. \tag{5.123}$$

In this case, the analytical solution of the direct problem for $u(x,t)$ is not available, thus the energy $E(t)$ is not available either. In such a situation, the data (5.77) is simulated numerically by solving first the direct problem (5.73), (5.74), (5.116) and (5.117) with known $r(t)$ given by (5.123), using the BEM with $N = N_0 = 40$. Further, this data is perturbed as in (5.114), where $\sigma = p \times \max_{t\in[0,T]}|E(t)| = 1.2 \times 10^{-5}p$. In this case, the system of equations (5.111) becomes

$$X\underline{r} = \underline{y}^{\epsilon}. \tag{5.124}$$

In order to avoid committing an inverse crime, the inverse problem is solved using the BEM with a different mesh size disretisation of $N = 40$ and $N_0 = 30$. Figure 5.6(a) shows the numerical results obtained without regularization, i.e. by the straightforward inversion of the system of equations (5.124), i.e.

$\underline{r} = X^{-1}\underline{y}^{\epsilon}$, for $p = 0$ (exact) and $p = 1\%$ (noisy) data. While for exact data the straightforward inversion produces very accurate results (RMSE(r)$= 2.9\times 10^{-4}$), for noisy data the numerical retrievals become highly oscillatory and unbounded. This is expected since the inverse problem under investigation is ill-posed. To overcome this instability, the Tikhonov regularization method of second-order is employed. This yields the stable and accurate numerical results (RMSE(r) $\in \{0.005, 0.013, 0.021\}$ for $p \in \{1, 3, 5\}\%$, respectively) illustrated in Figure 5.6(b). The regularization parameter was chosen according to the generalized cross validation (GCV) criterion, which is based on minimizing the GCV function [398], se also (1.16),

$$GCV(\lambda) = \frac{\|X(X^{\mathrm{T}}X + \lambda R_2^{\mathrm{T}}R_2)^{-1}X^{\mathrm{T}}\underline{y}^{\epsilon} - \underline{y}^{\epsilon}\|^2}{[\mathrm{Tr}(I - X(X^{\mathrm{T}}X + \lambda R_2^{\mathrm{T}}R_2)^{-1}X^{\mathrm{T}})]^2}. \tag{5.125}$$

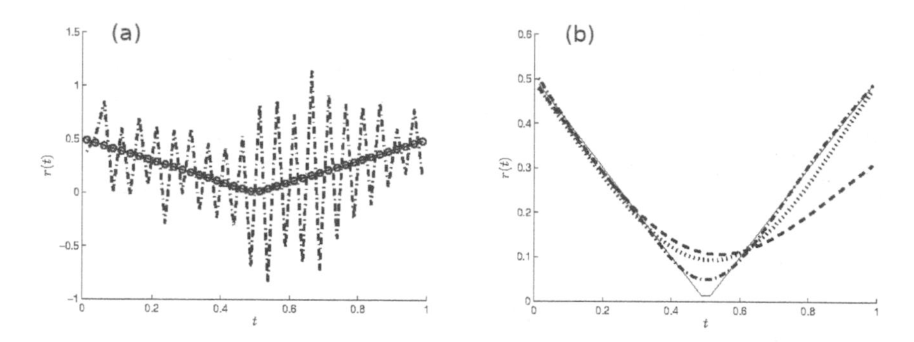

FIGURE 5.6: The analytical (—) solution (5.123) and the numerical results of $r(t)$: (a) with no regularization for exact data (∘∘∘) and noisy data $p = 1\%$ ($- \cdot -$); (b) with regularization for $p = 1\%$(∘ ∘ ∘), $p = 3\%(\cdots)$ and $p = 5\%(- - -)$, [165].

5.2.4 The fourth variant

A somewhere in between variant the previous two considers the nonlocal boundary conditions

$$u(0,t) + cu(1,t) = 0, \quad t \in [0,T] \tag{5.126}$$

and

$$u_x(0,t) + u_x(1,t) = 0, \quad t \in [0,T]. \tag{5.127}$$

Then, the auxiliary Sturm-Liouville spectral problem is

$$\begin{cases} y''(x) + \Lambda y(x) = 0, & x \in [0,1], \\ y(0) + cy(1) = 0, & y'(0) + y'(1) = 0. \end{cases} \tag{5.128}$$

In case $c = -1$, the periodic boundary condition (5.76) is re-obtained. In case $c = -1$, the boundary conditions in (5.128) are not regular, whilst in the case $b \neq -1$ they are regular but not strongly regular [314]. Also, in case $c = 1$ the boundary conditions in (5.128) are anti-periodic (self-adjoint). In what follows it is assumed that $c \neq \pm 1$, in which case each eigenvalue of the auxiliary Sturm-Liouville problem has algebraic multiplicity two. As in Theorem 5.11, the following well-posedness theorem for the inverse source problem (5.73), (5.74), (5.77), (5.126) and (5.127) holds [166].

Theorem 5.12. *Let $c \neq \pm 1$. Suppose that the following conditions hold:*
(A_1) $E \in C^1[0,T]$;
(A_2) $u_0 \in C^4[0,1]$, $u_0(0) = u_0(1) = u_0''(0) = u_0''(1) = 0$, $u_0' + u_0'(1) = u_0'''(0) + u_0'''(1) = 0$ and $E(0) = \int_0^1 u_0(x)dx$;
(A_3) $h \in C(\overline{Q}_T)$, $h(\cdot,t) \in C^4[0,1]$, $h(0,t) = h(1,t) = h_{xx}(0,t) = h_{xx}(1,t) = 0$, $h_x(0,t) + h_x(1,t) = h_{xxx}(0,t) + h_{xxx}(1,t) = 0$ and $\int_0^1 h(x,t)dx \neq 0$ for all $t \in [0,T]$.
Then, the inverse problem (5.73), (5.74), (5.77), (5.126), (5.116) *has a unique solution $(u(x,t), r(t)) \in (C^{2,1}(\Omega_T) \cap C^{1,0}(\overline{\Omega}_T)) \times C[0,T]$. Moreover, the solution depends continuously upon the input data (E, u_0, h) satisfying the assumptions $(A_1) - (A_3)$.*

5.2.4.1 BEM for the fourth variant

Application of the BEM to the problem (5.73), (5.76), (5.77), (5.126) and (5.127) results in the system of equations

$$-A_0\underline{q}_1 + A_1\underline{q}_1 + (cB_0 - B_1)\underline{u}_1 + C\underline{u}^0 + D_0\underline{r} = \underline{0},$$

$$\frac{1}{N_0}\sum_{k=1}^{N_0}\left(\left[(-A_{0,k}^{(1)} + A_{1,k}^{(1)})\middle|(cB_{0,k}^{(1)} - B_{1,k}^{(1)})\right]\begin{bmatrix}\underline{q}_1\\ \underline{u}_1\end{bmatrix} + C_k^{(1)}\underline{u}^0 + D_k^{(1)}\underline{r}\right) = \underline{E}.$$

Eliminating $\underline{u}_1$ and $\underline{q}_1$, the system (5.111) of $N \times N$ linear equations is obtained, where

$$X = \frac{1}{N_0}\sum_{k=1}^{N_0}\Bigg(\left[(-A_{0,k}^{(1)} + A_{1,k}^{(1)})\middle|(cB_{0,k}^{(1)} - B_{1,k}^{(1)})\right] \times \left[(A_0 - A_1)\middle|(-cB_0 + B_1)\right]^{-1} D_0 + D_k^{(1)}\Bigg),$$

$$\underline{y} = \underline{E} - \frac{1}{N_0}\sum_{k=1}^{N_0}\Bigg(\left[(-A_{0,k}^{(1)} + A_{1,k}^{(1)})\middle|(cB_{0,k}^{(1)} - B_{1,k}^{(1)})\right] \times \left[(A_0 - A_1)\middle|(-cB_0 + B_1)\right]^{-1} C + C_k^{(1)}\Bigg)\underline{u}^0.$$

As a numerical test example, consider the identification of a time-dependent discontinuous source function given by [398],

$$r(t) = \begin{cases} -1, & t \in [0, 0.2), \\ 1, & t \in [0.25, 0.5), \\ -1, & t \in [0.5, 0.75), \\ 1, & t \in [0.75, 1], \end{cases} \tag{5.129}$$

with the input data $T = 1$, $c = 2$, $u_0 = 0$ and $h = 1$, by using the BEM together with the Tikhonov regularization.

Data (5.77) is numerically simulated by solving first the direct problem (5.73), (5.74), (5.126), (5.127), with r known and given by (5.129). The numerical $E(t)$ simulated by solving the direct problem with $N = N_0 = 40$ is taken as the extra measured data for the inverse problem. To avoid committing an inverse crime, in the inverse problem the number of boundary elements is kept to be the same $N = 40$, but the number of cells is taken as $N_0 = 30$ to be different from 40 which was used in the direct problem.

In analyzing the inverse problem, first consider the linear BEM system of equations (5.111). The conditioned numbers of matrix X for $N = N_0 \in \{20, 40, 80\}$ are 291, 906 and 2737, respectively, which indicates some mild ill-conditioning.

For exact data, the numerical results for $r(t)$ obtained with the BEM for $N = 40$, $N_0 = 30$ are illustrated in Figure 5.7(a). These results have been obtained using the straightforward inversion of the linear system of equations (5.111), i.e. $\underline{r} = X^{-1}\underline{y}$. The very good agreement between the analytical (—) and numerical results ($\circ\circ\circ$) for $p = 0$ is observed.

Next, the energy data (5.77) is perturbed by $p = 1\%$ random Gaussian additive noise as in (5.114), where $\sigma = p \times \max_{t\in[0,T]} |E(t)| = 0.25p$. The numerical solution shown in Figure 5.7(a) by a dash-dot line ($- \cdot -$) has been obtained with no regularization, i.e. using the straightforward inversion of the noisy linear system, namely, $\underline{r}^\epsilon = X^{-1}\underline{y}^\epsilon$. From this figure it can be seen that the numerical solution for $r(t)$ is unstable, however, the results for $u(0,t)$ and $u_x(0,t)$ seem to remain stable (see Table 5.5). This is somewhat to be expected since the triplet solution $(r(t), u(0,t), u_x(0,t))$ is stable in the components of $u(0,t)$ and $u_x(0,t)$, but unstable in the source function $r(t)$. To overcome this instability, the first-order Tikhonov regularized solution

$$\underline{r}_\lambda = \left(X^{\mathrm{T}}X + \lambda R_1^{\mathrm{T}} R_1\right)^{-1} X^{\mathrm{T}} \underline{y}^\epsilon, \tag{5.130}$$

is constructed, where $\lambda > 0$ is chosen according to the GCV criterion. The numerically obtained results presented in Figure 5.7(b) and Table 5.5 illustrate that stable and accurate solutions have been obtained.

5.2.5 Other non-local variants

The existence and uniqueness of the solution $(u(x,t), r(t)) \in (C([0,T]; L^2(\Omega)) \cap L^\infty(0,T; H^1(\Omega))) \times L^2(0,T)$ with u_t, $\Delta u \in L^2(0,T; L^2(\Omega))$ and $\nabla\Delta u \in$

TABLE 5.5: The values of λ_{GCV} and RMSEs for $r(t)$, $u(0,t)$ and $u_x(0,t)$ obtained using the BEM together with the first-order Tikhonov regularization, for $p \in \{0,1,3,5\}\%$ noise [166].

p	λ	RMSE(r)	RMSE($u(0,t)$)	RMSE($u_x(0,t)$)
$p=0$	$\lambda=0$	1.42E-3	7.66E-5	2.55E-4
$p=1\%$	$\lambda=0$	7.36E-1	8.52E-3	6.13E-2
$p=1\%$	λ_{GCV} =1.4E-5	1.70E-1	7.08E-3	4.01E-2
$p=3\%$	λ_{GCV} =1.8E-4	2.28E-1	1.37E-2	6.51E-2
$p=5\%$	λ_{GCV} =1.0E-3	3.04E-1	2.47E-2	1.13E-1

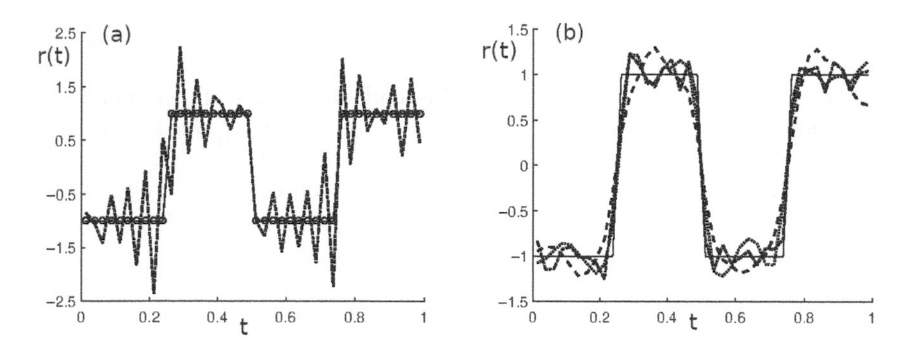

FIGURE 5.7: The analytical (——) and numerical results of $r(t)$ for: (a) exact data (∘∘∘) and $p=1\%$ noise (−·−) with no regularization ($\lambda=0$); and (b) for $p \in \{1(-\cdot-),\ 3(\cdots),\ 5(---)\}\%$ noisy with $\lambda_{GCV} \in \{1.4\text{E-}5, 1.8\text{E-}4, 1.0\text{E-}3\}$, respectively, in the first-order regularization [166].

$L^2(0,T;L^2(\Omega)^d)$ of the inverse source problem given by the semilinear heat equation

$$u_t - \Delta u(x,t) = r(t)h(x) + g(x,t) + \gamma(u), \quad (x,t) \in \Omega_T := \Omega \times (0,T),$$

in a bounded domain $\Omega \subset \mathbb{R}^d$, subject to initial condition

$$u(x,0) = u_0(x), \quad x \in \Omega, \tag{5.131}$$

the no-flux boundary condition

$$\frac{\partial u}{\partial n}(x,t) = 0, \quad (x,t) \in \partial\Omega \times (0,T), \tag{5.132}$$

and the extra boundary measurement

$$m(t) = \int_{\partial\Omega} u(x,t)dS, \quad t \in [0,T], \tag{5.133}$$

was established in [364], under the assumptions that $g \in C([0,T];H^1(\Omega))$, $h \in H^1(\Omega)$ with $\int_{\partial\Omega} h(x)dS \neq 0$, $m \in C^1([0,T])$, $u_0 \in V_1 := \{\phi : \Omega \to$

$\mathbb{R};\ \|\phi\| + \|\nabla\phi\| + \|\Delta\phi\| + \|\nabla\Delta\phi\| < \infty\}$, where the norms are in $L^2(\Omega)$, and γ is either (i) linear or (ii) $\gamma \in C^2(\mathbb{R})$ is nonlinear and Lipschitz continuous, and $|\nabla u| \leq C$ a.e. in Ω_T.

The same type of boundary measurement (5.133) was considered in [365], but for the linear heat equation

$$u_t - \Delta u(x,t) = r(t)h(x), \quad (x,t) \in \Omega_T := \Omega \times (0,T), \tag{5.134}$$

and the mixed nonlinear dynamic boundary conditions

$$\begin{aligned} -\frac{\partial u}{\partial n}(x,t) = u_t(x,t) + \sigma(u(x,t)), \quad (x,t) \in \Gamma_N \times (0,T), \\ u(x,t) = 0, \quad (x,t) \in \Gamma_D \times (0,T), \end{aligned} \tag{5.135}$$

where $\Gamma_D \cup \Gamma_N = \partial\Omega$, $\Gamma_D \cap \Gamma_N = \emptyset$, instead of (5.132). Then, under the assumptions that Γ_D is of non-zero measure, $h \in H_0^1(\Omega)$, $\int_\Omega h d\Omega \neq 0$, $m'' \in L^2(0,T)$, σ is Lipschitz continuous and $u_0 \in V_2 := \{\phi : \Omega \to \mathbb{R};\ \|\phi\| + \|\nabla\phi\| + \|\Delta\phi\| + \|\partial_n\phi\|_{L^2(\Gamma_N)} < \infty\}$, there exists a unique solution $(u(x,t), r(t))$ with $u \in C([0,T]; L^2(\Omega)) \cap L^\infty(0,T; V_2)$, $u_t \in L^2(0,T; H^1(\Omega))$, $r \in L^2(0,T)$ to the inverse source problem (5.131), (5.133)–(5.135), [365].

Chapter 6

Space- and Time-Dependent Sources

Consider a practical situation in which a fertilizer from a field is carried into a stream by rain in the form of run-off which in turn affects aquatic life. Then, in this application of water pollution, the inhomogeneous source term in the governing model is unknown and needs to be determined. As such, inverse source problems for the heat (diffusion) equation have attracted considerable interest in recent years, see [3, 200, 393, 399, 401] to name just a few. These studies, as well as the previous Chapters 4 and 5, have sought a coefficient source function depending on either space- or time-dependent variables. Aside from this, the reconstruction of a source depending on both space and time variables from a finite set of measurement data was attempted in [145, 403].

In this chapter, inverse heat source functions depending on both space and time, but which are either additively or multiplicatively separable are considered [146, 163, 164].

6.1 Additive Space- and Time-Dependent Heat Sources

Let $l > 0$ and $T > 0$ be fixed numbers and consider the inverse problem of finding the temperature $u(x,t) \in H^{2+\varepsilon,1+\varepsilon/2}(\overline{\Omega}_T)$ with $\varepsilon \in (0,1)$ and $\Omega_T := (0,l) \times (0,T)$, the time-dependent heat source $r(t) \in H^{\varepsilon/2}[0,T]$ and the space-dependent heat source $f(x) \in H^{\varepsilon}[0,l]$ satisfying the heat equation

$$u_t = u_{xx} + r(t)h_1(x,t) + f(x)h_2(x,t) + g(x,t), \quad (x,t) \in \Omega_T, \tag{6.1}$$

subject to the initial condition

$$u(x,0) = u_0(x), \quad x \in [0,l], \tag{6.2}$$

the Dirichlet boundary conditions

$$u(0,t) = \mu_0(t), \quad u(l,t) = \mu_l(t), \quad t \in (0,T), \tag{6.3}$$

and the over-determination conditions

$$u(X_0,t) = \chi(t), \quad t \in [0,T], \tag{6.4}$$

DOI: 10.1201/9780429400629-6

$$\int_0^T u(x,t)dt = \psi(x), \quad x \in [0,l], \tag{6.5}$$

$$f(X_0) = S_0, \tag{6.6}$$

where $h_1(x,t)$, $h_2(x,t)$, $g(x,t)$, $u_0(x)$, $\mu_0(t)$, $\mu_l(t)$, $\chi(t)$ and $\psi(x)$ are given functions, X_0 is a given sensor location within the interval $(0,l)$ and S_0 is a given value of the source function f at the given point X_0.

The time-average temperature measurement (6.5) represents a non-local condition/measurement that is convenient to use in practical situations where a local measurement of the temperature at a fixed time $\overline{T} \in (0,T]$, namely,

$$u(x,\overline{T}) =: \psi_{\overline{T}}(x), \quad x \in [0,l] \tag{6.7}$$

contains a large amount of noise. This may be due to harsh external conditions, or to the fact that many space measurements can, in fact, never be recorded at a fixed time instantaneously. If this is the case, one can have a selection of such very noisy local temperature measurements, but which on average produce the less noisy nonlocal measurement (6.5).

The inverse problem (6.1)–(6.6) has a unique local solution, as follows.

Theorem 6.1. ([205]) *Assume that the following conditions are satisfied:*
(i) $u_0(x)$, $\psi(x) \in H^{2+\varepsilon}[0,l]$, $\mu_0(t)$, $\mu_l(t)$, $\chi(t) \in H^{1+\varepsilon/2}[0,T]$, $g(x,t) \in H^{\varepsilon,\varepsilon/2}(\overline{\Omega}_T)$, h_1 independent of t and $h_1(x) \in H^{\varepsilon}[0,l]$, h_2 independent of x and $h_2(t) \in H^{\varepsilon/2}[0,T]$;
(ii) $h_1(X_0) \neq 0$, $\int_0^T h_2(t)dt \neq 0$, $\frac{h_2(t)}{\int_0^T h_2(\tau)d\tau} \geq 0$ for all $t \in [0,T]$;
(iii) $u_0(0) = \mu_0(0)$, $u_0(l) = \mu_l(0)$, $u_0(X_0) = \chi(0)$, $\int_0^T \chi(t)dt = \psi(X_0)$, $\psi(0) = \int_0^T \mu_0(t)dt$, $\psi(l) = \int_0^T \mu_l(t)\,dt$.
Then, for sufficiently small $T > 0$, there exists a unique solution of the inverse problem (6.1)–(6.6).

Condition (6.6) was omitted in [205], but it should be included in order to avoid the following non-uniqueness counterexample.

Counterexample 6.1. (non-uniqueness of solution of problem (6.1)–(6.5)) If h_1 is independent of t and h_2 is independent of x, as in assumption (i) of Theorem 6.1, then for an arbitrary constant c, the identity

$$r(t)h_1(x) + f(x)h_2(t) = (r(t) + ch_2(t))h_1(x) + (f(x) - ch_1(x))h_2(t)$$

means that the heat equation (6.1) is satisfied by both sets of solutions $(r(t), f(x))$ and $(r(t) + ch_2(t), f(x) - ch_1(x))$. This shows the non-uniqueness of solution of the incompletely formulated inverse source problem (6.1)–(6.5). Accordingly, condition (6.6) needs to be imposed in order to specify a particular solution. Theorem 6.1 also holds in higher dimensions. However, although the inverse problem (6.1)–(6.6) is uniquely solvable, the problem is still ill-posed because the continuous dependence on the input data (6.4) and (6.5) is violated. This instability can be seen from the following counterexample.

Counterexample 6.2. (instability of problem (6.1)–(6.6))
Let $l = \pi$, $X_0 = \pi/2$ and

$$u_n(x,t) = \frac{(1 - e^{-4n^2 t})\sin(2nx)}{n^{3/2}} + \frac{x(\pi - x)\sin(nt)}{n^{1/2}}, \tag{6.8}$$

for n positive integer. Then the initial and boundary conditions (6.2) and (6.3) are all homogeneous. The over-determination conditions (6.4) and (6.5) are

$$u\left(\frac{\pi}{2}, t\right) = \chi(t) = \frac{\pi^2 \sin(nt)}{4n^{1/2}}, \quad t \in [0,T],$$

$$\int_0^T u(x,t)dt = \psi(x) = \frac{\sin(2nx)}{n^{3/2}}\left[T + \frac{e^{-4n^2 T} - 1}{4n^2}\right]$$

$$+\frac{x(\pi - x)(1 - \cos(nT))}{n^{3/2}}, \quad x \in [0,\pi],$$

which tend to zero as $n \to \infty$. Take also

$$h_1(x,t) = x(\pi - x),\ h_2(x,t) = 1,\ g(x,t) = \frac{2\sin(nt)}{n^{1/2}},\ f(\pi/2) = S_0 = 0,$$

and observe that g tends to zero as $n \to \infty$. The above input satisfies the hypotheses of Theorem 6.1 and thus the inverse problem (6.1)–(6.6) is locally uniquely solvable. In fact, it has the unique (global) solution

$$r_n(t) = n^{1/2}\cos(nt), \quad f_n(x) = -4n^{1/2}\sin(2nx), \tag{6.9}$$

and $u_n(x,t)$ given by (6.8). However, one can observe that the solution (6.9) is unstable since it becomes unbounded, as $n \to \infty$, in any reasonable norm.

As in [350], it appears that the stability of solution can be restored under stronger assumptions, e.g. if $\|\chi\|_{H^1[0,T]} \to 0$, $\|\psi\|_{H^2[0,l]} \to 0$ then $\|f\|_{L^2[0,l]} \to 0$ and $\|r\|_{L^2[0,T]} \to 0$. Next, the inverse heat source problem (6.1)–(6.6) is solved using a regularized BEM.

6.1.1 BEM for additive source

As in the previous sections 4.1.4 and 5.2.1.1, by means of the fundamental solution (3.23) and the Green's formula, equation (6.1) implies the boundary integral equation,

$$\eta(x)u(x,t) = \int_0^t \left[G(x,t;\xi,\tau)\frac{\partial u}{\partial n}(\xi,\tau) - u(\xi,\tau)\frac{\partial G}{\partial n(\xi)}(x,t;\xi,\tau)\right]_{\xi\in\{0,l\}} d\tau$$
$$+ \int_0^l G(x,t;y,0)u(y,0)dy + \int_0^l \int_0^t G(x,t;y,\tau)r(\tau)h_1(y,\tau)d\tau dy$$
$$+ \int_0^l \int_0^t G(x,t;y,\tau)f(y)h_2(y,\tau)d\tau dy$$
$$+ \int_0^l \int_0^t G(x,t;y,\tau)g(y,\tau)d\tau dy, \quad (x,t)\in[0,l]\times(0,T). \tag{6.10}$$

Applying the BEM, as in subsection 5.2.1.1, results in (see also (5.28)),

$$\eta(x)u(x,t) = \sum_{j=1}^{N} [A_{0j}(x,t)q_{0j} + A_{lj}(x,t)q_{lj} - B_{0j}(x,t)\mu_{0j} - B_{lj}(x,t)\mu_{lj}]$$
$$+ \sum_{k=1}^{N_0} C_k(x,t)u_{0,k} + d_0(x,t) + d_1(x,t) + d_2(x,t), \tag{6.11}$$

for $(x,t) \in [0,l]\times(0,T)$, where $\mu_{0j} = \mu_0(\tilde{t}_j)$, $\mu_{lj} = \mu_l(\tilde{t}_j)$, $A_{\xi j}$, $B_{\xi j}$ for $\xi \in \{0,l\}$ and C_k are given by (5.29), and

$$d_0(x,t) = \int_0^l \int_0^t G(x,t;y,\tau)r(\tau)h_1(y,\tau)d\tau dy,$$
$$d_1(x,t) = \int_0^l \int_0^t G(x,t;y,\tau)g(y,\tau)d\tau dy,$$
$$d_2(x,t) = \int_0^l \int_0^t G(x,t;y,\tau)f(y)h_2(y,\tau)d\tau dy,$$

With the piecewise constant approximations

$$h_1(x,t) = h_1(x,\tilde{t}_j), \quad h_2(x,t) = h_2(\tilde{x}_k,t), \quad g(x,t) = g(x,\tilde{t}_j),$$
$$r(t) = r(\tilde{t}_j) =: r_j, \quad f(x) = f(\tilde{x}_k) =: f_k,$$
$$t\in(t_{j-1},t_j],\ x\in(x_{k-1},x_k],\ j=\overline{1,N},\ k=\overline{1,N_0},$$

these integrals can be approximated as, see also (5.31)–(5.34),

$$d_0(x,t) = \int_0^t r(\tau) \int_0^l G(x,t;y,\tau)h_1(y,\tau)dyd\tau = \sum_{j=1}^{N} D_{0,j}(x,t)r_j,$$
$$d_1(x,t) = \int_0^t \int_0^l G(x,t;y,\tau)g(y,\tau)dyd\tau = \sum_{j=1}^{N} D_{1,j}(x,t),$$
$$d_2(x,t) = \int_0^l f(y) \int_0^t G(x,t;y,\tau)h_2(y,\tau)d\tau dy = \sum_{k=1}^{N_0} D_{2,k}(x,t)f_k,$$

where

$$D_{0,j}(x,t) = \int_0^l h_1(y,\tilde{t}_j)A_{yj}(x,t)dy,$$
$$D_{1,j}(x,t) = \int_0^l g(y,\tilde{t}_j)A_{yj}(x,t)dy,$$
$$D_{2,k}(x,t) = \frac{1}{2}\int_0^t h_2(\tilde{x}_k,t)\left[\operatorname{erf}\left(\frac{x-x_{k-1}}{2\sqrt{t-\tau}}\right) - \operatorname{erf}\left(\frac{x-x_k}{2\sqrt{t-\tau}}\right)\right]d\tau.$$

These integrals are evaluated using Simpson's rule for numerical integration. With these approximations, the integral equation (6.11) becomes

$$\eta(x)u(x,t)$$
$$= \sum_{j=1}^{N}\left[A_{0j}(x,t)q_{0j} + A_{lj}(x,t)q_{lj} - B_{0j}(x,t)\mu_{0j} - B_{lj}(x,t)\mu_{lj}\right]$$
$$+ \sum_{k=1}^{N_0} C_k(x,t)u_{0,k} + \sum_{j=1}^{N} D_{0,j}(x,t)r_j + \sum_{k=1}^{N_0} D_{2,k}(x,t)f_k$$
$$+ \sum_{j=1}^{N} D_{1,j}(x,t), \quad (x,t)\in[0,l]\times(0,T). \tag{6.12}$$

Applying (6.12) at the boundary nodes $(0,\tilde{t}_i)$ and $(l,\tilde{t}_i)$ for $i = \overline{1,N}$, yields the system of $2N$ linear equations

$$A\underline{q} - B\underline{\mu} + C\underline{u}^0 + D_0\underline{r} + D_2\underline{f} + \underline{d} = \underline{0}, \tag{6.13}$$

where A, B, C, D_0, $\underline{q}$, $\underline{u}^0$ and $\underline{r}$ have been defined in (5.37) and

$$D_2 = \begin{bmatrix} D_{2,k}(0,\tilde{t}_i) \\ D_{2,k}(l,\tilde{t}_i) \end{bmatrix}_{2N\times N_0}, \quad \underline{d} = \begin{bmatrix} \sum_{j=1}^{N} D_{1,j}(0,\tilde{t}_i) \\ \sum_{j=1}^{N} D_{1,j}(l,\tilde{t}_i) \end{bmatrix}_{2N},$$
$$\underline{\mu} = \begin{bmatrix} \mu_{0j} \\ \mu_{lj} \end{bmatrix}_{2N}, \quad \underline{f} = \left[f_k\right]_{N_0},$$

To determine $\underline{r}$ and $\underline{f}$, conditions (6.4)–(6.6) are imposed. To impose (6.4), equation (6.12) is applied at the interior points $(X_0, \tilde{t}_i)$ for $i = \overline{1,N}$, giving rise to the following linear system of N equations:

$$A^I\underline{q} - B^I\underline{\mu} + C^I\underline{u}^0 + D_0^I\underline{r} + D_2^I\underline{f} + \underline{d}^I = \underline{\chi}, \tag{6.14}$$

where

$$A^I = \left[A_{0j}(X_0,\tilde{t}_i) \quad A_{lj}(X_0,\tilde{t}_i)\right]_{N\times 2N},$$
$$B^I = \left[B_{0j}(X_0,\tilde{t}_i) \quad B_{lj}(X_0,\tilde{t}_i)\right]_{N\times 2N}, \quad C^I = \left[C_k(X_0,\tilde{t}_i)\right]_{N\times N_0},$$
$$D_0^I = \left[D_{0,j}(X_0,\tilde{t}_i)\right]_{N\times N}, \quad D_2^I = \left[D_{2,k}(X_0,\tilde{t}_i)\right]_{N\times N_0},$$
$$\underline{d}^I = \left[\textstyle\sum_{j=1}^N D_{1,j}(X_0,\tilde{t}_i)\right]_N, \quad \underline{\chi} = \left[\chi_i\right]_N,$$

where $\chi_i := \chi(\tilde{t}_i) = u(X_0, \tilde{t}_i)$ for $i = \overline{1,N}$. The time-integral condition (6.5) is approximated by the midpoint rule

$$\psi(x) = \int_0^T u(x,t)dt = \frac{T}{N}\sum_{i=1}^N u(x,\tilde{t}_i), \quad x \in [0,l].$$

Applying this equation at the midpoint $\tilde{x}_k$ for $k = \overline{1,N_0}$ gives

$$\frac{T}{N}\sum_{i=1}^N u(\tilde{x}_k,\tilde{t}_i) = \psi(\tilde{x}_k) =: \psi_k. \tag{6.15}$$

Using (6.12), equation (6.15) yields

$$\frac{T}{N}\sum_{i=1}^N \left[A_i^{II}\underline{q} - B_i^{II}\underline{\mu} + C_i^{II}\underline{u}^0 + D_{0,i}^{II}\underline{r} + D_{2,i}^{II}\underline{f} + \underline{d}_i^{II}\right] = \underline{\psi}, \tag{6.16}$$

where

$$A_i^{II} = \left[A_{0j}(\tilde{x}_k,\tilde{t}_i) \quad A_{lj}(\tilde{x}_k,\tilde{t}_i)\right]_{N_0\times 2N},$$
$$B_i^{II} = \left[B_{0j}(\tilde{x}_k,\tilde{t}_i) \quad B_{lj}(\tilde{x}_k,\tilde{t}_i)\right]_{N_0\times 2N}, \quad C_i^{II} = \left[C_k(\tilde{x}_k,\tilde{t}_i)\right]_{N_0\times N_0},$$
$$D_{0,i}^{II} = \left[D_{0,j}(\tilde{x}_k,\tilde{t}_i)\right]_{N_0\times N}, \quad D_{2,i}^{II} = \left[D_{2,k}(\tilde{x}_k,\tilde{t}_i)\right]_{N_0\times N_0},$$
$$\underline{d}_i^{II} = \left[\textstyle\sum_{j=1}^N D_{1,j}(\tilde{x}_k,\tilde{t}_i)\right]_{N_0}, \quad \underline{\psi} = \left[\psi_k\right]_{N_0}.$$

Finally, consider condition (6.6). Since the space midpoint discretization was employed, S_0 at the given point $X_0 \in (0,l)$ is approximated as

$$S_0 = s(X_0) \approx \frac{s(\tilde{x}_{k^*}) + s(\tilde{x}_{k^*+1})}{2}, \tag{6.17}$$

where k^* is the number in $\{1,\ldots,N_0-1\}$ which satisfies $\tilde{x}_{k^*} \le X_0 < \tilde{x}_{k^*+1}$.

Now $\underline{r}$ and $\underline{f}$ can be found by eliminating $\underline{q}$ from (6.13), and combining expressions (6.14), (6.16) and (6.17), a linear system of $(N+N_0+1)$ equations with $(N+N_0)$ unknowns is obtained as [163]:

$$X\underline{w} = \underline{y}, \tag{6.18}$$

where

$$X = \begin{bmatrix} A^I A^{-1} D_0 - D_0^I & A^I A^{-1} D_2 - D_2^I \\ \frac{T}{N}\sum_{i=1}^{N}(A_i^{II} A^{-1} D_0 - D_{0,i}^{II}) & \frac{T}{N}\sum_{i=1}^{N}(A_i^{II} A^{-1} D_2 - D_{2,i}^{II}) \\ 0\ldots 0 & 0\ldots 0\ \frac{1}{2}\ \frac{1}{2}\ 0\ldots 0 \end{bmatrix},$$

$$\underline{w} = \begin{bmatrix} \underline{r} \\ \underline{f} \end{bmatrix},$$

$$\underline{y} = \begin{bmatrix} -\underline{\chi} + A^I A^{-1}(B\underline{\mu} - C\underline{u}^0 - \underline{d}) - B^I\underline{\mu} + C^I\underline{u}^0 + \underline{d}^I \\ -\underline{\psi} + \frac{T}{N}\sum_{i=1}^{N}\left(A_i^{II} A^{-1}(B\underline{\mu} - C\underline{u}^0 - \underline{d}) - B_i^{II}\underline{\mu} + C_i^{II}\underline{u}^0 + \underline{d}_i^{II}\right) \\ S_0 \end{bmatrix}.$$

In practice, the measured data is contaminated by errors. To numerically simulate these errors, noise is added in the input functions $\chi(t)$ and $\psi(x)$ as

$$\underline{\chi}^p = \underline{\chi} + random(Normal, 0, \sigma_\chi, 1, N), \tag{6.19}$$

$$\underline{\psi}^p = \underline{\psi} + random(Normal, 0, \sigma_\psi, 1, N_0), \tag{6.20}$$

where $\sigma_\chi = p \times \max_{t\in[0,T]}|\chi(t)|$ and $\sigma_\psi = p \times \max_{x\in[0,l]}|\psi(x)|$, where p represents the percentage of noise. Note that the measurement (6.6) is already contaminated by error due to the approximation made in (6.17). Thus, if $\underline{y}$ is contaminated with noise such that $\|\underline{y}^\epsilon - \underline{y}\| \approx \epsilon$, then the direct least-squares solution $\underline{w} = (X^{\mathrm{T}}X)^{-1}X^{\mathrm{T}}\underline{y}^\epsilon$ of the perturbed system of equations

$$X\underline{w} = \underline{y}^\epsilon, \tag{6.21}$$

will be unstable. To overcome this instability, regularization methods such as the truncated singular value decomposition (TSVD) or the Tikhonov regularization method needs to be employed.

6.1.2 Truncated singular value decomposition (TSVD)

As described in section 1.3.2, consider the decomposition $X = U\Sigma V^{\mathrm{T}}$, where $U = [\underline{U}_1, \underline{U}_2, \ldots, \underline{U}_{N+N_0}]$ and $V = [\underline{V}_1, \underline{V}_2, \ldots, \underline{V}_{N+N_0}]$ are $(N+N_0+1) \times (N+N_0)$ matrices with columns, $\underline{U}_j$ and $\underline{V}_j$ for $j = \overline{1,(N+N_0)}$, such that $U^{\mathrm{T}}U = I = V^{\mathrm{T}}V$, and $\Sigma = \mathrm{diag}(\sigma_1, \sigma_2, \ldots, \sigma_{N+N_0})$ is an $(N+N_0)\times(N+N_0)$ diagonal matrix containing the singular values of the matrix X, σ_j for $j = \overline{1,(N+N_0)}$, in decreasing order $\sigma_1 \geq \sigma_2 \geq \cdots \geq \sigma_{N+N_0} \geq 0$. Then, the SVD solution of the system of equations (6.21) is given by

$$\underline{w} = \left(\sum_{j=1}^{N+N_0} \frac{1}{\sigma_j}\underline{V}_j \cdot \underline{U}_j^{\mathrm{T}}\right)\underline{y}^\epsilon. \tag{6.22}$$

In MATLAB, this decomposition is operated using the command svds$(X, N+N_0)$. For obtaining a stable solution, the sum in expression (6.22) should be truncated by omitting its last $(N+N_0) - N_t$ small singular values, where N_t denotes the truncation level. This way, the regularized solution is given by

$$\underline{w}_{N_t} = \left(\sum_{j=1}^{N_t} \frac{1}{\sigma_j} \underline{V}_j \cdot \underline{U}_j^{\mathrm{T}} \right) \underline{y}^{\epsilon}. \tag{6.23}$$

To indicate the appropriate truncation level N_t, the L-curve criterion, the GCV method, and the discrepancy principle can be utilized. The L-curve method is based on plotting the residual $\|X\underline{w}_{N_t} - \underline{y}^{\epsilon}\|$ versus the norm of the solution $\|\underline{w}_{N_t}\|$, for various values of N_t, [137, 265], whilst the GCV method estimates the truncation number N_t by minimizing the GCV function [25]

$$GCV(N_t) = \frac{\|X\underline{w}_{N_t} - \underline{y}^{\epsilon}\|^2}{[\mathrm{Tr}(I - XX^*)]^2} = \frac{\|XX^*\underline{y}^{\epsilon} - \underline{y}^{\epsilon}\|^2}{[\mathrm{Tr}(I - XX^*)]^2}, \tag{6.24}$$

where $X^* = V\Sigma^*_{N_t}U^{\mathrm{T}}$ and $\Sigma^*_{N_t} = \mathrm{diag}(\frac{1}{\sigma_1}, \frac{1}{\sigma_2}, \ldots, \frac{1}{\sigma_{N_t}})$. Finally, the discrepancy principle, based on the knowledge of the noise level

$$\epsilon = \|\underline{y}^{\epsilon} - \underline{y}\| = \sqrt{\|\underline{\chi}^p - \underline{\chi}\|^2 + \|\underline{\psi}^p - \underline{\psi}\|^2}, \tag{6.25}$$

selects the truncation number N_t such that

$$\|X\underline{w}_{N_t} - \underline{y}^{\epsilon}\| \approx \epsilon. \tag{6.26}$$

6.1.3 Tikhonov regularization

The Tikhonov regularization is another way of obtaining a stable solution of the ill-conditioned system of equations (6.21). This method is based on minimizing the regularized linear least-squares objective function

$$\|X\underline{w} - \underline{y}^{\epsilon}\|^2 + \lambda_1\|Q_1\underline{r}\|^2 + \lambda_2\|Q_2\underline{f}\|^2, \tag{6.27}$$

where Q_1 and Q_2 are any of the (differential) regularization matrices defined in (1.11)–(1.13). Minimizing (6.27) one obtains the regularized solution

$$\underline{w}_{\lambda_1,\lambda_2} = \left(X^{\mathrm{T}}X + R^{\mathrm{T}}R\right)^{-1} X^{\mathrm{T}}\underline{y}^{\epsilon}, \tag{6.28}$$

where the matrix $R = R(\lambda_1, \lambda_2)$ is the block matrix of upper-left subblock $\lambda_1 Q_1$ and lower-right subblock $\lambda_2 Q_2$.

Initially, one can take $\lambda := \lambda_1 = \lambda_2$ and consider the L-curve criterion, the GCV method and the discrepancy principle as choices for indicating the single regularization parameter λ. The L-curve method is based on plotting the graph of the residual $\|X\underline{w}_{\lambda} - \underline{y}^{\epsilon}\|$ versus the solution norm $\|\underline{w}_{\lambda}\|$ for a range of values of λ, [137], whereas the GCV method suggests choosing the parameter λ by minimizing the following GCV function,

$$GCV(\lambda) = \frac{\|X\underline{w}_\lambda - \underline{y}^\epsilon\|^2}{[\mathrm{Tr}(I - X(X^{\mathrm{T}}X + R^{\mathrm{T}}R)^{-1}X^{\mathrm{T}})]^2} = \frac{\|X(X^{\mathrm{T}}X + R^{\mathrm{T}}R)^{-1}X^{\mathrm{T}}\underline{y}^\epsilon - \underline{y}^\epsilon\|^2}{[\mathrm{Tr}(I - X(X^{\mathrm{T}}X + R^{\mathrm{T}}R)^{-1}X^{\mathrm{T}})]^2}, \tag{6.29}$$

where here $R = R(\lambda)$. Both the L-curve and the GCV are heuristic because they do not require the knowledge of the level of noise ϵ. More rigorously, one can use the discrepancy principle [304], which selects λ such that

$$\|X\underline{w}_\lambda - \underline{y}^\epsilon\| \approx \epsilon. \tag{6.30}$$

If general multiple regularization parameters λ_1 and λ_2 are allowed in (6.27) then, for their selection one could employ the L-surface criterion [30], which plots the residual $\|X\underline{w}_{\lambda_1,\lambda_2} - \underline{y}^\epsilon\|$ versus $\|Q_1\underline{r}\|$ and $\|Q_2\underline{f}\|$ for various positive values of λ_1 and λ_2.

6.1.4 Numerical example

To test the accuracy of the numerical solutions, let us introduce the RSMEs (5.60) and

$$\mathrm{RMSE}(f) = \sqrt{\frac{l}{N_0}\sum_{k=1}^{N_0}\left(f_{exact}(\tilde{x}_k) - f_{numerical}(\tilde{x}_k)\right)^2}. \tag{6.31}$$

Consider a test example with $T = l = 1$ and the input data

$$u_0(x) = u(x,0) = x^2, \quad \mu_0(t) = u(0,t) = 0, \quad \mu_1(t) = u(1,t) = e^t,$$
$$\chi(t) = u(X_0,t) = X_0^2 e^t, \quad \psi(x) = \int_0^1 u(x,t)dt = x^2(e-1),$$
$$S_0 = f(X_0) = \sin(\pi X_0), \quad h_1(x,t) = e^x,$$
$$h_2(x,t) = t+1, \quad g(x,t) = (x^2-2)e^t - t^2 e^x - (t+1)\sin(\pi x).$$

The conditions of Theorem 6.1 are satisfied hence the inverse source problem (6.1)–(6.6) with the above data has the unique solution

$$u(x,t) = x^2 e^t, \quad r(t) = t^2, \quad f(x) = \sin(\pi x). \tag{6.32}$$

As shown in Counterexample 6.2, the inverse source problem (6.1)–(6.6) is ill-posed since small errors in the measured data (6.4)–(6.6) cause large errors in the solution. In order to quantify the degree of ill-conditioning, the condition numbers of the matrix X for $N = N_0 \in \{20, 40, 80\}$ and $X_0 \in \{\frac{1}{4}, \frac{1}{2}, \frac{3}{4}\}$

are given in Table 6.1, indicating that the system of equations (6.18) is ill-conditioned. Looking at the columns of Table 6.1 it can be seen that the condition number only slightly decreases as X_0 increase, hence the numerical results are not expected to be significantly influenced by the choice of X_0 within some interval $[\frac{1}{4}, \frac{3}{4}]$ away from the end points $x = 0$ and $x = l = 1$. Of course, as X_0 gets closer to the boundary point $x = 0$ or $x = l$ then the specification of the interval temperature measurement (6.4) resembles a heat flux prescription. This newly generated inverse problem in which Cauchy data are specified at $x = 0$ or $x = l$ would be an interesting future work. In what follows, the discretization $N = N_0 = 40$ and $X_0 = 1/2$ are fixed.

TABLE 6.1: The condition numbers of the matrix X in (6.18), for various $N = N_0 \in \{20, 40, 80\}$ and $X_0 \in \{\frac{1}{4}, \frac{1}{2}, \frac{3}{4}\}$.

$N = N_0$	20	40	80
$X_0 = 1/4$	1.94E+3	9.07E+3	5.04E+4
$X_0 = 1/2$	1.97E+3	7.51E+3	4.06E+4
$X_0 = 3/4$	1.93E+3	6.50E+3	3.33E+4

TABLE 6.2: The RMSEs, obtained using the SVD, the TSVD, and the Tikhonov regularization of orders zero, one and two, for $p = 1\%$ noise.

Method	Criterion	Parameter	RMSE(r)	RMSE(f)
SVD	-	-	16.2	101
TSVD	discrepancy principle	$N_t = 14$	0.204	0.177
Zeroth		$\lambda_1 = \lambda_2 = 1.3E-3$	0.187	0.183
First		$\lambda_1 = \lambda_2 = 2.8E-2$	0.128	0.350
Second		$\lambda_1 = \lambda_2 = 1.5$	0.097	0.265
Second	L-curve	$\lambda_1 = \lambda_2 = 10$	0.161	0.423
Second	L-surface	$\lambda_1 = 10,\ \lambda_2 = 1$	0.079	0.218
Second	trial & error	$\lambda_1 = 8,\ \lambda_2 = 5.2E-2$	0.001	0.053

The input functions $\chi(t)$ and $\psi(x)$ are perturbed by $p = 1\%$ noise, as in (6.19) and (6.20), respectively. The L-curve method and the discrepancy principle are employed as criteria for choosing the regularization parameters, and the suggested parameters are given in Table 6.2. Figure 6.1 presents all results obtained using the TSVD and the Tikhonov regularization of orders zero, one, and two with the regularization parameters suggested by the discrepancy principle. Looking more closely at Figure 6.1(a), it can be seen that the approximate solutions for $r(t)$ obtained by the first- and the second-order Tikhonov regularization are reasonably stable, whereas the numerical solution for $f(x)$, as shown in Figure 6.1(b), is rather inaccurate. The second-order Tikhonov regularization with the regularization parameter suggested by the L-curve method $\lambda = 10$ yields the results shown in Figure 6.3. After analyzing this numerical solution, it can be clearly observed that accurate solutions

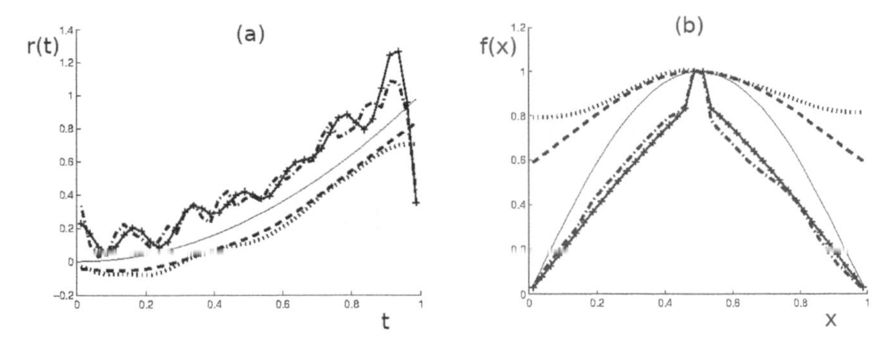

FIGURE 6.1: The analytical (——) and numerical results of (a) $r(t)$ and (b) $f(x)$, obtained using the TSVD ($-+-$) with $N_t = 14$, and the Tikhonov regularization of orders zero ($-\cdot-$), one ($\cdots$) and two ($---$), with discrepancy principle regularization parameters, for $p = 1\%$ noise [163].

for both r and f cannot be obtained by using $\lambda_1 = \lambda_2$. Therefore, the case $\lambda_1 \neq \lambda_2$ is considered and the L-surfaces are shown in Figure 6.2, which suggest $\lambda_1 = 10$, $\lambda_2 = 1$. However, the numerical results obtained using these values of the regularization were found still inaccurate. Alternatively, using trial and error to seek out the appropriated regularization parameters, it is found that regularization parameters $\lambda_1 = 8$ and $\lambda_2 = 5.2E{-}2$ yield an accurate and stable numerical solution, see Figure 6.3. Nevertheless, more research has to be undertaken in the future for the selection of appropriate multiple regularization parameters [78].

This section is closed by mentioning that an iterative process of regularization, whose stopping criterion based on the single choice of the iteration number does not involve the choice of two Tikhonov regularization parameters, was developed in [146]. Next, another inverse formulation concerned with the identification of an additively separable source is presented.

6.2 Additive Space- and Time-Dependent Heat Sources. Integral Observations

Let Ω be a bounded domain in $\mathbb{R}^d$ and T a given positive number. Denote by $\Omega_T := \Omega \times (0, T)$ and consider the problem of determining the temperature $u(x,t)$ and the heat source components $r(t)$ and $f(x)$ satisfying

$$u_t = \nabla^2 u + r(t)h_1(x) + f(x)h_2(t) + g(x,t), \quad (x,t) \in \Omega_T, \tag{6.33}$$

$$u(x,0) = u_0(x), \quad x \in \overline{\Omega}, \tag{6.34}$$

$$u = \mu \quad \text{on } \partial\Omega \times (0,T) \tag{6.35}$$

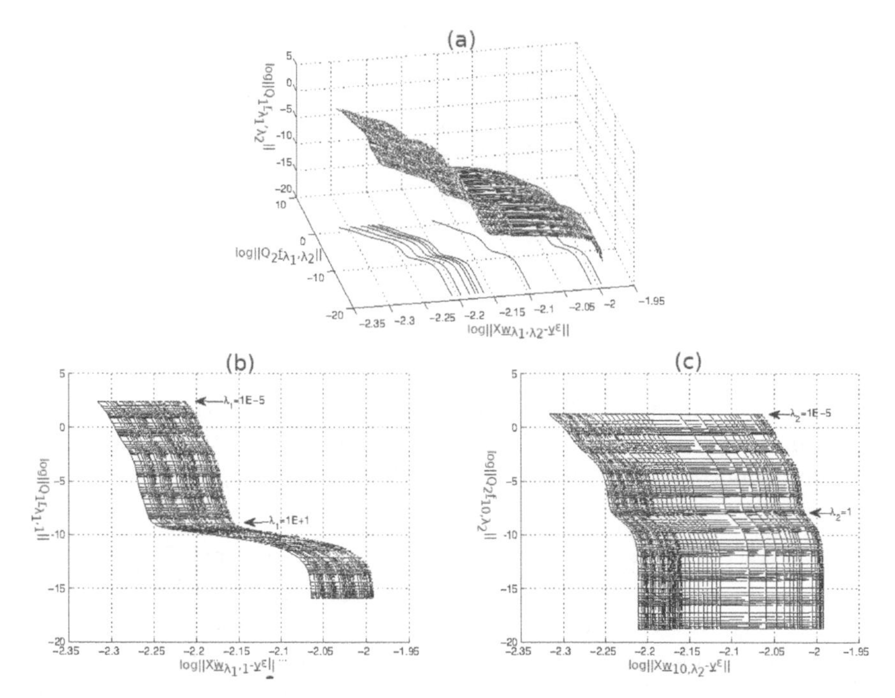

FIGURE 6.2: The L-surface on (a) a three-dimensional plot of $\log\|X\underline{w}_{\lambda_1,\lambda_2} - \underline{y}^{\epsilon}\|$ versus $\log\|Q_1\underline{r}_{\lambda_1,\lambda_2}\|$ and $\log\|Q_2\underline{f}_{\lambda_1,\lambda_2}\|$, (b) plane of $\log\|X\underline{w}_{\lambda_1,1} - \underline{y}^{\epsilon}\|$ versus $\log\|Q_1\underline{r}_{\lambda_1,1}\|$, and (c) plane of $\log\|X\underline{w}_{10,\lambda_2} - \underline{y}^{\epsilon}\|$ versus $\log\|Q_2\underline{f}_{10,\lambda_2}\|$, using the second-order Tikhonov regularization, for $p = 1\%$ noise [163].

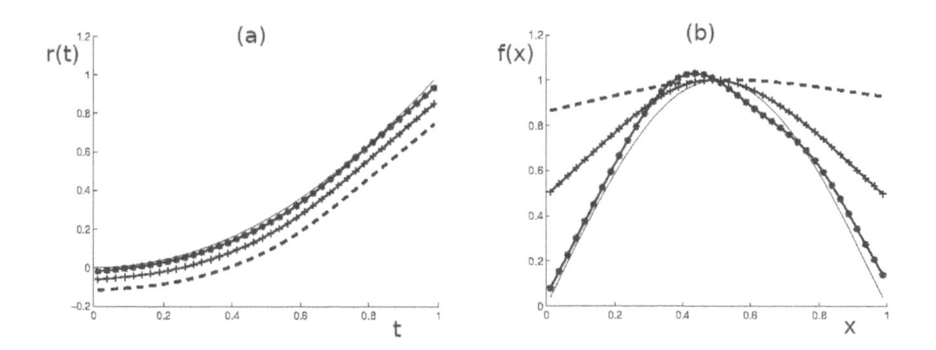

FIGURE 6.3: The analytical (—) and numerical results of (a) $r(t)$ and (b) $f(x)$, obtained using the second-order Tikhonov regularization with regularization parameters suggested by the L-curve criterion $\lambda = \lambda_1 = \lambda_2 = 10$ $(- - -)$, the L-surface method $(\lambda_1, \lambda_2) = (10, 1)$ $(- + -)$, and the trial and error $(\lambda_1, \lambda_2) = (8, 5.2E - 2)$ $(- * -)$, for $p = 1\%$ noise [163].

and some extra measurements. In the previous section, the measurement (6.4) at a fixed point $X_0 \in \Omega$ was considered. However, since this cannot be defined in the usual weak variational framework, it is replaced by the weighted mass/energy measurement

$$\int_{\Omega} \omega_1(x)u(x,t)dx = E(t), \quad t \in [0,T], \tag{6.36}$$

where ω_1 is a given function. Moreover, this is particularly advantageous in practical applications where local point measurement contains too large errors and then the use of non-local average measurements appears more realistic and reliable. Consider also the additional time-average temperature measurement

$$\int_0^T \omega_2(t)u(x,t)dt = U_T(x), \quad x \in \overline{\Omega}, \tag{6.37}$$

where ω_2 is a given function. The quantity

$$\int_{\Omega} \omega_1(x)f(x)dx = C_0. \tag{6.38}$$

is also assumed available. Then, the following local existence and uniqueness theorem holds, see also [146, 205].

Theorem 6.2. *Suppose that the following conditions are satisfied:*

(A1) $h_1 \in C(\overline{\Omega})$, $h_2 \in C[0,T]$, $u_0 \in H^{2+\varepsilon}(\overline{\Omega})$, $U_T \in H^{2+\varepsilon}(\overline{\Omega})$, $\mu \in H^{2+\varepsilon,1+\varepsilon/2}(\partial\Omega \times (0,T))$, $g \in H^{\varepsilon,\varepsilon/2}(\overline{\Omega}_T)$, $E \in H^{1+\varepsilon/2}[0,T]$, $\omega_1 \in H^{2+\varepsilon}(\Omega)$, $\omega_2 \in H^{1+\varepsilon/2}[0,T]$, $\partial\Omega \in H^{2+\varepsilon}$, where $\varepsilon \in (0,1)$;

(A2) $u_0|_{\partial\Omega} = \mu(\cdot,0)|_{\partial\Omega}$, $U_T|_{\partial\Omega} = \int_0^T \omega_2(t)\mu(\cdot,t)|_{\partial\Omega}dt$, $E(0) = \int_{\Omega}\omega_1(x)u_0(x)dx$, $\int_{\Omega}\omega_1(x)U_T(x)dx = \int_0^T \omega_2(t)E(t)dt$;

(A3) $\mathrm{A} := \int_0^T \omega_2(t)h_2(t)dt \neq 0$, $\mathrm{B} := \int_{\Omega}\omega_1(x)h_1(x)dx \neq 0$, $s(t) := \frac{h_2(t)}{\int_0^T \omega_2(\tau)h_2(\tau)d\tau} \geq 0$ for $t \in [0,T]$, and $\int_0^T s(t)dt \leq 1$.

Then, for sufficiently small $T > 0$ there exists a unique solution $(r,f,u) \in H^{\varepsilon/2}[0,T] \times H^{\varepsilon}(\overline{\Omega}) \times H^{2+\varepsilon,1+\varepsilon/2}(\overline{\Omega}_T)$ of the inverse problem (6.33)–(6.38).

Proof. Differentiating (6.36) yields

$$\begin{aligned} r(t) = \frac{1}{\mathrm{B}}\Bigg[E'(t) - \int_{\Omega}(\omega_1(x)g(x,t) + u(x,t)\nabla^2\omega_1(x))dx - C_0h_2(t) \\ + \int_{\partial\Omega}\left(\mu(x,t)\frac{\partial\omega_1}{\partial n}(x,t) - \omega_1(x)\frac{\partial u}{\partial n}(x,t)\right)dS\Bigg], \quad t \in [0,T]. \end{aligned} \tag{6.39}$$

Integrating (6.39) and using (6.37),

$$\int_0^T \omega_2(t)r(t)dt = \frac{1}{\mathrm{B}}\left[\int_0^T E'(t)\omega_2(t)dt - \int_Q \omega_1(x)\omega_2(t)g(x,t)dxdt \right.$$
$$\left. - \int_\Omega \omega_1(x)\nabla^2 U_T(x)dx - C_0\int_0^T \omega_2(t)h_2(t)dt\right] =: F_1. \quad (6.40)$$

Taking the Laplacian of (6.37),

$$\nabla^2 U_T(x) = \int_0^T \omega_2(t)\nabla^2 u(x,t)dt = \int_0^T \omega_2(t)u_t(x,t)dt - \int_0^T \omega_2(t)g(x,t)dt$$
$$-f(x)\int_0^T \omega_2(t)h_2(t)dt - h_1(x)\int_0^T \omega_2(t)r(t)dt.$$

Using (6.40) and rearranging give

$$f(x) = \frac{1}{\mathrm{A}}\left[\int_0^T \omega_2(t)u_t(x,t)dt - \int_0^T \omega_2(t)g(x,t)dx \right.$$
$$\left. -\nabla^2 U_T(x) - F_1 h_1(x)\right], \quad x \in \overline{\Omega}. \quad (6.41)$$

Let $\mathrm{U}_0 \in C^{2,1}(\Omega_T)\cap C^{1,0}(\overline{\Omega}_T)$ be the solution of the direct problem (6.33) with $r = f = 0$ subject to (6.34) and (6.35). Then, if G is the Green function of the Dirichlet problem for the heat equation (6.33) with $r = f = g = 0$, the solution $u(x,t)$ has the representation

$$u(x,t) = \mathrm{U}_0(x,t) + \int_0^t \int_\Omega \mathrm{G}(x,t;\xi,\tau)\left[f(\xi)h_2(\tau) + r(\tau)h_1(\xi)\right]d\xi d\tau,$$
$$(x,t) \in \overline{\Omega}_T. \quad (6.42)$$

Applying (6.42) at $t = T$ and substituting into (6.41) yield

$$f(x) = f_1(x) + \frac{\omega_2(T)}{\mathrm{A}}\int_0^T \int_\Omega \mathrm{G}(x,T;\xi,\tau)r(\tau)h_1(\xi)d\xi d\tau$$
$$-\frac{1}{\mathrm{A}}\int_0^T \omega_2'(t)\left(\int_0^t \int_\Omega \mathrm{G}(x,t;\xi,\tau)r(\tau)h_1(\xi)d\xi d\tau\right)dt$$
$$+\frac{1}{\mathrm{A}}\int_\Omega \left[\omega_2(T)K(x;\xi,T) + L(x;\xi)\right]f(\xi)d\xi, \quad (6.43)$$

where

$$f_1(x) := \frac{1}{\mathrm{A}}\left[\int_0^T \omega_2(t)\nabla^2 \mathrm{U}_0(x,t)dt - \Delta U_T(x) - F_1 h_1(x)\right],$$

$$K(x;\xi,t) := \int_0^t \mathrm{G}(x,t;\xi,\tau)h_2(\tau)d\tau, \quad L(x;\xi) := \int_0^T \omega_2'(t)K(x;\xi,t)dt.$$

Also, taking the gradient of (6.42) multiplied with the outward unit normal $\underline{n}$ and substituting into (6.39) yield

$$r(t) = \frac{1}{\mathrm{B}}\left\{E'(t) - C_0 h_2(t) - \int_\Omega \omega_1(x)g(x,t)dx + \int_{\partial\Omega} \mu(x,t)\frac{\partial \omega_1}{\partial n}(x)dS \right.$$
$$- \int_{\partial\Omega} \omega_1(x)\left[\frac{\partial \mathrm{U}_0}{\partial n}(x,t)\right.$$
$$\left. + \int_0^t \int_\Omega \frac{\partial \mathrm{G}}{\partial n}(x,t;\xi,\tau)\left[f(\xi)h_2(\tau) + r(\tau)h_1(\xi)\right] d\xi d\tau\right] dS$$
$$- \int_\Omega \nabla^2\omega_1(x)\left[\mathrm{U}_0(x,t)\right.$$
$$\left.\left. + \int_0^t \int_\Omega \mathrm{G}(x,t;\xi,\tau)\left[f(\xi)h_2(\tau) + r(\tau)h_1(\xi)\right] d\xi d\tau\right] dx\right\},$$

or

$$r(t) = r_1(t)$$
$$-\frac{1}{\mathrm{B}}\int_0^t \int_\Omega \left(\int_\Omega \omega_1(x)\nabla_x^2 \mathrm{G}(x,t;\xi,\tau)dx\right)\left[f(\xi)h_2(\tau) + r(\tau)h_1(\xi)\right] d\xi d\tau, \quad (6.44)$$

where

$$r_1(t) := \frac{1}{\mathrm{B}}\left[E'(t) - \int_\Omega \omega_1(x)g(x,t)dx + \int_{\partial\Omega} \mu(x,t)\frac{\partial \omega_1}{\partial n}(x)dS \right.$$
$$\left. -C_0 h_2(t) - \int_{\partial\Omega} \omega_1(x)\frac{\partial \mathrm{U}_0}{\partial n}(x,t)dS - \int_\Omega \nabla^2\omega_1(x)\mathrm{U}_0(x,t)dx\right]$$
$$= \frac{1}{\mathrm{B}}\left[E'(t) - C_0 h_2(t) - \int_\Omega \omega_1(x)\frac{\partial \mathrm{U}_0}{\partial t}(x,t)dx\right]. \quad (6.45)$$

Now the problem is equivalent to the coupled system formed with the Fredholm integral equation (6.43) and the Volterra integral equation (6.44), where $f_1(x)$ and $r_1(t)$ are known from the data of the problem. Consider first the homogeneous form of the Fredholm integral equation (6.43),

$$f_{\mathrm{h}}(x) = \frac{1}{\mathrm{A}}\int_\Omega M(x;\xi)f_{\mathrm{h}}(\xi)d\xi, \quad x \in \overline{\Omega}, \quad (6.46)$$

where $M(x;\xi) = \omega_2(T)K(x;\xi,T) + L(x;\xi)$. To show that this has only the trivial solution, as in [205] the inverse source problem consisting of finding the pair solution $(f(x), v(x,t))$ of the problem

$$\begin{cases} v_t = \nabla^2 v + f_{\mathrm{h}}(x)h_2(t), & (x,t) \in \Omega_T, \\ v(x,0) = 0, & x \in \overline{\Omega}, \\ v(x,t) = 0, & (x,t) \in \partial\Omega \times (0,T), \\ \int_0^T \omega_2(t)v(x,t)dt = 0, & x \in \overline{\Omega}. \end{cases} \quad (6.47)$$

is considered. As before, it can be obtained that f_{h} satisfies (6.46), and that

$$v(x,T) = \int_\Omega K(x;\xi,T) f_{\mathrm{h}}(\xi)d\xi, \quad x \in \overline{\Omega}.$$

Substituting these into the governing equation in (6.47) yields

$$v_t = \nabla^2 v + s(t)v(x,T),$$

which together with the zero intial and boundary conditions in (6.47), and the fact that from assumption (A_3) it follows that $s \geq 0$ with $\int_0^T s(t)dt \leq 1$, give that $v \equiv 0$ and $f_{\mathrm{h}} \equiv 0$, [205]. This also implies that the equation (6.46) has only the trivial solution. By the Fredholm alternative there exists a unique solution to (6.43), represented in terms of the resolvent $\Gamma(x;\xi)$ as:

$$f(x) = \int_\Omega \Gamma(x;y)\Bigg\{ f_1(y) + \frac{\omega_2(T)}{\mathrm{A}} \int_0^T \int_\Omega \mathrm{G}(y,T;\xi,\tau) r(\tau) h_1(\xi) d\xi d\tau - \frac{1}{\mathrm{A}} \int_0^T \omega_2'(t) \left(\int_0^t \int_\Omega \mathrm{G}(y,t;\xi,\tau) r(\tau) h_1(\xi) d\xi d\tau \right) dt \Bigg\} dy, \quad x \in \overline{\Omega}. \tag{6.48}$$

Introducing (6.48) into (6.44) gives the following integral equation:

$$\begin{aligned} r(t) = r_2(t) - &\frac{1}{\mathrm{B}} \int_0^t \int_\Omega \left(\int_\Omega \omega_1(x) \nabla_x^2 \mathrm{G}(x,t;\xi,\tau) dx \right) r(\tau) h_1(\xi) d\xi d\tau \\ &- \frac{1}{\mathrm{AB}} \int_\Omega \left[\int_0^t \left(\int_\Omega \omega_1(x) \nabla_x^2 \mathrm{G}(x,t;\xi,\tau) dx \right) h_2(\tau) d\tau \right] \\ &\times \Bigg[\int_\Omega \Gamma(\xi;y) \Bigg\{ \omega_2(T) \int_0^T \int_\Omega \mathrm{G}(y,T;\eta,\tau) h_1(\eta) r(\tau) d\eta d\tau \\ &- \int_0^T \omega_2'(t) \left(\int_0^t \int_\Omega \mathrm{G}(y,t;\eta,\tau) h_1(\eta) r(\tau) d\eta d\tau \right) dt \Bigg\} dy \Bigg] d\xi, \end{aligned} \tag{6.49}$$

with respect to $r(t)$, where

$$r_2(t) := r_1(t) - \frac{1}{\mathrm{B}} \int_\Omega \left(\int_0^t \left(\int_\Omega \omega_1(x) \nabla_x^2 \mathrm{G}(x,t;\xi,\tau) dx \right) h_2(\tau) d\tau \right) \left(\int_\Omega \Gamma(\xi;y) f_1(y) dy \right) d\xi.$$

The right-hand side of (6.49) is composed of a Volterra operator and a Fredholm operator. In summary, the equation (6.49) is of Fredholm type and, therefore, existence and uniqueness of solution hold if the norm of its kernel is sub-unitary. This is satisfied for $t \in [0, T_0]$, where $T_0 \in (0, T]$ is sufficiently small. Then, the system of integral equations (6.43) and (6.44) has a unique solution in $\overline{\Omega}_{T_0} = \overline{\Omega} \times [0, T_0]$.

6.3 Multiplicative Space- and Time-Dependent Source

In the previous sections the reconstruction of an additive space- and time-dependent source has been analyzed. In this section, a multiplicative source of the form $r(t)f(x)$, in which both $r(t)$ and $f(x)$ are unknown functions, is considered. In contrast to the previously investigated linear reconstruction of the additive source, this new inverse source problem formulation is more difficult to solve because the problem is now nonlinear. Moreover, its ill-posedness with respect to small errors in the input data being blown up in the output source solution adds even further difficulty.

Consider the following inverse initial-boundary value problem of finding the temperature $u(x,t)$ and the multiplicatively separable source function $F(x,t) := r(t)f(x)$, which satisfy the equation

$$u_t(x,t) = u_{xx}(x,t) + r(t)f(x), \quad (x,t) \in \Omega_T = (0,l) \times (0,T), \tag{6.50}$$

subject to the initial condition

$$u(x,0) = u_0(x), \quad x \in [0,l], \tag{6.51}$$

the homogeneous Neumann boundary conditions

$$u_x(0,t) = u_x(l,t) = 0, \quad t \in (0,T), \tag{6.52}$$

and the additional temperature measurement

$$u(X_0,t) = \chi(t), \quad t \in [0,T], \tag{6.53}$$

at a fixed sensor location $X_0 \in (0,l)$, and

$$u(x,T) = u_T(x), \quad x \in [0,l], \tag{6.54}$$

at the 'upper-base' final time $t = T$. Equation (6.52) expresses adiabatic boundary conditions at the ends $\{0,l\}$ of the finite slab $(0,l)$.

To avoid trivial non-uniqueness represented by the identity $r(t)f(x) = \frac{r(t)}{c} \cdot cf(x)$, with c arbitrary non-zero constant, a fixing condition, say

$$f(X_0) = S_0 \tag{6.55}$$

is imposed. The functions u_0, χ, u_T and the constant S_0 are given, whilst the functions $r(t)$, $f(x)$ and $u(x,t)$ are unknown. The specified interior and the final time temperature measurements (6.53) and (6.54) are time- and space-dependent functions $\chi(t)$ and $u_T(x)$ and they are meant to supply the missing information to identify the time- and space-dependent components $r(t)$ and $f(x)$, respectively. However, because of the nonlinearity represented by the product $r(t)f(x)$, the system of equations (6.50)–(6.55) cannot be decoupled to identify separately $r(t)$ and $f(x)$ from (6.53) and (6.54), respectively. It is further assumed that the conditions (6.51)–(6.54) are compatible, i.e.,

$$u_0'(0) = u_0'(l) = u_T'(0) = u_T'(l) = 0, \ \chi(0) = u_0(X_0), \ \chi(T) = u_T(X_0). \quad (6.56)$$

The unique solvability of the inverse problem (6.50)–(6.55), was established in [355]. With some slight corrections, this theorem reads as follows.

Theorem 6.3. ([164]) *Suppose that* $u_0(x) \in H^4(0,l)$, $u_T(x) \in H^4(0,l)$ *and* $\chi(t) \in H^2(0,T)$ *satisfy* (6.56) *and that* $S_0 \neq 0$. *Also, assume that:*
(i) $\mathrm{d} := \chi'(0) - u_0''(X_0) \neq 0$, $m := \frac{\chi'(T) - u_T''(X_0)}{\mathrm{d}} \neq 0$;
(ii) $u_0'''(0) = u_0'''(l) = u_T'''(0) = u_T'''(l) = 0$;
(iii) $\zeta_1 < 1$, $4\zeta_2\zeta_3 - (1-\zeta_1)^2 \leq 0$, $\zeta_4 < 1$,
where $\zeta_1 := \frac{2}{m^2\mathrm{d}^2} \max\left\{\mathrm{d}^2 + \frac{4l^2\|\chi''\|^2}{\pi^2}, 4l\|\theta\|^2 + lm^2\|u_0'''\|^2\right\}$,
$\zeta_2 := \frac{2}{\mathrm{d}^2} \max\left\{\frac{4l^6}{\pi^4 m^4}, 1\right\}$, $\zeta_3 := \frac{2\|\theta'\|^2}{m^2} + \frac{4\|\chi''\|^2}{\mathrm{d}^2}\left(\frac{2\|\theta\|^2}{m^2} + \|u_0'''\|^2\right)$,
$\zeta_4 := \frac{1}{m^2\mathrm{d}^2} \max\left\{\mathrm{d}^2 + 2l^3 z_0 + 4l^2\|\chi''\|^2, 4l^3 z_0 + 4l^3\|\theta'\|^2 + 2lm^2\|u_0'''\|^2\right\}$,
$\theta(x) = u_T'''(x) - m u_0'''(x)$, $\quad z_0 = \frac{1-\zeta_1}{2\zeta_2}$.
Then the inverse problem given by equations (6.50)–(6.55) *has a unique solution* $u(x,t) \in H^{4,2}(\Omega_T) \cap C(0,T;H^4(0,l)) \cap C(0,L;H^2(0,T))$, $r(t) \in H^1(0,T)$ *and* $f(x) \in H^2(0,l)$.

Remark 6.1. (i) In the above, the norms are in $L^2(0,T)$ or $L^2(0,l)$, and

$$H^{4,2}(\Omega_T) := \{u \in L^2(\Omega_T) | \partial_x^j u \in L^2(0,l) \text{ for } j = \overline{1,4}, \\ \text{and } \partial_t^i u \in L^2(0,T) \text{ for } i = 1,2\}.$$

(ii) In deriving the expressions for ζ_1, ζ_4 and z_0 above (which are different from those given in [355, 361]) use has been made of the Hölder and Wirtinger inequalities for the function $W := u_{tx}$, satisfying $W(0,t) = W(l,t) = 0$, i.e.,

$$|W_x(X_0,t)|^2 \leq l\|W_{xx}(\cdot,t)\|^2, \quad \forall t \in [0,T], \quad \|W(\cdot,T)\| \leq \frac{l}{\pi}\|W_x(\cdot,T)\|.$$

Although the inverse problem (6.50)–(6.55) has a unique solution it is still ill-posed because it violates the continuous dependence upon the input data (6.53) and (6.54). This can be seen from the following example of instability. Let $l = \pi$, $X_0 = \pi/2$ and for $n \in \mathbb{N}^*$ take

$$u_n(x,t) = \frac{[1 - \exp(-(2n+1)^2 t)]\cos((2n+1)x)}{n^{3/2}}, \quad (6.57)$$

which satisfies the heat equation (6.50) with the sources

$$r(t) = \frac{(2n+1)}{n^{3/4}}, \quad f(x) = \frac{(2n+1)\cos((2n+1)x)}{n^{3/4}}. \quad (6.58)$$

The initial condition (6.51) and the Neumann boundary conditions (6.52) are all homogeneous and the measured data (6.53)–(6.55) are given by

$$S_0 = \chi(t) = 0, \ u_T(x) = \frac{[1 - \exp(-(2n+1)^2 T)] \cos((2n+1)x)}{n^{3/2}} \to 0, \quad (6.59)$$

as $n \to \infty$. However, the sources $r(t)$ and $f(x)$ given by (6.58) become unbounded as $n \to \infty$, in any reasonable norm.

6.3.1 BEM for multiplicative source

The BEM is based on the following boundary integral equation:

$$\eta(x)u(x,t) = \int_0^t \left[G(x,t;\xi,\tau)\frac{\partial u}{\partial n}(\xi,\tau) - u(\xi,\tau)\frac{\partial G}{\partial n(\xi)}(x,t;\xi,\tau) \right]_{\xi\in\{0,l\}} d\tau$$
$$+ \int_0^l G(x,t;y,0)u(y,0)dy + \int_0^l \int_0^T G(x,t;y,\tau)r(\tau)f(y)d\tau dy,$$
$$(x,t) \in [0,l] \times (0,T), \quad (6.60)$$

Applying the homogeneous Neumann boundary condition (6.52) and using the constant BEM interpolations, as described in subsection 5.2.1.1, along with the piecewise constant approximations

$$r(t) = r(\tilde{t}_j) =: r_j, \quad t \in (t_{j-1}, t_j], \ j = \overline{1,N},$$
$$f(x) = f(\tilde{x}_k) =: f_k, \quad x \in (x_{k-1}, x_k], \ k = \overline{1,N_0},$$

equation (6.60) becomes [164], (see also (5.28)),

$$\eta(x)u(x,t) = \sum_{j=1}^{N} [-B_{0j}(x,t)u_{0j} - B_{lj}(x,t)u_{lj}]$$
$$+ \sum_{k=1}^{N_0} [C_k(x,t)u_{0,k} + D_k^r(x,t)f_k], \quad (6.61)$$

where the coefficients $B_{\xi j}$ for $\xi \in \{0,l\}$ and C_k are defined in (5.29) and

$$D_k^r(x,t) = \int_{x_{k-1}}^{x_k} \int_0^t G(x,t;y,\tau)r(\tau)d\tau dy = \sum_{j=1}^{N} d_{j,k}(x,t)r_j, \quad (6.62)$$

where $d_{j,k}(x,t) = \int_{x_{k-1}}^{x_k} \int_{t_{j-1}}^{t_j} G(x,t;y,\tau)d\tau dy$ are given by $d_{j,k}(x,t) = 0$ for $t \leq t_{j-1}$ and

$$d_{j,k}(x,t) = J(x,t,x_{k-1},t_{j-1}) - J(x,t,x_k,t_{j-1})$$
$$- \begin{cases} -\frac{(x-x_{k-1})^2}{4} + \frac{(x-x_k)^2}{4}, & t_{j-1} < t \leq t_j,\ x \leq x_{k-1}, \\ \frac{(x-x_{k-1})^2}{4} + \frac{(x-x_k)^2}{4}, & t_{j-1} < t \leq t_j,\ x_{k-1} < x \leq x_k, \\ \frac{(x-x_{k-1})^2}{4} - \frac{(x-x_k)^2}{4}, & t_{j-1} < t \leq t_j,\ x > x_k, \\ J(x,t,x_{k-1},t_j) - J(x,t,x_k,t_j), & t > t_j, \end{cases}$$

for $j = \overline{1,N}$, $k = \overline{1,N_0}$, where

$$J(x,t,x_k,t_j) = \left(\frac{(x-x_k)^2}{4} + \frac{t-t_j}{2}\right) \operatorname{erf}\left(\frac{x-x_k}{2\sqrt{t-t_j}}\right) + \frac{\sqrt{t-t_j}}{2\sqrt{\pi}}(x-x_k)\exp\left(-\frac{(x-x_k)^2}{4(t-t_j)}\right).$$

By applying (6.61) at the boundary element nodes $(0,\tilde{t}_i)$ and $(l,\tilde{t}_i)$ for $i = \overline{1,N}$, the system of $2N$ equations

$$-B\underline{u} + C\underline{u}^0 + D^r\underline{f} = \underline{0}, \tag{6.63}$$

is obtained, where B, C and $\underline{u}^0$ have been defined in (5.37) and

$$D^r = \begin{bmatrix} \sum_{j=1}^N d_{j,k}(0,\tilde{t}_i)r_j \\ \sum_{j=1}^N d_{j,k}(l,\tilde{t}_i)r_j \end{bmatrix}_{2N\times N_0}, \quad \underline{u} = \begin{bmatrix} u_{0j} \\ u_{Lj} \end{bmatrix}_N, \quad \underline{f} = \left[f_k\right]_{N_0}.$$

For the direct problem, one can find now the boundary temperatures $u(0,\tilde{t}_i) = u_{0i}$ and $u(l,\tilde{t}_i) = u_{li}$ from (6.63) as

$$\underline{u} = B^{-1}(C\underline{u}^0 + D^r\underline{f}). \tag{6.64}$$

Furthermore, the temperatures $u(X_0,\tilde{t}_i)$ for $i = \overline{1,N}$ and $u(\tilde{x}_k,T)$ for $k = \overline{1,N_0}$ are explicitly given from (6.61) as

$$\left[u(X_0,\tilde{t}_i)\right]_N = -B^I\underline{u} + C^I\underline{u}^0 + D^{rI}\underline{f} = \left[\chi(\tilde{t}_i)\right]_N, \tag{6.65}$$
$$\left[u(\tilde{x}_k,T)\right]_{N_0} = -B^{II}\underline{u} + C^{II}\underline{u}^0 + D^{rII}\underline{f} = \left[u_T(\tilde{x}_k)\right]_{N_0}, \tag{6.66}$$

where

$$B^I = \left[B_{0j}(X_0,\tilde{t}_i) \quad B_{lj}(X_0,\tilde{t}_i)\right]_{N\times 2N}, \quad C^I = \left[C_k(X_0,\tilde{t}_i)\right]_{N\times N_0},$$
$$D^{rI} = \left[\sum_{j=1}^N d_{j,k}(X_0,\tilde{t}_i)r_j\right]_{N\times N_0}, \quad B^{II} = \left[B_{0j}(\tilde{x}_k,T) \quad B_{lj}(\tilde{x}_k,T)\right]_{N_0\times 2N},$$
$$C^{II} = \left[C_k(\tilde{x}_k,T)\right]_{N_0\times N_0}, \quad D^{rII} = \left[\sum_{j=1}^N d_{j,k}(\tilde{x}_k,T)r_j\right]_{N_0\times N_0}.$$

6.3.2 Solution of the inverse problem

In this section, the unknown components $r(t)$ and $f(x)$ of the multiplicative source term in the inverse problem (6.50)–(6.55) are obtained simultaneously by using the BEM together with a classical minimization process which is commonly used when numerically solving inverse problems. The conditions (6.53)–(6.55) are imposed by minimizing the nonlinear least-squares function

$$\mathbb{F}_0(r, f) := \sum_{i=1}^{N} \left(u(X_0, \tilde{t}_i) - \chi(\tilde{t}_i)\right)^2$$
$$+ \sum_{k=1}^{N_0} \left(u(\tilde{x}_k, T) - u_T(\tilde{x}_k)\right)^2 + (f(X_0) - S_0)^2. \tag{6.67}$$

Here, the approximated temperatures $u(X_0, t)$ and $u(x, T)$, as introduced earlier in (6.14) and (6.16), respectively, are now employed into the above objective function with the initial guesses r_0 and f_0. Whereas $f(X_0)$ is approximated as $f(X_0) \approx (f(\tilde{x}_{N_k}) + f(\tilde{x}_{N_k+1}))/2$, where $1 \le N_k < N_0$ is the largest number for which $x_{N_k} \le X_0$. Then, introducing into (6.67) yield

$$\begin{aligned} \mathbb{F}_0(\underline{r}, \underline{f}) = \| &- B^I B^{-1}(C\underline{u}^0 + D^r \underline{f}) + C^I \underline{u}^0 + D^{rI} \underline{f} - \underline{\chi}\|^2 \\ +\| &- B^{II} B^{-1}(C\underline{u}^0 + D^r \underline{f}) + C^{II} \underline{u}^0 + D^{rII} \underline{f} - \underline{u}_T\|^2 \\ &+ (f(X_0) - S_0)^2. \end{aligned} \tag{6.68}$$

The minimization of (6.68) is performed using the *lsqnonlin* routine from the MATLAB Optimization Toolbox. This routine attempts to find the minimum of a sum of squares by starting from an arbitrary initial guesses $\underline{r}_0$, $\underline{f}_0$ for $\underline{r}$, $\underline{f}$, respectively, using the Trust-Region-Reflective (TRR) algorithm [87, 88, 201].

6.3.3 Numerical results and discussion

Consider a test example with $T = 1$, $l = 1/10$, $X_0 = 1/20$,

$$u_0(x) = u(x, 0) = 0, \quad x \in [0, l], \tag{6.69}$$

$$\chi(t) = u(1/20, t) = -(t - t^2)e^t, \quad u_T(x) = u(x, 1) = 0,$$
$$S_0 = f(1/20) = -40. \tag{6.70}$$

The analytical solution is given by

$$r(t) = -\frac{e^t}{40}\left(400\pi^2 t^2 - 400\pi^2 t + t^2 + t - 1\right), \quad f(x) = 40\cos(20\pi x), \tag{6.71}$$

$$u(x, t) = e^t(t - t^2)\cos(20\pi x), \quad (x, t) \in \overline{\Omega}_T. \tag{6.72}$$

Taking $l = 1/10$ sufficiently small such that $S_0 = -40 \neq 0$, $\mathrm{d} = -1 \neq 0$, $m = -e \neq 0$, $\theta(x) = u_T(x) = u_0(x) \equiv 0$, $\zeta_1 = 0.296 < 1$, $\zeta_2 = 2$, $\zeta_3 = 0$,

$4\zeta_2\zeta_3 - (1-\zeta_1)^2 = -0.495 \leq 0$, $z_0 = 0.175$ and $\zeta_4 = 0.261 < 1$, the conditions (i)–(iii) of Theorem 6.2 are satisfied guaranteeing the existence and uniqueness of the solution. The numerical solution is obtained, as described in Section 6.3.2, by minimizing the objective function (6.67). Preliminary numerical investigations showed that the initial guesses r_0 and f_0 cannot be so arbitrary for the minimization process to converge. Therefore, the initial guess

$$\underline{r}_0 = \underline{r} + random('Normal', 0, \sigma_r, N, 1),$$
$$\underline{f}_0 = \underline{f} + random('Normal', 0, \sigma_f, N_0, 1), \tag{6.73}$$

is chosen, where $\underline{r}$ and $\underline{f}$ are given by (6.71), $\sigma_r = p_0 \times \max_{t\in[0,T]} |r(t)|$, $\sigma_f = p_0 \times \max_{x\in[0,l]} |f(x)|$ where p_0 is a percentage of random deviation. The intention here was that the initial guesses (6.73) are understood as some general priors for the unknowns presumed to have been obtained experimentally from some simple but inaccurate practical measurement. Hereafter, unless otherwise specified, results obtained with $p_0 = 100\%$ perturbed initial guess (which is far from the exact solution (6.71)) and $N = N_0 = 20$ are presented.

When no regularization was imposed the numerical results for $r(t)$ and $f(x)$ have been obtained inaccurate and partially unstable. To improve stability, a regularization process is applied based on minimizing the objective function

$$\mathbb{F}_\lambda(\underline{r}, \underline{f}) := \mathbb{F}_0(\underline{r}, \underline{f}) + \lambda\left(\|Q_1\underline{r}\|^2 + \|Q_2\underline{f}\|^2\right), \tag{6.74}$$

where the regularization term is as in (6.27) with $\lambda = \lambda_1 = \lambda_2$. Initially, the first- and second-order regularizations based on minimizing the objective function (6.74) as

$$\mathbb{F}_\lambda(\underline{r}, \underline{f}) = \mathbb{F}_0(\underline{r}, \underline{f}) + \lambda\left(\sum_{i=1}^{N-1}(r_{i+1} - r_i)^2 + \sum_{k=1}^{N_0-1}(f_{k+1} - f_k)^2\right), \tag{6.75}$$

$$\mathbb{F}_\lambda(\underline{r}, \underline{f}) = \mathbb{F}_0(\underline{r}, \underline{f})$$
$$+\lambda\left(\sum_{i=2}^{N-1}(r_{i+1} - 2r_i + r_{i-1})^2 + \sum_{k=2}^{N_0-1}(f_{k+1} - 2f_k + f_{k-1})^2\right), \tag{6.76}$$

respectively, were applied. By trial and error, among various regularization parameters $\lambda \in \{10^{-9}, ..., 10^2\}$, those results obtained with $\lambda = 10^{-5}$ are shown in Figure 6.4. As it can be seen, applying orders one or two regularizations (6.75) or (6.76) yield stable, but rather inaccurate results, especially near the endpoints of the intervals of definition. To improve on these inaccuracies, a hybrid combination of first- and second- order regularizations given by

$$\mathbb{F}_\lambda(\underline{r}, \underline{f}) = \mathbb{F}_0(\underline{r}, \underline{f})$$
$$+\lambda\Bigg((r_2 - r_1)^2 + (r_N - r_{N-1})^2 + \sum_{i=2}^{N-1}(r_{i+1} - 2r_i + r_{i-1})^2$$
$$+(f_2 - f_1)^2 + (f_{N_0} - f_{N_0-1})^2 + \sum_{k=2}^{N_0-1}(f_{k+1} - 2f_k + f_{k-1})^2\Bigg) \tag{6.77}$$

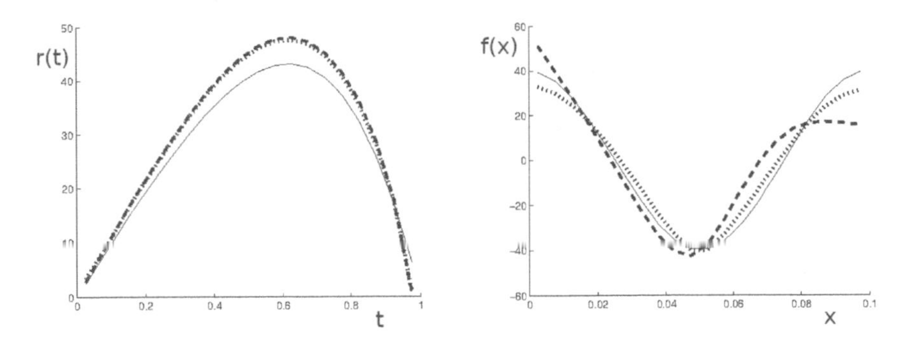

FIGURE 6.4: The numerical results for $r(t)$ and $f(x)$ obtained with the first-order regularization $(\cdots)$ and the second-order regularization $(- - -)$ with regularization parameter $\lambda = 10^{-5}$, for exact data. The corresponding analytical solutions are shown by continuous line (—), [164].

is applied. An appropriate regularization parameter λ which balances accuracy and stability needs to be chosen. Here, the L-curve method is applied by plotting the solution norm $\sqrt{\|Q_1\underline{r}\|^2 + \|Q_2\underline{f}\|^2}$ versus the residual norm $\sqrt{\mathbb{F}_0(\underline{r}, \underline{f})}$ for various values of λ. Although not illustrated, it is reported that the corner of the L-curve occurs nearby $\lambda = 10^{-5}$, with other appropriate values between the wide range 10^{-6} to 10^{-4}, [164]. With this value of the regularization parameter the numerical results are shown in Figure 6.5 and, compared to the previous Figure 6.4, very good agreement between the exact and the numerical solutions is now obtained. Results are summarized in terms of the RMSE (5.60) and (6.31) in Table 6.3 for various initial guesses (6.73) with $p_0 \in \{40, 60, 80, 100\}\%$ in (6.73). It can be seen that whilst the choice of the initial guess seems to matter for the accuracy of the unregularized solution, this restriction disappears when regularization with $\lambda = 10^{-5}$ is imposed. This shows that the regularization method employed is robust with respect to the independence on the initial guess.

TABLE 6.3: The RMSEs, for exact data.

p_0	λ	RMSE(r)	RMSE(f)
40%	0	6.34	18.4
	10^{-5}	1.52	0.81
60%	0	9.75	26.7
	10^{-5}	1.51	0.76
80%	0	25.8	44.9
	10^{-5}	1.52	0.81
100%	0	53.7	54.2
	10^{-5}	1.52	0.81

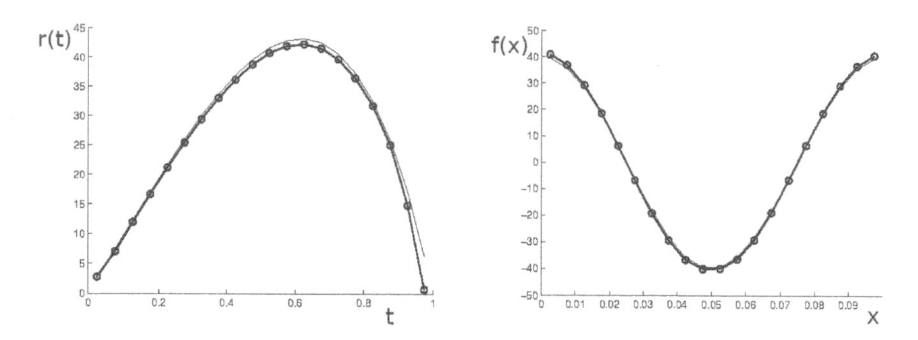

FIGURE 6.5: The numerical results $(-\circ-)$ for $r(t)$ and $f(x)$, obtained with the hybrid-order regularization (6.77) with the L-curve regularization parameter $\lambda = 10^{-5}$, for exact data. The corresponding analytical solutions are shown by continuous line (—), [164].

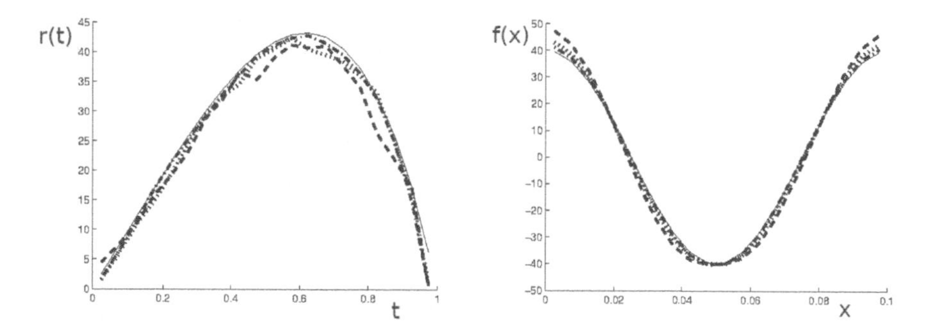

FIGURE 6.6: The numerical results for $r(t)$ and $f(x)$ obtained with the hybrid-order regularization (6.77) with regularization parameter $\lambda = 10^{-5}$ for $P \in \{1(-\cdot-), 3(\cdots), 5(---)\}\%$ noisy data. The corresponding analytical solutions are shown by continuous line (—), [164].

To test the stability of the nonlinear regularization, the inverse problem is solved when random noises $\underline{\epsilon}_{\chi}^{P} = random('Normal', 0, \sigma_{\chi}^{P}, N, 1)$ and $\underline{\epsilon}_{u_T}^{P} = random('Normal', 0, \sigma_{u_T}^{P}, N_0, 1)$ are added to the input functions $\chi(t)$ and $u_T(x)$, respectively, where P represents the percentage of noise, $\sigma_{\chi}^{P} = P \times \max_{t\in[0,T]} |\chi(t)|$ and $\sigma_{u_T}^{P} = P \times \max_{x\in[0,l]} |u_T(x)|$. Then, $\underline{\chi}$ and $\underline{u}_T$ in (6.68) are replaced by their noisy perturbations $\underline{\chi}^P = \underline{\chi} + \underline{\epsilon}_{\chi}^{P}$ and $\underline{u}_T^P = \underline{u}_T + \underline{\epsilon}_{u_T}^{P}$. The numerical results obtained with $\lambda = 10^{-5}$, (suggested by the L-curve), illustrated in Figure 6.6 show that stable and accurate numerical results are obtained for all $P \in \{1, 3, 5\}\%$ noises.

The identification of both multiplicative $r(t)f(x)$ and additive $\phi(t) + \Phi(x)$ components of sources of the form $r(t)f(x)+\phi(t)+\Phi(x)$ can also be considered [355], but its numerical implementation is yet to be realized.

Chapter 7

Inverse Wave Force Problems

The wave equation governs many physical problems such as the vibrations of a spring or membrane, acoustic scattering, etc. When it comes to mathematical modeling, probably the most investigated are the direct and inverse acoustic scattering problems, see e.g., [91]. On the other hand, inverse source/force problems for the hyperbolic wave equation have been less investigated. The forcing function is assumed to depend only upon the space variable(s) or the time variable in order to ensure uniqueness of solution.

7.1 Determination of a Space-Dependent Force in the One-Dimensional Wave Equation from Cauchy Data

The governing equation for a vibrating string of length $l > 0$ acted upon a space-dependent force $f(x)$ is given by the one-dimensional wave equation

$$u_{tt} = c^2 u_{xx} + f(x), \quad x \in (0, l) \times (0, \infty), \tag{7.1}$$

where u represents the displacement and $c > 0$ is the wave speed. Equation (7.1) has to be solved subject to the initial conditions

$$u(x, 0) = u_0(x), \quad x \in [0, l], \tag{7.2}$$

$$u_t(x, 0) = v_0(x), \quad x \in [0, l], \tag{7.3}$$

where u_0 and v_0 represent the initial displacement and velocity, respectively, and to the boundary conditions

$$u(0, t) = p_0(t), \quad t \in (0, \infty), \tag{7.4}$$

$$\mu u(l, t) + (1 - \mu) u_x(l, t) = p_l(t), \quad t \in (0, \infty), \tag{7.5}$$

where $\mu \in \{0, 1\}$, with $\mu = 1$ for the Dirichlet boundary condition and $\mu = 0$ for the Neumann boundary condition. Robin boundary conditions can also be considered. In (7.4) and (7.5), p_0 and p_l are given functions satisfying the compatibility conditions

$$p_0(0) = u_0(0), \quad p_l(0) = \mu u_0(l) + (1 - \mu) u_0'(l). \tag{7.6}$$

DOI: 10.1201/9780429400629-7

If the force f is given, then equations (7.1)–(7.6) form a direct well-posed problem for $u(x,t)$ which can be solved using the BEM, [32]. However, if the force function f is unknown, then the above equations are not sufficient to determine the pair solution $(u(x,t), f(x))$. Then, as suggested in [61], the above system of equations can be supplied with the measurement of the flux tension of the string at the end $x = 0$, namely,

$$-u_x(0,t) = q_0(t), \quad t \in [0,T], \tag{7.7}$$

where q_0 is a given function over a time of interest. Then, the inverse force problem requires determining the pair solution$(u(x,t), f(x))$ satisfying equations (7.1)–(7.7). In the above, f had to be restricted to depend on x only since otherwise, if f depends on both x and t, one can always add to $u(x,t)$ any function of the form $t^2x^2(x-l)^2U(x,t)$ with arbitrary $U \in C^{2,1}([0,l] \times [0,\infty))$ and still obtain another solution satisfying (7.1)–(7.7). Note that the unknown force $f(x)$ depends on the space variable x, whilst the measurement (7.7) of the flux tension $q_0(t)$ depends on the time variable t.

The problem (7.1)–(7.7) has at most one solution [61]. Moreover, the solution depends continuously on the input data if $f \in C^2[0,l]$ with $\text{supp}(f) \subset (0,l)$ and

$$\max_{x\in[0,l]}\{|f(x)|, |f'(x)|, |f''(x)|\} \leq \text{K}, \tag{7.8}$$

where K is a known positive constant. Due to the linearity of the inverse problem (7.1)–(7.7) it is convenient to split it into the form [61],

$$u = v + w, \tag{7.9}$$

where v satisfies the well-posed direct problem

$$v_{tt} = c^2 v_{xx}, \quad (x,t) \in (0,l) \times (0,\infty), \tag{7.10}$$

$$v(x,0) = u_0(x), \quad x \in [0,l], \tag{7.11}$$

$$v_t(x,0) = v_0(x), \quad x \in [0,l], \tag{7.12}$$

$$v(0,t) = p_0(t), \quad t \in (0,\infty), \tag{7.13}$$

$$\mu v(l,t) + (1-\mu)v_x(l,t) = p_l(t), \quad t \in (0,\infty). \tag{7.14}$$

and (w, f) satisfies the ill-posed inverse problem

$$w_{tt} = c^2 w_{xx} + f(x), \quad (x,t) \in (0,l) \times (0,\infty), \tag{7.15}$$

$$w(x,0) = w_t(x,0) = 0, \quad x \in [0,l], \tag{7.16}$$

$$w(0,t) = 0, \quad t \in (0,\infty), \tag{7.17}$$

$$\mu w(l,t) + (1-\mu)w_x(l,t) = 0, \quad t \in (0,\infty), \tag{7.18}$$

$$w_x(0,t) = -q_0(t) - v_x(0,t), \quad t \in [0,T]. \tag{7.19}$$

Observe that one could also control the Dirichlet data (7.4) instead of the Neumann data (7.7). The solution of the direct and well-posed problem (7.10)–(7.14) is obtained using the BEM, as described in the next section.

7.1.1 BEM for solving the direct problem

The BEM for the one-dimensional wave equation (7.10) is based on the application of integration by parts and the use of the fundamental solution,

$$u^*(x,t;\xi,\tau) = -\frac{1}{2c}\mathrm{H}(c(t-\tau) - |x-\xi|), \tag{7.20}$$

where H is the Heaviside function, to result in [32],

$$\begin{aligned} 2v(\xi,t) &= v(\xi - ct, 0) + v(\xi + ct, 0) \\ &+ \frac{1}{c}\int_{\xi-ct}^{\xi+ct} v_t(x,0)dx + v(l, t-(l-\xi)/c) \\ +c\int_0^{t-(l-\xi)/c} & v_x(l,\tau)d\tau + v(0, t-\xi/c) - c\int_0^{t-\xi/c} v_x(0,\tau)d\tau, \\ &(\xi,t) \in (0,l) \times (0,\infty). \end{aligned} \tag{7.21}$$

Equation (7.21) is valid if

$$v(0,0) = v(l,0) = 0, \text{ i.e., } u_0(0) = u_0(l) = 0, \tag{7.22}$$

otherwise, one can work with the modified function

$$\widetilde{v}(x,t) = v(x,t) - \frac{u_0(l) - u_0(0)}{l}x - u_0(0), \tag{7.23}$$

which satisfies $\widetilde{v}(0,0) = \widetilde{v}(l,0) = 0$. It is important to remark that in expression (7.21) the time and space coordinates must be within the domain $[0,l]\times[0,\infty)$ and the integrals must have their lower limit of integration smaller than the upper one. If any of these conditions are not satisfied the integrals are taken to be zero. Equation (7.21) yields the interior solution $v(\xi,t)$ for $(\xi,t) \in (0,l) \times (0,\infty)$ of (7.10) in terms of the initial and boundary data. In general, at a boundary point only one boundary condition is imposed and the first step of the BEM requires the evaluation of the unspecified boundary data. For this, first evaluate the boundary integral equation (7.21) at the end points $\xi \in \{0,l\}$ to yield [32],

$$\begin{aligned} v(0,t) &= v(ct,0) + \frac{1}{c}\int_0^{ct} v_t(x,0)dx + v(l,t-l/c) \\ &+c\left[\int_0^{t-l/c} v_x(l,\tau)d\tau - \int_0^t v_x(0,\tau)d\tau\right], \quad t\in(0,\infty), \end{aligned} \tag{7.24}$$

$$\begin{aligned} v(l,t) &= v(0,t-l/c) + v(l-ct,0) + \frac{1}{c}\int_{l-ct}^{l} v_t(x,0)dx \\ &+c\left[\int_0^t v_x(l,\tau)d\tau - \int_0^{t-l/c} v_x(0,\tau)d\tau\right], \quad t\in(0,\infty). \end{aligned} \tag{7.25}$$

These equations also hold under the assumption (7.22). Since $v_x(0,t)$ is of interest only for $t \in [0,T]$, let us restrict the boundary integral equations (7.24) and (7.25) to the time interval $[0,T]$. For the numerical discretization of (7.24) and (7.25), divide the time interval $[0,T]$ into a series of N uniform boundary elements $[t_{j-1},t_j]$ for $j = \overline{1,N}$, where $t_j = jT/N$ for $j = \overline{0,N}$. Similarly, divide the space interval $[0,l]$ into a series of M uniform cells $[x_{i-1},x_i]$ for $i = \overline{1,M}$, where $x_i = il/M$ for $i = \overline{0,M}$. Then, approximate the boundary and initial values as

$$v(0,\tau) = \sum_{j=1}^{N} \phi^j(\tau)v_j^0, \quad v(l,\tau) = \sum_{j=1}^{N} \phi^j(\tau)v_j^l, \quad \tau \in [0,T], \tag{7.26}$$

$$v_x(0,\tau) = \sum_{j=1}^{N} \theta^j(\tau)v_j'^0, \quad v_x(l,\tau) = \sum_{j=1}^{N} \theta^j(\tau)v_j'^l, \quad \tau \in [0,T], \tag{7.27}$$

$$v(x,0) = \sum_{i=1}^{M} \psi_i(x)v_0^i, \quad v_t(x,0) = \sum_{i=1}^{M} \psi_i(x)v_0^i, \quad x \in [0,l], \tag{7.28}$$

where

$$v_j^0 := v(0,t_j),\ v_j^l := v(l,t_j),\ v_j'^0 := v_x(0,t_j),\ v_j'^l := v_x(l,t_j),\ j = \overline{1,N}, \tag{7.29}$$

$$u_0^i := v(x_i,0),\ v_0^i := v_t(x_i,0),\ i = \overline{1,M}. \tag{7.30}$$

The functions ϕ^j, θ^j and ψ_i are interpolants, e.g. piecewise polynomial functions chosen such that $\phi^j(t_n) = \theta^j(t_n) = \delta_{jn}$ for $j,n = \overline{1,N}$, $\psi_i(x_m) = \delta_{im}$ for i, $m = \overline{1,M}$, where δ_{jn} is the Kronecker delta symbol. For example, if $\theta^j(\tau)$ is a piecewise constant function then, $\theta^j(\tau) = \chi_{(t_{j-1},t_j]}(\tau)$ is the characteristic function of $(t_{j-1},t_j]$. Then, $v_x(0,\tau) = v_j'^0$ for $\tau \in (t_{j-1},t_j]$, etc. one also has that $\int_0^{t_n} \theta^j(\tau)d\tau = t_j - t_{j-1}$ for $j = \overline{1,n}$. Using the approximations (7.26)–(7.30) into (7.24) and (7.25), one obtains, for $n = \overline{1,N}$, [184],

$$\begin{aligned} &v_n^0 + cv_n'^0 \int_0^{t_n} \theta^n(\tau)d\tau - \phi^n(t_n - l/c)v_n^l - cv_n'^l \int_0^{t_n - l/c} \theta^n(\tau)d\tau \\ &= \sum_{j=1}^{n-1} \phi^j(t_n - l/c)v_j^l + \sum_{i=1}^{M} \psi_i(ct_n)u_0^i + c\sum_{j=1}^{n-1} v_j'^l \int_0^{t_n - l/c} \theta^j(\tau)d\tau \\ &\quad - c\sum_{j=1}^{n-1} v_j'^0 \int_0^{t_n} \theta^j(\tau)d\tau + \frac{1}{c}\sum_{i=1}^{M} v_0^i \int_0^{ct_n} \psi_i(x)dx =: \mathrm{F} \end{aligned} \tag{7.31}$$

and

$$v_n^l - cv_n'^l \int_0^{t_n} \theta^n(\tau)d\tau - \phi^n(t_n - l/c)v_n^0 + cv_n'^0 \int_0^{t_n - l/c} \theta^n(\tau)d\tau$$
$$= \sum_{j=1}^{n-1} \phi^j(t_n - l/c)v_j^0 + \sum_{i=1}^{M} \psi_i(l - ct_n)u_0^i + c\sum_{j=1}^{n-1} v_j'^l \int_0^{t_n} \theta^j(\tau)d\tau$$
$$-c\sum_{j=1}^{n-1} v_j'^0 \int_0^{t_n - l/c} \theta^j(\tau)d\tau + \frac{1}{c}\sum_{i=1}^{M} v_0^i \int_{l-ct_n}^{l} \psi_i(x)dx =: \mathrm{G}. \quad (7.32)$$

Denoting

$$A = c\int_0^{t_n} \theta^n(\tau)d\tau, \quad B = \phi^n(t_n - l/c), \quad D = c\int_0^{t_n - l/c} \theta^n(\tau)d\tau, \quad (7.33)$$

equations (7.31) and (7.32) can be rewritten as

$$\begin{cases} v_n^0 + Av_n'^0 - Bv_n^l - Dv_n'^l = \mathrm{F}, \\ v_n^l - Av_n'^l - Bv_n^0 + Dv_n'^0 = \mathrm{G}. \end{cases} \quad (7.34)$$

At each time t_n for $n = \overline{1, N}$, the system of equations (7.34) is a time marching BEM in which the values of F and G are expressed in terms of the previous values of the solution at the times t_0,...,t_{n-1}. Note that upon the imposition of the initial conditions (7.11) and (7.12),

$$u_0^i = v(x_i, 0) = u_0(x_i), \quad v_0^i = v_t(x_i, 0) = v_0(x_i), \quad i = \overline{1, M} \quad (7.35)$$

are known. The system of equations (7.34) contains 2 equations with 4 unknowns. Two more equations are known from (7.13) and (7.14), as

$$v_n^0 = v(0, t_n) = p_0(t_n) =: p_0^n, \ n = \overline{1, N}, \quad (7.36)$$

$$\mu v_n^l + (1-\mu)v_n'^l = \mu v(l, t_n) + (1-\mu)v_x(l, t_n) = p_l(t_n) =: p_l^n, \quad n = \overline{1, N}. \quad (7.37)$$

The solution of the system of equations (7.34), (7.36) and (7.37) can be expressed explicitly at each time step t_n for $n = \overline{1, N}$, as follows:

(a) For $\mu = 1$, i.e. the Dirichlet problem (7.10)–(7.14) in which (7.14) is

$$v(l, t) = p_l(t), \quad t \in (0, \infty), \quad (7.38)$$

and equation (7.37) yields

$$v_n^l = p_l^n, \quad n = \overline{1, N}, \quad (7.39)$$

the unspecified boundary values are the Neumann flux values. Introduction of (7.36) and (7.39) into (7.34) yields the simplified system of two equations with two unknowns given by

$$\begin{cases} Av_n'^0 - Dv_n'^l = F - p_0^n + Bp_l^n =: \widetilde{\mathrm{F}}, \\ Dv_n'^0 - Av_n'^l = G - p_l^n + Bp_0^n =: \widetilde{\mathrm{G}}. \end{cases} \tag{7.40}$$

Application of Cramer's rule immediately yields the solution

$$v_n'^0 = \frac{D\widetilde{\mathrm{G}} - A\widetilde{\mathrm{F}}}{D^2 - A^2}, \quad v_n'^l = \frac{A\widetilde{\mathrm{G}} - D\widetilde{\mathrm{F}}}{D^2 - A^2}. \tag{7.41}$$

(b) For $\mu = 0$, i.e. the mixed problem (7.10)–(7.14) in which (7.14) is

$$v_x(L,t) = p_l(t), \quad t \in (0,\infty) \tag{7.42}$$

and equation (7.37) yields

$$v_n'^l = p_l^n, \quad n = \overline{1,N}, \tag{7.43}$$

the unspecified boundary values are the Neumann data at $x = 0$ and the Dirichlet data at $x = l$. Inserting (7.36) and (7.43) into (7.34) yields

$$\begin{cases} Av_n'^0 - Bv_n^l = F - p_0^n + Dp_l^n =: \widetilde{\widetilde{\mathrm{F}}}, \\ Dv_n'^0 + v_n^l = G + Bp_0^n + Ap_l^n =: \widetilde{\widetilde{\mathrm{G}}}. \end{cases} \tag{7.44}$$

This gives the solution

$$v_n'^0 = \frac{\widetilde{\widetilde{\mathrm{F}}} + B\widetilde{\widetilde{\mathrm{G}}}}{A + DB}, \quad v_n^l = \frac{A\widetilde{\widetilde{\mathrm{G}}} - D\widetilde{\widetilde{\mathrm{F}}}}{A + DB}. \tag{7.45}$$

Alternatively, instead of employing a time-marching BEM it is possible to employ a global BEM by assembling (7.31) and (7.32) as a full system of $2N$ linear equations with $4N$ unknowns v_j^0, v_j^l, $v_j'^0$, $v_j'^l$ for $j = \overline{1,N}$, namely,

$$\sum_{j=1}^{n} \left[cv_j'^0 \int_0^{t_n} \theta^j(\tau)d\tau - cv_j'^l \int_0^{t_n - l/c} \theta^j(\tau)d\tau - \phi^j(t_n - l/c)v_j^l \right]$$
$$= \sum_{i=1}^{M} \left[\psi_i(ct_n)u_0^i + \frac{1}{c}v_0^i \int_0^{ct_n} \psi_i(x)dx \right] - v_n^0, \quad n = \overline{1,N}, \tag{7.46}$$

and

$$\sum_{j=1}^{n} \left[cv_j'^l \int_0^{t_n} \theta^j(\tau)d\tau - cv_j'^0 \int_0^{t_n - l/c} \theta^j(\tau)d\tau + \phi^j(t_n - l/c)v_j^0 \right]$$
$$= -\sum_{i=1}^{M} \left[\psi_i(l - ct_n)u_0^i + \frac{1}{c}v_0^i \int_{l-ct_n}^{l} \psi_i(x)dx \right] + v_n^l, \quad n = \overline{1,N}. \tag{7.47}$$

The other $2N$ equations are given by (7.36) and (7.37). Introduction of (7.36) and (7.37) into (7.46) and (7.47) finally results in a linear system of $2N$ equations with $2N$ unknowns, which can be solved using a Gaussian elimination procedure. Once all the boundary values have been determined, the interior solution can be obtained explicitly using (7.21). This gives

$$\begin{aligned} 2v(\xi, t_n) = \sum_{j=1}^{n} \Big[\phi^j(t_n - (l-\xi)/c) v_j^l + \phi^j(t_n - \xi/c) v_j^0 \Big] \\ + c \sum_{j=1}^{n} \left[v_j'^l \int_0^{t_n-(l-\xi)/c} \theta^j(\tau) d\tau - v_j'^0 \int_0^{t_n-\xi/c} \theta^j(\tau) d\tau \right] \\ + \sum_{i=1}^{M} \Big[\psi_i(\xi - ct_n) + \psi_i(\xi + ct_n) \Big] u_0^i \\ + \frac{1}{c} \sum_{i=1}^{M} v_0^i \int_{\xi-ct_n}^{\xi+ct_n} \psi_i(x) dx, \quad n = \overline{1, N}, \ \xi \in (0,1). \end{aligned} \tag{7.48}$$

In (7.48), for the piecewise constant interpolations (7.29) and (7.30),

$$\int_0^{t_n-\xi/c} \theta^j(\tau) d\tau = \mathrm{H}(t_n - t_{j-1} - \xi/c)(t_j - t_{j-1}),$$

$$\int_0^{t_n-(l-\xi)/c} \theta^j(\tau) d\tau = \mathrm{H}(t_n - t_{j-1} - (l-\xi)/c)(t_j - t_{j-1}).$$

As experienced in [32], by comparing the BEM with the method of characteristics one observes that by choosing $\Delta t = T/N$ and $\Delta x = l/M$ such that the Courant number $c\Delta t/\Delta x = cTM/(Nl) = 1$ then, the grid is set to fit the characteristic net and the error involved is made arbitrarily small by reducing interpolation between points.

The flux $v_x(0,t)$ obtained numerically using the BEM is then introduced into (7.19) and the inverse problem (7.15)–(7.19) for the pair solution $(w(x,t), f(x))$ is solved using the method described next.

7.1.2 Method for solving the inverse problem

Based on the separation of variables, the approximate solution to the inverse force problem (7.15)-(7.19) is given by [61],

$$w_K(x, t; \underline{b}) = \frac{\sqrt{2}}{c^2} \sum_{k=1}^{K} \frac{b_k}{\lambda_k^2} (1 - \cos(c\lambda_k t)) \sin(\lambda_k x),$$
$$(x,t) \in [0,l] \times [0, \infty), \tag{7.49}$$

$$f_K(x) = \sqrt{2} \sum_{k=1}^{K} b_k \sin(\lambda_k x), \quad x \in (0,l), \tag{7.50}$$

where K is a truncation number and

$$\lambda_k = \begin{cases} \frac{k\pi}{l} & \text{if } \mu = 1, \\ \frac{(k-\frac{1}{2})\pi}{L} & \text{if } \mu = 0. \end{cases} \tag{7.51}$$

The coefficients $\underline{b} = (b_k)_{k=\overline{1,K}}$ are to be determined by imposing the additional boundary condition (7.19). This results in

$$-q_0(t) - v_x(0,t) =: g(t) = \frac{\partial w_K}{\partial x}(0, t; \underline{b})$$
$$= \frac{\sqrt{2}}{c^2} \sum_{k=1}^{K} \frac{b_k}{\lambda_k}(1 - \cos(c\lambda_k t)), \quad t \in [0, T]. \tag{7.52}$$

In practice, the additional observation (7.7) comes from measurement which is inherently contaminated with errors. This is modelled by replacing the exact data $q_0(t)$ with the noisy data

$$q_0^\epsilon(t_n) = q_0(t_n) + \epsilon, \quad n = \overline{1, N}, \tag{7.53}$$

where ϵ are N random noisy variables generated (using the Fortran NAG routine G05DDF) from a Gaussian normal distribution with mean zero and standard deviation $\sigma = p\% \times \max_{t\in[0,T]}|q_0(t)|$, where $p\%$ represents the percentage of noise. The noisy data (7.53) also induces noise in g as given by

$$g^\epsilon(t_n) = -q_0^\epsilon(t_n) - v_x(0, t_n) = g(t_n) - \epsilon, \quad n = \overline{1, N}. \tag{7.54}$$

Then, apply the condition (7.52) with g replaced by g^ϵ in a least-squares penalized sense by minimizing the Tikhonov functional

$$\mathcal{J}(\underline{b}) := \sum_{n=1}^{N} \left[\frac{\sqrt{2}}{c^2} \sum_{k=1}^{K} \frac{b_k}{\lambda_k}(1 - \cos(c\lambda_k t_n)) - g^\epsilon(t_n) \right]^2 + \lambda \sum_{k=1}^{K} b_k^2, \tag{7.55}$$

where $\lambda \geq 0$ is a regularization parameter to be prescribed. Denoting

$$\underline{g}^\epsilon = (g^\epsilon(t_n))_{n=\overline{1,N}}, \; Q_{nk} = \frac{\sqrt{2}(1 - \cos(c\lambda_k t_n))}{c^2 \lambda_k}, \; n = \overline{1, N}, \; k = \overline{1, K}, \tag{7.56}$$

expression (7.55) can be recast in a compact form as

$$\mathcal{J}(\underline{b}) = \| Q\underline{b} - \underline{g}^\epsilon \|^2 + \lambda \| \underline{b} \|^2. \tag{7.57}$$

In general, $N \geq K$ and the minimization of (7.57) yields the solution

$$\underline{b}_\lambda = (Q^{\mathrm{T}}Q + \lambda I)^{-1} Q^{\mathrm{T}} \underline{g}^\epsilon. \tag{7.58}$$

Once $\underline{b}$ has been found, the spacewise dependent force function is obtained from (7.50). Also, the displacement solution is obtained from (7.9) and (7.49).

TABLE 7.1: Condition number of the matrix Q given by equation (7.64).

K	$N = 20$	$N = 40$	$N = 80$
5	83	82	82
10	372	367	366
20	1420	1550	1540

7.1.3 Numerical example

In this section, numerical results obtained using the combined BEM and Tikhonov regularization are presented and discussed [184]. For simplicity, take $c = l = T = \mu = 1$ and in the BEM use constant time and space interpolation functions. Consider an analytical solution given by

$$u(x,t) = \sin(\pi x) + t + t^2/2, \quad (x,t) \in [0,1] \times [0,\infty), \tag{7.59}$$

$$f(x) = 1 + \pi^2 \sin(\pi x), \quad x \in [0,1]. \tag{7.60}$$

This generates the input data (7.2)–(7.5) and (7.7) given by

$$u(x,0) = u_0(x) = \sin(\pi x), \quad u_t(x,0) = v_0(x) = 1, \quad x \in [0,1], \tag{7.61}$$

$$u(0,t) = p_0(t) = t + t^2/2, \quad u(1,t) = p_l(t) = t + t^2/2, \quad t \in [0,\infty), \tag{7.62}$$

$$-u_x(0,t) = q_0(t) = -\pi, \quad t \in [0,1]. \tag{7.63}$$

Condition (7.22) is satisfied by the initial displacement $u_0(x)$ in (7.61) hence, there is no need to employ the modified function (7.23). The numerical solution for $v_x(0,t)$ at the points $(t_n)_{n=\overline{1,N}}$ obtained using the BEM by solving the direct problem (with known force given by (7.60)) is then input into (7.52) to determine the values for $g(t_n)$ and its noisy counterpart $g^\epsilon(t_n)$ given by (7.54) for $n = \overline{1,N}$, which are needed for solving the inverse problem (7.15)–(7.19). Since this problem is ill-posed, the matrix Q in (7.56) having the entries

$$Q_{nk} = \frac{\sqrt{2}(1 - \cos(\frac{k\pi n}{N}))}{k\pi}, \quad n = \overline{1,N}, \ k = \overline{1,K}, \tag{7.64}$$

will be ill-conditioned. The condition number defined as the ratio between the largest to the smallest singular values of the matrix Q is calculated in MATLAB using the command cond(Q) and is presented in Table 7.1 for various $N \in \{20, 40, 80\}$ and $K \in \{5, 10, 20\}$. The condition number is not affected by the increase in the number of measurements N, but it increases rapidly as the number K of basis functions increases.

Let us fix $N = 80$ and now proceed to solving the inverse problem (7.15)–(7.19), which, based on the BEM, has been reduced to solving the linear, but ill-conditioned system of equations

$$Q\underline{b} = \underline{g}^\epsilon. \tag{7.65}$$

Using the Tikhonov regularization one obtains a stable solution given explicitly by (7.58), provided that the regularization parameter λ is suitably chosen.

Exact data

First, consider the case of exact data, i.e. $p = 0$ and hence $\epsilon = 0$ in (7.53) and (7.54). Then $\underline{g}^{\epsilon} = \underline{g}$ and the system of equations (7.65) becomes

$$Q\underline{b} = \underline{g}. \tag{7.66}$$

Although there is no random noise added to the data q_0, there still exists some numerical noise in the data g in (7.52). This is given by the small discrepancy between the unavailable exact solution $v_x(0, t)$ of the direct problem and its numerical BEM solution. Figure 7.1 shows the retrieved coefficient vector $\underline{b} = (b_k)_{k=\overline{1,K}}$ for $K = 20$ obtained using no regularization, i.e. $\lambda = 0$, in which case, equation (7.58) produces the least-squares solution

$$\underline{b} = (Q^{\mathrm{T}}Q)^{-1}Q^{\mathrm{T}}\underline{g} \tag{7.67}$$

of the system of equations (7.66). Note that the analytical values for the sine Fourier series coefficients are given by

$$\begin{aligned} b_k &= \sqrt{2}\int_0^1 f(x)\sin(k\pi x)dx = \sqrt{2}\int_0^1 (1+\pi^2\sin(\pi x))\sin(k\pi x)dx \\ &= \begin{cases} \frac{2\sqrt{2}}{\pi} + \frac{\pi^2}{\sqrt{2}} \simeq 7.8791 & \text{if } k = 1, \\ 0 & \text{if } k = \text{even}, \\ \frac{2\sqrt{2}}{k\pi} & \text{if } k = \text{odd} \geq 3. \end{cases} \end{aligned} \tag{7.68}$$

By inspecting Figure 7.1 it appears that the leading term b_1 is the most significant in the series expansions (7.49) and (7.50), which, via (7.9), give the solutions for $f(x)$ and $u(x, t)$. The numerical solution for the force plotted in Figure 7.2 shows good agreement with the analytical solution (7.60).

Noisy data

To investigate the stability of the numerical solution, some ($p\% = 1\%$) noise is included into the input data (7.7), as given by (7.53). The numerical solutions for $f(x)$ obtained for various values of $K \in \{5, 10, 20\}$ and no regularization are plotted in Figure 7.3. As illustrated elsewhere [184], there was little difference between the results for $u(x, t)$ obtained with and without noise which also were in very good agreement with the exact solution (7.59). It also means that the numerical solution for the displacement $u(x, t)$ is stable with respect to noise added in the input data (7.7). In contrast, in Figure 7.3 the unregularized numerical solution for $f(x)$ manifests instabilities as K increases. For small K the numerically retrieved solution is quite stable showing that taking a small number of basis functions in the series expansion (7.50) has a regularizing effect. However, as K increase to 10 or 20 it can be clearly seen that oscillations start to appear. Eventually, these oscillations will become

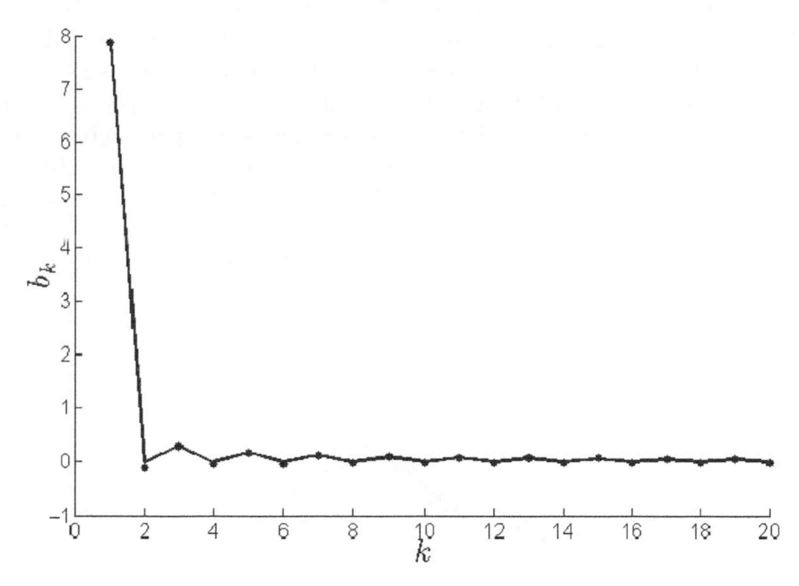

FIGURE 7.1: The numerical solution ($\cdots$) for $(b_k)_{k=\overline{1,K}}$ for $K = 20$, $N = 80$, obtained with no regularization, i.e. $\lambda = 0$, for exact data, in comparison with the exact solution (7.68) (—), [184].

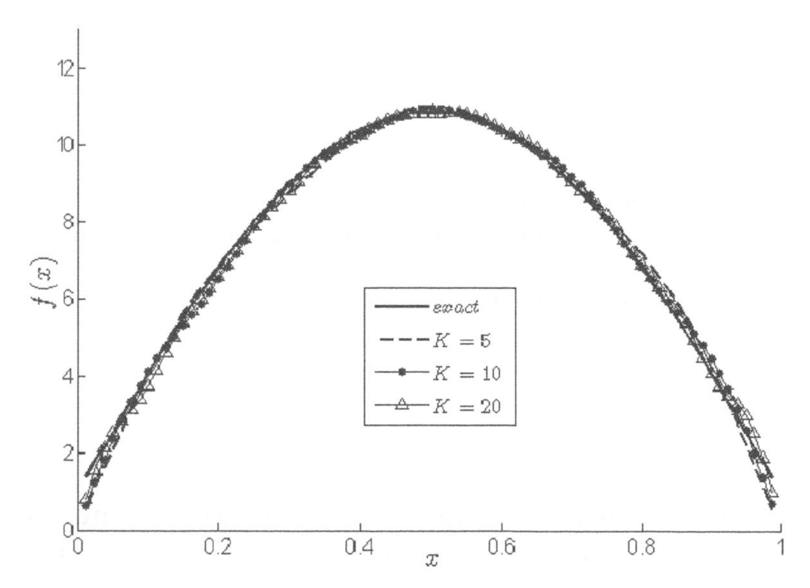

FIGURE 7.2: The exact solution (7.60) in comparison with the numerical solution (7.50) for $K \in \{5, 10, 20\}$, no regularization, for exact data [184].

highly unbounded, as K increases even further. To deal with this instability, the Tikhonov regularization is employed. Therefore, fix $K = 20$ and aim to alleviate the instability of the numerical solution for $f(x)$ shown by (—△—) in Figure 7.3 obtained with no regularization, i.e. $\lambda = 0$, for $p\% = 1\%$ noisy data. Including regularization with various regularization parameters, the numerical results shown in Figure 7.4 indicate that the value of $\lambda = 10^0$ is too large and the solution is oversmooth, whilst the value of $\lambda = 10^{-2}$ is too small and the solution is unstable. One could therefore select the value of $\lambda = 10^{-1}$ and conclude that this is a reasonable choice of the regularization parameter which balances the smoothness with the instability of solution.

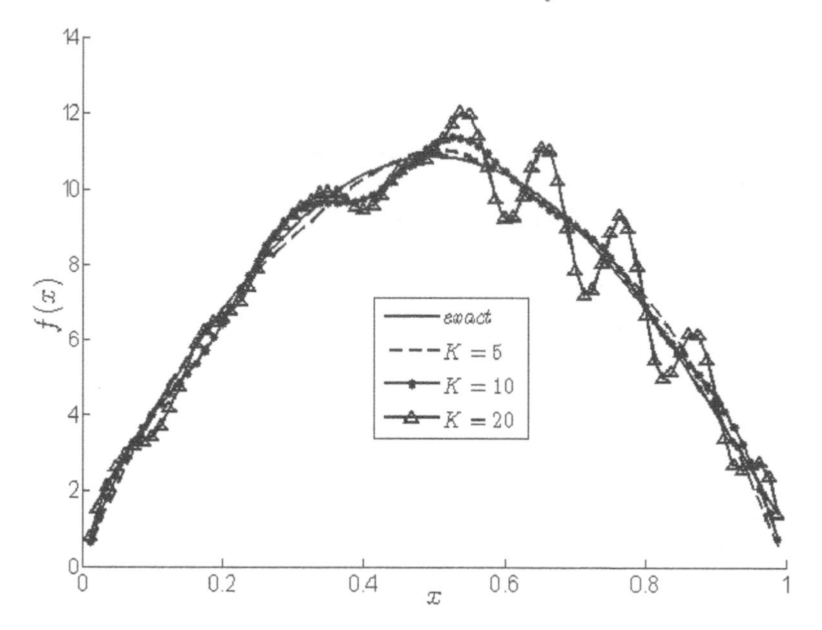

FIGURE 7.3: The exact solution (7.60) in comparison with the numerical solution (7.50) for $K \in \{5, 10, 20\}$, no regularization, for $p\% = 1\%$ noise [184].

The next section, formulates the inverse force identification problem in multiple-dimensions.

7.2 Determination of the Force Function in the Multi-Dimensional Wave Equation from Cauchy Data

The governing equation for a vibrating bounded structure $\Omega \subset \mathbb{R}^d$, acted upon by a force $F(\underline{x}, t)$ is given by the wave equation

$$u_{tt}(\underline{x}, t) = \nabla^2 u(\underline{x}, t) + F(\underline{x}, t), \quad (\underline{x}, t) \in \Omega \times (0, T) =: \Omega_T, \qquad (7.69)$$

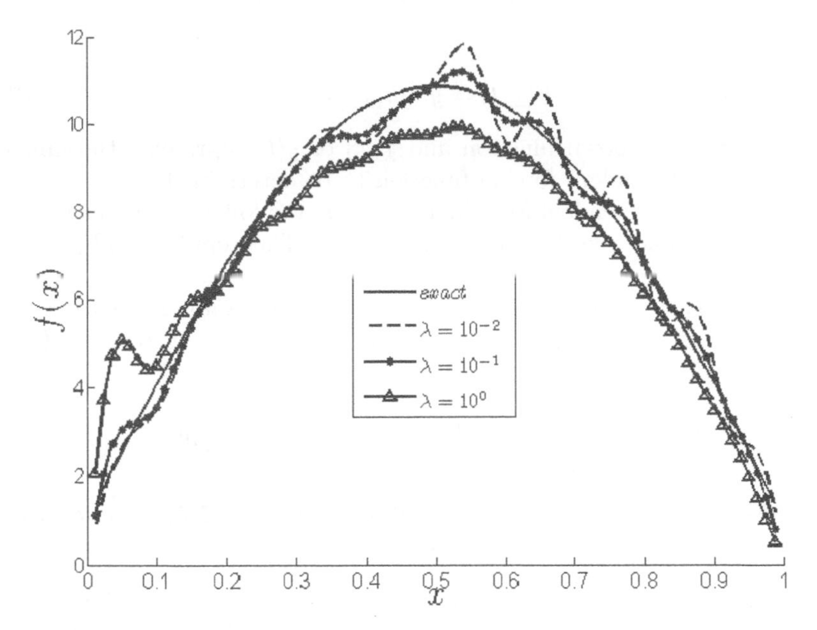

FIGURE 7.4: The exact solution (7.60) in comparison with the numerical solution (7.50), for $K = 20$, $p\% = 1\%$ noise, and regularization parameters $\lambda \in \{10^{-2}, 10^{-1}, 10^{0}\}$, [184].

where $u(\underline{x}, t)$ represents the displacement and, for simplicity, the wave speed has been taken equal to unity. Equation (7.69) has to be solved subject to the initial conditions

$$u(\underline{x}, 0) = u_0(\underline{x}), \quad u_t(\underline{x}, 0) = v_0(\underline{x}), \quad \underline{x} \in \Omega, \tag{7.70}$$

and the Dirichlet boundary condition

$$u(\underline{x}, t) = P(\underline{x}, t), \quad (\underline{x}, t) \in \partial\Omega \times (0, T). \tag{7.71}$$

If the force $F(\underline{x}, t)$ is given, then equations (7.69)–(7.71) form a direct well-posed problem. However, if the force function $F(\underline{x}, t)$ is unknown, then consider the additional measurement given by the flux tension

$$\frac{\partial u}{\partial n}(\underline{x}, t) = q(\underline{x}, t), \quad (x, t) \in \Gamma \times (0, T), \tag{7.72}$$

where $\Gamma \subset \partial\Omega$, or by the mass integral

$$\int_{\Omega} \omega(\underline{x}) u(\underline{x}, t) d\Omega = \psi(t), \quad t \in [0, T], \tag{7.73}$$

where ω is a given weight function. To ensure a unique solution it is further assumed that

$$F(\underline{x}, t) = f(\underline{x}) h(\underline{x}, t), \tag{7.74}$$

or

$$F(\underline{x}, t) = g(t)h(\underline{x}, t), \tag{7.75}$$

where $h(\underline{x}, t)$ is a known function and $f(\underline{x})$ or $g(t)$ represents the unknown space- or time-dependent forcing function to be determined.

For the space-dependent force identification, the following uniqueness Theorems 7.1–7.3 hold, see [114, Theorem 9], [232, Theorem 3.8] and [395].

Theorem 7.1. *Assume that $\Omega \subset \mathbb{R}^d$ is a bounded star-shaped domain with sufficiently smooth boundary such that $T > diam(\Omega)$. Let $h \in H^2(0, T; L^\infty(\Omega))$ be such that $h(.,0) \in L^\infty(\Omega)$, $h_t(.,0) \in L^\infty(\Omega)$ and*

$$H := \frac{||h_{tt}||_{L^2(0,T;L^\infty(\Omega))}}{inf_{\underline{x}\in\Omega}|h(\underline{x}, 0)|} \quad \text{is sufficiently small.} \tag{7.76}$$

If $\Gamma = \partial\Omega$, then the inverse problem (7.69)–(7.72) and (7.74) has at most one solution $(u(\underline{x}, t), f(\underline{x}))$ in the class of functions

$$u \in L^2(0, T; H^1(\Omega)),\ u_t \in L^2(0, T; L^2(\Omega)),\ u_{tt} \in L^2(0, T; (H^1(\Omega))'),$$
$$f \in L^2(\Omega). \tag{7.77}$$

Theorem 7.2. *Assume that $\Omega \subset \mathbb{R}^d$ is a bounded domain with piecewise smooth boundary. Let $h \in C^3(\overline{\Omega} \times [0, T])$ be such that $h(\underline{x}, 0) \neq 0$ for $\underline{x} \in \overline{\Omega}$. Then, if $\Gamma = \partial\Omega$, the inverse problem (7.69)–(7.72) and (7.74) has at most one solution $(u(\underline{x}, t), f(\underline{x})) \in C^3(\overline{\Omega} \times [0, T]) \times C(\overline{\Omega})$.*

Theorem 7.3. *Assume that $\Omega \subset \mathbb{R}^d$ is a bounded star-shaped domain with smooth boundary such that $T > diam(\Omega)$. Let $h \in C^1[0, T]$ be independent of $\underline{x}$ in (7.74) such that equation (7.69) becomes*

$$u_{tt}(\underline{x}, t) = \nabla^2 u(\underline{x}, t) + f(\underline{x})h(t), \quad (\underline{x}, t) \in \Omega \times (0, T), \tag{7.78}$$

and assume further that $h(0) \neq 0$. Then the inverse problem (7.70)–(7.72) and (7.78) has at most one solution in the class of functions

$$u \in C^1([0, T]; H^1(\Omega)) \cap C^2([0, T]; L^2(\Omega)), \quad f \in L^2(\Omega). \tag{7.79}$$

For the time-dependent force identification it is assumed, without loss of generality, that the Dirichlet boundary data in (7.71) is homogeneous, i.e. $P \equiv 0$, and that the input data satisfy the regularity conditions

$$u_0 \in \overset{\circ}{W}{}_2^1(\Omega),\ v_0 \in L_2(\Omega),\ \psi \in C^2[0, T],\ h \in C([0, T]; L_2(\Omega)),$$
$$\omega \in \overset{\circ}{W}{}_2^1(\Omega), \tag{7.80}$$

the compatibility conditions

$$\int_\Omega \omega(\underline{x})u_0(\underline{x})d\Omega = \psi(0), \quad \int_\Omega \omega(\underline{x})v_0(\underline{x})d\Omega = \psi'(0), \tag{7.81}$$

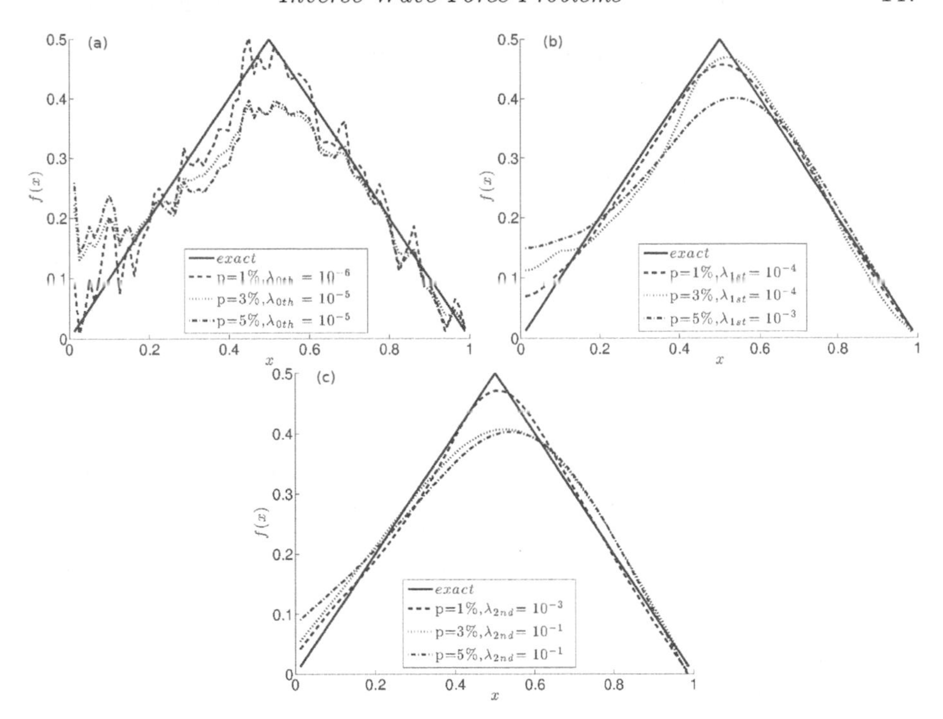

FIGURE 7.5: The exact solution (7.86) for the force $f(x)$ in comparison with the Tikhonov regularized solution (7.87) of various orders (a) zeroth-order, (b) first-order and (c) second-order, for $p \in \{1, 3, 5\}\%$ noise [185].

and the identifiably condition

$$\int_{\Omega} h(\underline{x}, t)\omega(\underline{x})d\Omega \neq 0, \quad t \in [0, T]. \tag{7.82}$$

Then, from the Corollary 9.2.1 of [336], the following unique solvability theorem of the inverse problem (7.69)–(7.71), (7.73) and (7.75), can be obtained.

Theorem 7.4. *Let* $P \equiv 0$ *in* (7.71). *If the input data satisfy conditions* (7.80)–(7.82), *then there exists a unique solution* $(u(\underline{x}, t), g(t))$ *of the inverse problem* (7.69)–(7.71), (7.73) *and* (7.75) *in the class of functions*

$$u \in C([0, T]; H^1(\Omega)) \cap C^1([0, T]; L_2(\Omega)), \quad g \in C[0, T]. \tag{7.83}$$

7.2.1 Numerical results

In this section, numerical results are presented only for the inverse space-dependent force problem (7.70)–(7.72) and (7.78). For the inverse time-dependent force problem (7.69)–(7.71), (7.73) and (7.75) numerical results can be found in [186].

Take the one-dimensional case, $\Omega = (0, l)$ with $l = 1$, $T = 1$, $h(x, t) = 1+t$, and consider first the direct problem (7.70), (7.71) and (7.78) with the input data

$$u(x,0) = u_0(x) = 0, \quad u_t(x,0) = v_0(x) = 0, \quad x \in [0,1], \tag{7.84}$$

$$u(0,t) = p_0(t) = 0, \quad u(1,t) = p_l(t) = 0, \quad t \in (0,1], \tag{7.85}$$

$$f(x) = \begin{cases} x & \text{if } x \in [0, 1/2], \\ 1 - x & \text{if } x \in (1/2, 1]. \end{cases} \tag{7.86}$$

The inverse problem is given by (7.78) with $h(t) = 1 + t$, (7.84), (7.85) and (7.72) with $q_0(t) = -u_x(0, t)$ numerically simulated by solving the direct problem (7.78), (7.84)–(7.86) using the finite-difference method (FDM) with a uniform space-time mesh of $N = M = 80$ discretizing the intervals $[0, l = 1]$ and $[0, T = 1]$. Further, this flux is perturbed by $p \in \{1, 3, 5\}\%$ additive random noise with mean zero and standard deviation equal to $p \times \max_{t\in[0,1]}|q_0(t)|$. The numerical FDM solution for $f(x)$ obtained with no regularization has been found highly oscillatory and unstable. To deal with this instability the Tikhonov regularization of various orders is employed, giving the solution [329],

$$\underline{f}_\lambda = (A^{\mathrm{T}}A + \lambda R_k^{\mathrm{T}} R_k)^{-1} A^{\mathrm{T}} \underline{b}^\epsilon, \tag{7.87}$$

where A represents the matrix resulting from the FDM discretization of the inverse problem and $\underline{b}^\epsilon$ represents the right-hand side vector which is contaminated with noise. Figure 7.5 shows the regularized numerical solutions (7.87) for various orders of regularization $k \in \{0, 1, 2\}$ in comparison with the exact solution (7.86). From this figure it can be seen that the numerical results are stable and they become more accurate as the amount of noise p decreases. Also, the first- and second-order regularization methods produce smoother and more accurate numerical solutions than the zeroth-order regularization.

7.3 Determination of a Space-Dependent Force Function from Final or Time-Averaged Displacement Data

In the previous sections, the reconstruction of a space-dependent force in the wave equation from Cauchy data measurements of both displacement and its normal derivative on the boundary has been attempted, see also [61, 184, 185]. This inverse formulation is, as expected, ill-posed because the unknown output force $f(\underline{x})$ depends on $\underline{x}$ in the domain Ω, whilst the known input data, say u

and $\partial_n u$, depend on $(\underline{x}, t)$ on the boundary $\partial\Omega \times (0, T)$. Although the uniqueness of solution holds [114, 232, 395], it seems more natural to measure instead the displacement $u(\underline{x}, t)$ for $\underline{x} \in \Omega$ and time $t = T$, or the time-averaged displacement $\int_0^T u(\underline{x}, t)dt$ for $\underline{x} \in \Omega$. This way, the output-input mapping satisfies the meta-theorem that the overposed data and the unknown force function lie in the same direction [330]. This spacewise-dependent force $f(\underline{x})$ identification from the upper-base spacewise dependent displacement measurement $u(\underline{x}, T)$ was investigated theoretically in Section 8.2 of [336], where the uniqueness of solution was proved. For other wave-related force identification studies, which use the final time displacement data, refer to [154], which employs a weak solution approach for a quite general inverse problem with a non-unique solution, and to [363] which nicely introduces a quasi nonlinearity in the governing wave equation to resolve the non-uniqueness of solution. The other inverse problem generated by the measurement of the time-averaged displacement $\int_0^T u(\underline{x}, t)dt$ was investigated in [274], and is also presented below.

In summary, the inverse force problem formulated in this section requires finding the displacement $u(\underline{x}, t)$ and the force $f(\underline{x})$ in the wave equation

$$u_{tt} - \mathcal{L}u = f(\underline{x})h(\underline{x}, t) =: F(\underline{x}, t) \quad \text{in} \quad \Omega_T, \tag{7.88}$$

where h is a given function, and, in general, for a homogeneous medium $\mathcal{L} = \nabla^2$ is the Laplacian operator. For inhomogeneous media, $\mathcal{L}u = c^2(\underline{x})\nabla^2 u$ or $\nabla \cdot (\mathbb{K}(\underline{x})\nabla u)$, where $\mathbb{K}$ and c are given positive material properties [84]. Equation (7.88) has to be solved subject to prescribed initial conditions

$$u(\underline{x}, 0) = u_0(\underline{x}) \quad u_t(\underline{x}, 0) = v_0(\underline{x}) \quad \underline{x} \in \Omega, \tag{7.89}$$

homogenous Dirichlet boundary conditions,

$$u(\underline{x}, t) = 0, \quad (\underline{x}, t) \in \partial\Omega \times (0, T), \tag{7.90}$$

and the additional final displacement measurement

$$u(\underline{x}, T) = u_T(\underline{x}), \quad \underline{x} \in \Omega, \tag{7.91}$$

or, the time-average displacement observation

$$\int_0^T \omega(t)u(\underline{x}, t)dt = U_T(\underline{x}), \quad \underline{x} \in \Omega, \tag{7.92}$$

where ω is a given function. The additional measurement $u_t(\underline{x}, T) = v_T(\underline{x})$ for $\underline{x} \in \Omega$, [154], does not provide sufficient extra information to the displacement specification (7.91) to ensure the uniqueness of solution.

7.3.1 Direct and adjoint problems

Provided that $u_0 \in H_0^1(\Omega)$, $v_0 \in L^2(\Omega)$ and $F \in L^2(\Omega_T)$, the well-posedness of the direct problem (7.88)–(7.90) when f is given, holds in the space of weak solutions $u \in L^2(0,T;H_0^1(\Omega))$ with $u_t \in L^2(0,T;L^2(\Omega))$ and $u_{tt} \in L^2(0,T;H^{-1}(\Omega))$, [274]. This implies $u \in C([0,T];L^2(\Omega))$ and $u_t \in C([0,T];H^{-1}(\Omega))$, which, in particular, yield that the restrictions at $t=0$ of the solution and its derivative make sense. Moreover, the estimate

$$\max_{t\in[0,T]}\left(||u(.,t)||_{H_0^1(\Omega)} + ||u_t(.,t)||_{L^2(\Omega)}\right) + ||u_{tt}||_{L^2(0,T;H^{-1}(\Omega))}$$
$$\leq C\left(||F||_{L^2(0,T;L^2(\Omega))} + ||u_0||_{H_0^1(\Omega)} + ||v_0||_{L^2(\Omega)}\right), \quad (7.93)$$

holds, for some positive constant C. Similar considerations can be done for the formally adjoint backward hyperbolic problem to (7.88)–(7.90),

$$v_{tt} - \mathcal{L}^* v = G(\underline{x},t), \quad (\underline{x},t) \in \Omega_T, \quad (7.94)$$
$$v(\underline{x},T) = \zeta(\underline{x}), \quad v_t(\underline{x},T) = \xi(\underline{x}), \quad \underline{x} \in \Omega, \quad (7.95)$$
$$v(\underline{x},t) = 0, \quad (\underline{x},t) \in \partial\Omega \times (0,T), \quad (7.96)$$

where $\mathcal{L}^*$ is the adjoint of $\mathcal{L}$. Furthermore, using integration by parts, the following Green's formula holds:

$$\int_\Omega [u_t(\underline{x},T)\zeta(\underline{x}) - v_0(\underline{x})v(\underline{x},0)]\, d\underline{x}$$
$$- \int_\Omega [u(\underline{x},T)\xi(\underline{x}) - u_0(\underline{x})v_t(\underline{x},0)]\, d\underline{x}$$
$$= \int_0^T \int_\Omega F(\underline{x},t)v(\underline{x},t)d\underline{x}dt - \int_0^T \int_\Omega G(\underline{x},t)u(\underline{x},t)d\underline{x}dt. \quad (7.97)$$

It is possible to formulate the direct problem (7.88)–(7.90) in a more general abstract setting, as described in Chapter 8 of [336], by considering the problem

$$u'' = \mathcal{L}u(t) + F(t), \quad t \in [0,T], \quad (7.98)$$

$$u(0) = u_0, \quad u'(0) = v_0, \quad (7.99)$$

where $\mathcal{L}$ is a closed linear operator with a dense domain of definition $D(\mathcal{L})$ on a Hilbert space $\mathcal{X}$. Assume that $\mathcal{L}$ generates a strongly continuous cosine function $C(t) = \cos(\sqrt{-\mathcal{L}}t)$, such that equation (7.98) is hyperbolic. In the case of a self-adjoint operator, the previous assumption is equivalent to say that $\mathcal{L}$ is semi-bounded from above [336, p.525]. According to [336, p.538], if

$$u_0 \in D(\mathcal{L}) := \{u \in X | C(t)u \in C^2(\mathbb{R})\}, \quad v_0 \in \mathrm{E} := \{u \in X | C(t)u \in C^1(\mathbb{R})\},$$
$$F \in C^1([0,T];X) \oplus C([0,T];D(\mathcal{L})),$$

then the direct problem (7.98) and (7.99) has a unique solution $u \in C^2([0,T];X) \cap C([0,T];D(\mathcal{L}))$, which can be represented explicitly as

$$u(t) = C(t)u_0 + S(t)v_0 + \int_0^t S(t-s)F(s)ds, \quad (7.100)$$

where $S(t) = \int_0^t C(s)ds = \frac{1}{\sqrt{-\mathcal{L}}}\sin(\sqrt{-\mathcal{L}}t)$ is the associated sine function.

7.3.2 Inverse problem

Consider first, for simplicity, the one-dimensional case and take $\Omega = (0,l)$, where $l > 0$ represents the length of a vibrating string. Let us also take $h(x,t) = 1$ and $\mathcal{L} = \partial^2/\partial x^2$. Then, in [61] it was remarked that the inverse force problem (7.88)–(7.91) has a unique solution if and only if $T/l \notin \mathbb{Q}$. This follows immediately from the separation of variables, whereas for $u_0 = v_0 = 0$ and $h = 1$ the solution of the inverse problem

$$u_{tt} - u_{xx} = f(x), \quad (x,t) \in (0,l) \times (0,T), \tag{7.101}$$

$$u(x,0) = u_t(x,0) = 0, \quad x \in [0,l], \tag{7.102}$$

$$u(0,t) = u(l,t) = 0, \quad t \in (0,T), \tag{7.103}$$

$$u(x,T) = 0, \quad x \in [0,l], \tag{7.104}$$

is given by

$$u(x,t) = \frac{\sqrt{2}}{\pi^2} \sum_{k=1}^{\infty} \frac{c_k}{k^2} \left(1 - \cos\left(\frac{k\pi t}{l}\right)\right) \sin\left(\frac{k\pi x}{l}\right), \tag{7.105}$$

$$f(x) = \sum_{k=1}^{\infty} c_k \sin\left(\frac{k\pi x}{l}\right), \tag{7.106}$$

where

$$c_k = \frac{\sqrt{2}}{l} \int_0^l f(x) \sin\left(\frac{k\pi x}{l}\right) dx, \quad k \geq 1. \tag{7.107}$$

Now, to impose (7.104), apply (7.105) at $t = T$ to obtain

$$0 = \frac{\sqrt{2}}{\pi^2} \sum_{k=1}^{\infty} \frac{c_k}{k^2} \left(1 - \cos\left(\frac{k\pi T}{l}\right)\right) \sin\left(\frac{k\pi x}{l}\right), \quad x \in [0,l]. \tag{7.108}$$

Observe that $c_k = 0$ for all $k \geq 1$, and hence from (7.105) and (7.106), $u = f = 0$ if and only if $T/l \notin \mathbb{Q}$. Moreover, this condition cannot be removed even if one additionally prescribe $u_t(x,T)$, as it can be easily seen by differentiating (7.105) with respect to t. However, if the time-average displacement measurement (7.92) (with say $\omega = 1$) is considered instead of (7.91), by integrating (7.105) with respect to t and make it zero, yield

$$0 = \frac{\sqrt{2}}{\pi^2} \sum_{k=1}^{\infty} \frac{c_k}{k^2} \left(T - \frac{l}{k\pi} \sin\left(\frac{k\pi T}{l}\right)\right) \sin\left(\frac{k\pi x}{l}\right), \quad x \in [0,l]. \tag{7.109}$$

Since $\frac{k\pi T}{l} > \sin\left(\frac{k\pi T}{l}\right)$ for all $k \in \mathbb{N}^*$, it follows that $c_k = 0$ for all $k \in \mathbb{N}^*$ and hence, from (7.105) and (7.106), that $u = f = 0$. Thus, the inverse problem (7.101)–(7.103) together with the integral condition

$$\int_0^T u(x,t)dt = 0, \quad x \in [0,l], \tag{7.110}$$

has only the trivial solution, which in turn implies that the solution of the inverse problem given by (7.101)–(7.103) and the time-average displacement

$$\int_0^T u(x,t)dt = U_T(x), \quad x \in [0,l], \tag{7.111}$$

being measured, is unique, with no restriction on the ratio T/l being irrational number or not. In the case of an arbitrary integrable weight function $\omega(t)$ in (7.92), the necessary and sufficient condition for uniqueness becomes

$$\int_0^T \omega(t)\left(1 - \cos\left(\frac{k\pi t}{l}\right)\right) dt \neq 0, \quad \forall\, k \in \mathbb{N}^*. \tag{7.112}$$

Returning to the abstract setting, consider the inverse problem of determining the displacement u and the force f satisfying

$$u'' = \mathcal{L}u(t) + h(t)f, \quad t \in [0,T], \tag{7.113}$$

subject to the initial conditions (7.99) and the additional measurement

$$u(T) = u_T \tag{7.114}$$

or

$$\int_0^T \omega(t)u(t)dt = U_T. \tag{7.115}$$

First, the following theorem gives the unique solvability of the inverse problem (7.99), (7.113) and (7.114).

Theorem 7.5. ([274]) *Let $\mathcal{L}$ be a self-adjoint and semi-bounded from above operator in the Hilbert space $\mathcal{X}$ and assume that*

$$u_0 \in D(\mathcal{L}), \quad v_0 \in \mathrm{E}, \quad h \in C^1([0,T];\mathcal{X}), \quad u_T \in D(\mathcal{L}).$$

Then the inverse problem (7.99), (7.113) and (7.114) has a unique solution $u \in C^2([0,T];\mathcal{X}) \cap C([0,T];D(\mathcal{L}))$, $f \in \mathcal{X}$ if, see Corollary 8.2.7 of [336], h is non-negative and strictly increasing, as a function of $t \in [0,T]$, or if, see Corollary 8.2.8 of [336], $h \equiv 1$ and $1 \notin \overline{Z}$, where

$$Z := \{\cos(\sqrt{-\Lambda}T) | \Lambda \in \Sigma(\mathcal{L}) \backslash \{0\}\}, \tag{7.116}$$

and $\Sigma(\mathcal{L})$ denotes the spectrum of $\mathcal{L}$. In the above, the value $h(s)$ is identified with the operator of multiplication by the number $h(s)$ in the space $\mathcal{X}$.

Remark that in the previous one-dimensional setting of the inverse problem (7.101)–(7.104) the condition that $1 \notin \overline{Z}$ recasts as $1 - \cos(\sqrt{-\Lambda}\ T) \neq 0$, where $\Lambda = -n^2\pi^2/l^2$ for $n \in \mathbb{N}^*$, which is equivalent to say that $T/l \notin \mathbb{Q}$.

Note that for $\mathcal{L} = \nabla^2$, $\Omega \in C^1$, u_0, v_0, $u_T \in C^1(\Omega)$ satisfying compatibility conditions and $h \in C^1(\Omega_T)$ satisfying $h(\underline{x},t) > 0$, $h_t(\underline{x},t) > 0$, $\forall(\underline{x},t) \in \overline{\Omega}_T$, the existence of solution, i.e. the solvability of the inverse problem (7.88)–(7.91) was also established in [9] in the classes of functions $f \in L^2(\Omega)$, $u \in L^2(0,T;H^1(\Omega))$, $u_{tt} \in L^2(Q_T)$ and $\nabla^2 u \in L^2(Q_T)$.

Second, for the inverse problem (7.99), (7.113) and (7.115) the existence and uniqueness of solution hold, as given by the following theorem.

Theorem 7.6. ([274]) *In the abstract setting of Theorem 7.5 and assuming*

$$u_0 \in D(\mathcal{L}), \quad v_0 \in \mathrm{E}, \quad h \in C^1([0,T];\mathcal{X}), \quad \omega = 1, \quad U_T \in D(\mathcal{L}), \tag{7.117}$$

the inverse problem (7.99), (7.113) *and* (7.115) *has a unique solution* $u \in C^2([0,T];\mathcal{X}) \cap C([0,T];D(\mathcal{L}))$, $f \in \mathcal{X}$, *if* $0 \not\equiv h$ *is non-negative, as a function of* $t \in [0,T]$.

Remark 7.1. On applying (7.92) to (7.100) results in

$$\Phi(\mathcal{L})f = U_T - S(T)u_0 - \frac{(1-C(T))}{-\mathcal{L}}v_0, \tag{7.118}$$

where

$$\Phi(\mathcal{L}) = \int_0^T \int_0^t h(s)S(t-s)dsdt$$
$$= \int_0^T \int_0^t h(s)\frac{\sin(\sqrt{-\mathcal{L}}(t-s))}{\sqrt{-\mathcal{L}}}dsdt. \tag{7.119}$$

In the particular case $h = 1$, equation (7.118) simplifies as

$$\left(\frac{T-S(T)}{-\mathcal{L}}\right) f = U_T - S(T)u_0 - \frac{(1-C(T))}{-\mathcal{L}}v_0$$

which, since $T > S(T)$, it yields the solution

$$f = \frac{-\mathcal{L}U_T + S(T)\mathcal{L}u_0 - (1-C(T))v_0}{T-S(T)}. \tag{7.120}$$

The displacement is given by (7.100) which, for $h = 1$, simplifies as

$$u(t) = C(t)u_0 + S(t)v_0 + \left(\frac{1-C(t)}{-\mathcal{L}}\right) f = C(t)u_0$$
$$+S(t)v_0 + \left(\frac{1-C(t)}{T-S(T)}\right)\left(U_T - S(T)u_0 - \frac{(1-C(t))}{-\mathcal{L}}v_0\right). \tag{7.121}$$

Even if the solution exists and is unique, both inverse problems (7.88)–(7.91) and (7.88)–(7.90), (7.92) are still ill-posed since the continuous dependence upon the data (7.91) or (7.92) is violated. This can be seen from the following example of instability. Let $\Omega = (0, l = \pi)$ and, for $n \in \mathbb{N}^*$ take

$$u_n(x,t) = \frac{(1-\cos(nt))\sin(nx)}{n^{3/2}}, \quad (x,t) \in (0,\pi) \times (0,T)$$

which satisfies the wave equation with homogenous initial and Dirichlet boundary conditions,

$$u_{nT}(x) = u_n(x,T) = \frac{(1-\cos(nT))\sin(nx)}{n^{3/2}}, \quad x \in [0,\pi],$$

$$U_{nT}(x) = \int_0^T u_n(x,t)dt = \frac{\sin(nx)}{n^{3/2}}\left(T - \frac{\sin(nT)}{n}\right), \quad x \in [0,\pi]$$

and the force

$$f_n(x) = n^{1/2}\sin(nx), \quad x \in (0,\pi).$$

One can observe that whilst all the input data tends to zero, the force $f_n(x)$ becomes oscillatory and unbounded, as $n \to \infty$.

7.3.3 Variational formulation of the inverse problem

For the inverse problem (7.88)–(7.91), define $A : L^2(\Omega) \to L^2(\Omega)$,

$$Af = u_f(\cdot, T), \tag{7.122}$$

where $u_f(x,t)$ is the unique weak solution of the direct problem (7.88)–(7.90) corresponding to the given force f. The operator A is bounded and affine. By A_0, denote the similar linear operator defined for $u_0 = v_0 = 0$. The inverse problem (7.88)–(7.91) recasts as

$$Af = u_T. \tag{7.123}$$

Since in practice u_T is contaminated with random noisy errors it is convenient to minimize the least-squares cost functional $J : L^2(\Omega) \to \mathbb{R}_+$ defined by

$$J(f) = \frac{1}{2}||Af - u_T||^2_{L^2(\Omega)}. \tag{7.124}$$

It can be shown that J is weakly continuous on closed and convex subsets of $L^2(\Omega)$, which in turn, due to Weierstrass' theorem implies that there exists a solution to the minimization of (7.124), [154]. Next it is proved that J is Frechet differentiable and derive its gradient. For this, let u_χ solve (7.88)–(7.90) with $f = \chi \in L^2(\Omega)$ and $u_0 = v_0 = 0$. Then,

$$\begin{aligned} J(f+\chi) - J(f) &= \frac{1}{2}||Af + A_0\chi - u_T||^2_{L^2(\Omega)} - \frac{1}{2}||Af - u_T||^2_{L^2(\Omega)} \\ &= \int_\Omega (Af(\underline{x}) - u_T(\underline{x}))A_0\chi(\underline{x})d\Omega + \frac{1}{2}\int_\Omega (A_0\chi(\underline{x}))^2 d\Omega. \end{aligned} \tag{7.125}$$

The first term in the right-hand side can, using the definition of the operator A_0, be rewritten as

$$\int_\Omega (Af(\underline{x}) - u_T(\underline{x}))A_0\chi(\underline{x})d\Omega = \int_\Omega (u_f(\underline{x},T) - u_T(\underline{x}))\, u_\chi(\underline{x},T)d\Omega. \quad (7.126)$$

Let v_1 be the solution to the adjoint problem (7.94)-(7.96) with $\zeta = G = 0$ and $\xi = Af - u_T = u_f(.,T) - u_T$. From formula (7.97) applied to v_1 and u_χ,

$$\begin{aligned}
&\int_\Omega (u_f(\underline{x},T) - u_T(\underline{x}))u_\chi(\underline{x},T)d\Omega \\
&= -\int_0^T \int_\Omega \chi(\underline{x})h(\underline{x},t)v_1(\underline{x},t)d\Omega dt \\
&= -\int_\Omega \chi(\underline{x}) \left(\int_0^T h(\underline{x},t)v_1(\underline{x},t)dt \right) d\Omega. \quad (7.127)
\end{aligned}$$

From (7.125) and (7.127), and since $||A_0\chi||^2_{L^2(\Omega)}$ can be estimated using $||\chi||^2_{L^2(\Omega)}$ by (7.93), see also [154] for one-dimensional explicit estimates, it follows that the functional J is Frechet differentiable and its gradient is given by

$$J'(f) = -\int_0^T h(\underline{x},t)v_1(\underline{x},t)dt. \quad (7.128)$$

To show that J is in fact twice Frechet differentiable and convex, let v_2 be the solution of the adjoint problem (7.94)–(7.96) with $\zeta = G = 0$ and $\xi = u_\chi(\cdot,T)$. From formula (7.97) applied to the functions u_χ and v_2,

$$-\int_\Omega u_\chi^2(\underline{x},T)d\Omega = \int_0^T \int_\Omega \chi(\underline{x})h(\underline{x},t)v_2(\underline{x},t)d\Omega dt. \quad (7.129)$$

Following a similar argument as above

$$J''(f)h = -\int_0^T h(\underline{x},t)v_2(\underline{x},t)dt. \quad (7.130)$$

Then, from (7.129) and (7.130),

$$(J''(f)\chi,\chi)_{L^2(\Omega)} = ||u_\chi(.,T)||^2_{L^2(\Omega)} \geq 0, \quad (7.131)$$

which implies that J is convex. Similarly, for the inverse problem (7.88)–(7.90), (7.92), define the operator $\tilde{A} : L^2(\Omega) \to L^2(\Omega)$ by

$$\tilde{A}f = \int_0^T \omega(t)u_f(.,t)dt, \quad (7.132)$$

which is bounded and affine and by $\tilde{A}_0$ denote its linear part. Then, the inverse problem (7.88)–(7.90), (7.92) recasts as

$$\tilde{A}f = U_T. \quad (7.133)$$

As in the previous case, a quasi-solution to (7.133) is sought in the form of minimizing the least-squares cost functional $\tilde{J} : L^2(\Omega) \to \mathbb{R}_+$ defined by

$$\tilde{J}(f) := \frac{1}{2}||\tilde{A}f - U_T||^2_{L^2(\Omega)}. \tag{7.134}$$

As in (7.125) and (7.127),

$$\tilde{J}(f+\chi) - \tilde{J}(f) = \frac{1}{2}\int_\Omega (\tilde{A}_0\chi(\underline{x}))^2 d\Omega + \int_\Omega \left(\int_0^T \omega(t) u_\chi(\underline{x},t)dt\right) \times \left(\int_0^T \omega(t) u_f(\underline{x},t)dt - U_T(\underline{x})\right) d\Omega. \tag{7.135}$$

Let us discuss first the particular case when the weight function $\omega(t)$ is a non-zero constant, say equal to unity, and then the general case.

(i) In the particular case when $\omega(t) \equiv 1$, i.e. (7.92) recasts as

$$\int_0^T u(\underline{x},t)dt = U_T(\underline{x}), \quad \underline{x} \in \Omega, \tag{7.136}$$

expression (7.135) becomes

$$\tilde{J}(f+\chi) - \tilde{J}(f) = \frac{1}{2}\int_\Omega (\tilde{A}_0\chi(\underline{x}))^2 d\Omega + \int_\Omega \left(\int_0^T u_\chi(\underline{x},t)dt\right)\left(\int_0^T u_f(\underline{x},t)dt - U_T(\underline{x})\right) d\Omega. \tag{7.137}$$

Let $\tilde{v}_1$ be the solution of the adjoint problem (7.94)–(7.96) with $\zeta = G = 0$ and $\xi = \tilde{A}f - U_T = \int_0^T u_f(\cdot,t)dt - U_T$. Then,

$$w_\chi(\underline{x},t) := \int_0^t u_\chi(\underline{x},s)ds, \tag{7.138}$$

satisfies the wave equation with homogenous initial and boundary conditions and right-hand side equal to $\left(\int_0^T h(\underline{x},s)ds\right)\chi(\underline{x})$. Then, from Green's formula (7.97) applied to w_χ and $\tilde{v}_1$ it follows from (7.137) that

$$\tilde{J}(f+\chi) - \tilde{J}(f) = \frac{1}{2}\int_\Omega (\tilde{A}_0\chi(\underline{x}))^2 d\Omega - \int_\Omega\int_0^T \chi(\underline{x})\left(\int_0^t h(\underline{x},s)ds\right)\tilde{v}_1(\underline{x},t)dtd\Omega. \tag{7.139}$$

From this it follows that

$$\tilde{J}'(f) = -\int_0^T \left(\int_0^t h(\underline{x},s)ds\right)\tilde{v}_1(\underline{x},t)dt. \tag{7.140}$$

(ii) In the general case, rewrite (7.135) as

$$\tilde{J}(f+\chi) - \tilde{J}(f) = \frac{1}{2}\int_\Omega (\tilde{A}\chi(\underline{x}))^2 d\Omega$$

$$+\int_\Omega \int_0^T \omega(t) u_\chi(\underline{x}, t) \left(\int_0^T \omega(\tau) u_f(\underline{x}, \tau) d\tau - U_T(\underline{x}) \right) dt d\Omega. \quad (7.141)$$

Let $\tilde{V}_1$ be the solution of the adjoint problem (7.94)–(7.96) with $\xi = \zeta = 0$ and $G(\underline{x}, t) = \omega(t) \left(\int_0^T \omega(\tau) u_f(\underline{x}, \tau) d\tau - U_T(\underline{x}) \right)$. Then, Green's formula (7.97) applied to $\tilde{V}_1$ and u_χ implies that the last term in (7.141) is equal to $\int_0^T \int_\Omega \chi(\underline{x}) h(\underline{x}, t) \tilde{V}_1(\underline{x}, t) d\Omega dt$, and consequently,

$$\tilde{J}'(f) = \int_0^T h(\underline{x}, t) \tilde{V}_1(\underline{x}, t) dt. \quad (7.142)$$

Let us finally show that (7.142) reduces to (7.140) when $\omega(t) \equiv 1$. In such a situation, the problems for $\tilde{v}_1$ and $\tilde{V}_1$ are given by

$$\begin{cases} \tilde{v}_{1tt} - \mathcal{L}^* \tilde{v}_1 = 0 \quad \text{in } \Omega_T, \\ \tilde{v}_1(\underline{x}, T) = 0, \quad \tilde{v}_{1t}(\underline{x}, T) = \int_0^T u_f(\underline{x}, t) dt - U_T(\underline{x}), \quad \underline{x} \in \Omega, \\ \tilde{v}_1(\underline{x}, t) = 0, \quad (\underline{x}, t) \in \partial\Omega \times (0, T), \end{cases}$$

$$\begin{cases} \tilde{V}_{1tt} - \mathcal{L}^* \tilde{V}_1 = \int_0^T u_f(\underline{x}, t) dt - U_T(\underline{x}) \quad \text{in } \Omega_T, \\ \tilde{V}_1(\underline{x}, T) = 0, \quad \tilde{V}_{1t}(\underline{x}, T) = 0, \quad \underline{x} \in \Omega \\ \tilde{V}_1(\underline{x}, t) = 0, \quad (\underline{x}, t) \in \partial\Omega \times (0, T). \end{cases}$$

One can observe that $\tilde{V}_{1t}(\underline{x}, t) = \tilde{v}_1(\underline{x}, t)$. Then, starting from (7.142), using integration by parts, it can be derived that

$$\begin{aligned} \tilde{J}'(f) = \int_0^T h(\underline{x}, t) \tilde{V}_1(\underline{x}, t) dt &= \left(\int_0^T h(\underline{x}, s) ds \right) \tilde{V}_1(\underline{x}, t) \Big|_{t=0}^{t=T} \\ &- \int_0^T \left(\int_0^t h(\underline{x}, s) ds \right) \tilde{V}_{1t}(\underline{x}, t) dt \\ &= -\int_0^T \left(\int_0^t h(\underline{x}, s) ds \right) \tilde{v}_1(\underline{x}, t) dt. \end{aligned} \quad (7.143)$$

Hence, (7.140) and (7.142) coincide in the case $\omega(t) \equiv 1$.

7.3.4 Landweber-Fridman method (LFM)

Once the gradient of the functional J (or $\tilde{J}$) has been derived, one can apply the iterative Landweber-Fridman method (LFM), e.g. [113], for obtaining a stable solution to the inverse problem, as follows:

Step 1. Choose an arbitrary function $f_0 \in L^2(\Omega)$. Let u^0 be the solution of the direct problem (7.88)–(7.90) with $f = f_0$.

Step 2. Assume that f_k and u^k have been constructed. For the inverse problem (7.88)–(7.91), let v^k solve the adjoint problem (7.94)–(7.96) with $\zeta = G = 0$ and

$$\xi_k(\underline{x}) = u^k(\underline{x}, T) - u_T(\underline{x}), \quad \underline{x} \in \Omega, \tag{7.144}$$

and calculate the gradient (7.128) given by

$$z_k(\underline{x}) = -\int_0^T h(\underline{x}, t) v^k(\underline{x}, t) dt, \quad \underline{x} \in \Omega. \tag{7.145}$$

For the inverse problem (7.88)–(7.90) and (7.136) let $\tilde{v}_k$, solve the adjoint problem (7.94)–(7.96) with $\zeta = G = 0$ and

$$\xi_k(\underline{x}) = \int_0^T u^k(\underline{x}, t) dt - U_T(\underline{x}), \quad \underline{x} \in \Omega, \tag{7.146}$$

and calculate the gradient (7.140) given by

$$z_k(\underline{x}) = -\int_0^T \left(\int_0^t h(\underline{x}, s) ds \right) \tilde{v}_k(\underline{x}, t) dt, \quad \underline{x} \in \Omega. \tag{7.147}$$

Step 3. Construct the new iterate for the force given by

$$f_{k+1}(\underline{x}) = f_k(\underline{x}) - \gamma z_k(\underline{x}), \quad \underline{x} \in \Omega, \tag{7.148}$$

where $0 < \gamma < \frac{2}{||A||^2}$ (respectively $\frac{2}{||\tilde{A}||^2}$) is a relaxation factor to be prescribed and the spectral norm of the operator A (respectively $\tilde{A}$) is defined as

$$||A|| = \sup_{f \in L^2(\Omega) \setminus \{0\}} \frac{||Af||_{L^2(\Omega)}}{||f||_{L^2(\Omega)}}. \tag{7.149}$$

Let u^{k+1} be the solution of the direct problem (7.88)–(7.90) with $f = f_{k+1}$.

Step 4. Repeat steps 2 and 3 until convergence is achieved in the case of exact data u_T (or U_T). In the case of noisy data

$$||u_T - u_T^\epsilon||_{L^2(\Omega)} \leq \epsilon \quad \text{or} \quad ||U_T - U_T^\epsilon||_{L^2(\Omega)} \leq \epsilon \tag{7.150}$$

one can use the Morozov discrepancy principle [110, 113] to terminate the iterations at the smallest $k = k(\epsilon)$ for which

$$||u^k(.,T) - u_T^\epsilon||_{L^2(\Omega)} \leq \tau\epsilon \text{ or } \left\| \int_0^T \omega(t) u^k(.,t) - U_T^\epsilon \right\|_{L^2(\Omega)} \leq \tau\epsilon, \tag{7.151}$$

where $\tau > 1$ is some constant to be prescribed. According to (7.124) and (7.134), criterion (7.151) can be rewritten as

$$J(f_k) \leq \tau^2 \frac{\epsilon^2}{2} \quad \text{or} \quad \tilde{J}(f_k) \leq \tau^2 \frac{\epsilon^2}{2}. \tag{7.152}$$

7.3.5 Numerical results and discussion

In all examples $T = 1$, $\Omega = (0, l)$ with $l = 1$ and $\mathcal{L} = \nabla^2 = \partial^2/\partial x^2$ is the Laplacian operator. Take the initial guess arbitrary such as $f_0 \equiv 0$.

Example 7.1 Consider first the direct problem (7.88)–(7.90) given by

$$u_{tt} - u_{xx} = f(x)h(x,t), \quad (x,t) \in (0,1) \times (0,1), \tag{7.153}$$

with $h \equiv 1$, subject to

$$u(x,0) = u_0(x) = 2\sin(\pi x), \quad u_t(x,0) = v_0(x) = 0, \quad x \in [0,1], \tag{7.154}$$

$$u(0,t) = u(1,t) = 0, \quad t \in (0,1), \tag{7.155}$$

when the force is given by

$$f(x) = \pi^2 \sin(\pi x), \quad x \in (0,1). \tag{7.156}$$

The exact solution of this direct problem is given by

$$u(x,t) = \sin(\pi x)(\cos(\pi t) + 1), \quad (x,t) \in [0,1] \times [0,1]. \tag{7.157}$$

The final displacement is given by

$$u(x,T) = u(x,1) = u_T(x) = 0, \quad x \in [0,1], \tag{7.158}$$

and the time-average displacement (for $\omega = 1$) is

$$\int_0^T u(x,t)dt = \int_0^1 u(x,t)dt = U_T(x) = \sin(\pi x), \quad x \in [0,1], \tag{7.159}$$

The discrete finite-difference form of problem (7.153)-(7.155) is as follows. Divide the solution domain $(0,l) \times (0,T)$ into M and N subintervals of equal space length Δx and time-step Δt, where $\Delta x = l/M$ and $\Delta t = T/N$. Denote $u_{i,j} := u(x_i, t_j)$, where $x_i = i\Delta x$, $t_j = j\Delta t$, and $f_i := f(x_i)$, and $h_{i,j} := h(x_i, t_j)$ for $i = \overline{0, M}$, $j = \overline{0, N}$. Then, a central finite difference approximation to (7.153)–(7.155) at the mesh points $(x_i, t_j) = (i\Delta x, j\Delta t)$ of the rectangular mesh covering the solution domain $(0,l) \times (0,T)$ is,

$$u_{i,j+1} = r^2 u_{i+1,j} + 2(1 - r^2)u_{i,j} + r^2 u_{i-1,j} - u_{i,j-1} + (\Delta t)^2 f_i h_{i,j},$$
$$i = \overline{1,(M-1)}, \quad j = \overline{1,(N-1)}, \tag{7.160}$$

$$u_{i,0} = u_0(x_i),\ i = \overline{0,M}, \quad \frac{u_{i,1} - u_{i,-1}}{2\Delta t} = v_0(x_i),\ i = \overline{1,(M-1)}, \tag{7.161}$$

$$u_{0,j} = 0, \quad u_{M,j} = 0, \quad j = \overline{1,N}, \tag{7.162}$$

where $r = \Delta t/\Delta x$. Equation (7.160) represents an explicit FDM which is stable if $r \leq 1$, giving approximate values for the solution at mesh points along $t = 2\Delta t, 3\Delta t, ...$, as soon as the solution at the mesh points along $t = \Delta t$ has been determined. Putting $j = 0$ in equation (7.160) and using (7.161),

$$u_{i,1} = \frac{1}{2}r^2 u_0(x_{i+1}) + (1-r^2)u_0(x_i) + \frac{1}{2}r^2 u_0(x_{i-1}) + (\Delta t)v_0(x_i) + \frac{1}{2}(\Delta t)^2 f_i h_{i,0}, \quad i = \overline{1,(M-1)}. \tag{7.163}$$

For finding the numerical solution to (7.158), put $j = N-1$ in (7.160). And for (7.159), use the trapezoidal rule approximation

$$\int_0^T u(x_i,t)dt = \frac{\Delta t}{2}\left(u_0(x_i) + 2\sum_{j=1}^{N-1} u(x_i,t_j) + u(x_i,t_N)\right), \quad i = \overline{1, M-1}. \tag{7.164}$$

Consider next solving the inverse problem with fixed $N = M = 80$ and $\gamma = 1$. Since $h \equiv 1$ and also since $T/l = 1 \in \mathbb{Q}$, the uniqueness of solution of the inverse problem (7.153)–(7.155) does not hold when measuring the final displacement (7.158). Therefore, for Example 7.1 the inverse problem (7.153)–(7.155) is only considered with the time-average displacement measurement (7.159), which has a unique solution given by equations (7.156) and (7.157). The objective function (7.134) given by

$$\tilde{J}(f_k) = \frac{1}{2}||\xi_k||^2 = \frac{1}{2}\sum_{i=1}^{M-1} \xi_k^2(x_i), \tag{7.165}$$

where ξ_k is given by (7.146), plotted in Figure 7.6 shows that the convergence of $\tilde{J}$ is achieved after about 300 iterations. The figure also shows that the error between the exact solution f and numerical solution f_k defined by

$$E(f_k) = ||f_{exact} - f_k|| = \sqrt{\sum_{i=1}^{M-1}(f(x_i) - f_k(x_i))^2} \tag{7.166}$$

decreases monotonically to zero, as a function of the number of iterations k.

Figure 7.7 shows the monotonic increasing convergence of the numerical solution f_k towards the exact solution (7.156).

The additional observation (7.92) is perturbed by noise as

$$U_T^{\epsilon}(x_i) = U_T(x_i) + \epsilon_i, \quad i = \overline{1,(M-1)}, \tag{7.167}$$

where $(\epsilon_i)_{i=\overline{1,M-1}}$ are random noisy variables generated from a Gaussian normal distribution with mean zero and standard deviation $\sigma = p \times$

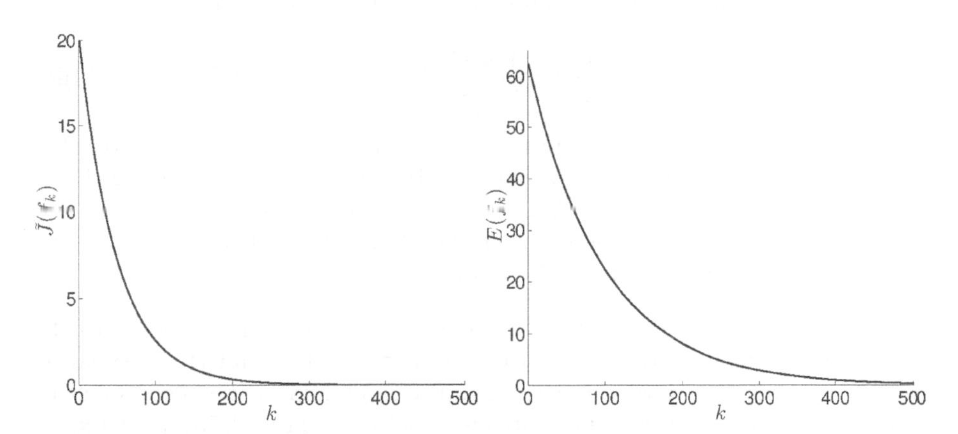

FIGURE 7.6: The objective function $\tilde{J}(f_k)$ and the accuracy error $E(f_k)$, versus the number of iterations $k = \overline{1,500}$, no noise, for Example 7.1, [274].

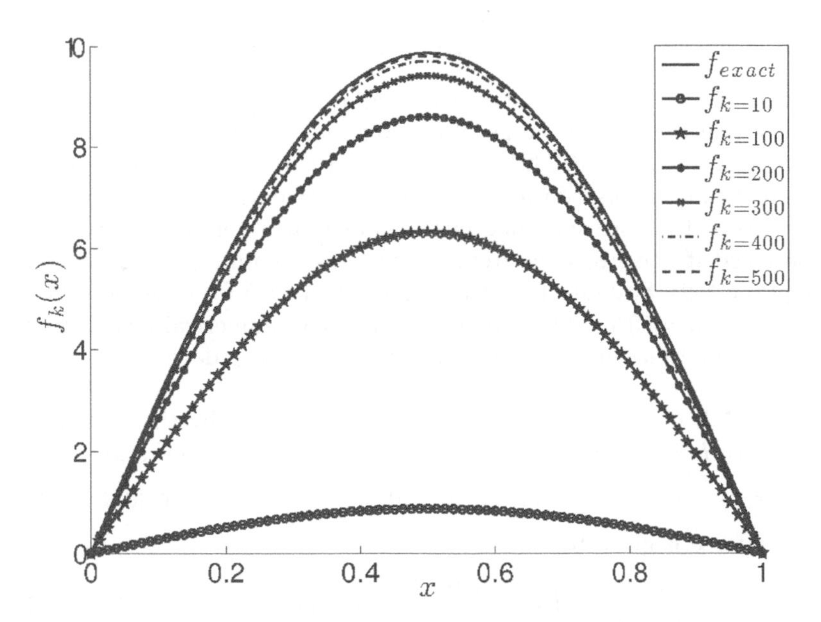

FIGURE 7.7: The numerical solution f_k at various iteration numbers k, in comparison with the exact solution (7.156), no noise, for Example 7.1, [274].

$\max_{x\in[0,L]} |U_T(x)|$, where p is the percentage of noise. The total amount of noise introduced in the objective functional (7.134) is then given by

$$\frac{1}{2}\epsilon^2 = \frac{1}{2}\sum_{i=1}^{M-1} \epsilon_i^2. \tag{7.168}$$

For various percentages of noise p, the values of k_{discr} and k_{opt} together with the corresponding accuracy errors (7.166) are given in Table 7.2. From this table it can be seen that there is not much difference between k_{opt} and k_{discr} for all percentages of noise p considered and this adds to the robustness of the numerical iterative method employed.

TABLE 7.2: The stopping iteration number k_{discr} chosen according to the discrepancy principle criterion (7.152) (with $\tau = 1.15$), and the optimal iteration number k_{opt} chosen according to the minimum of the accuracy error function (7.166), for various percentages of noise $p \in \{10, 30, 50\}\%$ for Example 7.1. The corresponding accuracy errors $E(f_{k_{discr}})$ and $E(f_{k_{opt}})$ are also included.

p	10%	30%	50%
k_{opt}	373	276	232
$E(f_{k_{opt}})$	1.84	4.21	6.01
k_{discr}	300	245	205
$E(f_{k_{discr}})$	2.52	4.52	6.43

Figures 7.8(a) and 7.8(b) show the regularized numerical solution for $f(x)$ obtained with the various values of the iteration numbers listed in Table 7.2, namely, $k_{opt} \in \{373, 276, 232\}$ and $k_{discr} \in \{240, 225, 220\}$, respectively, for $p \in \{10, 30, 50\}\%$ noisy data. It can be seen that there is not much difference between the corresponding curves and that stable numerical solutions are obtained if one stops the iteration process according to the discrepancy principle (7.152). Stability is further maintained even for large percentages of noise such as $p = 50\%$. Furthermore, as expected, numerical results in Figure 7.8 become more accurate as the percentage of noise p decreases.

Example 7.2 Consider first the direct problem given by the wave equation (7.153) with $h(x,t) = 1 + t$, the boundary conditions (7.155) and

$$u(x,0) = u_0(x) = \sin(\pi x), \quad u_t(x,0) = v_0(x) = 0, \quad x \in [0,1], \tag{7.169}$$

when the force is given by

$$f(x) = \frac{1}{\tilde{\sigma}\sqrt{2\pi}} \exp\left(-\frac{(x-\mu)^2}{2\tilde{\sigma}^2}\right), \quad \text{with } \tilde{\sigma} = 0.1,\ \mu = 0.5. \tag{7.170}$$

The force (7.170) is a Gaussian normal function with mean μ and standard deviation $\tilde{\sigma}$. As $\tilde{\sigma} \to 0$, expression (7.170) mimics the Dirac delta distribution

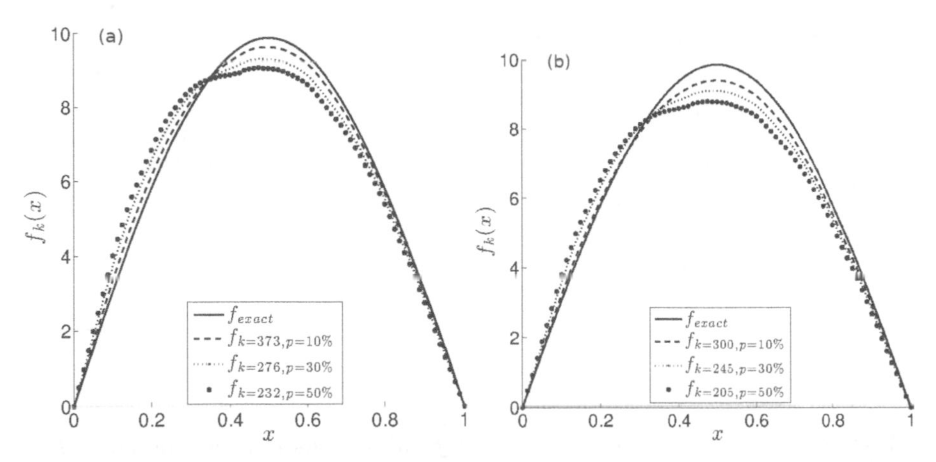

FIGURE 7.8: The exact solution f in comparison with the numerical solution f_k for (a) $k_{opt} \in \{373, 276, 232\}$ and (b) $k_{discr} \in \{300, 245, 205\}$, for $p \in \{10, 30, 50\}\%$ noise, for Example 7.1, [274].

$\delta(x - \mu)$. Unlike in the previous example, for the above direct problem an explicit analytical solution for the displacement $u(x,t)$ is not readily available and thus, the values (7.91) and (7.92) of $u(x,1)$ and $\int_0^1 u(x,t)dt$ (for $\omega(t) = 1$), respectively, are obtained numerically using the FDM. The function $h(x,t) = 1 + t$ satisfies $h(x,t) \geq 0$, $h_t(x,t) > 0$, $\forall(x,t) \in \overline{\Omega}_T$ and hence, according to Theorems 7.5 and 7.6, both the inverse problems (7.153), (7.155), (7.169) with the measurement (7.91) or (7.92), have unique solutions.

Take $M = N = 80$ in the FDM and $\gamma = 1$. First, consider the case without noise, i.e. $p = 0$. Figure 7.9 shows the convergence of the numerical solutions, as the number of iterations increases. It can be seen that the number of iterations necessary to achieve a high level of accuracy is large of $O(10^5)$. It is much larger than in the previous Example 7.1 because the force function (7.170) to be retrieved has a small standard deviation $\tilde{\sigma}$ and therefore a sharper peak centred at the mean value $\mu = 0.5$ than the trigonometric function (7.156). By comparing the results in Figures 7.9(a) and 7.9(b) one can also observe that the convergence for the inverse problem with the displacement measurement (7.91) is much faster (and for some number of iterations more accurate) than that with time-average displacement measurement (7.111).

Next, add some $p \in \{1, 3, 5\}\%$ noise in the data (7.91) and (7.111). Figure 7.10 is analogous to Figure 7.8 of Example 7.1 and similar conclusions can be drawn in terms of comparing the inverse problems with either the displacement measurement (7.91) or with the time-average measurement (7.111). Of course, since more iterations are required for Example 7.2 than for Example 7.1, the thresholds k_{discr} and k_{opt} are much higher (and also more different between themselves). Furthermore, the accuracy of the numerical results in Figure 7.8 for Example 7.1 is much

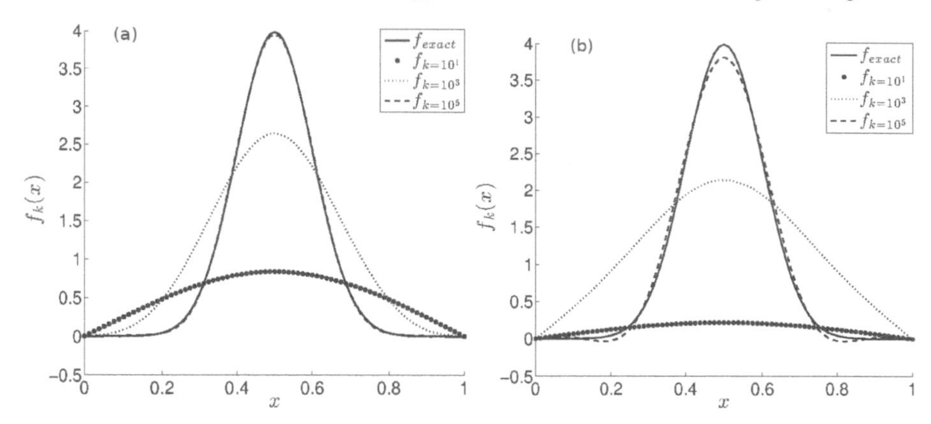

FIGURE 7.9: The numerical solution f_k at various iteration numbers $k \in \{10^1, 10^3, 10^5\}$, in comparison with the exact solution (7.170), no noise for Example 7.2 with (a) the displacement measurement (7.91) and (b) the time-average displacement measurement (7.111), [274].

higher than that in Figure 7.10 for Example 7.2, as expected since the trigonometric source (7.156) is less complicated than the Gaussian normal bell-shaped function (7.170).

Example 7.3 Consider first the direct problem given by the wave equation (7.153) with $h(x,t) = 1 + t$, the boundary conditions (7.155) and

$$u(x,0) = u_0(x) = 0, \quad u_t(x,0) = v_0(x) = 0, \quad x \in [0,1], \tag{7.171}$$

when the force is given by

$$f(x) = \begin{cases} x & \text{if } x \in [0, 1/2], \\ 1 - x & \text{if } x \in (1/2, 1]. \end{cases} \tag{7.172}$$

This force has a triangular shape, being continuous but non-differentiable at the peak $x = 1/2$. This example also does not possess an explicit analytical solution for the displacement $u(x,t)$ being readily available. The numerically simulated $u(x,1)$ is used as input data (7.91) in the inverse problem (7.153), (7.155), (7.171) whose existence and uniqueness of solution is guaranteed from Theorem 7.6 since $h(x,t) = 1+t$ satisfies $h(x,t) \geq 0$, $h_t(x,t) > 0$, $\forall (x,t) \in \overline{\Omega}_T$. As before, fix $N = M = 80$ and $\gamma = 1$.

It was observed in Example 7.2 and elsewhere that the convergence of the iterative LFM can become prohibitively show. One way to increase the rate of convergence is to increase the value of the relaxation parameter γ in (7.148) and this effect will be investigated in the next Example 7.4. Alternatively, one can speed up the convergence of the minimization of the least-squares functional (7.124) or (7.134) by employing the convergent and regularizing conjugate gradient method (CGM). In addition, the CGM does not require any choice of a relaxation parameter γ, as the LFM does, in order to iterate in formula (7.148).

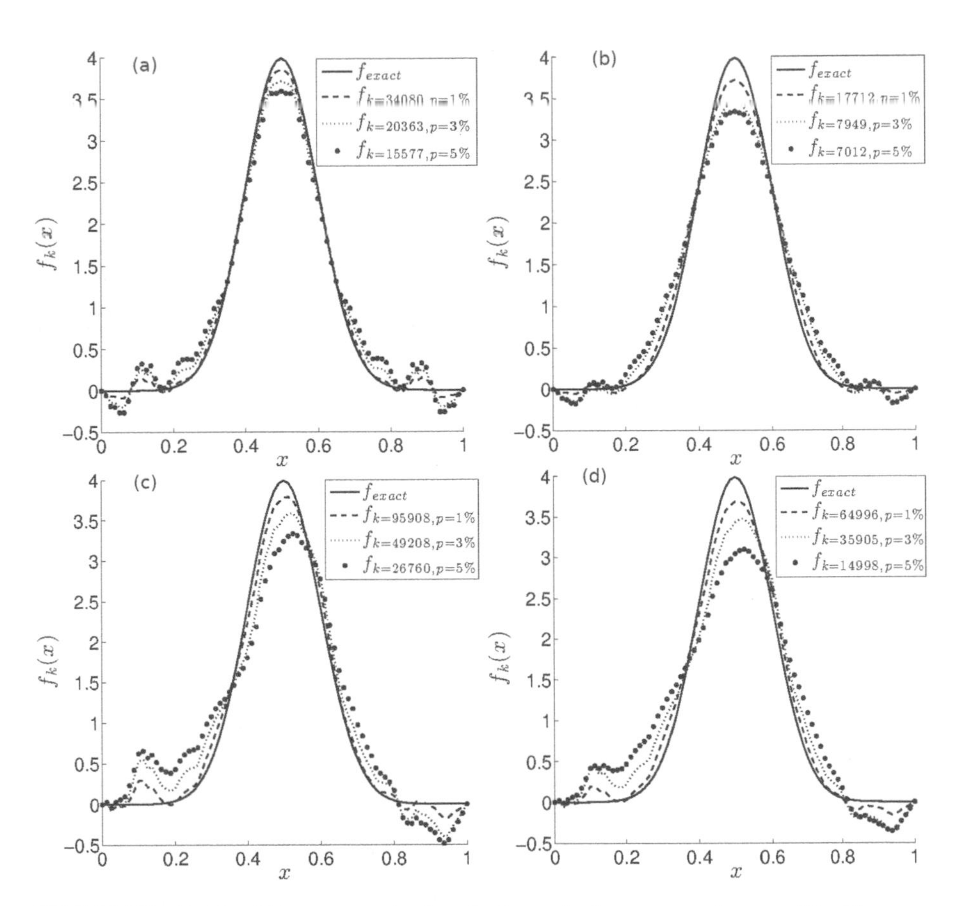

FIGURE 7.10: The numerical solution f_k in comparison with the exact solution (7.170), for $p \in \{1, 3, 5\}\%$ noise, for Example 7.2 with the displacement measurement (7.91) for (a) $k_{opt} \in \{34080, 20363, 15577\}$, (b) $k_{discr} \in \{17712, 7949, 7012\}$ with $\tau = 1.2$, and with the time-average displacement measurement (7.111) for (c) $k_{opt} \in \{95908, 49208, 26760\}$, (d) $k_{discr} \in \{64996, 35905, 14998\}$ with $\tau = 1.1$, [274].

Let steps 1 and 2 be the same as in the LFM of subsection 7.3.4. The next steps are as follows [274]:

Calculate

$$d_k(x) = -z_k(x) + \beta_{k-1} d_{k-1}(x), \tag{7.173}$$

with the convention that $\beta_{-1} = 0$ and

$$\beta_{k-1} = \frac{||z_k||^2_{L^2(\Omega)}}{||z_{k-1}||^2_{L^2(\Omega)}}, \quad k \geq 1. \tag{7.174}$$

Solve the direct problem (7.88)–(7.90) with $u_0 = v_0 = 0$ and $f = d_k$ to determine $A_0 d_k$, set the direction search

$$\alpha_k = \frac{||z_k||^2_{L^2(\Omega)}}{||A_0 d_k||^2_{L^2(\Omega)}}, \quad k \geq 0, \tag{7.175}$$

and pass to the new iteration by letting

$$f_{k+1}(x) = f_k(x) + \alpha_k d_k(x). \tag{7.176}$$

Let u^{k+1} solve the direct problem (7.88)−(7.90) with $f = f_{k+1}$ and repeat the above steps until the stopping criterion (7.152) is satisfied.

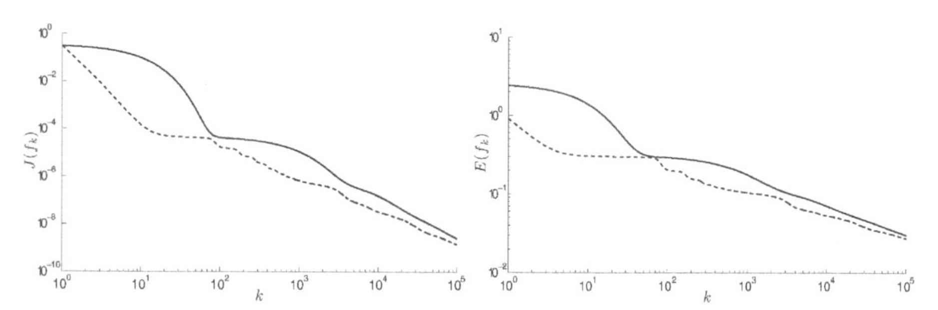

FIGURE 7.11: The objective function $J(f_k)$ and the accuracy error $E(f_k)$, versus the number of iterations $k = \overline{1, 10^5}$, obtained using the LFM (—) and the CGM (- - -), no noise, for Example 7.3, [274].

The objective function (7.124), the accuracy error (7.166) and the numerical solution for the force at various iteration numbers obtained using the LFM and the CGM are plotted in Figures 7.11 and 7.12. It can be seen that it takes a large number of iterations of $O(10^5)$ to converge with a good accuracy to the exact solution (7.172), similarly to what happened for Example 7.2, when the LFM is employed. In comparison to the previous Examples 7.1 and 7.2 this is to be expected because the force function (7.172) to be retrieved is non-smooth possessing a sharp corner at the peak $x = 1/2$. Moreover, the behaviour of the convergence is similar to that of Example 7.2 for which the

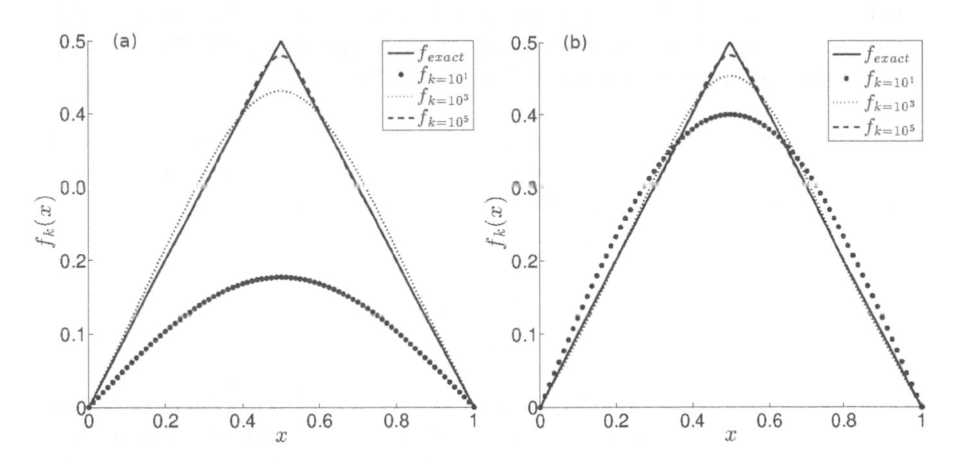

FIGURE 7.12: Numerical solution f_k for various iteration numbers $k \in \{10^1, 10^3, 10^5\}$, in comparison with the exact solution (7.172), obtained using (a) the LFM and (b) the CGM, no noise, for Example 7.3, [274].

TABLE 7.3: The stopping iteration numbers k_{discr} chosen according to the discrepancy principle (7.152) (with $\tau = 1.15$) and the optimal iteration numbers k_{opt} chosen according to the minimum of the accuracy error (7.166), for various percentages of noise $p \in \{1, 3, 5\}\%$ for Example 7.3, obtained using the LFM and CGM. The CGM results are included in brackets. The corresponding accuracy errors $E(f_{k_{discr}})$ and $E(f_{k_{opt}})$ are also included.

p	1%	3%	5%
k_{opt}	10171	3429	68
	(2908)	(295)	(12)
$E(f_{k_{opt}})$	0.120	0.300	0.319
	(0.109)	(0.294)	(0.316)
k_{discr}	2995	95	70
	(130)	(21)	(11)
$E(f_{k_{discr}})$	0.150	0.310	0.319
	(0.234)	(0.310)	(0.316)

Gaussian normal force function (7.170) to be retrieved, although smooth, it possesses also a sharp peak at $x = 1/2$. On the other hand, the convergence is much faster when the CGM is employed. When $p \in \{1, 3, 5\}\%$ noise is added to the input data (7.91), the numerical details included in Table 7.3 show that the CGM is about 10 times faster than the LFM.

Example 7.4 The previous example investigated a severe test given by the non-smooth triangular shape force function (7.172). In this example, an even more severe test example given by the discontinuous force

$$f(x) = \begin{cases} 0 & \text{if } 0 \leq x < \frac{1}{3}, \\ 1 & \text{if } \frac{1}{3} \leq x \leq \frac{2}{3}, \\ 0 & \text{if } \frac{2}{3} < x \leq 1. \end{cases} \tag{7.177}$$

is considered. Take the same input data as in Example 7.3. Then, on solving the direct problem given by (7.153), (7.155), (7.171) with the forcing term given by the product of the functions $h(x,t) = 1+t$ and $f(x)$ given by (7.177), the numerically simulated data for $\int_0^1 u(x,t)dt$ is used as input (7.111) in the inverse problem. Again, as in Example 7.2, the function $h(x,t) = 1+t$ satisfies $0 \not\equiv h(x,t) \geq 0$, $\forall (x,t) \in \overline{\Omega}_T$ and hence, according to Theorem 7.6, the inverse problem has a unique solution. As expected, for exact data a very slow convergence of the objective function (7.134) is encountered by the LFM because the force function (7.177) to be retrieved is discontinuous at the points $x \in \{1/3, 2/3\}$. In fact, the relaxation factor γ had to be increased from $\gamma = 1$ to higher values to achieve convergence in a reasonable number of iterations.

Figure 7.13 shows the objective function (7.134) and the accuracy error (7.166), versus the number of iterations $k = \overline{1, 10^5}$, for various values of the relaxation parameter $\gamma \in \{1, 5, 15\}$. From this figure it can be seen that the rate of convergence increases as γ is increased from 1 to 5 and then to 15. The corresponding numerical solutions for the force $f_k(x)$ are shown in Figure 7.14 for various numbers of iterations $k \in \{10^1, 10^3, 10^5\}$, and again more accurate results are obtained as k and/or γ are/is increased.

To investigate the stability of the numerical solution, $p \in \{1, 3, 5\}\%$ noise is included into the input data, as given by (7.167), and the numerical results obtained with $\gamma = 15$ are presented in Figures 7.15 and 7.16. Figure 7.15 justifies the choice of the stopping iteration number k_{discr} and furthermore, the numerical solutions illustrated in Figure 7.16 show that stable and reasonably accurate results are obtained for recovering the severely discontinuous force function (7.177). Extensions to two-dimensions are also available [274].

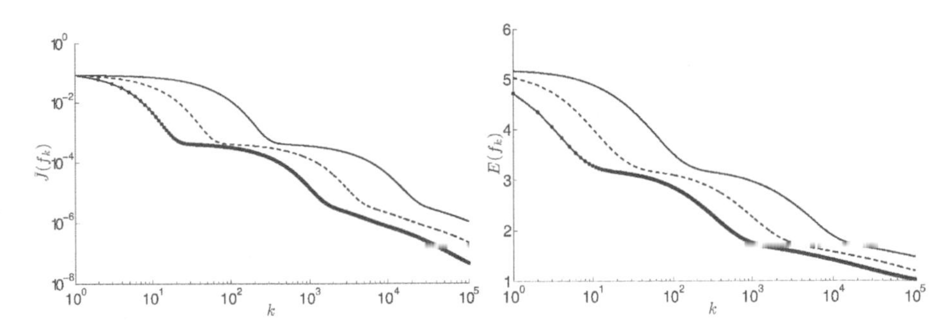

FIGURE 7.13: The objective function $\tilde{J}(f_k)$ and the accuracy error $E(f_k)$, as functions of the number of iterations $k = 1, 10^5$, for various $\gamma = 1$ (—), $\gamma = 5$ (- - -) and $\gamma = 15$ (– • –), no noise, for Example 7.4, [274].

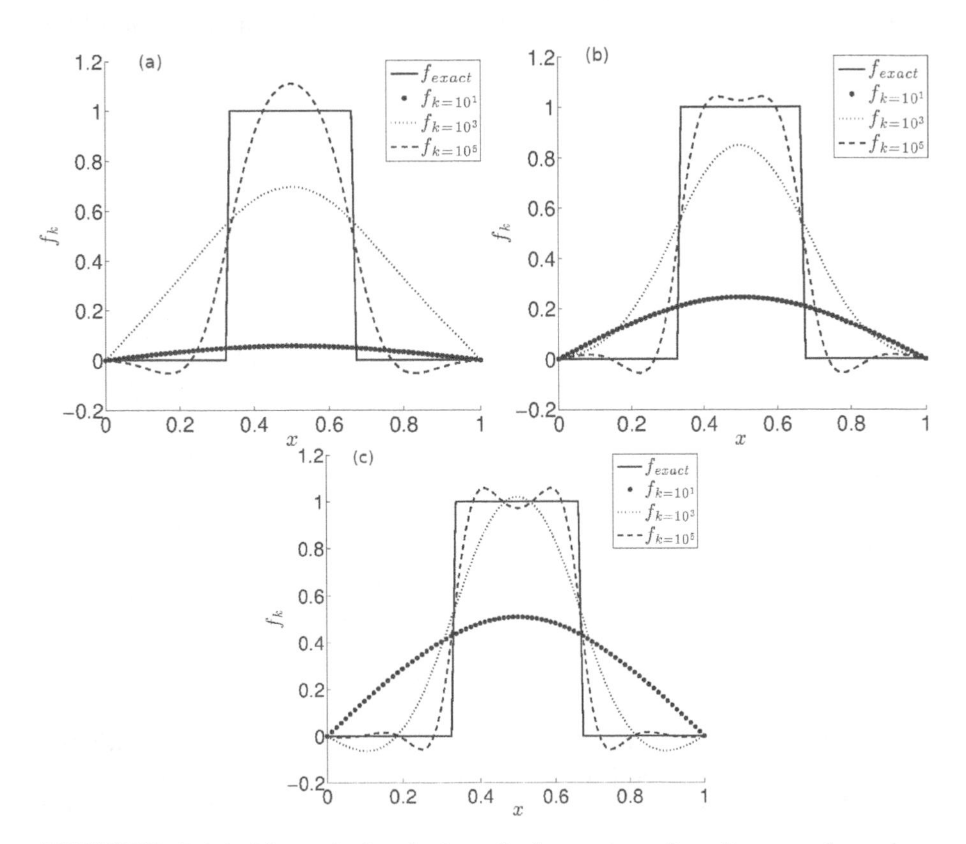

FIGURE 7.14: Numerical solution f_k for various iteration numbers $k \in \{10^1, 10^3, 10^5\}$, in comparison with the exact solution (7.177), for (a) $\gamma = 1$, (b) $\gamma = 5$ and (c) $\gamma = 15$, no noise, for Example 7.4, [274].

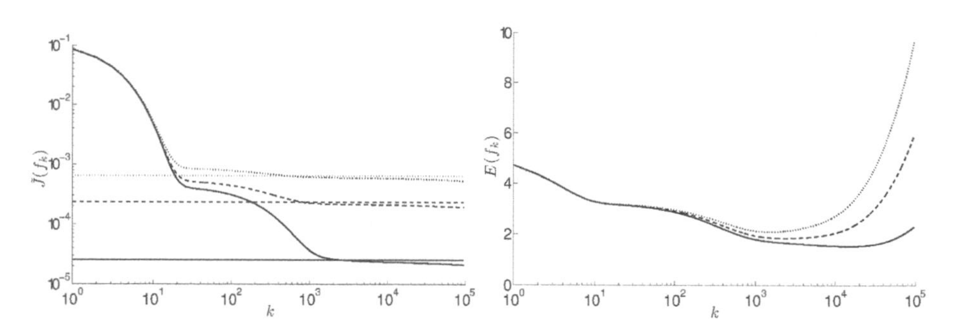

FIGURE 7.15: The objective function $\tilde{J}(f_k)$ and the accuracy error $E(f_k)$, versus the number of iterations $k = \overline{1, 10^5}$, for $p = 1\%$ (—), $p = 3\%$ (- - -) and $p = 5\%$ ($\cdots$) noise with $\gamma = 15$, for Example 7.4. The horizontal lines represents the threshold $\tau^2 \frac{\epsilon^2}{2}$ with $\tau = 1.1$, [274].

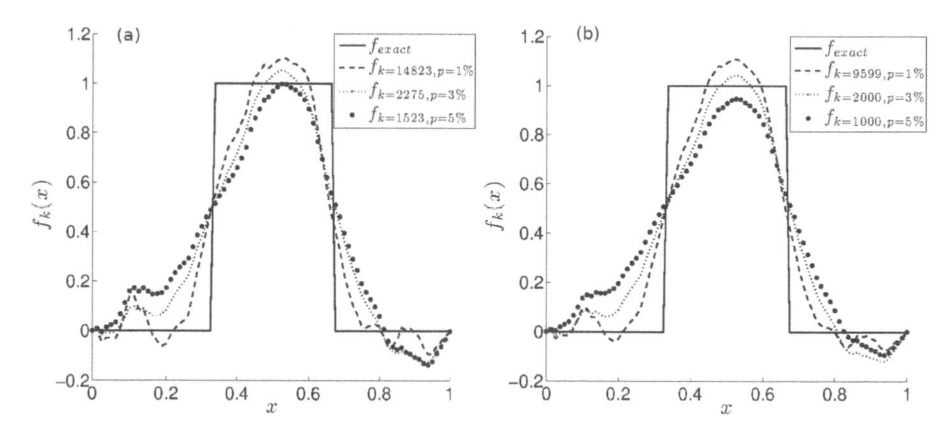

FIGURE 7.16: The exact solution (7.177) for f in comparison with the numerical solution f_k for (a) $k_{opt} \in \{14823, 2275, 1523\}$ and (b) $k_{discr} \in \{9599, 2000, 1000\}$ with $\tau = 1.1$, for $p \in \{1, 3, 5\}\%$ noise with $\gamma = 15$, for Example 7.4, [274].

Chapter 8

Reconstruction of Interfacial Coefficients

The knowledge of heat transfer behaviour in composite conductors requires characterization of the heat transfer coefficient at the contact interfaces between constituent materials. This chapter is devoted to such an inverse problem in which a generalized interface condition contains an unknown space- and time-varying interface coefficient that is to be determined from non-invasive temperature measurements on the boundary [409]. The uniqueness of solution holds, but the problem does not depend continuously on the input measured temperature data. A new preconditioned conjugate gradient method (CGM) is utilized to deal with this ill-posedness. In comparison with the standard CGM with no preconditioning, this method has the merit that the gradient of the objective functional does not vanish at the final time, which restores accuracy and stability when the input data is contaminated with noise and when the initial guess is not close to the true solution.

8.1 Introduction

For many multi-layer composite materials, the thermal behaviour is difficult to predict, due to the fact that the temperature at the interfaces is discontinuous. An important problem is the accurate prediction of the heat transfer coefficients (HTCs) at the solid-solid thermal contacting interfaces where the prescribed conditions can be linear [174] or nonlinear [400]. In this context, the knowledge of the thermal contact conductance (TCC) for a linear contact and the Stefan-Boltzmann coefficient (SBC) [80] for a nonlinear contact are essential. TCC characterizes the thermal resistance at the contact region between two materials, owing to the effect of surface roughness. In addition, for some high-temperature applications [80, 172], if there exists a tiny air layer between the mated surfaces, the Stefan-Boltzmann radiation condition should be applied. The parameter characterizing the effect of thermal radiation on the temperature drop is the SBC. Accurate estimations of TCC and SBC are important for quality control and monitoring in various fields, e.g., metal casting

DOI: 10.1201/9780429400629-8

[174], heat exchangers [175], quenching [70], plasma-facing components [119] and blood perfusion measurement [283].

For the determination of TCC, some explicit expressions were proposed through direct analysis of deformation of asperities [303] and thermomechanical simulation [362]. However, deviations between the analytical predictions and experimental measurements were observed [390]. Experimentally, the most direct method for the measurement of TCC is to use a series of thermocouples to measure the temperature profile along the centreline, and then extrapolate it to the interface to obtain the temperature drop across the interface [362, 390]. This type of method has some drawbacks of being sensitive to noise and time-consuming. Accordingly, many works were devoted to retrieve the solution to an associated inverse heat conduction problem (IHCP) from few temperature measurements based on steady state [122, 85] or transient heat transfer [86, 174, 249]. In the context of IHCPs, different techniques were applied to address the ill-posedness, e.g., the CGM [45, 174, 175, 283], the Gauss-Newton algorithm [249], the function specification method [70], the reciprocity functional [323] and the Tikhonov regularization [122].

When the boundary condition at the interface is nonlinear, Hu et al. [172] reconstructed an inaccessible boundary of a three-layer composite material from Cauchy data on an accessible boundary and Stefan-Boltzmann radiation conditions at the interfaces. Wei [384] also investigated the boundary identification nonlinear problem with Stefan-Boltzmann interface condition, and showed the uniqueness of a moving boundary for this inverse problem. Murio [310] studied the numerical identification of an interface source function in a generalized nonlinear boundary condition by a stable space-marching finite difference method in conjunction with mollification. In the aforementioned works, the SBC was a constant, but this assumption is not always appropriate, as it can vary spatially along the interface and even temporally, due to damages or thin coatings on the surface. For this reason, Cheng et al. [80] studied an inverse problem of determining the space-dependent SBC and the defect area from temperature measurements on an observation surface.

In this chapter, the generalized inverse problem of reconstructing the space and time-dependent interfacial coefficient (TCC or SBC) at the interface of a bi-material from temperature measurements on an accessible portion of the exterior boundary of the bodies placed in contact is considered. This is formulated in section 8.2, where the uniqueness and the discontinuous dependence on the data of the solution are also briefly discussed. Compared to the one-dimensional (1D) inverse problems investigated in [174, 175], the present setting is formulated in any dimension hence, enabling practical 2D and 3D problems to be considered. Moreover, the thermal properties are assumed space-dependent to enable application to functionally graded materials [321]. The uniqueness result that is sketched in subsection 8.2.1 specifies sufficient data that can be measured, e.g., in the 1D case a boundary temperature measurement with one thermocouple at one end of the finite slab is sufficient, whilst Huang et al. [174] considered an extra intrusive internal thermocouple that is not actually needed for the uniqueness of solution, although, of course, adding more non-redundant information improves stability. In subsection 8.3.3, a preconditioned CGM is developed for overcoming the vanishing of the gradient at final time, encountered in the conventional CGM [67, 174, 175].

8.2 Mathematical Formulation

Consider a two-layer composite heat conductor $\Omega = \Omega_1 \cup \Omega_2$, where Ω_i is the ith subdomain with thermal conductivity k_i and heat capacity per unit volume C_i for $i = 1, 2$. The thermal properties (k_i and C_i) are assumed to be heterogeneous space-dependent known quantities. Let Γ be the interface between these two subdomains and $\partial\Omega_i$ the boundary of the subdomain Ω_i. Thus, $\partial\Omega_1 = \Gamma_1 \cup \Gamma$ and $\partial\Omega_2 = \Gamma_2 \cup \Gamma$.

The heat conduction model is given by the heat equations

$$C_i(\underline{x})\frac{\partial u_i}{\partial t} = \nabla \cdot (k_i(\underline{x})\nabla u_i), \quad (\underline{x}, t) \in \Omega_i \times (0, T), \quad i = 1, 2 \tag{8.1}$$

where $T > 0$ is the final time, with the Neumann boundary conditions,

$$-k_i(\underline{x})\frac{\partial u_i}{\partial n_i} = q_i(\underline{x}, t), \quad (\underline{x}, t) \in \Gamma_i \times (0, T), \quad i = 1, 2, \tag{8.2}$$

the general interface condition on $S_T := \Gamma \times (0, T)$, namely,

$$-k_1(\underline{x})\frac{\partial u_1}{\partial n_1} = k_2(\underline{x})\frac{\partial u_2}{\partial n_2} = \phi(\underline{x}, t)[f(u_1) - f(u_2)], \quad (\underline{x}, t) \in S_T \tag{8.3}$$

and initial conditions,

$$u_i(\underline{x}, 0) = a_i(\underline{x}), \quad \underline{x} \in \Omega_i, \quad i = 1, 2, \tag{8.4}$$

where u_i is the temperature in the domain Ω_i, $\underline{n}_i$ is the outward unit normal to the boundary $\partial\Omega_i$, q_i is the heat flux on the boundary Γ_i, $a_i(\underline{x})$ is the initial temperature in the domain Ω_i for $i = 1, 2$, and, for simplicity, heat sources have been neglected. The materials Ω_1 and Ω_2 can be anisotropic in which case the scalars $k_1(\underline{x})$ and $k_2(\underline{x})$ become symmetric and positive definite tensors. Also, the Neumann boundary conditions (8.2) can be replaced by more general Robin boundary conditions allowing for convection through the boundaries Γ_1 and Γ_2. Equation (8.3) characterises a general condition at the interface Γ between the two materials. If $f(u) = u$, the coefficient $\phi(\underline{x}, t)$ represents the TCC. If $f(u) = u^4$, then equation (8.3) is referred to as the Stefan-Boltzmann interface condition, and $\phi(\underline{x}, t)$ is the SBC, [80]. Both of these two conditions can cause a temperature discontinuity across the interface Γ.

The direct problem consists in the determination of the temperatures u_1 and u_2 in the domains Ω_1 and Ω_2, from the knowledge of the geometry, thermal properties, boundary conditions, initial conditions and the interface coefficient. On the contrary, in the inverse problem, the space- and time-dependent coefficient $\phi(\underline{x},t)$ at the interface Γ and the temperatures u_1 and u_2 are unknown and have to be determined. Thus, the objective is to determine $\phi(\underline{x},t)$, $u_1(\underline{x},t)$ and $u_2(\underline{x},t)$, given Ω_i, k_i, C_i, q_i, a_i, $i=1,2$, and the boundary temperature measured with a thermal camera on an accessible portion $\Gamma_0 \subset \Gamma_1$,

$$u_1(\underline{x}_i,t) = Y(\underline{x}_i,t), \quad (\underline{x}_i,t) \in \Gamma_0 \times (0,T), \quad i = \overline{1,N_m}, \tag{8.5}$$

where $\underline{x} = \underline{x}_i$ for $i = \overline{1,N_m}$ are N_m measurement points on Γ_0.

8.2.1 Ill-posedness of the inverse problem

The inverse problem (8.1)-(8.5) has at most one solution, as follows. First, the uniqueness of solution of the Cauchy problem for u_1 in $\Omega_1 \times (0,T)$ given by (8.1), (8.2) and (8.4) for $i=1$, and (8.5) follows from the Holmgren theorem [172]. In fact, the initial condition (8.4) with $i=1$ is not needed for this uniqueness analytic continuation argument. As a byproduct, it follows that $u_1|_{\Gamma\times(0,T)}$ and $k_1\frac{\partial u_1}{\partial n_1}|_{\Gamma\times(0,T)}$ are uniquely determined. Then, the first identity in (8.3) yields that $k_2\frac{\partial u_2}{\partial n_2}|_{\Gamma\times(0,T)}$ is also known and this, together with (8.2) and (8.4) for $i=2$, form a direct well-posed Neumann problem for the heat equation for u_2 in $\Omega_2 \times (0,T)$. In particular, it yields $u_2|_{\Gamma\times(0,T)}$ and finally, the interface coefficient is uniquely determined from (8.3) as

$$\phi = \frac{-k_1\frac{\partial u_1}{\partial n_1}}{f(u_1) - f(u_2)} \quad \text{on } S_T, \tag{8.6}$$

provided that the denominator is non-zero. Even if the uniqueness holds, the inverse problem (8.1)-(8.5) is still ill-posed since the interface coefficient ϕ does not depend continuously on the input measured data (8.5). This can easily be seen from the following example of instability. Consider a square domain Ω and take $\Omega_1 = (0,x_c)\times(0,l)$, $\Omega_2 = (x_c,l)\times(0,l)$, $\Gamma_0 = \{0\}\times(0,l)$, $\Gamma_1 = \{0\}\times(0,l) \cup (0,x_c)\times\{0,l\}$, $\Gamma_2 = \{l\}\times(0,l)\cup(x_c,l)\times\{0,l\}$, $\Gamma = \{x_c\}\times(0,l)$, where $l=1/4$ and x_c is a fixed value in $(0,l)$. The thermal properties of the materials are taken constant $k_1=k_2=c_1=c_2=1$. Take the function f in (8.3) to be $f(u)=u^4$. With the initial temperatures $a_1(x,y) = p_1\cos(\alpha x)\cos(\pi y)$, $a_2(x,y) = p_2\cos(\alpha(x-l))\cos(\pi y)$, the overspecified data

$u_1(0,y,t) = Y(y,t) = p_1 \cos(\pi y)\, e^{-mt}$, and the heat fluxes

$$q_1(x,y,t) = \begin{cases} 0, & \text{on } x = 0 \text{ or } y = 0, \\ \frac{\pi p_1}{\sqrt{2}} \cos(\alpha x) e^{-mt}, & \text{on } y = l, \end{cases}$$
$$(x,y,t) \in \Gamma_1 \times (0,T),$$

$$q_2(x,y,t) = \begin{cases} 0, & \text{on } x = l \text{ or } y = 0, \\ \frac{\pi p_2}{\sqrt{2}} \cos(\alpha(x-l)) e^{-mt}, & \text{on } y = l, \end{cases}$$
$$(x,y,t) \in \Gamma_2 \times (0,T),$$

where $p_1 = 1/[\alpha \sin(\alpha x_c)]$, $p_2 = 1/[\alpha \sin(\alpha(x_c - l))]$, $\alpha = \sqrt{m - \pi^2}$, $m \in \mathbb{N}$, $m \geq 10$, the solution (u_1, u_2, ϕ) satisfying (8.1)–(8.5) is given by

$$u_1(x,y,t) = p_1 \cos(\alpha x) \cos(\pi y) e^{-mt}, \; (x,y,t) \in \Omega_1 \times (0,T),$$
$$u_2(x,y,t) = p_2 \cos(\alpha(x-l)) \cos(\pi y) e^{-mt}, \; (x,y,t) \in \Omega_2 \times (0,T),$$
$$\phi(y,t) = \frac{e^{3mt}}{\cos^3(\pi y)} \left[p_1^4 \cos^4(\alpha x_c) - p_2^4 \cos^4(\alpha(x_c - l)) \right]^{-1},$$
$$(y,t) \in [0,l] \times (0,T).$$

As $m \to \infty$, the temperatures u_1 and u_2 tend to zero, as well as the input data $Y(y,t)$, while the interface coefficient ϕ becomes unbounded.

8.3 Conjugate Gradient Method (CGM)

The inverse problem of finding the triplet functions (u_1, u_2, ϕ) satisfying (8.1)–(8.5) is solved using the CGM by minimizing the least-squares functional [409]

$$J(\phi) = \frac{1}{2} \sum_{i=1}^{N_m} \| u_1(\underline{x}_i, t; \phi) - Y(\underline{x}_i, t) \|^2_{L^2(0,T)}, \tag{8.7}$$

where $u(\underline{x}_i, t; \phi)$ is the solution of the direct problem for a given particular function ϕ. The CGM is based on the recurrence relationship

$$\phi^{n+1}(\underline{x}, t) = \phi^n(\underline{x}, t) - \beta^n d^n(\underline{x}, t), \quad n \in \mathbb{N}, \tag{8.8}$$

where n is the iteration number and ϕ^0 is an initial guess, β^n is the search step size and $d^n(\underline{x}, t)$ is the direction of descent defined recurrently as,

$$d^0(\underline{x}, t) = J'(\phi^0), \quad d^n(\underline{x}, t) = J'(\phi^n) + \gamma^n d^{n-1}(\underline{x}, t), \quad n \in \mathbb{N}^*, \tag{8.9}$$

where $J'(\phi^n)$ is the gradient of J with respect to ϕ, and γ^n is the conjugate coefficient. Although there are many choices for γ^n, the Polak-Ribiere method is recommended, due to its computational performance [319],

$$\gamma^n = \frac{\langle J'(\phi^n), J'(\phi^n) - J'(\phi^{n-1})\rangle_{L^2(S_T)}}{\|J'(\phi^{n-1})\|^2_{L^2(S_T)}}, \quad n \in \mathbb{N}^*, \tag{8.10}$$

where $\langle \cdot, \cdot \rangle_{L^2}$ denotes the L^2-inner product. Further, the search step size β^n is chosen as the one that minimizes the objective functional J at each iteration,

$$\beta^n = \arg\min_\beta J(\phi^n - \beta d^n). \tag{8.11}$$

By following a similar analysis to that of [322], β^n is obtained as,

$$\beta^n = \frac{\sum_{i=1}^{N_m} \langle u_1(\underline{x}_i, t; \phi^n) - Y(\underline{x}_i, t), \Delta u_1^n(\underline{x}_i, t)\rangle_{L^2(0,T)}}{\sum_{i=1}^{N_m} \|\Delta u_1^n(\underline{x}_i, t)\|^2_{L^2(0,T)}}, \quad n \in \mathbb{N}, \tag{8.12}$$

where $\Delta u_1^n(\underline{x}_i, t) = \Delta u_1(\underline{x}_i, t; d^n)$, $i = \overline{1, N_m}$, is the solution to a sensitivity problem (subsection 8.3.1) with $\Delta\phi^n = d^n$.

8.3.1 Sensitivity problem

To obtain the search step size via Eq.(8.12), a sensitivity problem is constructed. By adding a perturbation $\varepsilon\Delta\phi(\underline{x}, t)$ to $\phi(\underline{x}, t)$, the subsequent responses $u_1(\underline{x}, t)$ and $u_2(\underline{x}, t)$ are perturbed by $\varepsilon\Delta u_1(\underline{x}, t)$ and $\varepsilon\Delta u_2(\underline{x}, t)$, respectively, where ε is a small parameter. By replacing u_1, u_2 and ϕ in the direct problem (8.1)–(8.4) by $(u_1 + \varepsilon\Delta u_1)$, $(u_2 + \varepsilon\Delta u_2)$ and $(\phi + \varepsilon\Delta\phi)$, respectively, and subtracting the resulting formulation from the original direct problem, one obtains the following sensitivity problem:

$$\begin{cases} C_i(\underline{x})\frac{\partial(\Delta u_i)}{\partial t} = \nabla \cdot (k_i(\underline{x})\nabla(\Delta u_i)), & (\underline{x}, t) \in \Omega_i \times (0, T), \\ -k_i(\underline{x})\frac{\partial(\Delta u_i)}{\partial n_i} = 0, & (\underline{x}, t) \in \Gamma_i \times (0, T), \\ \Delta u_i(\underline{x}, 0) = 0, & \underline{x} \in \Omega_i, \quad i = 1, 2, \end{cases} \tag{8.13}$$

with the interface condition,

$$-k_1(\underline{x})\frac{\partial(\Delta u_1)}{\partial n_1} = k_2(\underline{x})\frac{\partial(\Delta u_2)}{\partial n_2} = \phi(\underline{x}, t)[f'(u_1)\Delta u_1 - f'(u_2)\Delta u_2]$$
$$+\Delta\phi(\underline{x}, t)[f(u_1) - f(u_2)], \quad (\underline{x}, t) \in S_T, \tag{8.14}$$

where the terms of order ε^2 have been neglected and use has been made of the first-order approximation $f(u + \varepsilon\Delta u) \approx f(u) + \varepsilon f'(u)\Delta u$.

8.3.2 Adjoint problem

Due to the constraint that $u_1(\underline{x}, t; \phi)$ in the objective functional (8.7) is the solution of the direct problem, two Lagrange multipliers $\lambda_1(\underline{x}, t)$ and $\lambda_2(\underline{x}, t)$

are introduced to construct the constrained objective functional,

$$J(\phi) = \frac{1}{2}\sum_{i=1}^{N_m}\int_0^T [u_1(\underline{x}_i, t; \phi) - Y(\underline{x}_i, t)]^2 \, dt$$

$$+\sum_{j=1}^{2}\int_{\Omega_j}\int_0^T \lambda_j \left[C_j \frac{\partial u_j}{\partial t} - \nabla \cdot (k_j \nabla u_j)\right] dt \, d\Omega_j. \tag{8.15}$$

The functional $J(\phi)$ has a variation $\Delta J(\phi)$ corresponding to the perturbation of ϕ. Note that $\Delta J(\phi)$ is the directional derivative of $J(\phi)$ in the direction of $\Delta\phi$, [322], and thus can be derived from (8.15), as follows:

$$\Delta J(\phi) = \sum_{i=1}^{N_m}\int_{\Omega_1}\int_0^T \Delta u_1 \, [u_1(\underline{x}, t; \phi) - Y(\underline{x}_i, t)] \, \delta(\underline{x} - \underline{x}_i) \, dt \, d\Omega_1$$

$$+\sum_{j=1}^{2}\int_{\Omega_j}\int_0^T \lambda_j \left[C_j \frac{\partial(\Delta u_j)}{\partial t} - \nabla \cdot (k_j \nabla(\Delta u_j))\right] dt \, d\Omega_j, \tag{8.16}$$

where $\delta(\cdot)$ is the Dirac delta function. Using integration by parts, the second and third integrals on the right-hand side of (8.16) become

$$I_j = \int_{\Omega_j}\int_0^T \lambda_j \left[C_j \frac{\partial(\Delta u_j)}{\partial t} - \nabla \cdot (k_j \nabla(\Delta u_j))\right] dt \, d\Omega_j$$

$$= \int_{\Omega_j} C_j \lambda_j \Delta u_j \Big|_{t=0}^{t=T} d\Omega_j - \int_{\Omega_j}\int_0^T C_j \Delta u_j \frac{\partial \lambda_j}{\partial t} \, dt \, d\Omega_j$$

$$-\int_{\partial\Omega_j}\int_0^T k_j \lambda_j \frac{\partial(\Delta u_j)}{\partial n} \, dt \, dS + \int_{\partial\Omega_j}\int_0^T k_j \Delta u_j \frac{\partial \lambda_j}{\partial n} \, dt \, dS$$

$$-\int_{\Omega_j}\int_0^T \Delta u_j \nabla \cdot (k_j \nabla \lambda_j) \, dt \, d\Omega_j, \quad j = 1, 2. \tag{8.17}$$

Substituting (8.17), and the boundary and initial conditions of the sensitivity problem (8.13) into (8.16), give

$$\Delta J(\phi) = \sum_{i=1}^{N_m}\int_{\Omega_1}\int_0^T \Delta u_1 \, [u_1(\underline{x}, t; \phi) - Y(\underline{x}_i, t)] \, \delta(\underline{x} - \underline{x}_i) \, dt \, d\Omega_1$$

$$+\sum_{j=1}^{2}\left[\int_{\Omega_j} C_j \lambda_i(\underline{x}, T) \Delta u_j \, d\Omega_j\right.$$

$$-\int_{\Omega_j}\int_0^T \Delta u_j \left(C_j \frac{\partial \lambda_j}{\partial t} + \nabla \cdot (k_j \nabla \lambda_j)\right) dt \, d\Omega_j$$

$$\left.-\int_{\Gamma}\int_0^T k_j \lambda_j \frac{\partial(\Delta u_j)}{\partial n} \, dt \, dS + \int_{\partial\Omega_j}\int_0^T k_j \Delta u_j \frac{\partial \lambda_j}{\partial n} \, dt \, dS\right].$$

Let the terms containing Δu_1 and Δu_2 vanish and utilize the interface condition (8.14), to obtain the adjoint problems,

$$\begin{cases} C_1(\underline{x})\frac{\partial \lambda_1}{\partial t} + \nabla\cdot(k_1(\underline{x})\nabla\lambda_1) \\ = \sum_{i=1}^{N_m}[u_1(\underline{x},t;\phi) - Y(\underline{x}_i,t)]\delta(\underline{x}-\underline{x}_i), \quad (\underline{x},t)\in\Omega_1\times(0,T), \\ \frac{\partial\lambda_1}{\partial n_1} = 0, \quad (\underline{x},t)\in\Gamma_1\times(0,T), \\ \lambda_1(\underline{x},T) = 0, \quad \underline{x}\in\Omega_1, \\ -k_1(\underline{x})\frac{\partial\lambda_1}{\partial n_1} = \phi(\underline{x},t)f'(u_1)[\lambda_1(\underline{x},t)-\lambda_2(\underline{x},t)], \quad (\underline{x},t)\in S_T, \end{cases} \tag{8.18}$$

$$\begin{cases} C_2(\underline{x})\frac{\partial \lambda_2}{\partial t} + \nabla\cdot(k_2(\underline{x})\nabla\lambda_2) = 0, \quad (\underline{x},t)\in\Omega_2\times(0,T), \\ \frac{\partial\lambda_2}{\partial n_2} = 0, \quad (\underline{x},t)\in\Gamma_2\times(0,T), \\ \lambda_2(\underline{x},T) = 0, \quad \underline{x}\in\Omega_2, \\ k_2(\underline{x})\frac{\partial\lambda_2}{\partial n_2} = \phi(\underline{x},t)f'(u_2)[\lambda_1(\underline{x},t)-\lambda_2(\underline{x},t)], \quad (\underline{x},t)\in S_T. \end{cases} \tag{8.19}$$

Consequently,

$$\Delta J(\phi) = \int_\Gamma\int_0^T \Delta\phi[f(u_1(\underline{x},t)) - f(u_2(\underline{x},t))][\lambda_1(\underline{x},t)-\lambda_2(\underline{x},t)]\,dt\,ds, \tag{8.20}$$

and thus the L^2-gradient of the functional $J(\phi)$ is,

$$J'_L(\phi) = [f(u_1(\underline{x},t)) - f(u_2(\underline{x},t))][\lambda_1(\underline{x},t)-\lambda_2(\underline{x},t)], \quad (\underline{x},t)\in S_T. \tag{8.21}$$

8.3.3 Preconditioning

The usual gradient (8.21) defined in $L^2(S_T)$ may be too rough and result in a poor rate of convergence [346]. The standard CGM can be improved by using operator preconditioning [319], which is widely used for minimizing nonlinear least-squares problems [345]. The Sobolev gradient is smoother than the L^2-gradient and was used to generate preconditioners for the steepest descent method [317]. Moreover, it was applied to the solution of inverse problems, e.g., in electrical impedance tomography [235], for the Robin inverse problem [211, 212] and in the parameter identification for the bio-heat equation [67]. Below, a preconditioner employed in [409] is introduced.

Let $H^{0,1}(S_T) := \{\phi\in L^2(S_T) | \phi_t\in L^2(S_T)\}$ be the Hilbert space of functions endowed with the norm

$$\|\phi\|_{H^{0,1}(S_T)} = \left[\int_0^T \left(\|\phi(\cdot,t)\|^2_{L^2(\Gamma)} + \|\phi_t(\cdot,t)\|^2_{L^2(\Gamma)}\right)dt\right]^{1/2}.$$

Let $J'_H(\phi)$ denote the gradient of $J(\phi)$ defined in the space $H^{0,1}(S_T)$, and $\mathcal{H}^{0,1}_\kappa(S_T)$ the corresponding weighted inner product defined by [212],

$$\Delta J(\phi) := \langle J'_H(\phi), \Delta\phi\rangle_{\mathcal{H}^{0,1}_\kappa(S_T)} = \int_\Gamma\int_0^T \left(J'_H(\phi)\Delta\phi + \kappa\frac{\partial J'_H(\phi)}{\partial t}\frac{\partial(\Delta\phi)}{\partial t}\right)dt\,d\Gamma,$$

where κ is a positive constant to be prescribed. In particular, in one-dimension, ϕ only depends on time, and thus J'_H is referred to as the Sobolev gradient [317] in the space $H^1(0,T)$. Integrating by parts,

$$\Delta J(\phi) = \kappa\Delta\phi \left.\frac{\partial J'_H}{\partial t}\right|_0^T + \int_\Gamma\int_0^T \left(J'_H - \kappa\frac{\partial^2 J'_H}{\partial t^2}\right)\Delta\phi\, dt\, d\Gamma. \tag{8.22}$$

If J'_H satisfies the conditions,

$$\left.\frac{\partial J'_H}{\partial t}\right|_{t=0} = \left.\frac{\partial J'_H}{\partial t}\right|_{t=T} = 0, \tag{8.23}$$

then, from (8.20)–(8.22),

$$J'_H - \kappa\frac{\partial^2 J'_H}{\partial t^2} = J'_L \quad \text{on } S_T. \tag{8.24}$$

In other words, J'_H is obtained from J'_L via $M^{-1} := (I - \kappa\nabla_t^2)^{-1}$, where I is an identity operator and $\nabla_t^2 = \frac{\partial^2}{\partial t^2}$ is a Laplacian in time. The operator M is viewed as a preconditioner for accelerating the convergence and improving the accuracy of solution [212, 319, 345]. Its inverse M^{-1} is essentially an integral operator and has a smoothing effect on the gradient of the objective functional [389]. In addition, κ is regarded as an additional regularization parameter [212, 389]. The preconditioned CGM is implemented by substituting the gradient $J'_H(\phi)$ into (8.9) to obtain a different direction of descent from the standard CGM. Also, the conjugate coefficient in (8.10) is replaced by [235]:

$$\gamma^n = \frac{\langle J'_L(\phi^n), J'_H(\phi^n) - J'_H(\phi^{n-1})\rangle_{L^2(S_T)}}{\langle J'_L(\phi^{n-1}), J'_H(\phi^{n-1})\rangle_{L^2(S_T)}}, \quad n = 1, 2, \ldots. \tag{8.25}$$

Note that the L^2-gradient in (8.21) always vanishes at the final time $t = T$ according to the adjoint problems (8.18) and (8.19), for which $\lambda_1(\underline{x}, T) = \lambda_2(\underline{x}, T) = 0$. Therefore, if the initial guess does not match the exact value of the unknown function ϕ at $t = T$, a direct application of the L^2-gradient J'_L fails to update the estimation of ϕ at $t = T$. This can be avoided by employing the preconditioned gradient J'_H, satisfying the imposed conditions (8.23). For this reason, the presented preconditioned CGM possesses merits of improved accuracy for an arbitrary initial guess in comparison with the standard CGM.

8.3.4 Stopping criterion

The noisy temperature measurements are numerically simulated as

$$Y^{\text{noise}}(\underline{x}_i, t) = Y(\underline{x}_i, t) + \epsilon_i(t), \quad (\underline{x}_i, t) \in \Gamma_0 \times (0,T),\ i = \overline{1, N_m}, \tag{8.26}$$

where $\epsilon_i(t)$ are random variables drawn from a normal distribution with zero mean and standard deviations $\sigma_i = p\% \times \max_{t\in[0,T]} |u(\underline{x}_i, t)|$ for $i = \overline{1, N_m}$, where p is the percentage of noise. As shown in subsection 8.2.1, the inverse problem (8.1)-(8.5) is ill-posed and small errors in the measured temperature (8.5) cause large oscillations in the interface coefficient. On the other hand, the CGM is semi-convergent, i.e., convergence at the beginning of the iterations, but divergence of solution as the iterations proceed [131]. Thus, the iterative procedure is stopped at the first iteration number $k = k(\epsilon)$ for which

$$J(\phi^k) \leq \mathcal{E} = \frac{1}{2}\sum_{i=1}^{N_m} \|\epsilon_i(t)\|^2_{L^2(0,T)}\,. \tag{8.27}$$

8.3.5 Algorithm

Step 1. Set $n = 0$ and choose an arbitrary initial guess $\phi^0(\underline{x}, t)$ for $\phi(\underline{x}, t)$.

Step 2. Solve the direct problem (equations (8.1)–(8.4)) to obtain $u_1^n = u_1(\underline{x}, t; \phi^n)$ and $u_2^n = u_2(\underline{x}, t; \phi^n)$, and calculate the objective functional $J(\phi^n)$. If $J(\phi^n)$ satisfies the stopping criterion (8.27), then stop, else go to step 3.

Step 3. Solve the adjoint problem (equations (8.18) and (8.19)) to calculate $\lambda_1(\underline{x}, t; \phi^n)$ and $\lambda_2(\underline{x}, t; \phi^n)$, and the gradient $J'_L(\phi^n)$ by (8.21).

Step 4. Solve (8.23)–(8.24) to calculate $J'_H(\phi^n)$ from $J'_L(\phi^n)$.

Step 5. Substitute $J'_H(\phi^n)$ into (8.25) and (8.9) to obtain the conjugate coefficient γ^n and the direction of descent d^n, respectively.

Step 6. Solve the sensitivity problem (equations (8.13) and (8.14)) to obtain $\Delta u_1(\underline{x}, t; \phi^n)$ with the condition $\Delta\phi^n = d^n$, and then calculate the search step size β^n using (8.12).

Step 7. Obtain $\phi^{n+1}(\underline{x}, t)$ via (8.8). If $J(\phi^{n+1})$ satisfies the stopping criterion (8.27), then stop, else set $n = n + 1$ and go to step 2.

8.4 Numerical Results and Discussions

The direct, sensitivity and adjoint problems are solved by the alternating-direction implicit (ADI) method [327]. The integrals involved in the implementation of CGM are approximated by the trapezium rule, and the Dirac delta function in (8.18) is approximated by

$$\delta(\underline{x} - \underline{x}_i) \approx \frac{1}{c\sqrt{\pi}} e^{-|\underline{x}-\underline{x}_i|^2/c^2}, \quad i = \overline{1, N_m},$$

where c is a small positive constant, such as 10^{-3}.

Two different types of interface condition will be considered in the following examples; one is the thermal contact condition with unknown TCC and the other is the Stefan-Boltzmann radiation condition with unknown SBC.

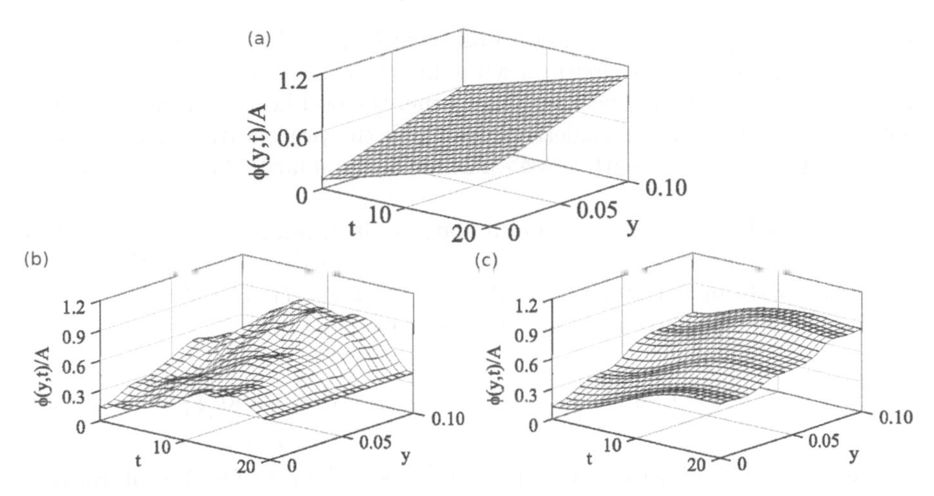

FIGURE 8.1: (a) The exact $\phi(y,t)$ given by (8.28) and the numerical solutions for $p\% = 5\%$ noise with (b) $\kappa = 0$, (c) $\kappa = 10^2$, for Example 8.1, [409].

Example 8.1 Consider the following input data:

$$T = 20s, \quad L_x = L_y = 0.1m, \quad k_1 = 54W/(^\circ C\ m), \quad k_2 = 213W/(^\circ C\ m),$$
$$C_1 = 3.66 \times 10^6 J/(^\circ C\ m^3), \quad C_2 = 3.27 \times 10^6 J/(^\circ C\ m^3)$$
$$x_c = 0.01m, \quad a_1 = 30^\circ C, \quad a_2 = 900^\circ C, \quad q_1 = q_2 = 0,$$

where k_1 and C_1 are thermal properties of AISI 1050 steel [86], whilst k_2 and C_2 correspond to the material considered in [174] for determining the TCC during metal casting. Both Ω_1 and Ω_2 are rectangular, with $\Omega_1 = (0, x_c) \times (0, L_y)$ and $\Omega_2 = (x_c, L_x) \times (0, L_y)$, corresponding to the bi-material being segmented at the vertical interface $x = x_c$ along the y-axis with $y \in (0, L_y)$. The two components are both homogeneous and possess different thermal properties. The boundaries $\partial\Omega_1 \backslash \Gamma$ and $\partial\Omega_2 \backslash \Gamma$ are kept insulated. At the interface $\Gamma = \{x_c\} \times (0, L_y)$, for the linear law $f(u) = u$, the unknown interface coefficient $\phi(y,t)$ represents a TCC, which is time-dependent and varies along the y-direction. The exact solution is taken as the linear function

$$\phi(y,t) = 0.5A\frac{y}{L_y} + 0.5A\frac{t}{T} + 0.1A, \quad (y,t) \in [0, L_y] \times [0, T], \tag{8.28}$$

with the scaling constant $A = 1 \times 10^3 W/(^\circ C\ m^2)$. Assume that the temperature measurements (8.5) are taken at $N_m \geq 2$ points on $\Gamma_0 = \{0\} \times [0, L_y]$, uniformly distributed with a step size $L_y/(N_m - 1)$. Take the initial guess $\phi^0(y,t) = 0.4A$, the numbers of ADI nodes as $N_x = 102$, $N_y = 41$ and $N_t = 101$, and the number of measurement points as $N_m = 10$. For noisy data ($p = 5$), the retrieved solutions for $\kappa = 0$ stopped after 8 interactions and for $\kappa = 10^2$ stopped after 15 iterations according to the criterion (8.27), are shown in Figure 8.1 in comparison with the exact solution (8.28). Besides the improvement of accuracy near the final time, the oscillations in the numerical solutions are smoothed considerably by using the preconditioned CGM with $\kappa = 10^2$, in comparison with those obtained by standard CGM with $\kappa = 0$.

Example 8.2. Take the same input data as in Example 8.1, but the heat transfer at the interface $\Gamma = \{x_c\} \times (0, L_y)$ is governed by a Stefan-Boltzmann radiation condition with unknown SBC, $\phi(y,t)$. Thus, $f(u) = u^4$ and the exact solution of $\phi(y,t)$ is given by the trigonometric function

$$\phi(y,t) = 0.5A \cdot \sin\left(\frac{2\pi y}{L_y}\right) + A, \quad (y,t) \in [0, L_y] \times [0, T], \tag{8.29}$$

where the scaling constant $A = 1 \times 10^{-7} W/(^\circ C^4\ m^2)$. The initial guess is taken as $\phi^0(y,t) = 0.4A$, which is far from the analytical solution at the final time $t = T = 20s$, to see further the merits of the preconditioned CGM over its standard version. The numbers of nodes are taken as $N_x = 102$, $N_y = 41$ and $N_t = 101$, and the number of measurement points $N_m = 10$. For noisy data $(p = 5)$, Figure 8.2 shows $\phi(y,t)$ obtained after 5 iterations for $\kappa = 0$ and 4 iterations for $\kappa = 10^3$, determined by the discrepancy principle (8.27). The oscillations induced by the noise in the input data are removed substantially by using the preconditioned CGM with $\kappa = 10^3$. Moreover, the inaccuracies occurring near the final time are also reduced. The rigorous choice of the smoothing parameter κ in (8.24) is still open to further investigations.

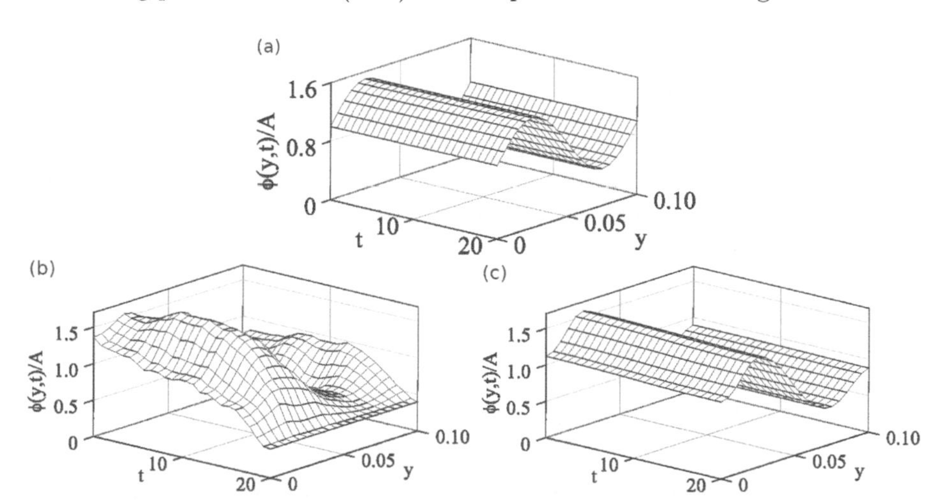

FIGURE 8.2: (a) The exact $\phi(y,t)$ given by (8.29) and the numerical solutions for $p\% = 5\%$ noise with (b) $\kappa = 0$, (c) $\kappa = 10^3$, for Example 8.2, [409].

Chapter 9

Identification of Constant Parameters in Diffusion

This chapter considers the identification of constant parameters in diffusion processes arising from the modelling of the fluid flow through a porous medium or of the heat flow through a conductor.

9.1 Homogeneous and Isotropic Diffusion

The diffusion of a Newtonian fluid through a homogeneous and isotropic porous medium can be obtained by combining the conservation of mass, the state equation of fluid's compressibility and the Darcy's law to result in the diffusion equation

$$\frac{\partial u}{\partial t} = s\nabla^2 u, \tag{9.1}$$

where $s = \kappa/(\phi\mu c\rho_0)$ is the diffusivity, κ and ϕ are the permeability and porosity of the porous medium, respectively, μ is the kinematic viscosity of the fluid, c is the fluid's compressibility, ρ_0 is a reference density, and the dependent variable u can be either the pressure p, or the density ρ (in heat conduction, u represents the temperature). In general, the porous medium hydraulic properties ϕ, κ, the fluid viscosity μ and its small compressibility c may depend upon time t and location $\underline{x}$, but for a homogeneous and isotropic diffusion they remain constant. In (9.1), for simplicity, it was assumed that any sources or sinks were absent. Moreover, if the medium is removing fluid or the fluid is being absorbed by other means, it will be lost at a constant rate $\lambda \geq 0$ proportional to u, and equation (9.1) modifies to include the additional absorption term as

$$\frac{\partial u}{\partial t} = s\nabla^2 u - \lambda u. \tag{9.2}$$

For an infnite medium occupying the whole space $\mathbb{R}^d$, the infinity and initial conditions are assumed uniform and taken to be equal to a constant u_0 and written as

$$\lim_{\|(\underline{x},t)\|\to\infty} u(\underline{x}, t) = u_0, \tag{9.3}$$

DOI: 10.1201/9780429400629-9

and

$$\lim_{t \searrow t_0} = u(\underline{x}, t) = u_0. \tag{9.4}$$

Assuming that the diffusion starts at the initial time $t = t_0$ with the total amount $a > 0$ of fluid concentrated at a location $\underline{x}_0$, then u satisfies the generalized form of equation (9.2), namely,

$$\frac{\partial u}{\partial t} = s\nabla^2 u - \lambda u + \delta((\underline{x}, t) - (\underline{x}_0, t_0)), \tag{9.5}$$

where δ is the Dirac-delta function, and the normalization condition [42, 318]

$$\int_{\mathbb{R}^d} (u(\underline{x}, t) - u_0) d\underline{x} = a e^{-\lambda(t-t_0)}, \quad t > t_0. \tag{9.6}$$

The solution of the direct problem given by (9.3)–(9.6) is [318],

$$u(\underline{x}, t) = u_0 + \frac{a e^{-\lambda(t-t_0)}}{[4\pi s(t-t_0)]^{d/2}} \exp\left(-\frac{\|\underline{x} - \underline{x}_0\|^2}{4s(t-t_0)}\right), \quad t > t_0, \quad \mathbf{x} \in \mathbb{R}^d. \tag{9.7}$$

As for the inverse coefficient problem, this is concerned with finding the constants a, s, λ, t_0 and $\underline{x}_0$ from additional measurements of u at some locations $\underline{x}_i$ and times t_j for $i, j = 1, 2, ...$ Taking the logarithm in (9.7) yields

$$\ln(u(\underline{x}, t) - u_0) = -\frac{\|\underline{x} - \underline{x}_0\|^2}{4s(t-t_0)} - \lambda(t - t_0) - \frac{d}{2}\ln(t - t_0) + C, \tag{9.8}$$

where $C := \ln(a) - \frac{d}{2}\ln(4\pi s)$. Denoting $\psi_{i,j} := \ln(u(\underline{x}_i, t_j) - u_0)$, equation (9.8) applied at $(\underline{x}, t) = (\underline{x}_i, t_j)$ yields

$$\psi_{i,j} = -\frac{\|\underline{x}_i - \underline{x}_0\|^2}{4s(t_j - t_0)} - \lambda(t_j - t_0) - \frac{d}{2}\ln(t_j - t_0) + C. \tag{9.9}$$

The following theorem gives the identifiability of the parameters t_0, λ and C.

Theorem 9.1. ([42, 318]) *Four measurements of $\psi_{i,j}$ at two different times $t_1 \neq t_2 > t_0$ common at each of two distinct points $\underline{x}_1 \neq \underline{x}_2 \in \mathbb{R}^d$, suffice to obtain in closed form the initial time t_0 provided that $\psi_{2,1} - \psi_{1,1} \neq \psi_{2,2} - \psi_{1,2}$. One additional measurement at the location $\underline{x}_1$ or $\underline{x}_2$, but at a distinct time $t_0 < t_3 \notin \{t_1, t_2\}$, namely $\psi_{1,3}$ or $\psi_{2,3}$, lead to a closed form solution for the absorption rate λ and the auxiliary parameter C.*

Proof. The proof of [318] is paraphrased and accommodated, as described in [42]. By using the four measurements $\psi_{i,j}$, $i, j = 1, 2$, the value of t_0 can be found by firstly subtracting $\psi_{1,1}$ from $\psi_{2,1}$, and $\psi_{2,1}$ from $\psi_{2,2}$ giving

$$\psi_{2,1} - \psi_{1,1} = \frac{\|\underline{x}_1 - \underline{x}_0\|^2 - \|\underline{x}_2 - \underline{x}_0\|^2}{4s(t_1 - t_0)},$$

$$\psi_{2,2} - \psi_{1,2} = \frac{\|\underline{x}_1 - \underline{x}_0\|^2 - \|\underline{x}_2 - \underline{x}_0\|^2}{4s(t_2 - t_0)},$$

which yield the identity

$$(\psi_{2,1} - \psi_{1,1})(t_1 - t_0) = (\psi_{2,2} - \psi_{1,2})(t_2 - t_0). \tag{9.10}$$

Upon re-arrangement, equation (9.10) yields explicitly the value of

$$t_0 = \frac{(\psi_{2,1} - \psi_{1,1})t_1 - (\psi_{2,2} - \psi_{1,2})t_2}{(\psi_{2,1} - \psi_{1,1}) - (\psi_{2,2} - \psi_{1,2}))}. \tag{9.11}$$

Expression (9.11) exists because, by the assumption of the theorem, $\psi_{2,1} - \psi_{1,1} \neq \psi_{2,2} - \psi_{1,2}$. Remark that when $\psi_{2,1} - \psi_{1,1} = \psi_{2,2} - \psi_{1,2}$ then, $\|\underline{x}_1 - \underline{x}_0\| = \|\underline{x}_2 - \underline{x}_0\|$. This shows that if the position of measurements lie on the surface of a sphere, then t_0 cannot be determined.

Let us now prove the second part of the theorem. Manipulations of equation (9.9) by taking $\psi_{1,1}(t_1 - t_0) - \psi_{1,k}(t_k - t_0)$ for $k = 2, 3$, lead to the following linear system of two equations in the two unknowns λ and C:

$$\begin{cases} \lambda\Lambda_2 + C\Gamma_2 = \Phi_2, \\ \lambda\Lambda_3 + C\Gamma_3 = \Phi_3, \end{cases} \tag{9.12}$$

where

$$\Phi_k = (t_k - t_0)\left[\psi_{1,k} + \frac{d}{2}\ln(t_k - t_0)\right] - (t_1 - t_0)\left[\psi_{1,1} + \frac{d}{2}\ln(t_1 - t_0)\right],$$
$$\Lambda_k = (t_1 - t_0)^2 - (t_k - t_0)^2, \quad \Gamma_k = t_k - t_1, \quad k = 2, 3.$$

This linear system of equations has the determinant

$$\begin{vmatrix} \Lambda_2 & \Gamma_2 \\ \Lambda_3 & \Gamma_3 \end{vmatrix} = (t_2 - t_3)(t_1 - t_2)(t_3 - t_1),$$

which is non-zero because the times t_1, t_2 and t_3 are mutually distinct. Thus, the solution of the system of equations (9.12) is unique and given by

$$\lambda = \frac{\begin{vmatrix} \Phi_2 & \Gamma_2 \\ \Phi_3 & \Gamma_3 \end{vmatrix}}{\begin{vmatrix} \Lambda_2 & \Gamma_2 \\ \Lambda_3 & \Gamma_3 \end{vmatrix}}, \quad C = \frac{\begin{vmatrix} \Lambda_2 & \Phi_2 \\ \Lambda_3 & \Phi_3 \end{vmatrix}}{\begin{vmatrix} \Lambda_2 & \Gamma_2 \\ \Lambda_3 & \Gamma_3 \end{vmatrix}}. \tag{9.13}$$

The following theorem gives the identifiability of the parameters a, s and $\underline{x}_0$.

Theorem 9.2. ([42, 318]) *A number of $(d+2)$ measurements of $\psi_{i,1}$ at a common time $t_1 > t_0$ at each of $(d+2)$ distinct points $\underline{x}_i \neq \underline{x}_0$ for $i = \overline{1,(d+2)}$, in general position in $\mathbb{R}^d$ suffice to obtain in closed form the centre of diffusion $\underline{x}_0$ and the parameters a and s.*

Proof. The proof of [318] is completed and accommodated, as described in [42]. Having found t_0, C and λ, they can be substituted into the rearranged form of (9.9) to yield

$$\frac{\|\underline{x}_i - \underline{x}_0\|^2}{s} = -4(t_1 - t_0)\left[\psi_{i,1} + \lambda(t_1 - t_0) + \frac{d}{2}\ln(t_1 - t_0) - C\right], \quad i = \overline{1,(d+2)}. \quad (9.14)$$

Since $s > 0$ and $\underline{x}_i \neq \underline{x}_0$ for $i = \overline{1,(d+2)}$, it follows that the right hand side of (9.14) is positive. Denote by

$$r_{i,j}^2 := \frac{\|\underline{x}_i - \underline{x}_0\|^2}{\|\underline{x}_j - \underline{x}_0\|^2} = \frac{\psi_{i,1} + \lambda(t_1 - t_0) + \frac{d}{2}\ln(t_1 - t_0) - C}{\psi_{j,1} + \lambda(t_1 - t_0) + \frac{d}{2}\ln(t_1 - t_0) - C} > 0, \quad i \neq j = \overline{1,(n+2)}. \quad (9.15)$$

If $r_{i,j} \neq 1$, then it can be easily seen that the first equality in (9.15) constraints $\underline{x}_0$ to lie on the surface of a d-dimensional sphere centred at the point $\underline{P}_{i,j} := \frac{\underline{x}_i - r_{i,j}^2\underline{x}_j}{1 - r_{i,j}^2}$ of radius $R_{i,j} := \frac{r_{i,j}\|\underline{x}_i - \underline{x}_j\|}{|r_{i,j,}^2 - 1|}$. On the other hand, if $r_{i,j} = 1$, then the first equality in (9.15) constraints $\underline{x}_0$ to lie on the hyperplane

$$\underline{x}_0 \cdot (\underline{x}_i - \underline{x}_j) = \frac{1}{2}\left(\|\underline{x}_i\|^2 - \|\underline{x}_j\|^2\right),$$

with normal vector $\underline{x} = \underline{x}_i - \underline{x}_j$. Taking $j = 1$ in the above expression and subtracting the equations of a sphere for $i = 2$ and $i = \overline{3,(d+2)}$, one obtains the following d linear equations:

$$2\underline{x}_0 \cdot (\underline{P}_{2,1} - \underline{P}_{i,1}) = R_{i,1}^2 - R_{2,1}^2 + \|\underline{P}_{2,1}\|^2 - \|\underline{x}_{i,1}\|^2, \quad i = \overline{3,(d+2)} \quad (9.16)$$

to determine the d components of the centre of diffusion $\underline{x}_0$. For $(d+2)$ distinct points $\underline{x}_i \neq \underline{x}_0$ for $i = \overline{1,(d+2)}$, in general position in $\mathbb{R}^d$, this determined system of equations has a unique solution, which can be determined using the Cramer rule, for example. Once $\underline{x}_0$ has been determined, equation (9.14) yields the diffusion coefficient s. Finally, knowing C and s yields $a = (4\pi s)^{d/2}e^C$.

Example 9.1 The inverse diffusion model previously described can be applied to invert the experimental data from a tumour dosage [318]. These measurements are inherently contaminated with noisy errors are therefore, apart from the uniqueness of solution, the instability phenomenon may also appear. This experiment concerns the three-dimensional, i.e. $d = 3$, unsteady diffusion of a dopamine fluid injected at the initial time $t_0 = 0$ in the brain of a monkey with a dimensionless (normalized) amount $a = 1000$, where $u_0 = 0$. Twelve measurements of the concentration $u(\underline{x}_i, t_1)$ are taken at the same instant $t_1 = 1$ hour at various locations $\underline{x}_i \in \mathbb{R}^3$ for $i = \overline{1,12}$, see Table 9.1. In this table the values of $\psi_{i,1} = \ln(u(\mathbf{x}_i, t_1))$ for $i = \overline{1,12}$, are also given, as they are needed in (9.9). Note that these data already contain both random errors

TABLE 9.1: The concentration measurements $u_i := u(\underline{x}_i, t_1)$, $\psi_{i,1} = \ln(u_i)$ and the locations $\underline{x}_i$ for $i = \overline{1, 12}$, [42, 318].

$\underline{x}_i$	u_i	$\psi_{i,1}$
$\underline{x}_1 = (1.5, 1.9, 1.5)$	43	3.761
$\underline{x}_2 = (3.5, 1.9, 1.5)$	78	4.356
$\underline{x}_3 = (1.5, 2.66, 1.5)$	12	2.484
$\underline{x}_4 = (3.5, 2.66, 1.5)$	42	3.737
$\underline{x}_5 = (1.5, 1.9, 3.5)$	19	2.944
$\underline{x}_6 = (3.5, 1.9, 3.5)$	26	3.258
$\underline{x}_7 = (1.5, 2.66, 3.5)$	4	1.386
$\underline{x}_8 = (3.5, 2.66, 3.5)$	15	2.708
$\underline{x}_9 = (2.5, 1.9, 1.5)$	77	4.343
$\underline{x}_{10} = (2.5, 2.66, 1.5)$	55	4.007
$\underline{x}_{11} = (2.5, 1.9, 3.5)$	53	3.970
$\underline{x}_{12} = (2.5, 2.66, 3.5)$	22	3.091

from measurements, as well as additional round-off errors caused by taking only the first few decimals in the logarithmic values for $\psi_{i,1}$ for $i = \overline{1, 12}$.

Since the practical measurement data is given at a single fixed time t_1 we cannot apply Theorem 9.1, but Theorem 9.2. The data $\psi_{i,1}$ for $i = \overline{1, 12}$ can be applied to equation (9.14) to see if one can determine the centre of diffusion $\underline{x}_0$, the rate of absorption λ and the diffusivity s. Given that $t_1 = 1$, $t_0 = 0$, $d = 3$ and $a = 1000$, equation (9.14) yields

$$\|\underline{x}_i - \underline{x}_0\|^2 = -4s\left[\psi_{i,1} + \lambda + \frac{3}{2}\ln(s) + 3\ln\left(\frac{\sqrt{\pi}}{5}\right)\right], \quad i = \overline{1, 12} \tag{9.17}$$

or, on eliminating λ,

$$\underline{x}_0 \cdot (\underline{x}_1 - \underline{x}_i) + 2s(\psi_{i,1} - \psi_{1,1}) = \frac{1}{2}\left(\|\underline{x}_1\|^2 - \|\underline{x}_i\|^2\right), \quad i = \overline{2, 12}.$$

This produces a system of 11 linear equations in the four unknowns $\underline{x}_0 = (x_{0,1}, x_{0,2}, x_{0,3})$ and s, which, when solved using an ordinary least-squares method, yields $\underline{x}_0 = (2.657, 1.695, 2.217)$ and $s = 0.260$. Finally, the value of λ can be obtained from (9.17) applied for $i = 1$ to give $\lambda = 0.897$, [42].

9.2 A Two-Dimensional Tracer Dispersion Problem

Tracer experiments have been widely used to characterize dispersion of fluids in flowing systems [39], and this section will investigate how a column of fluid diffuses into an orthotropic medium. A practical application of this is that of oil diffusing from a well. In tracer experiments, the liquid phase flow and mixing fields are essentially two-dimensional in the plane of the tray [302]. The experiment is modelled as a point source injection at a rate equal to the Dirac delta function of non-volatile tracer into an infinite plane $\delta((\underline{x},t)-(\underline{x}_0,t_0))$, [171]. If one considers a sufficiently small region in which the gradients in the froth density are not too large such that we can neglect the hydrodynamic and turbulent contributions, one can approximate the flow field as uniform with constant intensity dispersion tensor in the two-dimensional plane. If it is further assumed that only the diagonal components D_1 and D_2 of this tensor are important, i.e. the medium is orthotropic, and that the liquid density is constant, then the governing parabolic partial differential equation for the average, liquid phase tracer concentration is given by [121],

$$\frac{\partial u}{\partial t}+V_1\frac{\partial u}{\partial x}+V_2\frac{\partial u}{\partial y}-D_1\frac{\partial^2 u}{\partial x^2}-D_2\frac{\partial^2 u}{\partial y^2}+\lambda u=\delta((x,y,t)-(x_0,y_0,t_0)), \quad (9.18)$$

where $\lambda \geq 0$ is the rate at which the fluid is being absorbed from the medium, and V_1 and V_2 are the components of fluid velocity. Equation (9.18) accounts for dispersion, advection and absorption. Applying the conditions (9.3), (9.4) and (9.6) from the isotropic model when $d=2$ and $u_0=0$,

$$\lim_{\|(x,y,t)\|\to\infty} u(x,y,t)=0, \quad \lim_{t\searrow t_0} = u(x,y,t)=0, \quad (9.19)$$

$$\int\int_{\mathbb{R}^2} u(x,y,t)dxdy = ae^{-\lambda(t-t_0)}. \quad (9.20)$$

Note that by employing the transformation

$$u(x,y,t)=v(x,y,t)\ \exp(\Delta(x,y,t)), \quad (9.21)$$

where

$$\Delta(x,y,t):=\frac{V_1x}{2D_1}+\frac{V_2y}{2D_2}+t\left(\frac{V_1^2}{4D_1}+\frac{V_2^2}{4D_2}\right) \quad (9.22)$$

the advection term can be eliminated such that equation (9.18) reduces to

$$\begin{aligned}&\frac{\partial v}{\partial t}-D_1\frac{\partial^2 v}{\partial x^2}-D_2\frac{\partial^2 v}{\partial y^2}+\left(\lambda+\frac{V_1^2}{2D_1}+\frac{V_2^2}{2D_2}\right)v\\&\quad=\delta((x,y,t)-(x_0,y_0,t_0))\exp(-\Delta(x,y,t)).\end{aligned} \quad (9.23)$$

Employing further the change of coordinates [73],

$$\xi = x(\sqrt{D_1D_2}/D_1)^{1/2}, \quad \eta = y(\sqrt{D_1D_2}/D_2)^{1/2}, \quad (9.24)$$

the orthotropic equation (9.23) takes the isotropic form

$$\begin{aligned}&\frac{\partial v}{\partial t}-\sqrt{D_1D_2}\left(\frac{\partial^2 v}{\partial \xi^2}+\frac{\partial^2 v}{\partial \eta^2}\right)+\left(\lambda+\frac{V_1^2}{2D_1}+\frac{V_2^2}{2D_2}\right)v\\&=\delta[(\xi(\sqrt{D_1D_2}/D_1)^{-1/2},\eta(\sqrt{D_1D_2}/D_1)^{-1/2},t)-(x_0,y_0,t_0)]\\&\quad\times\exp[-\Delta(\xi(\sqrt{D_1D_2}/D_1)^{-1/2},\eta(\sqrt{D_1D_2}/D_1)^{-1/2},t)].\end{aligned} \quad (9.25)$$

One can easily check that

$$u(x,y,t) = \frac{a e^{-\lambda(t-t_0)}}{4\pi(t-t_0)\sqrt{D_1 D_2}} \exp(-A(x,y,t)), \tag{9.26}$$

where

$$A(x,y,t) := \frac{D_2[x - x_0 - (t-t_0)V_1]^2 + D_1[y - y_0 - (t-t_0)V_2]^2}{4(t-t_0)D_1 D_2},$$

satisfies the direct problem (9.18)–(9.20).

As far as the inverse coefficient problem is concerned, this requires finding the constants t_0, a and λ, while assuming that the dispersion coefficients D_1 and D_2, and the fluid velocity components V_1 and V_2 are known constants, and that the centre of diffusion/dispersion (x_0, y_0) is also known. First, take the logarithm in (9.26) to obtain

$$\ln(4\pi u(x,y,t)) = -A(x,y,t) - \lambda(t-t_0) - \ln(t-t_0) + \beta, \tag{9.27}$$

where $\beta := \ln(a) - \frac{1}{2}\ln(D_1 D_2)$. Then, apply (9.27) at the measurement points (x_i, y_i, t_j) and denote $\phi_{i,j} := \ln(4\pi u(x_i, y_i, t_j))$ to obtain

$$\phi_{i,j} = -A(x_i, y_i, t_j) - \lambda(t_j - t_0) - \ln(t_j - t_0) + \beta. \tag{9.28}$$

Theorem 9.3. ([42]) *The initial time t_0 can be found using four measurements of u at two different positions (x_1, y_1) and (x_2, y_2), and at two different times $t_1 \neq t_2 > t_0$, provided that $\phi_{2,1} - \phi_{1,1} \neq \phi_{2,2} - \phi_{1,2}$. Furthermore, one additional measurement of $u(x_1, y_1, t_3)$ or $u(x_2, y_2, t_3)$ at the same location (x_1, y_1) or (x_2, y_2), but at a third different time $t_0 < t_3 \notin \{t_1, t_2\}$ leads to a closed form unique retrieval of the absorption rate λ and the auxiliary parameter β.*

Proof. Subtract $\phi_{2,1}$ from $\phi_{1,1}$, and $\phi_{2,2}$ from $\phi_{1,2}$, to obtain

$$\begin{aligned}
\phi_{2,1} - \phi_{1,1} &= A(x_1, y_1, t_1) - A(x_2, y_2, t_1) \\
&= \frac{(x_1 - x_2)[x_1 + x_2 - 2x_0 - 2(t_1 - t_0)V_1]}{4D_1(t_1 - t_0)} \\
&\quad + \frac{(y_1 - y_2)[y_1 + y_2 - 2y_0 - 2(t_1 - t_0)V_2]}{4D_2(t_1 - t_0)}, \\
\phi_{2,2} - \phi_{1,2} &= A(x_1, y_1, t_2) - A(x_2, y_2, t_2) \\
&= \frac{(x_1 - x_2)[x_1 + x_2 - 2x_0 - 2(t_2 - t_0)V_1]}{4D_1(t_2 - t_0)} \\
&\quad + \frac{(y_1 - y_2)[y_1 + y_2 - 2y_0 - 2(t_2 - t_0)V_2]}{4D_2(t_2 - t_0)}.
\end{aligned} \tag{9.29}$$

From these

$$\begin{aligned}
&4D_1 D_2[(\phi_{2,2} - \phi_{1,2})(t_2 - t_0) - (\phi_{2,1} - \phi_{1,1})(t_1 - t_0)] \\
&\quad = 2(t_1 - t_2)[D_2 V_1(x_1 - x_2) + D_1 V_2(y_1 - y_2)].
\end{aligned} \tag{9.30}$$

The rearrangement of (9.30) in terms of t_0 gives

$$t_0 = \frac{(t_1 - t_2)[D_2V_1(x_1 - x_2) + D_1V_2(y_1 - y_2)]}{2D_1D_2(\phi_{2,1} - \phi_{1,1}) - (\phi_{2,2} - \phi_{1,2})} + \frac{t_1(\phi_{2,1} - \phi_{1,1}) - t_2(\phi_{2,2} - \phi_{1,2})}{(\phi_{2,1} - \phi_{1,1}) - (\phi_{2,2} - \phi_{1,2})}. \tag{9.31}$$

This expression is well-defined since the denominator is non-zero from the hypothesis of the theorem. If $\chi := \phi_{2,1} - \phi_{1,1} = \phi_{2,2} - \phi_{1,2}$ then we cannot apply directly equation (9.31). However, equation (9.30) gives

$$4D_1D_2(t_2 - t_1)\chi = 2(t_1 - t_2)[D_2V_1(x_1 - x_2) + D_1V_2(y_1 - y_2)], \tag{9.32}$$

or, upon simplifying with $t_1 - t_2 \neq 0$ yields

$$-2D_1D_2\chi = D_2V_1(x_1 - x_2) + D_1V_2(y_1 - y_2). \tag{9.33}$$

Further, from equations (9.29) and (9.33)

$$D_2(x_1 - x_2)(x_1 + x_2 - 2x_0) = D_1(y_1 - y_2)(y_1 + y_2 - 2y_0),$$

or, rearranging

$$\frac{(x_1 - x_0)^2}{D_1} + \frac{(y_1 - y_0)^2}{D_2} = \frac{(x_2 - x_0)^2}{D_1} + \frac{(y_2 - y_0)^2}{D_2}. \tag{9.34}$$

This rearrangement shows that if the positions of the measurements conform to lie on the boundary of this ellipse then, t_0 cannot be determined.

Consider now the second part of the theorem for the determination of the absorption rate λ and the auxiliary parameter β. Using equation (9.28), subtract $(t_1 - t_0)\phi_{1,1}$ from $(t_k - t_0)\phi_{1,k}$ with $k = 2, 3$, to get

$$(t_k - t_0)\phi_{1,k} - (t_1 - t_0)\phi_{1,1} = \beta(t_k - t_1) + \lambda[(t_1 - t_0)^2 - (t_k - t_0)^2]$$
$$+(t_1 - t_0)[\ln(t_1 - t_0) + A(x_1, y_1, t_1)] - (t_k - t_0)[\ln(t_k - t_0) + A(x_1, y_1, t_k)].$$

This can be rearranged as the system of equations (9.12), where

$$\Phi_k = (t_k - t_0)[\phi_{1,k} + \ln(t_k - t_0) + A(x_1, y_1, t_k)]$$
$$-(t_1 - t_0)[\phi_{1,1} + \ln(t_1 - t_0) + A(x_1, y_1, t_1)], \ k = 2, 3.$$

The final solution for λ and β is then given in closed form by (9.13).

9.3 Determination of Constant Thermal Properties

Inverse problems of determining unknown constant parameters in heat conduction were treated previously in some detail in [12]. Usually, they involve the determination of a single parameter from overspecified boundary data. In some applications, however, it is desirable to determine more than one parameter [59]. Consider for example, the one-dimensional conduction of heat in a homogeneous medium of finite length. Using certain experiments, the thermal conductivity and heat capacity of a homogeneous heat conductor can uniquely be determined from a single measurement (at two instants) of the heat flux [56]. In [56], the case of a semi-infinite homogeneous heat conductor was also investigated and it was shown that further data, must be specified to obtain a unique solution. In [56, 59], the determination of the two thermophysical property coefficients was accomplished by means of an intersecting graph technique. However, this method is restricted to the identification of only two parameters and cannot easily be extended to more than two unknowns. Therefore, in this section, an iterative nonlinear least-squares BEM is employed for the numerical inversion. Numerically, it is shown that the thermal properties can uniquely and stably be retrieved from two measurements containing at least one heat flux for finite homogeneous heat conductors, whilst theoretically, for semi-infinite heat conductors it is shown that one heat flux and one internal temperature measurement are necessary and sufficient [255].

9.3.1 Finite homogeneous conductor

Consider a homogeneous finite slab heat conductor of length $l > 0$ initially at a uniform temperature. The upstream face of the sample $x = 0$ is heated, whilst the downstream face $x = l$ is kept insulated. Then, the inverse problem consists of determining a pair of positive constants (C, K) and a function $u \in C^1([0, l] \times [0, T]) \cap C^2((0, l) \times (0, T))$ such that

$$\begin{cases} C\frac{\partial u}{\partial t}(x,t) = K\frac{\partial^2 u}{\partial x^2}(x,t), & (x,t) \in (0,l) \times (0,T), \\ u(x,0) = 0, & x \in (0,l), \\ u(0,t) = \mu_0(t), & t \in (0,T), \\ \frac{\partial u}{\partial x}(l,t) = 0, & t \in (0,T), \end{cases} \tag{9.35}$$

plus any two of the following additional boundary measurements:

$$(a)\ -K\frac{\partial u}{\partial x}(0,\bar{t}_1) = q_1, \quad (b)\ -K\frac{\partial u}{\partial x}(0,\bar{t}_2) = q_2,$$
$$(c)\ u(l,\bar{t}_1) = g_1, \quad (d)\ u(l,\bar{t}_2) = g_2, \tag{9.36}$$

where $0 < \bar{t}_1 < \bar{t}_2 \leq T$ are given instants, μ_0 is a strictly monotone increasing continuously differentiable given function satisfying $\mu_0(0) = 0$, and $q_1 > 0$, $q_2 > 0$ are given numbers. In (9.35), u is the temperature, and C and K are the heat capacity and thermal conductivity, respectively. It is not necessary that the times at which we record the heat flux in (9.36)(a,b) and the temperature in (9.36)(c,d) measurements be the same. Also, temperature measurements at internal locations $(x_i, \bar{t}_i)$, $i = 1, 2$, with $0 < x_1 \leq x_2 < l$, may replace the boundary temperature measurements in (9.36)(c,d).

The case of a single unknown parameter $\alpha = K/C$ representing the thermal diffusivity of the heat conductor was investigated in [52], and the more general case of a single unknown time-dependent diffusivity $\alpha(t) = K(t)/C(t)$ in [51], see also Chapter 10. However, in some situations, it is desirable to determine both C and K. Using the Green function for the initial boundary value formulation (9.35), it can be shown that the solution (C, K, u) of the inverse problem given by equations (9.35), (9.36)(a,b) is unique [56]. However, for inhomogeneous conditions the theoretical analysis becomes more difficult [103] and numerical methods become necessary. For this, remark that the boundary integral equation associated to (9.35) is given by

$$0.5u(x,t) = \alpha \int_0^t \left[\mu_0(\tau) \frac{\partial G}{\partial \xi}(x,t;0,t\tau) - u(l,\tau)\frac{\partial G}{\partial \xi}(x,t;l,\tau) \right.$$
$$\left. -G(x,t;0,\tau)\frac{\partial u}{\partial \xi}(0,\tau) \right] dt\tau, \quad t \in (0,T),\ x \in \{0,l\}, \tag{9.37}$$

where the fundamental solution (3.23) is slightly modified to account for the thermal diffusivity α, namely,

$$G(x,t;\xi,\tau) = \frac{\mathrm{H}(t-\tau)}{\sqrt{4\pi\alpha(t-\tau)}} \exp\left(-\frac{(x-\xi)^2}{4\alpha(t-\tau)} \right).$$

The BEM, based on discretizing (9.37) using N constant uniform boundary elements on the time interval $[0,T]$ at each of the boundaries $x = 0$ and $x = l$, is employed, as described in the previous chapters related to the heat equation.

Consider the same example investigated in [59] using a graphical intersecting finite-difference technique, and therefore take $l = 1$, $\mu_0(t) = t$, $T = 12$, $\bar{t}_1 = 2$ and $\bar{t}_2 = 10$. The data q_1, q_2, g_1 and g_2 in (9.36) is generated numerically by solving using the BEM, [255], the direct mixed well-posed problem given by (9.35) in which $C = 0.7$ and $K = 0.2$ are assumed to be known.

For inversion, minimize the nonlinear least-squares function

$$S(C,K) := \lambda_1|q_1 - (-K\partial_x u(0,\bar{t}_1))^c|^2 + \lambda_2|q_2 - (-K\partial_x u(0,\bar{t}_2))^c|^2$$
$$+\lambda_3|g_1 - u^c(l,\bar{t}_1)|^2 + \lambda_4|g_2 - u^c(l,\bar{t}_2)|^2, \tag{9.38}$$

where the superscript ^{c} stands for the numerically calculated quantities from iteratively solving using the BEM the mixed direct problem (9.35). The minimization of the function S defined in (9.38) is performed subject to the simple bounds on the variables $0 < m_1 \leq C \leq M_1 < \infty$ and $0 < m_2 \leq K \leq M_2 < \infty$. In expression (9.38), the constants λ_i for $i = \overline{1,4}$, are chosen zero if the corresponding quantity with which they are multiplied is not measured, and $\lambda_i = 1$ for $i = \overline{1,3}$, $\lambda_4 = 10^{-2}$ otherwise. The value of 10^{-2} instead of unity for λ_4 was included for scaling purposes only, because of the different order of magnitudes of the measurements q_1, q_2, g_1 and g_2. If other values of λ_4 such unity, 10^{-3}, or 10^{-4} are used then the convergence of the iterative process will still be achieved, but in a larger number of iterations. Based on this discussion, six

inverse problems are formulated given by (9.35) together with either of the following additional data:
(i) (9.36)(a,b), i.e. $\lambda_1 = \lambda_2 = 1$, $\lambda_3 = \lambda_4 = 0$;
(ii) (9.36)(a,c), i.e. $\lambda_1 = \lambda_3 = 1$, $\lambda_2 = \lambda_4 = 0$;
(iii) (9.36)(a,d), i.e. $\lambda_1 = 1$, $\lambda_4 = 10^{-2}$, $\lambda_2 = \lambda_3 = 0$;
(iv) (9.36)(b,c), i.e. $\lambda_2 = \lambda_3 = 1$, $\lambda_1 = \lambda_4 = 0$;
(v) (9.36)(b,d), i.e. $\lambda_2 = 1$, $\lambda_4 = 10^{-2}$, $\lambda_1 = \lambda_3 = 0$;
(vi) (9.36)(c,d), i.e. $\lambda_3 = 1$, $\lambda_4 = 10^{-2}$, $\lambda_1 = \lambda_2 = 0$.

The inverse formulation (vi) given by equations (9.35), (9.36)(c,d) involve only the thermal diffusivity parameter $\alpha = K/C$, and therefore in this case only α can be retrieved. In contrast, all the inverse formulations (i)–(v) contain both C and K since always a flux measurement (9.36)(a) or (9.36)(b) is taken into account, and in these situations both properties K and C are expected to be retrieved. Note that in case (vi) the inverse problem is ill-posed since the solution is not unique, and obviously unstable, whilst in the cases (i)-(v) the corresponding inverse problems are well-posed and therefore no regularization is necessary. The commonly issue of inverse problems being ill-posed is not present in these latter formulations since only a small number of parameters are to be identified, and thus there is no need to further penalize the least-squares functional (9.38). Nevertheless, regularization would be necessary if, for example, the heat conductor is inhomogeneous.

In a first instance, since only two parameters C and K are to be estimated, one can simply plot the graph of the function $S : [m_1, M_1] \times [m_2, M_2] \to \mathbb{R}_+$ given by (9.38) and search for its minimum. At least for the formulation (i), it is known that the function $S(C, K)$ has a unique global minimum [56]. However, as it also happens with the graphical intersecting technique of [56], this approach is restricted to two-parameters identification only, since in higher dimensions the manifolds generated by the graph of the function S become impossible to be graphically illustrated. Furthermore, the graphical inspection of local minima which may occur, especially when the input data is noisy, becomes difficult and not rigorous. Finally, as it also happens with the simulated annealing or genetic algorithm techniques, the interval $[m_1, M_1] \times [m_2, M_2]$ in which the minimal solution (C, K) is sought, ought to be quite tight in order for a feasible time computation to be possible. Therefore, to overcome these restrictions, a constrained minimization of the function S, given by the NAG routine E04UCF, is employed. This routine is based on an iterative sequential programming technique which is designed to minimize an arbitrary smooth function subject to simple bounds on the variables and linear and/or nonlinear constraints. The constraints that the solution (C, K) should be positive and finite are numerically implemented by taking the lower bound as a very small positive number such as $m_i = 10^{-10}$ and the upper bound as a very large number such as $M_i = 10^{10}$ for $i = 1, 2$.

The BEM is employed with $N = 120$ boundary elements. The gradient of the least-squares function (9.38) is calculated using forward finite differences with a step size h, i.e.,

$$\frac{\partial S}{\partial C}(C,K) = \frac{S(C+h,K) - S(C,K)}{h}, \quad \frac{\partial S}{\partial K}(C,K) = \frac{S(C,K+h) - S(C,K)}{h}.$$

For the accurate evaluation of the gradient ∇S, $h = 10^{-3}$ is used for the formulations (i) and (v), $h = 10^{-4}$ for the formulation (iii), and $h = 10^{-5}$ for the formulations (ii), (iv), and (vi).

An arbitrary initial guess $(C_0, K_0) = (1, 1)$ was chosen. Several other guesses were considered and it was found that the numerical results did not differ significantly for all the formulations (i)$-$(v), showing that the constrained minimization approach proposed is robust. In formulation (vi), different initial guesses gave different values for the retrieved coefficients K and C, however the retrieval of the ratio $\alpha = K/C$ was found to be independent of the initial guess, as expected.

The stability of the numerical solution has been investigated by introducing noise in the measured data given by equations (9.36) using the truncated values $q_1 = 0.56$, $q_2 = 0.69$, $g_1 = 0.69$ and $g_2 = 8.2$, of the direct problem BEM solution.

The optimal solution has been reached in 15 iterations in the formulation (i), 8 iterations in the formulations (ii) and (v), 10 iterations in the formulations (iii) and (iv), and 4 iterations in the formulation (vi). Table 9.2 shows the numerical retrieved values of C and K for the formulations (i)$-$(vi) obtained when both exact and noisy data are inverted. The minimum values of the function S and the CPU time (in seconds) are also included. From Table 9.2 it can be seen that in all the inverse formulations (i)$-$(v) accurate and stable approximations to the exact solution $(C, K) = (0.7, 0.2)$ are obtained, whilst in the inverse formulation (vi), as expected, only the thermal diffusivity $\alpha = K/C = 0.2/0.7 \approx 0.2857$ can be retrieved.

9.3.2 Semi-infinite homogeneous conductor

It is interesting to consider the semi-infinite analog of problem (9.35) and (9.36) for which a similar theoretical analysis to that performed in [56] can be developed [255]. That is, find two positive constants C and K, and a bounded function $u \in C^1([0,\infty) \times [0,\infty)) \cap C^2((0,\infty) \times (0,\infty))$ such that

$$\begin{cases} C\frac{\partial u}{\partial t}(x,t) = K\frac{\partial^2 u}{\partial x^2}(x,t), & (x,t) \in (0,\infty) \times (0,\infty), \\ u(x,0) = 0, & x \in [0,\infty), \\ u(0,t) = \mu_0(t), & t \in (0,\infty), \end{cases} \tag{9.39}$$

plus the following two additional boundary measurements:

$$(a) \; -K\frac{\partial u}{\partial x}(0,\overline{\overline{t}}_1) = \overline{q}_1, \quad (b) \; u(x_1,\overline{\overline{t}}_1) = \overline{g}_1, \tag{9.40}$$

where $\overline{\overline{t}}_1 > 0$, $x_1 > 0$ are given, μ_0 is a strictly monotone increasing continuously differentiable given function satisfying $\mu_0(0) = 0$, and $\overline{q}_1 > 0$ and $\overline{g}_1$

TABLE 9.2: The numerically retrieved values of C and K for the formulations (i)-(vi) obtained when exact data q_1, q_2, g_1, g_2 and noisy data $q_1 = 0.56$, $q_2 = 0.69$, $g_1 = 0.69$, $g_2 = 8.2$ are inverted.

Formulation	C	K	$S(C,K)$
(i) noiseless	0.6999	0.1999	$1.7E-19$
(i) noisy	0.6903	0.2036	$2.6E-19$
(ii) noiseless	0.6999	0.1999	$2.6E-21$
(ii) noisy	0.6982	0.1993	$4.4E-21$
(iii) noiseless	0.6999	0.1999	$5.3E-17$
(iii) noisy	0.7052	0.1956	$1.8E-22$
(iv) noiseless	0.6999	0.1999	$2.2E-20$
(iv) noisy	0.6904	0.1971	$3.4E-22$
(v) noiseless	0.6999	0.1999	$4.0E-17$
(v) noisy	0.6905	0.1916	$2.1E-16$
(vi) noiseless	1.5440	0.4411	$1.9E-18$
(vi) noisy	1.5449	0.4402	$2.3E-5$

are given numbers. As in the finite conductor case, it is not necessary that the time $\bar{\bar{t}}_1$ at which we record the heat flux (9.40)(a) and the temperature (9.40)(b) measurements be the same. The imposition of other measurement conditions such as (9.36)(b) or (9.36)(d) is redundant, as discussed below.

If it exists, the bounded solution u of (9.39) is given by [55]

$$u(x,t) = -2\alpha \int_0^t \mu_0(t') \frac{\partial G}{\partial x'}(x,t;0,t')dt', \quad (x,t) \in (0,\infty) \times (0,\infty). \quad (9.41)$$

and [56]

$$\overline{q}_1 = \sqrt{KC} \int_0^{\bar{\bar{t}}_1} \frac{\mu_0'(\tau)}{\sqrt{\pi(\bar{\bar{t}}_1 - \tau)}} d\tau. \quad (9.42)$$

Equation (9.42) is well-defined since $\overline{q}_1 > 0$ and $\mu_0'(t) > 0$ for $t \in (0,\infty)$, and the following relationship between C and K holds:

$$KC = \overline{q}_1^2 \left[\int_0^{\bar{\bar{t}}_1} \frac{\mu_0'(\tau)}{\sqrt{\pi(\bar{\bar{t}}_1 - \tau)}} d\tau \right]^{-2} =: A(\bar{\bar{t}}_1, \overline{q}_1, f). \quad (9.43)$$

This relationship is explicit and it must be satisfied by the exact values of C and K, regardless of the choice of $\bar{\bar{t}}_1 > 0$ and μ_0 strictly monotone increasing continuously diffentiable function satisfying $\mu_0(0) = 0$. Since $\overline{q}_1 = \overline{q}_1(\bar{\bar{t}}_1)$, it follows that $A(\bar{\bar{t}}_1, q_1, f)$ must be a constant. This in turn, implies that in the semi-infinite case the two flux conditions (9.36)(a,b) will coincide, and therefore only one of them, as given by (9.40)(a), can be used as additional data measurement in the inversion. The same data redundancy happens with the condition (9.40)(b) and a possible additional temperature measurement, say $u(x_2, \bar{\bar{t}}_2) = \overline{g}_2$, where $x_2 > 0$, $\bar{\bar{t}}_2 > 0$, since the right-hand side of equation (9.41) depends only on $\alpha = K/C$. Applying (9.41) at $(x_1, \bar{\bar{t}}_1)$ and using (9.40)(b) the following relation involving α is obtained:

$$\overline{g}_1 = W(\alpha) := \int_0^{\bar{\bar{t}}_1} \frac{x_1 \mu_0(\tau)}{2\sqrt{\pi\alpha(\bar{\bar{t}}_1 - \tau)^3}} \exp\left(-\frac{x_1^2}{4\alpha(\bar{\bar{t}}_1 - \tau)}\right) d\tau. \tag{9.44}$$

Since $\mu_0(0) = 0$ and $\mu_0'(t) > 0$ it follows that $\mu_0(t) > 0$ for $t \in (0, \infty)$ and

$$0 \leq W(\alpha) \leq \mu_0(\bar{\bar{t}}_1) \operatorname{erfc}\left(\frac{x_1}{2\sqrt{\alpha\bar{\bar{t}}_1}}\right),$$

$$\lim_{\alpha \to 0} W(\alpha) = 0, \quad 0 \leq L := \lim_{\alpha \to \infty} W(\alpha) \leq \mu_0(\bar{\bar{t}}_1). \tag{9.45}$$

Differenting (9.44) yields

$$W'(\alpha) = \int_0^{\bar{\bar{t}}_1} \frac{x_1 \mu_0(\tau)}{4\sqrt{\pi\alpha^3(\bar{\bar{t}}_1 - \tau)^3}} \left[\frac{x_1^2}{2\alpha(\bar{\bar{t}}_1 - \tau)} - 1\right] \exp\left(-\frac{x_1^2}{4\alpha(\bar{\bar{t}}_1 - \tau)}\right) d\tau.$$

Clearly, $W'(\alpha) > 0$ for $0 < \alpha \leq \frac{x_1^2}{2\bar{\bar{t}}_1}$. For $\alpha > \frac{x_1^2}{2\bar{\bar{t}}_1}$, by splitting the above integral from 0 to $\bar{\bar{t}}_0 := \frac{x_1^2}{2\alpha}$ and from $\bar{\bar{t}}_0$ to $\bar{\bar{t}}_1$, one obtains

$$W'(\alpha) \geq \mu_0(\bar{\bar{t}}_1 - \bar{\bar{t}}_0) \int_0^{\bar{\bar{t}}_1} \frac{x_1}{4\sqrt{\pi\alpha^3(\bar{\bar{t}}_1 - \tau)^3}} \left[\frac{x_1^2}{2\alpha(\bar{\bar{t}}_1 - \tau)} - 1\right]$$
$$\times \exp\left(-\frac{x_1^2}{4\alpha(\bar{\bar{t}}_1 - \tau)}\right) d\tau = \frac{x_1 \mu_0(\bar{\bar{t}}_1 - \bar{\bar{t}}_0)}{2\sqrt{\pi\alpha^3 \bar{\bar{t}}_1}} \exp\left(-\frac{x_1^2}{4\alpha\bar{\bar{t}}_1}\right) > 0.$$

Hence, $W'(\alpha) > 0$ for all $\alpha > 0$, implying that the function $W : (0, \infty) \to (0, \mu_0(\bar{\bar{t}}_1))$ is bijective and invertible, and therefore,

$$\alpha = K/C = W^{-1}(\overline{g_1}) > 0 \tag{9.46}$$

exists provided that $0 < \overline{g}_1 < \mu_0(\bar{\bar{t}}_1)$. The inversion of (9.44) needed in (9.46) can be performed numerically using, e.g., the NAG routine C05NCF.

In conclusion, the solution of the problem (9.39) and (9.40) exists and is unique if $q_1 > 0$, $0 < \overline{g}_1 < \mu_0(\bar{\bar{t}}_1)$, $\mu_0(0) = 0$ and $\mu_0'(t) > 0$ for $t \in (0, \infty)$.

Moreover, from (9.43) and (9.46) it is explicitly given by

$$C = \frac{\overline{q}_1}{\sqrt{W^{-1}(\overline{g_1})}} \left[\int_0^{\bar{\bar{t}}_1} \frac{\mu_0'(\tau)}{\sqrt{\pi(\bar{\bar{t}}_1 - \tau)}} d\tau \right]^{-1},$$

$$K = \overline{q}_1 \sqrt{W^{-1}(\overline{g_1})} \left[\int_0^{\bar{\bar{t}}_1} \frac{\mu_0'(\tau)}{\sqrt{\pi(\bar{\bar{t}}_1 - \tau)}} d\tau \right]^{-1}. \tag{9.47}$$

Using (9.47), similarly as in (9.42), the heat flux is obtained as

$$q(t) = -K\frac{\partial u}{\partial x}(0,t) = \overline{q}_1 \left[\int_0^{\bar{\bar{t}}_1} \frac{\mu_0'(\tau)}{\sqrt{\pi(\bar{\bar{t}}_1 - \tau)}} d\tau \right]^{-1} \left[\int_0^{t} \frac{\mu_0'(\tau)}{\sqrt{\pi(t - \tau)}} d\tau \right] \tag{9.48}$$

for $t > 0$. From this equation it can be seen that $q(t)$ can be determined from the heat flux measurement (9.40)(a) only. Finally, using (9.41) and (9.46), the interior temperature can be obtained as

$$u(x,t) = \int_0^t \frac{x\mu_0(\tau)}{2\sqrt{\pi W^{-1}(\overline{g}_1)(t-\tau)^3}} \exp\left(-\frac{x^2}{4W^{-1}(\overline{g}_1)(t-\tau)} \right) d\tau \tag{9.49}$$

for $x > 0$ and $t > 0$. From this equation it can be seen that $u(x,t)$ can be determined from the temperature measurement (9.40)(b) only.

Example 9.2 Let $\mu_0(t) = t$, $\overline{q}_1 > 0$ and $0 < \overline{g}_1 < \bar{\bar{t}}_1$ be given data which satisfy the conditions for the existence and uniqueness of the solution of the inverse problem. Then, $\int_0^{\bar{\bar{t}}_1} \frac{d\tau}{\sqrt{\pi(\bar{\bar{t}}_1 - \tau)}} = 2\sqrt{\bar{\bar{t}}_1/\pi}$ and equation (9.47) gives

$$C = \frac{\overline{q}_1\sqrt{\pi}}{2\sqrt{\bar{\bar{t}}_1 W^{-1}(\overline{T}_1)}}, \quad K = \frac{\overline{q}_1\sqrt{\pi W^{-1}(\overline{T}_1)}}{2\sqrt{\bar{\bar{t}}_1}}, \tag{9.50}$$

where from (9.44), $W : (0,\infty) \to (0, \bar{\bar{t}}_1)$ is given by

$$W(\alpha) := \int_0^{\bar{\bar{t}}_1} \frac{x_1\tau}{2\sqrt{\pi\alpha(\bar{\bar{t}}_1 - \tau)^3}} \exp\left(-\frac{x_1^2}{4\alpha(\bar{\bar{t}}_1 - \tau)} \right) d\tau$$

$$= \left(\frac{x_1^2 + 2\alpha\bar{\bar{t}}_1}{2\alpha} \right) \operatorname{erfc}\left(\frac{x_1}{2\sqrt{\alpha\bar{\bar{t}}_1}} \right) - \frac{x_1\sqrt{\bar{\bar{t}}_1}}{\sqrt{\pi\alpha}} \exp\left(-\frac{x_1^2}{4\alpha\bar{\bar{t}}_1} \right).$$

The heat flux can be obtained from equation (9.48) as $q(t) = \overline{q}_1\sqrt{t/\bar{\bar{t}}_1}$ for $t > 0$ and the temperature is obtained from (9.49) as

$$u(x,t) = \left(\frac{x^2 + 2W^{-1}(\overline{g}_1)t}{2W^{-1}(\overline{g}_1)}\right)\operatorname{erfc}\left(\frac{x}{2\sqrt{W^{-1}(\overline{g}_1)t}}\right)$$
$$-\frac{x\sqrt{t}}{\sqrt{\pi W^{-1}(\overline{g}_1)}}\exp\left(-\frac{x^2}{4W^{-1}(\overline{g}_1)t}\right), \quad x > 0,\ t > 0.$$

Chapter 10

Time-Dependent Conductivity

Determination of the leading coefficient or, the coefficient of highest-order derivative in the parabolic heat equation has been widely investigated [277, 168, 57, 382]. In this chapter, the identification of the time-dependent conductivity in the parabolic heat conduction equation is described using the various techniques developed in [178, 179, 180, 405].

10.1 Identification of the Time-Dependent Conductivity

In this section, the inverse problem of finding the temperature $u(x,t)$ of a one-dimensional heat conductor that is undertaking radioactive decay or damage, such that its unknown thermal conductivity/diffusivity $a(t)$ varies with the degree of decay, which can be related to time t, is considered. Since the unknown coefficient is time-dependent, the additional information required to render a unique solution is also temporal and it may be given by an interior temperature, the normal derivative of the temperature, the heat flux or, the energy output from the heat conducting body. These specifications give rise to several inverse problem formulations that are discussed next.

Consider the one-dimensional time-dependent parabolic heat equation

$$\frac{\partial u}{\partial t} = a(t)\frac{\partial^2 u}{\partial x^2}, \quad (x,t) \in (0,1) \times (0,T) =: Q_T, \tag{10.1}$$

with the initial condition

$$u(x,0) = u_0(x), \quad x \in [0,1], \tag{10.2}$$

and the Dirichlet boundary conditions

$$u(0,t) = \mu_0(t), \quad t \in (0,T), \tag{10.3}$$

$$u(1,t) = \mu_1(t), \quad t \in (0,T). \tag{10.4}$$

Equations (10.1)–(10.4) form the Dirichlet direct problem, which is linear and well-posed if the coefficient $a(t) > 0$ is given. The Neumann direct problem

DOI: 10.1201/9780429400629-10

given by (10.1), (10.2) and the heat flux boundary conditions

$$-a(t)\frac{\partial u}{\partial x}(0,t) = q_0(t), \quad t \in [0,T], \tag{10.5}$$

$$a(t)\frac{\partial u}{\partial x}(1,t) = q_1(t), \quad t \in [0,T]. \tag{10.6}$$

can also be considered. In contrast to the direct problem, in the inverse problem the thermal conductivity $a(t) > 0$ is unknown and additional information is required for a unique solution. With respect to what additional information can be provided, four inverse problems are distinguished, as follows.

(i) **Inverse Problem 1 (IP1)**. The additional measurement is the internal temperature in time at a fixed sensor location $X_0 \in (0,1)$,

$$u(X_0,t) = \chi(t), \quad t \in [0,T]. \tag{10.7}$$

(ii) **Inverse Problem 2 (IP2)**. The additional measurement is the normal derivative of u at one of the boundary points, say at $x = 0$,

$$-\frac{\partial u}{\partial x}(0,t) = h(t), \quad t \in [0,T]. \tag{10.8}$$

(iii) **Inverse Problem 3 (IP3)**. The additional measurement is the heat flux at one of the boundary ends, say at $x = 0$, as given by equation (10.5). The IP3 is more physical than IP2 since, in practice, one usually measures the heat flux (10.5) rather than the normal derivative of the temperature (10.8).

(iv) **Inverse Problem 4 (IP4)**. The additional measurement is the L^1-norm of u over the space domain $(0,1)$, i.e. the total energy output from the body,

$$\int_0^1 u(x,t)dx = E(t), \quad t \in [0,T]. \tag{10.9}$$

The solutions to the above inverse problems are sought as pairs $(a(t), u(x,t))$ with $u \in C(\overline{Q}_T) \cap C^{2,1}(Q_T)$ and $0 < a \in C[0,T]$. Moreover, in many cases it is customary and physically realistic to assume that

$$a \in \Lambda_{ad} := \{a \in C[0,T] \mid 0 < a_{min} \leq a(t) \leq a_{max} < \infty, \ \forall t \in [0,T]\},$$

where a_{min} and a_{max} are given positive numbers. Since $a > 0$, the function

$$\eta = \theta(t) = \int_0^t a(\tau)d\tau, \quad t \in [0,T], \tag{10.10}$$

is strictly increasing and $\theta(0) = 0$. Upon the substitution

$$U(x,\theta(t)) = u(x,t), \quad (x,t) \in [0,1] \times [0,T], \tag{10.11}$$

or, equivalently

$$U(x,\eta) = u(x,\theta^{-1}(\eta)), \quad (x,\eta) \in [0,1] \times [0,\theta(T)], \tag{10.12}$$

equation (1.1) transforms into the heat equation ([220] and Chapter 13 of [55]),

$$\frac{\partial U}{\partial \eta} = \frac{\partial^2 U}{\partial x^2}, \quad (x,\eta) \in (0,1) \times (0,\theta(T)). \tag{10.13}$$

Also, equations (10.2)–(10.9) transform into

$$U(x,0) = u_0(x), \quad x \in [0,1], \tag{10.14}$$

$$U(0,\eta) = \mu_0(\theta^{-1}(\eta)), \quad \eta \in (0,\theta(T)), \tag{10.15}$$

$$U(1,\eta) = \mu_1(\theta^{-1}(\eta)), \quad \eta \in (0,\theta(T)), \tag{10.16}$$

$$-a(\theta^{-1}(\eta))\frac{\partial U}{\partial x}(0,\eta) = q_0(\theta^{-1}(\eta)), \quad \eta \in (0,\theta(T)), \tag{10.17}$$

$$a(\theta^{-1}(\eta))\frac{\partial U}{\partial x}(1,\eta) = q_1(\theta^{-1}(\eta)), \quad \eta \in (0,\theta(T)), \tag{10.18}$$

$$U(X_0,\eta) = \chi(\theta^{-1}(\eta)), \quad \eta \in [0,\theta(T)], \tag{10.19}$$

$$-\frac{\partial U}{\partial x}(0,\eta) = h(\theta^{-1}(\eta)), \quad \eta \in [0,\theta(T)], \tag{10.20}$$

$$\int_0^1 U(x,\eta)dx = E(\theta^{-1}(\eta)), \quad \eta \in [0,\theta(T)]. \tag{10.21}$$

One has the following representation formula for the function $u(x,t)$ satisfying equations (10.1)–(10.4), [55]:

$$\begin{aligned}
u(x,t) = \frac{1}{2\sqrt{\pi\theta(t)}} \int_0^1 u_0(s) \sum_{n=-\infty}^{\infty} \Bigg[&\exp\left(-\frac{(x-s+2n)^2}{4\theta(t)}\right) \\
&- \exp\left(-\frac{(x+s+2n)^2}{4\theta(t)}\right) \Bigg] ds \\
+\frac{1}{2\sqrt{\pi}} \int_0^t \frac{\mu_0(\tau)a(\tau)}{(\theta(t)-\theta(\tau))^{3/2}} &\sum_{n=-\infty}^{\infty} (x+2n) \exp\left(-\frac{(x+2n)^2}{4(\theta(t)-\theta(\tau))}\right) d\tau \\
-\frac{1}{2\sqrt{\pi}} \int_0^t \frac{\mu_1(\tau)a(\tau)}{(\theta(t)-\theta(\tau))^{3/2}} &\sum_{n=-\infty}^{\infty} (x+2n-1) \exp\left(-\frac{(x+2n-1)^2}{4(\theta(t)-\theta(\tau))}\right) d\tau.
\end{aligned}$$

The IP1

Collecting the results of Theorem 13.3.1 of [55] and [221], for the IP1 given by equations (10.1)–(10.4) and (10.7), the following theorem gives the existence and uniqueness of its solution.

Theorem 10.1. *Let the Dirichlet boundary conditions* (10.3) *and* (10.4) *be homogeneous, i.e.* $\mu_0 = \mu_1 = 0$.
(a) Let $T = \infty$, $u_0 \in C^2[0,1]$, $u_0(0) = u_0(1) = 0$, $u_0'' < 0$, $\chi \in C^1[0,\infty)$, $\chi' < 0$, $\chi(0) = u_0(X_0)$ *and* $lim_{t\to T=\infty}\chi(t) = 0$. *Then, the solution of the IP1 exists and is unique.*
(b) Let $u_0 \equiv 1$. *Then, a necessary and sufficient condition for the IP1 to possess a solution is that* χ *is absolutely continuous and positive,* $\chi' < 0$ *and* $\chi(0) = 1$.

Although Theorem 10.1 shows that the solution of IP1 exists and is unique, the following example [55] shows that the IP1 is still ill-posed since it is unstable with respect to errors in the data (10.7). Let $X_0 = 0.5$, $\mu_0 = \mu_1 = 0$, $u_0(x) = \sin(\pi x)$ and $T = \infty$. One can easily check that the conditions on u_0 of Theorem 10.1(a) are satisfied. Further, the solution of IP1 is given by

$$u(x,t) = \sin(\pi x)\exp(-\pi^2\theta(t)), \quad a(t) = -\frac{\chi'(t)}{\pi^2\chi(t)}. \tag{10.22}$$

The solution for $a(t)$ exists and is positive if χ is continuously differentiable, $\chi' < 0$, $\chi(0) = 1$ and $\lim_{t\to T=\infty}\chi(t) = 0$. However, if χ contains errors the differentiation of a noisy function is an unstable procedure and therefore, $a(t)$ given by (10.22) is unstable. One can however mollify [311] the noisy data χ and a stable solution can then be obtained. Finite difference numerical solutions for the IP1 were investigated in [14, 278].

The IP2

For the IP2 given by equations (10.1)–(10.4) and (10.8), the following local well-posedness theorem holds [66].

Theorem 10.2. *Let* $\mu_0 = \mu_1 = 0$, $u_0 \in C^2[0,1]$, $u_0'' < 0$, $u_0(0) = u_0(1) = 0$. *Also, let* $h \in C^1[0,T]$, $h' > 0$, $h(0) = -u_0'(0)$. *Then, the IP2 is well-posed in a neighbourhood of* $t = 0$.

A numerical technique based on space decomposition in a reproducing kernel space for solving a version of the IP2 given by equations (10.1)–(10.3), (10.6) and (10.8) was investigated in [408].

The IP3

Collecting the results of [51, 220] and Theorem 13.3.2 of [55], for the IP3 given by equations (10.1)–(10.5), the following theorem gives the existence and uniqueness of its solution.

Theorem 10.3. *(a) Let* μ_0 *be a constant function and assume that* $(1-x)\mu_0 + x\mu_1(0) \not\equiv u_0(x) \in C^5[0,1]$, $u_0(0) = \mu_0(0)$, $u_0(1) = \mu_1(0)$, $u_0''(0) = u_0''(1) = 0$, $u_0' > 0$, $\mu_1 \in C^1[0,T]$, $0 > q_0 \in C^1[0,T]$. *Then, the IP3 is well-posed in a neighbourhood of* $t = 0$.

(b) Let $T = \infty$ and $\mu_0 = \mu_1 = 0$. If $u_0 \in C^1[0,1]$, $u_0(x) > 0$ for $x \in (0,1)$, $u_0(0) = u_0(1) = 0$, $u_0'(0) > 0$ and $0 > q_0 \in C[0,\infty)$ are such that $0 < -\int_0^t q_0(s)ds < \sum_{k=1}^{\infty} 2(k\pi)^{-1} \int_0^1 u_0(s)\sin(k\pi s)ds < \infty$, then the solution of IP3 exists and is unique.

(c) Let $u_0 = 0$. Assume that μ_0 and μ_1 are absolutely continuous on every compact subset of $[0,T)$, $\mu_0(0) = \mu_1(0) = 0$, $\mu_0' > 0$ and $\mu_1' \leq 0$ on $(0,T)$, μ_0', μ_1' and $\int_0^t \frac{\mu_1'(\tau)}{\sqrt{t-\tau}}d\tau$ are bounded on compact subsets of $(0,T)$. Assume also that $0 < q_0 \in C[0,T]$ and that the function

$$W(t) := \frac{\sqrt{\pi}q_0(t)}{\int_0^t \frac{\mu_0'(\tau)}{\sqrt{t-\tau}}d\tau}, \quad t \in (0,T]$$

satisfies $\lim_{t\searrow 0} W(t) =: W(0) > 0$. Then, the IP3 has a unique solution.

A numerical approximation for the IP3 was proposed in [221].

The IP4

The following theorem gives the local well-posedness of the IP4 given by equations (10.1)–(10.4) and (10.9).

Theorem 10.4. ([405]) *Let $\mu_0 = \mu_1 = 0$, $u_0 \in C^2[0,1]$, $u_0'' < 0$, $u_0(0) = u_0(1) = 0$. Also, let $E \in C^1[0,T]$, $E' < 0$, $E(0) = \int_0^1 u_0(x)dx$. Then, the IP4 is well-posed in a neighbourhood of $t = 0$.*

Proof. Following [66, 405], the solution of problem (10.13)–(10.16) with $\mu_0 = \mu_1 = 0$ is given by

$$U(x,\eta) = \sum_{n=1}^{\infty} b_n \sin(n\pi x)e^{-n^2\pi^2\eta}, \tag{10.23}$$

where

$$b_n = 2\int_0^1 u_0(x)\sin(n\pi x)dx = -\frac{2}{n^2\pi^2}\int_0^1 u_0''(x)\sin(n\pi x)dx, \quad n \geq 1.$$

Also, equations (10.21) and (10.23) give

$$E(\theta^{-1}(\eta)) = \int_0^1 U(x,\eta)dx = \sum_{n=1}^{\infty} \frac{b_n[1-(-1)^n]}{n\pi}e^{-n^2\pi^2\eta} =: D(\eta). \tag{10.24}$$

From the maximum principle applied to the function U_{xx}, the assumption $u_0'' < 0$ implies $U_{xx}(x,\eta) < 0$ and hence, from (10.13) it follows that $U_\eta(x,\eta) < 0$. Then, since $u_t = a(t)U_\eta$ and $a > 0$ it follows that $u_t(x,t) < 0$. Now from (10.9), $E'(t) = \int_0^1 u_t(x,t)dx < 0$ for $t \in [0,T]$. From this, one observes that condition $E \in C^1[0,T]$ with $E' < 0$ is a necessary condition for the solvability of the IP4. Moreover, since $D'(\eta) = E'(t)/a(t) < 0$ it follows that D is strictly decreasing and one-to-one function. The conditions of the theorem on u_0 implies that u_0 is strictly concave, it has a unique maximum point and $u_0(x) > 0$ for $x \in (0,1)$. From the strong minimium principle it follows that $U(x,\eta) > 0$ for $(x,\eta) \in (0,1) \times [0, \theta(T)]$ and thus $D(\eta) > 0$. This, combined with the fact that D is strictly decreasing, yield

$$0 < D(\theta(T)) < D(\eta) < D(0) = \sum_{n=1}^{\infty} \frac{b_n[1-(-1)^n]}{n\pi}, \quad \eta \in (0, \theta(T)). \quad (10.25)$$

Also, since E is continuous at $t = 0$, the range of E is included in the range of D. Therefore, equation (2.15) can be inverted to yield $\eta = D^{-1}(E(t))$ and then, by differentiation, $a(t)$ is uniquely determined.

To show the stability in a neighbourhood of $t = 0$, first remark that by differentiating (10.9) with respect to t,

$$E'(t) = a(t)(U_x(1,\eta) - U_x(0,\eta)). \quad (10.26)$$

From (10.23),

$$\Delta U(\eta) := U_x(1,\eta) - U_x(0,\eta) = \sum_{n=1}^{\infty} n\pi b_n[(-1)^n - 1]e^{-n^2\pi^2\eta}. \quad (10.27)$$

Since $E' < 0$, $a > 0$ it follows that $\Delta U < 0$. Let now a_1 and a_2 be two conductivities corresponding to two energy data E_1 and E_2, respectively. Then,

$$\begin{aligned}|\Delta U_1(\eta) - \Delta U_2(\eta)| &\leq 2\pi \sum_{n=1}^{\infty} n|b_n| \left| e^{-n^2\pi^2 \int_0^t a_1(\tau)d\tau} - e^{-n^2\pi^2 \int_0^t a_2(\tau)d\tau} \right| \\ &\leq 2\pi^3 \left(\int_0^t |a_1(\tau) - a_2(\tau)|d\tau \right) \sum_{n=1}^{\infty} n^3|b_n|e^{-n^2\pi^2 a_{min} t} \\ &\leq Ct \max_{\tau \in [0,t]} |a_1(\tau) - a_2(\tau)|,\end{aligned}$$

where $C = C(u_0)$ is some positive constant. Then, from (10.26),

$$\begin{aligned}|a_1(t) - a_2(t)| &= \left| \frac{E_1'(t)}{\Delta U_1(\eta)} - \frac{E_2'(t)}{\Delta U_2(\eta)} \right| \\ &= \left| \frac{E_1'(t)(\Delta U_2(\eta) - \Delta U_1(\eta)) + \Delta U_1(\eta)(E_1'(t) - E_2'(t))}{\Delta U_1(\eta)\Delta U_2(\eta)} \right|.\end{aligned}$$

Using the equations above yield

$$\|a_1 - a_2\|_t \leq C_1 t \|a_1 - a_2\|_t + C_2\|E_1' - E_2'\|_t, \quad (10.28)$$

for some positive constants C_1 and C_2, where $\|a_1 - a_2\|_t := \max_{\tau \in [0,t]} |a_1(\tau) -$

$a_2(\tau)|$ denotes the usual maximum norm defined on the space of continuous functions. From this inequality it can be seen that for sufficiently small $T_0 \in (0, T]$, the following stability estimate holds:

$$\|a_1 - a_2\|_{T_0} \leq C\|E_1' - E_2'\|_{T_0}. \tag{10.29}$$

When the boundary conditions are not homogeneous, the IP4 can be reduced to the following equation with respect to $a(t)$:

$$a(t) = \frac{E'(t)}{u_x(1,t) - u_x(0,t)}, \quad t \in [0, T],$$

where

$$u_x(x,t) = \int_0^1 G_2(x,t;\xi,\tau)u_0'(\xi)d\xi - \int_0^t G_2(x,t;0,\tau)g_0'(\tau)d\tau + \int_0^t G_2(x,t;1,\tau)g_1'(\tau)d\tau,$$

and

$$G_k(x,t,\xi,\tau) = \frac{\mathrm{H}(t-\tau)}{\sqrt{4\pi(\theta(t)-\theta(\tau))}} \sum_{n=-\infty}^{\infty} \left[\exp\left(-\frac{(x-\xi+2n)^2}{4(\theta(t)-\theta(\tau))}\right) + (-1)^k \exp\left(-\frac{(x+\xi+2n)^2}{4(\theta(t)-\theta(\tau))}\right)\right], \tag{10.30}$$

is the Green's function for the Dirichlet ($k = 1$) and Neumann ($k = 2$) initial boundary value problems for the heat equation (10.1). Then, for the IP4, the following existence and uniqueness theorems holds [207, 277].

Theorem 10.5. *Suppose the following assumptions:*
1) $u_0 \in C^2[0,1]$, $\mu_i \in C^1[0,T]$, $i = 0, 1$, $E \in C^1[0,T]$;
2) $u_0'(1-x) - u_0'(x) > 0$, $x \in [0, 1/2)$, $\mu_0'(t) + \mu_1'(t) \geq 0$, $E'(t) > 0$, $t \in [0,T]$;
3) $u_0(0) = \mu_0(0)$, $u_0(1) = \mu_1(0)$, $E(0) = \int_0^1 u_0(x)dx$.
Then, a solution to IP4 exists locally in time. Moreover, if, in addition the following condition holds:
4) $\frac{1}{\sqrt{t}} \int_0^{1/2} (u_0'(1-\xi) - u_0'(\xi)) \sum_{n=-\infty}^{\infty} (-1)^n \exp\left(-\frac{(\xi+n)^2}{4t}\right) d\xi \geq C_0 > 0$, $t \in [0,T]$, *for some positive constant* C_0, *the existence of solution of IP4 is global.*

Theorem 10.6. *Under the assumption* $E'(t) \neq 0$ *for* $t \in [0,T]$, *the IP4 cannot have more than one solution.*

Remark 10.1. (i) In practice the energy $E(t)$ will be measured with some noisy error and therefore the evaluation of the noisy derivative in (10.26) should be stabilized, e.g. using the mollification method [311].

(ii) Note that Theorems 10.2 and 10.4 for the IP2 and IP4, respectively, also hold in higher dimensions [66].

(iii) As in [66] for the IP2, the condition that u_0 is concave in Theorem 10.4 cannot, in general, be removed. This can be seen by taking $u_0(x) = \frac{1}{4}\sin(\pi x) - \frac{1}{9}\sin(3\pi x)$ for $x \in [0,1]$, and $E(t)$ any continuously differentiable strictly decreasing function satisfying $E(0) = \frac{23}{54\pi}$. Then, $u_0(0) = u_0(1) = 0$ and $u_0''(x) = \pi^2 \sin(\pi x)[11/4 - 4\sin^2(\pi x)]$ has mixed sign on $[0,1]$. Then, $u(x,t)$ satisfying (10.1)–(10.4) with $\mu_0 = \mu_1 = 0$ and $u_0(x)$ defined as above is given, in terms of the function $\psi(t) := e^{-\pi^2 \int_0^t a(\tau)d\tau}$, by

$$u(x,t) = \frac{1}{4}\sin(\pi x)\psi(t) - \frac{1}{9}\sin(3\pi x)\psi^9(t).$$

Then, (10.9) yields

$$E(t) = \frac{1}{2\pi}\psi(t) - \frac{2}{27\pi}\psi^9(t).$$

One can observe that $\psi(0) = 1$, $\psi'(t) < 0$ and $E'(t) = \frac{\psi'(t)}{\pi}\left[\frac{1}{2} - \frac{2}{3}\psi^8(t)\right]$. By letting $t \searrow 0$, it can be seen that $E'(t) > 0$ in a neighbourhood of $t = 0$ and this is in contradiction with the initial hypothesis that E is strictly decreasing function. Thus, in this case, the solution of IP4 does not exist.

10.1.1 Satisfier function and Ritz-Galerkin method

Suppose first that $a(t)$ is known and consider the direct problem given by (10.1)–(10.4). This problem is linear and well-posed and it can be solved to find $u(x,t)$ in terms of $a(t)$. In the Ritz-Galerkin method the approximation $\hat{u}(x,t)$ of the solution $u(x,t)$ is sought in the form of the truncated series

$$\hat{u}(x,t) = \sum_{i=0}^{n}\sum_{j=0}^{m} c_{ij}\psi_{ij}(x,t) + w(x,t), \quad (x,t) \in [0,1] \times [0,T], \tag{10.31}$$

where $\psi_{ij}(x,t) = x(x-1)t\phi_i(x)\phi_j(t)$, and $\{\phi_i\}_{i=0}^{\infty}$ are basis functions, e.g. Legendre polynomials in $[0,1]$ given by $\phi_0(x) = 1$, $\phi_1(x) = 2x - 1$, $\phi_2(x) = \frac{1}{2}(3(2x-1)^2 - 1)$, etc. Let us define the satisfier function $w(x,t)$ which satisfies the initial condition (10.2), the Dirichlet boundary conditions (10.3) and (10.4) and either of the additional measurement conditions (10.5), (10.7)–(10.9), as follows [405].

For **IP1:**

$$\begin{aligned} w(x,t) = u_0(x) + (1-x)\left(\mu_0(t) - \mu_0(0)\right) + x\left(\mu_1(t) - \mu_1(0)\right) \\ +x(x-1)\left[\frac{\mu_0(t) - \mu_0(0) - (\mu_1(t) - \mu_1(0))}{X_0}\right. \\ \left.+\frac{\chi(t) - \chi(0) - (\mu_1(t) - \mu_1(0))}{X_0(X_0 - 1)}\right], \end{aligned}$$

where the following compatibility conditions were assumed:

$$u_0(0) = \mu_0(0), \quad u_0(1) = \mu_1(0), \quad u_0(X_0) = \chi(0).$$

For **IP2:**

$$w(x,t) = u_0(x) - s(x,0) + s(x,t),$$

where

$$s(x,t) = \mu_0(t) - xh(t) + x^2(\mu_1(t) - \mu_0(t) + h(t)),$$

and the following compatibility conditions were assumed:

$$u_0(0) = \mu_0(0), \quad u_0(1) = \mu_1(0), \quad u_0'(0) = -h(0).$$

For **IP3:**

$$w(x,t) = u_0(x) - s(x,0) + s(x,t),$$

where

$$s(x,t) = \mu_0(t) - x\frac{q_0(t)}{a(t)} + x^2\left(\mu_1(t) - \mu_0(t) + \frac{q_0(t)}{a(t)}\right),$$

and the following compatibility conditions were assumed:

$$u_0(0) = \mu_0(0), \quad u_0(1) = \mu_1(0), \quad a(0)u_0'(0) = -q_0(0).$$

For **IP4:**

$$w(x,t) = u_0(x) - s(x,0) + s(x,t),$$

where

$$s(x,t) = (3x^2 - 4x + 1)\mu_0(t) + (3x^2 - 2x)\mu_1(t) + (-6x^2 + 6x)E(t),$$

and the following compatibility conditions were assumed:

$$u_0(0) = \mu_0(0), \quad u_0(1) = \mu_1(0), \quad E(0) = \int_0^1 u_0(x)dx.$$

Generally, the satisfier function is not unique. But in special cases one can have a unique satisfier function, e.g. if it is separable [405].

Returning now to the Ritz-Galerkin approximation (10.31), by substituting $\hat{u}(x,t)$ in equation (10.1) gives

$$\hat{u}_t(x,t) = a(t)\hat{u}_{xx}(x,t). \tag{10.32}$$

The expansion coefficients c_{ij} are determined by the Galerkin equations

$$\langle \hat{u}_t(x,t) - a(t)\hat{u}_{xx}(x,t), \phi_i(x)\phi_j(t)\rangle = 0, \quad i = \overline{0,n}, \quad j = \overline{0,m}, \tag{10.33}$$

where $\langle\cdot\rangle$ denotes the inner product defined by

$$\langle \hat{u}_t(x,t) - a(t)\hat{u}_{xx}(x,t), \phi_i(x)\phi_j(t)\rangle$$
$$= \int_0^1 \int_0^T (\hat{u}_t(x,t) - a(t)\hat{u}_{xx}(x,t))\phi_i(x)\phi_j(t)dtdx. \tag{10.34}$$

Equations (10.33) form a linear system of equations

$$K\underline{C} = \underline{b} \tag{10.35}$$

which can be solved for the elements of $\underline{C} = (c_{ij})_{j=\overline{0,m}}^{i=\overline{0,n}}$. In (10.35), the transpose of the rows of the matrix K is [405]

$$\begin{pmatrix}
\int_0^1 \int_0^T [(\psi_{00})_t(x,t) - a(t)(\psi_{00})_{xx}(x,t)]\phi_{S-[(k-1)(n+1)+1]}(x)\phi_{k-1}(t)dtdx \\
\int_0^1 \int_0^T [(\psi_{10})_t(x,t) - a(t)(\psi_{10})_{xx}(x,t)]\phi_{S-[(k-1)(n+1)+1]}(x)\phi_{k-1}(t)dtdx \\
\vdots \\
\int_0^1 \int_0^T [(\psi_{n0})_t(x,t) - a(t)(\psi_{n0})_{xx}(x,t)]\phi_{S-[(k-1)(n+1)+1]}(x)\phi_{k-1}(t)dtdx \\
\int_0^1 \int_0^T [(\psi_{01})_t(x,t) - a(t)(\psi_{10})_{xx}(x,t)]\phi_{S-[(k-1)(n+1)+1]}(x)\phi_{k-1}(t)dtdx \\
\int_0^1 \int_0^T [(\psi_{11})_t(x,t) - a(t)(\psi_{11})_{xx}(x,t)]\phi_{S-[(k-1)(n+1)+1]}(x)\phi_{k-1}(t)dtdx \\
\vdots \\
\int_0^1 \int_0^T [(\psi_{n1})_t(x,t) - a(t)(\psi_{n1})_{xx}(x,t)]\phi_{S-[(k-1)(n+1)+1]}(x)\phi_{k-1}(t)dtdx \\
\vdots \\
\int_0^1 \int_0^T [(\psi_{0m})_t(x,t) - a(t)(\psi_{0m})_{xx}(x,t)]\phi_{S-[(k-1)(n+1)+1]}(x)\phi_{k-1}(t)dtdx \\
\int_0^1 \int_0^T [(\psi_{1m})_t(x,t) - a(t)(\psi_{1m})_{xx}(x,t)]\phi_{S-[(k-1)(n+1)+1]}(x)\phi_{k-1}(t)dtdx \\
\vdots \\
\int_0^1 \int_0^T [(\psi_{nm})_t(x,t) - a(t)(\psi_{nm})_{xx}(x,t)]\phi_{S-[(k-1)(n+1)+1]}(x)\phi_{k-1}(t)dtdx
\end{pmatrix}$$

for $(k-1)(n+1)+1 \leq S \leq k(n+1)$, $k = \overline{1, m+1}$. Also,

$$\underline{C} = [c_{00}, ..., c_{n0}, c_{01}, ..., c_{n1}, ..., c_{0,m}, ..., c_{nm}]^{\mathrm{T}},$$

$$\underline{b} = [b_{00}, ..., b_{n0}, b_{01}, ..., b_{n1}, ..., b_{0,m}, ..., b_{nm}]^{\mathrm{T}},$$

where

$$b_{ij} = \int_0^1 \int_0^T [a(t)w_{xx}(x,t) - w_t(x,t)]\phi_i(x)\phi_j(t)dtdx, \quad i = \overline{0,n}, \quad j = \overline{0,m}.$$

By solving the linear system of equations (10.35), the elements c_{ij}, $i = \overline{0,n}$, $j = \overline{0,m}$, can be expressed in terms of $a(t)$ and then from (10.31) the approximation of $\hat{u}(x,t)$ in terms of x, t and $a(t)$ can be obtained.

Let us now assume that the diffusion coefficient $a(t)$ is unknown and consider the inverse problem. Let

$$\hat{a}(t) = \sum_{i=0}^{L} a_i\phi_i(t), \quad t \in [0,T] \tag{10.36}$$

be an approximation for $a(t)$. This way the infinite dimensional problem of estimating a continuous function $a(t)$ is reduced to a finite dimensional problem of determining the vector $\underline{A} = (a_i)_{i=\overline{0,L}}$. The additional data $\hat{u}(X_0,t)$, $-\frac{\partial \hat{u}}{\partial x}(0,t)$, $-a(t)\frac{\partial \hat{u}}{\partial x}(0,t)$ and $\int_0^1 \hat{u}(x,t)dx$ approximating (10.7), (10.8), (10.5) anf (10.9), respectively, are computed in terms of the (unknown) coefficient $\underline{A}$ using the symbolic computation package MATHEMATICA. After that, to find the coefficient $\underline{A}$, the minimization problem

$$\text{for IP1:} \quad \min_{\underline{A}\in\mathbb{R}^{L+1}} \left\{||\hat{u}(X_0,t) - \chi(t)||^2 + \lambda||\underline{A}||^2\right\}, \tag{10.37}$$

$$\text{for IP2:} \quad \min_{\underline{A}\in\mathbb{R}^{L+1}} \left\{\left|\left| -\frac{\partial \hat{u}}{\partial x}(0,t) - h(t)\right|\right|^2 + \lambda||\underline{A}||^2\right\}, \tag{10.38}$$

$$\text{for IP3:} \quad \min_{\underline{A}\in\mathbb{R}^{L+1}} \left\{\left|\left| -\hat{a}(t)\frac{\partial \hat{u}}{\partial x}(0,t) - q_0(t)\right|\right|^2 + \lambda||\underline{A}||^2\right\}, \tag{10.39}$$

$$\text{for IP4:} \quad \min_{\underline{A}\in\mathbb{R}^{L+1}} \left\{\left|\left| \int_0^1 \hat{u}(x,t)dx - E(t)\right|\right|^2 + \lambda||\underline{A}||^2\right\} \tag{10.40}$$

is solved, where $\lambda \geq 0$ is the regularization parameter and the norm $||\cdot||^2 = \int_0^T (.)^2 dt$. If $F(\underline{A})$ denotes the function which is minimized in the above expressions, then one has to solve the system of nonlinear equations

$$\frac{\partial F}{\partial a_i}(\underline{A}) = 0, \quad i = \overline{0,L}. \tag{10.41}$$

10.1.2 Numerical example

Since the dimension $L+1$ of the vector of unknowns $\underline{A}$ will be taken rather small, namely around 3 to 7 coefficients to be determined, the instability of solution does not really manifest and the numerical results obtained with $\lambda = 0$ in equations (10.37)–(10.40) seem stable.

Consider the following example from [55, 405] with $T = 1$,

$$u_0(x) = \sin(\pi x), \quad g_0(t) = 0, \quad g_1(t) = 0,$$

and the analytical solution

$$u(x,t) = \sin(\pi x)\exp\left[-\pi^2\left(1.01t + \frac{1-\cos(3\pi t)}{3\pi}\right)\right], \quad a(t) = 1.01 + \sin(3\pi t)$$

In the IP1, take $X_0 = 0.5$ and

$$\chi(t) = \exp\left[-\pi^2\left(1.01t + \frac{1-\cos(3\pi t)}{3\pi}\right)\right]. \tag{10.42}$$

In the IP2, take

$$h(t) = -\pi\exp\left[-\pi^2\left(1.01t + \frac{1-\cos(3\pi t)}{3\pi}\right)\right]. \tag{10.43}$$

In the IP3, take

$$q_0(t) = -\pi(1.01 + \sin(3\pi t)) \exp\left[-\pi^2\left(1.01t + \frac{1-\cos(3\pi t)}{3\pi}\right)\right], \quad (10.44)$$

In the IP4, take

$$E(t) = \frac{2}{\pi} \exp\left[-\pi^2\left(1.01t + \frac{1-\cos(3\pi t)}{3\pi}\right)\right]. \quad (10.45)$$

All four problems IP1–IP4 are solved with $m = n = 5$, $L = 6$ for $p = 3\%$ additive Gaussian noise added into the measured data (10.42)–(10.45). Figure 10.1 shows the absolute errors between the exact and approximate solutions of $a(t)$ for all the IP1–IP4. It can be seen that on most of the time interval $t \in [0, 1]$, the most accurate results are obtained for IP1, followed by IP2, IP3 and IP4. This order of accuracy is also confirmed by the L_2–errors of 0.0159, 0.0428, 0.0441 and 0.0578 obtained for IP1–IP4, respectively.

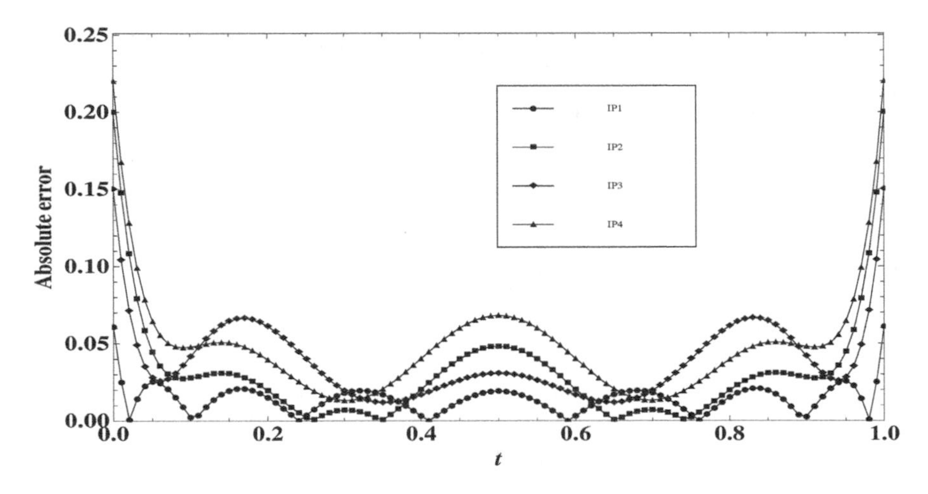

FIGURE 10.1: Absolute errors between the exact and approximate solutions of $a(t)$ for IP1–IP4, [405].

10.2 Identification of the Time-Dependent Conductivity of an Inhomogeneous Diffusive Material

In this section, the identification of the time-dependent conductivity of an inhomogeneous heat conductor, which has a space-varying known heat capacity, is considered. It is this later physically realistic feature that makes some of the methods of [66, 199, 201, 405] inapplicable. The same problem can be formulated in porous media by replacing the thermal properties with the corresponding hydraulic ones.

Consider the inverse problem of finding the time-dependent thermal conductivity $C[0,T] \ni a > 0$ and the temperature $u \in C^{2,1}(Q_T) \cap C^{1,0}(\overline{Q}_T)$, which satisfy the heat equation

$$c(x)\frac{\partial u}{\partial t}(x,t) = a(t)\frac{\partial^2 u}{\partial x^2}(x,t) + \tilde{F}(x,t), \ (x,t) \in Q_T, \tag{10.46}$$

where $c(x) > 0$ is the heat capacity and $\tilde{F}$ is a heat source, the initial condition

$$u(x,0) = u_0(x), \quad x \in [0,l], \tag{10.47}$$

the Dirichlet boundary conditions

$$u(0,t) = \mu_0(t), \quad u(l,t) = \mu_l(t), \quad t \in [0,T], \tag{10.48}$$

and the heat flux additional measurement

$$-a(t)u_x(0,t) = q_0(t), \quad t \in [0,T]. \tag{10.49}$$

Dividing equation (10.46) by $c(x)$ and denoting

$$b(x) = \frac{1}{c(x)}, \quad F(x,t) = \frac{\tilde{F}(x,t)}{c(x)} \tag{10.50}$$

yield

$$\frac{\partial u}{\partial t}(x,t) = a(t)b(x)\frac{\partial^2 u}{\partial x^2}(x,t) + F(x,t), \quad (x,t) \in Q_T. \tag{10.51}$$

The above inverse problem was investigated theoretically in Section 4.3 of [207] where its unique solvability was established, as given by the following Theorems 10.7 and 10.8.

Theorem 10.7. *Suppose that the following conditions hold:*
(i) (regularity conditions) $b \in C^1[0,l]$, $u_0 \in C^1[0,l]$, μ_0, $\mu_l \in C^1[0,T]$, $q_0 \in C[0,T]$, $F \in C^{1,0}(\overline{Q}_T)$;
(ii) (compatibility conditions) $u_0(0) = \mu_0(0)$, $u_0(l) = \mu_l(0)$;
(iii) (non-vanishing and monotonicity conditions) $u_0'(x) > 0$, $b(x) > 0$, $b'(x) \leq 0$ *for* $x \in [0,l]$, $q_0(t) < 0$, $\mu_0'(t) - F(0,t) \leq 0$, $\mu_l'(t) - F(l,t) \geq 0$ *for* $t \in [0,T]$, $F_x(x,t) \geq 0$ *for* $(x,t) \in \overline{Q}_T$.
Then, there exists a solution to the inverse problem (10.47)–(10.49), (10.51).

Theorem 10.8. *If* $0 < b \in C^1[0,l]$, $q_0(t) \neq 0$ *for* $t \in [0,T]$, *then the solution of the inverse problem (10.47)–(10.49), (10.51) is unique.*

Next, the stability of solution is established, as given by the following theorem.

Theorem 10.9. ([179]) *Suppose that the conditions of Theorem 10.7 are satisfied. Let* q_0 *and* $\tilde{q}_0$ *be two data in (10.49) satisfying* $q(t) \neq 0$ *and* $\tilde{q}(t) \neq 0$ *for* $t \in [0, T]$. *Let* $(a(t), u(x,t))$ *and* $(\tilde{a}(t), \tilde{u}(x,t))$ *be the corresponding solutions of the inverse problem (10.47)–(10.49) and (10.51). Then, for sufficiently small* T, *the following local stability estimate holds:*

$$\|a - \tilde{a}\|_{C[0,T]} \leq C \|q_0 - \tilde{q}_0\|_{C[0,T]}, \tag{10.52}$$

for some positive constant C.

Proof: First, observe that the pair of differences $\mathrm{A}(t) := a(t) - \tilde{a}(t)$, $U(x,t) := u(x,t) - \tilde{u}(x,t)$ is a solution of the following inverse problem:

$$U_t = a(t)b(x)U_{xx} + \mathrm{A}(t)b(x)\tilde{u}_{xx}, \quad (x,t) \in Q_T, \tag{10.53}$$

$$U(x,0) = 0, \quad x \in [0,l], \tag{10.54}$$

$$U(0,t) = U(l,t) = 0, \quad t \in (0,T), \tag{10.55}$$

$$-a(t)U_x(0,t) = \mathrm{A}(t)\tilde{u}_x(0,t) + q_0(t) - \tilde{q}_0(t), \quad t \in [0,T]. \tag{10.56}$$

The function $U(x,t)$ satisfying (10.53)–(10.55) can be expressed with the aid of its Green's function G_1 defined in (10.30), as

$$U(x,t) = \int_0^t \int_0^L G_1(x,t;\xi,\tau)\mathrm{A}(\tau)b(\xi)\tilde{u}_{\xi\xi}(\xi,\tau)d\xi d\tau, \quad (x,t) \in Q_T. \tag{10.57}$$

Differentiating (10.57) with respect to x and substituting into (10.56),

$$-a(t)\int_0^t \mathrm{A}(\tau)\left(\int_0^L G_{1x}(0,t;\xi,\tau)b(\xi)\tilde{u}_{\xi\xi}(\xi,\tau)d\xi\right)d\tau$$
$$= \mathrm{A}(t)\tilde{u}_x(0,t) + q_0(t) - \tilde{q}_0(t), \quad t \in [0,T]. \tag{10.58}$$

Since $\tilde{q}_0(t) \neq 0$, equation (10.49) implies that $\tilde{u}_x(0,t) \neq 0$ for $t \in [0,T]$. Thus, (10.58) is a linear Volterra integral equation of the second kind in $\mathrm{A}(t)$:

$$\mathrm{A}(t) = g(t) + \int_0^t \mathrm{D}(t,\tau)\mathrm{A}(\tau)d\tau, \quad t \in [0,T], \tag{10.59}$$

where

$$g(t) := \frac{\tilde{q}_0(t) - q_0(t)}{\tilde{u}_x(0,t)} \qquad t \in [0,T], \tag{10.60}$$

is a continuous function and the kernel

$$\mathrm{D}(t,\tau) := -\frac{a(t)}{\tilde{u}_x(0,t)}\int_0^L G_{1x}(0,t;\xi,\tau)b(\xi)\tilde{u}_{\xi\xi}(\xi,\tau)d\xi \tag{10.61}$$

has a weak integrable singularity since [207, p.147],

$$|G_{1x}(0,t;\xi,\tau)| \leq \frac{C_1}{\theta(t)-\theta(\tau)} \sum_{n=-\infty}^{\infty} \exp\left(-\frac{(\beta(\xi)+2n\beta(l))^2}{8(\theta(t)-\theta(\tau))}\right), \quad t>\tau,$$

for some positive constant C_1, where

$$\theta(t) = \int_0^t a(\tau)d\tau, \quad \beta(x) = \int_0^x \frac{d\xi}{\sqrt{b(\xi)}}. \tag{10.62}$$

Note that $\theta \in C^1[0,T]$ is a strictly increasing function with $\theta(0) = 0$ and $\theta'(t) = a(t) > 0$ for $t \in [0,T]$. Then, upon integration

$$\begin{aligned} \int_0^l |G_{1x}(0,t;\xi,\tau)|d\xi &\leq \frac{C_2}{\sqrt{\theta(t)-\theta(\tau)}} \\ &\leq \frac{C_2}{\min_{\rho\in[0,T]} a(\rho)} \frac{a(\tau)}{\sqrt{\theta(t)-\theta(\tau)}}, \quad t>\tau, \end{aligned} \tag{10.63}$$

for some positive constant C_2, which when used into (10.61) integrated produces the inequality

$$\int_0^t |\mathrm{D}(t,\tau)|d\tau \leq C_3 \int_0^t \frac{a(\tau)}{\sqrt{\theta(t)-\theta(\tau)}} d\tau = 2C_3\sqrt{\theta(t)}, \tag{10.64}$$

for some positive constant C_3. Now since the function $\sqrt{\theta(t)}$ is a monotonically increasing function and $\lim_{t\to 0}\sqrt{\theta(t)} = 0$, according to the theory of linear Volterra integral equations of the second kind, e.g. [55, Section 8.2], it follows that (10.59) is uniquely solvable. Furthermore, from (10.59) and (10.64),

$$\begin{aligned} |\mathrm{A}(t)| &\leq \|g\|_{C[0,T]} + \int_0^t |\mathrm{D}(t,\tau)||\mathrm{A}(\tau)|d\tau \\ &\leq \|g\|_{C[0,T]} + 2C_3\sqrt{\theta(T)}\|\mathrm{A}\|_{C[0,T]}, \quad t \in [0,T]. \end{aligned} \tag{10.65}$$

Since $\theta(0) = 0$ and θ is a monotonically increasing function, then, for sufficiently small T, it holds that $1 > 2C_3\sqrt{\theta(T)}$. Then, (10.60) and (10.65) yield that (10.52) holds, where it was used that from (10.60),

$$\|g\|_{C[0,T]} \leq \frac{\|\tilde{q}_0(t) - q_0(t)\|_{C[0,T]}}{\min_{t\in[0,T]} |\tilde{u}_x(0,t)|}. \tag{10.66}$$

A similar local stability can be obtained from (10.57) and (10.52) for $\|U\|_{C(\overline{Q}_T)} = \|u-\tilde{u}\|_{C(\overline{Q}_T)}$. Altogether, it has been established that the solution $(a(t), u(x,t))$ locally depends continuously upon the input data (10.49) in the maximum norm of $C[0,T]$.

The well-posedness established in Theorems 10.7–10.9 suggests that no regularization will be needed for obtaining a stable solution.

10.2.1 Numerical example

For the direct problem given by equations (10.47), (10.48) and (10.51), where $a(t)$, $b(x)$, $F(x,t)$, $u_0(x)$, $\mu_0(t)$ and $\mu_l(t)$ are known and $u(x,t)$ is the solution to be determined, the finite-difference method (FDM) with a Crank-Nicholson scheme using $M \times N$ subintervals of equal length discretizing the domain $\overline{Q}_T = [0,l] \times [0,T]$ is used [179].

For the inverse problem the functional

$$\mathrm{F}(a) := \left\| a(t)u_x(0,t) + q_0(t) \right\|^2_{L^2[0,T]} \tag{10.67}$$

is minimized. The discretization of (10.67) yields

$$\mathrm{F}(\underline{a}) = \sum_{j=0}^{N} \left[a_j u_x(0,t_j) + q_0(t_j) \right]^2, \tag{10.68}$$

where $\underline{a} = (a_j)_{j=\overline{0,N}}$, $a_j = a(t_j)$ and $t_j = jT/N$ for $j = \overline{0,N}$. In (10.68) at the first time step $j = 0$, the derivative $u_x(0,0)$ is obtained from the initial condition (10.47), as $u_x(0,0) = (4u_{0,1} - u_{0,2} - 3u_{0,0})M/(2l)$, where $u_{0,i} = u_0(x_i)$, $x_i = il/M$ for $i = \overline{0,M}$.

The minimization of the objective function (10.68) subjected to $a \in \Lambda_{ad}$ is accomplished using the Trust Region Reflective (TRR) optimization algorithm of the *lsqnonlin* MATLAB toolbox subject to the components of the vector $\underline{a}$ being constained in the interval $[10^{-10}, 10^3]$.

In the case of noisy data, $q_0(t_j)$ is replaced by $q_0^\epsilon(t_j)$ in (10.68). Numerically, the noisy data is simulated as $q_0^\epsilon(t_j) = q_0(t_j) + \epsilon_j$ for $j = \overline{0,N}$, where ϵ_j are random variables generated from a Gaussian normal distribution with mean zero and standard deviation $\sigma = p \times \max_{t\in[0,T]} |q_0(t)|$ and p represents the percentage of noise.

Consider the inverse problem (10.47)–(10.49) and (10.51) with the data

$$T = l = 1, \quad u_0(x) = u(x,0) = xe^x, \quad b(x) = 2 - x^2, \quad \mu_0(t) = u(0,t) = t^2,$$
$$\mu_l(t) = u(1,t) = e + t^2, \quad q_0(t) = -a(t)u_x(0,t) = -1 - \left| t - \frac{1}{2} \right|,$$
$$F(x,t) = 2t - \left(1 + \left| t - \frac{1}{2} \right| \right) (2 - x^2)(xe^x + 2e^x).$$

The conditions of Theorem 10.8 are satisfied, hence the uniqueness of the solution holds. With this data, the analytical solution is given by

$$a(t) = 1 + \left| t - \frac{1}{2} \right|, \quad u(x,t) = xe^x + t^2. \tag{10.69}$$

The numerical results for the time-dependent thermal conductivity $a(t)$ obtained with $M = N = 40$ and the initial guess $\underline{a}^{(0)} = \underline{1}$, for $p \in \{0, 2\%\}$ noise are presented in Figure 10.2. It can be seen that the numerical solution for the thermal conductivity is accurate and stable. The analyzed inverse problem is rather stable and in fact, as mentioned before, no regularization was needed to be included in the least-squares functional (10.68).

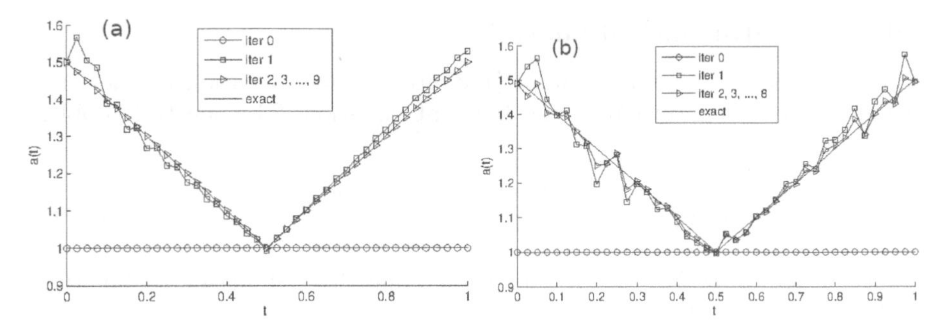

FIGURE 10.2: Results for $a(t)$ for (a) $p = 0$ and (b) $p = 2\%$ noise, [179].

10.3 Finding the Time-Dependent Diffusion Coefficient from an Integral Observation

Consider the inverse problem of finding the pair $(a(t), u(x,t)) \in C[0,T] \times (C^{2,1}(Q_T) \cap C^{1,0}(\overline{Q}_T))$ with $a > 0$, given by the heat equation

$$\frac{\partial u}{\partial t}(x,t) = a(t)\frac{\partial^2 u}{\partial x^2}(x,t), \quad (x,t) \in (0,1) \times (0,T) =: Q_T, \tag{10.70}$$

subject to the initial condition

$$u(x,0) = u_0(x), \quad x \in [0,1], \tag{10.71}$$

the periodic and adiabatic boundary conditions

$$u(0,t) = u(1,t), \quad t \in (0,T), \tag{10.72}$$

$$u_x(1,t) = 0, \quad t \in (0,T), \tag{10.73}$$

and the over-determination integral observation [308],

$$\rho(t)u(0,t) + \int_0^1 u(x,t)dx = E(t), \quad t \in [0,T], \tag{10.74}$$

with $\rho(t) = \alpha + \beta a^{-\gamma}(t)$, where α, β, $\gamma > 0$ are segregation coefficients. This problem is a mathematical model for the technological process of external guttering applied in cleaning admixtures from silicon chips [308]. In this case, $u_0(x)$ is the distribution of admixture in the chip for $x \in [0,1]$ at the initial time $t = 0$, while $u(x,t)$ is its distribution at time t. Condition (10.72) means that the admixtures in the left and right boundaries of the chip are the same. The adiabatic condition (10.73) means that the right boundary $x = 1$ of the chip is perfectly insulated. Condition (10.74) means that part of substance is segregated on the left side $x = 0$ of the chip [308]. When $\alpha = \beta = 0$, the resulting inverse problem was investigated in [199]. In this section, the non-trivial case when α and β are non-zero is addressed.

10.3.1 Mathematical analysis

The analysis is similar to that of [198] for the identification of the time-dependent blood perfusion in the bio-heat equation. The spectral problem

$$\begin{cases} X''(x) + \Lambda X(x) = 0, & x \in [0,1], \\ X(0) = X(1),\ X'(1) = 0 \end{cases} \tag{10.75}$$

has the eigenvalues $\Lambda_n = (2\pi n)^2$ for $n \in \mathbb{N}$, and the system of eigenfunctions and associated functions

$$X_0(x) = 2, \quad X_{2n-1}(x) = 4\cos(2\pi n x),$$
$$X_{2n}(x) = 4(1-x)\sin(2\pi n x), \quad n \in \mathbb{N}^*. \tag{10.76}$$

the system of functions $(X_n)_{n\in\mathbb{N}}$ is a basis in $L_2[0,1]$, [194]. The adjoint problem to (10.75) has the form

$$\begin{cases} Y''(x) + \Lambda Y(x) = 0, & x \in [0,1], \\ Y(0) = 0,\ Y'(0) = Y'(1). \end{cases} \tag{10.77}$$

Analogous to the problem (10.75), the system of eigenfunctions and associated functions of the problem (10.77) is given by

$$Y_0(x) = x, \quad Y_{2n-1}(x) = x\cos(2\pi n x), \quad Y_{2n}(x) = \sin(2\pi n x),$$
$$n \in \mathbb{N}^*. \tag{10.78}$$

The systems (10.76) and (10.78) form a bi-orthonormal system on the interval $[0,1]$. Denote $u_0^n := \int_0^1 u_0(x) Y_n(x) dx$ for $n \in \mathbb{N}$.

Theorem 10.9. ([180]) (Existence and uniqueness) *Let* $u_0 \in C^3[0,1]$ *and* $E \in C[0,T]$ *satisfy:*
$u_0(0) = u_0(1)$, $u_0'(1) = 0$, $u_0''(0) = u_0''(1)$; $u_0^{2k} \geq 0$, $u_0^{2k-1} \leq 0$, $k \in \mathbb{N}^*$; $u_0^0 + 2u_0^1 < 0$, $E(t) < 2u_0^0$, $\forall t \in [0,T]$.
Then, there exist positive numbers α_0 *and* γ_0 *determined by the input data such that the inverse problem (10.70)–(10.74) with the parameters* $\alpha < \alpha_0$ *and* $\gamma > \gamma_0$ *has a unique solution.*

Theorem 10.10. ([180]) (Continuous dependence upon data) *Let the input data* $\Phi = \{u_0, E\}$ *satisfy the assumptions of Theorem 10.9 with*

$$2u_0^0 - E_{\max} \geq N_1 > 0, \quad u_0^0 + 2u_0^1 \leq -N_2 < 0,$$
$$\|u_0\|_{C^3[0,1]} \leq N_3, \quad \|E\|_{C[0,T]} \leq N_4,$$

for some positive numbers N_i *for* $i = \overline{1,4}$. *Then, the solution* $(a(t), u(x,t))$ *of the inverse problem (10.70)–(10.74) depends continuously upon the data for sufficiently small* α *and large* γ.

10.3.2 Numerical solution

Consider the minimization of the least-squares objective function

$$F(a) := \left\| u(0,t)(\alpha + \beta a^{-\gamma}(t)) + \int_0^1 u(x,t)dx - E(t) \right\|^2_{L^2[0,T]}, \tag{10.79}$$

or its discretized form

$$F(\underline{a}) = \sum_{j=1}^{N} \left[u(0,t_j)(\alpha + \beta a_j^{-\gamma}) + \frac{1}{M} \sum_{i=0}^{M-1} u_{i,j}(\underline{a}) - E(t_j) \right]^2, \tag{10.80}$$

where $\underline{a} = (a_j)_{j=\overline{1,N}}$, $a_j = a(t_j)$, $t_j = jT/N$ for $j = \overline{1,N}$, the values $u_{i,j}(\underline{a})$ are computed from a FDM, [179], direct solver (with an $M \times N = 40 \times 40$ number of grid points discretizing the domain $\overline{Q}_T$) for given $\underline{a}$ and the time-step multiplier T/N has been dropped. If the compatibility condition $u(0,0) = u_0(0)$ is satisfied, then (10.74) applied at $t = 0$, yields

$$a(0) = \left(\frac{\beta u_0(0)}{E(0) - \int_0^1 u_0(x)dx - \alpha u_0(0)} \right)^{1/\gamma}. \tag{10.81}$$

The minimization of (10.80) subjected to $a \in \Lambda_{ad}$ with $a_{min} = 10^{-10}$ and $a_{max} = 10^3$ is accomplished using the MATLAB optimization toolbox routine *lsqnonlin*. In the case of noisy data, $E(t_j)$ is replaced by $E^\epsilon(t_j)$ in (10.80) given by

$$E^\epsilon(t_j) = E(t_j) + \epsilon_j, \quad j = \overline{1,N}, \tag{10.82}$$

where ϵ_j are random variables generated from a Gaussian normal distribution with mean zero and standard deviation $\sigma = p \times \max_{t \in [0,T]} |E(t)|$ and p represents the percentage of noise.

Consider the inverse problem (10.70)–(10.74) with $T = 1$, the data

$$u(x,0) = u_0(x) = -\frac{\cos(2\pi x)}{e}, \quad E(t) = -\left(1 + 8\pi^2\sqrt{1+t}\right)\exp(-\sqrt{1+t}),$$

and $\alpha = \beta = \gamma = 1$. One can easily check that the conditions of Theorems 10.9 and 10.10 are satisfied and that the analytical solution is given by

$$a(t) = \frac{1}{8\pi^2\sqrt{1+t}}, \quad u(x,t) = -\cos(2\pi x)\exp(-\sqrt{1+t}). \tag{10.83}$$

The initial guess for $a(t)$ is taken equal to the constant $a(0) = 1/(8\pi^2)$ which is known from expression (10.81). Both exact data, i.e. $p = 0$, as well as $p \in \{2\%, 20\%\}$ noisy data (10.82) are inverted. The objective function (10.80) and the numerically obtained results for $a(t)$ are presented in Figure 10.3. It can be seen that a very fast convergence of the objective function is achieved in 4 to 8 iterations to reach a very low value of $O(10^{-25})$. The agreement between the numerical results and the analytical solution for $a(t)$ is excellent for exact data, i.e. $p = 0$, and is consistent with the errors in the input data for $p > 0$. The numerical solutions for $a(t)$ converge to their corresponding exact solutions in (10.83), as p decreases from 20% to 2% and then to zero. The nonlinear least-squares minimization (10.80) produces good and consistent reconstructions of the solution even for a large amount of noise such as 20%.

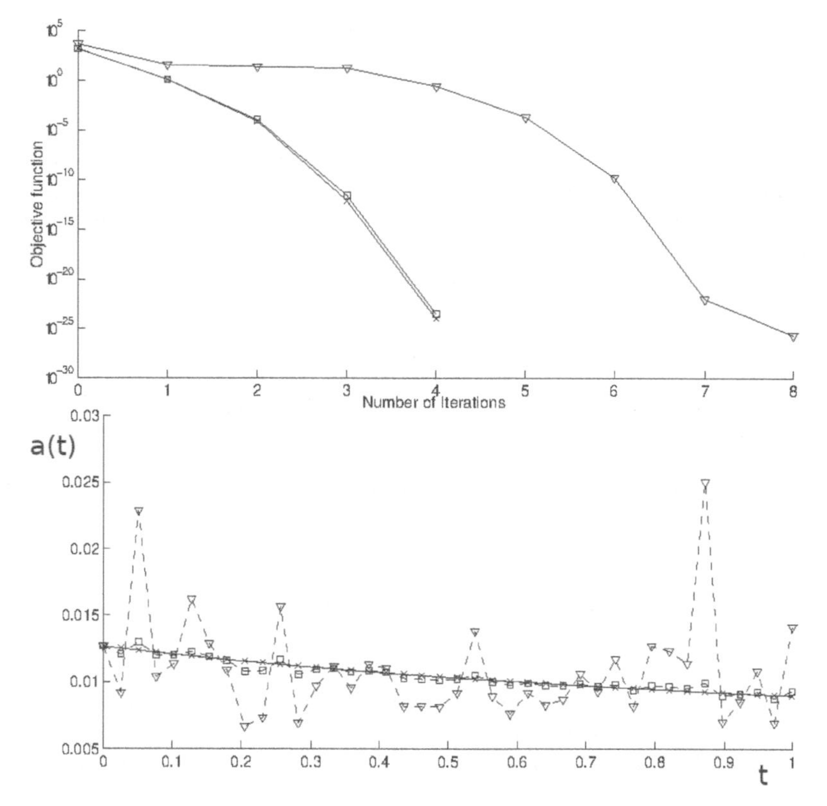

FIGURE 10.3: Objective function (10.80) and results for $a(t)$ for $p = 0$ $(-\times-)$, $p = 2\%$ $(-\Box-)$ and $p = 20\%$ $(-\nabla-)$ noise [180].

10.4 Determination of a Time-Dependent Thermal Conductivity and a Free Boundary

Many heat transfer applications can be modelled by the heat equation with a fixed boundary. However, there are numerous other problems for which the domain or the boundary varies with time and such problems are known as free boundary or Stefan problems. For instance, when a conductor melts and the liquid is drained away as it appears, the problem within the remaining solid involves the heat equation in a domain that is changing with time.

Consider the one-dimensional time-dependent heat equation

$$\frac{\partial u}{\partial t}(x,t) = a(t)\frac{\partial^2 u}{\partial x^2}(x,t) + F(x,t), \quad x \in (0, h(t)), \ t \in (0,T), \tag{10.84}$$

with unknown free smooth boundary $x = h(t) > 0$ and time-dependent thermal conductivity $a(t) > 0$. The initial condition is

$$u(x,0) = u_0(x), \quad 0 < x < h(0) =: h_0, \tag{10.85}$$

where $h_0 > 0$ is given, and the boundary conditions

$$u(0,t) = \mu_0(t), \quad u(h(t),t) = \mu_1(t), \quad t \in [0,T] \tag{10.86}$$

$$-a(t)u_x(0,t) = q_0(t), \quad \int_0^{h(t)} u(x,t)dx = E(t), \quad t \in [0,T]. \tag{10.87}$$

Note that μ_0 and q_0 represent Cauchy data at $x = 0$, whilst E represent the specification of the energy of the heat conducting moving boundary system [49, 267].

First, introduce the variable $y = x/h(t)$ to reduce (10.84)–(10.87) to the following problem for the unknowns $a(t)$, $h(t)$ and $v(y,t) := u(yh(t),t)$, [206]:

$$\frac{\partial v}{\partial t}(y,t) = \frac{a(t)}{h^2(t)}\frac{\partial^2 v}{\partial y^2}(y,t) + \frac{yh'(t)}{h(t)}\frac{\partial v}{\partial y}(y,t) + F(yh(t),t), \quad (y,t) \in Q_T, \tag{10.88}$$

in the fixed domain $Q_T = (0,1) \times (0,T)$,

$$v(y,0) = u_0(h_0 y), \quad y \in (0,1), \tag{10.89}$$

$$v(0,t) = \mu_0(t), \quad v(1,t) = \mu_1(t), \quad t \in [0,T], \tag{10.90}$$

$$-a(t)v_y(0,t) = q_0(t)h(t), \quad h(t)\int_0^1 v(y,t)dy = E(t), \quad t \in [0,T]. \tag{10.91}$$

The triplet $(0 < h(t), 0 < a(t), v(y,t))$ is called a solution to the inverse problem if it belongs to the class $C^1[0,T] \times C[0,T] \times C^{2,1}(\overline{Q})$ and satisfies the equations (10.88)–(10.91). For the input data, the following regularity and compatibility conditions are assumed:
(A) $0 < \mu_i(t) \in C^1[0,T]$ for $t \in [0,T]$, $i = 0,1$, $0 < E(t) \in C^1[0,T]$ for $t \in [0,T]$, $0 > q_0(t) \in C^1[0,T]$, $0 < u_0 \in C^2[0,h_0]$, $u_0'(x) > 0$ for $x \in [0,h_0]$, and $0 \leq F(x,t) \in C^{1,0}([0,H_1] \times [0,T])$, where

$$H_1 = \left(\min\left\{\min_{[0,h_0]} u_0(x), \min_{[0,T]} \mu_0(t), \min_{[0,T]} \mu_1(t)\right\}\right)^{-1} \times \max_{[0,T]} E(t);$$

(B) $u_0(0) = \mu_0(0)$, $u_0(h_0) = \mu_1(0)$ and $\int_0^{h_0} u_0(x)dx = E(0)$.

The existence and uniqueness of solution hold, as follows [206].

Theorem 10.11. (Local existence) *If the conditions* (A) *and* (B) *are satisfied, then there exists* $T_0 \in (0, T]$, *(defined by the input data) such that a solution to the problem (10.88)–(10.91) exists locally for* $(y, t) \in [0, 1] \times [0, T_0]$.

Theorem 10.12. (Uniqueness) *Suppose that:*
(i) $0 \le F(x, t) \in C^{1,0}([0, H_1] \times [0, T])$;
(ii) $u_0(x) > 0$ *for* $x \in [0, h_0]$, $\mu_0(t) > 0$, $\mu_1(t) > 0$, $q_0(t) < 0$ *and* $E(t) > 0$ *for* $t \in [0, T]$.
Then, a solution to the problem (10.88)–(10.91) is unique.

10.4.1 Numerical results and discussion

The inverse problem (10.88)–(10.91) is solved by minimizing the functional

$$\mathrm{F}(a, h) = \left\| -\frac{a(t)}{h(t)} v_y(0, t) - \mu_3(t) \right\|^2$$
$$+ \left\| h(t) \int_0^1 v(y, t)dy - \mu_4(t) \right\|^2 + \beta \left(\|a(t)\|^2 + \|h(t)\|^2 \right), \tag{10.92}$$

where $\beta \ge 0$ is a regularization parameter and the norm is the $L^2[0, T]$-norm. The discretization of (10.92) is

$$\mathrm{F}(\underline{a}, \underline{h}) = \sum_{j=0}^{N} \left[-\frac{a_j}{h_j} v_y(0, t_j) - \mu_3(t_j) \right]^2$$
$$+ \sum_{j=1}^{N} \left[h_j \int_0^1 v(y, t_j)dy - \mu_4(t_j) \right]^2 + \beta \left(\sum_{j=0}^{N} a_j^2 + \sum_{j=1}^{N} h_j^2 \right), \tag{10.93}$$

where $\underline{a} = (a_j)_{j=\overline{0,N}}$, $a_j = a(t_j)$, $\underline{h} = (h_j)_{j=\overline{1,N}}$, $h_j = h(t_j)$, $t_j = jT/N$ for $j = \overline{0, N}$, and v is the solution of the direct problem (10.88)–(10.90) (obtained using a FDM, [179], with the number of grids $M \times N = 40 \times 40$ discretizing the domain $\overline{Q}_T$) for given $a(t)$ and $h(t)$. The minimization of F subject to the physical constraints $\underline{a} > \underline{0}$ and $\underline{h} > \underline{0}$ is accomplished using the MATLAB toolbox routine *lsqnonlin*, [178].

At the first time step, i.e. $j = 0$, the derivative $v_y(0, 0)$ is approximated by $v_y(0, 0) = (4u_0^1 - u_0^2 - 3u_0^0)M/2$, where $u_0^i = u_0(h_0 y_i)$, $y_i = i/M$ for $i = \overline{0, M}$. In addition, the derivative of the free surface is approximated as

$$h'(0) = \frac{\mu_2'(0) - a(0)u_0''(h_0) - F(h_0, 0)}{u_0'(h_0)}, \quad h'(t_j) = \frac{h_j - h_{j-1}}{T/N}, \quad j = \overline{1, N}.$$

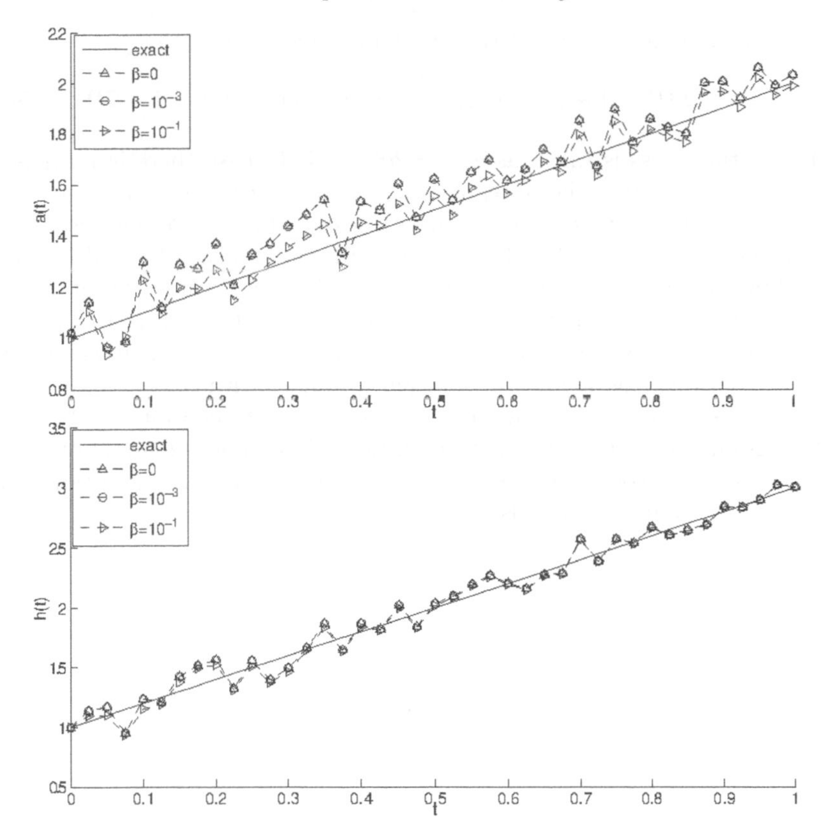

FIGURE 10.4: The thermal conductivity $a(t)$ and the free surface $h(t)$, with $p = 2\%$ noise and regularization [180].

If there is noise in the measured data (10.91), $q_0(t_j)$ and $E(t_j)$ in (10.93) are replaced by $q_0^{\epsilon 1}(t_j)$ and $E^{\epsilon 2}(t_j)$, given by

$$q_0^{\epsilon 1}(t_j) = q_0(t_j) + \epsilon 1_j, \quad E^{\epsilon 2}(t_j) = E(t_j) + \epsilon 2_j, \quad j = \overline{0, N}, \tag{10.94}$$

where $\epsilon 1_j$ and $\epsilon 2_j$ are random variables generated from a Gaussian normal distribution with mean zero and standard deviations $\sigma_1 = p \times \max_{t\in[0,T]} |q_0(t)|$ and $\sigma_2 = p \times \max_{t\in[0,T]} |E(t)|$, respectively.

Consider the inverse problem (10.84)–(10.87) with unknown coefficients $a(t)$ and $h(t)$, and solve this inverse problem with the following input data:

$$T = 1,\ h_0 = 1,\ \mu_0(t) = 1 + 8t,\ \mu_1(t) = (2 + 2t)^2 + 8t,\ q_0(t) = -2(1 + t),$$
$$E(t) = \frac{(2 + 2t)^3 - 1}{3} + 8t(1 + 2t),\ u_0(x) = (1 + x)^2,\ F(x, t) = 6 - 2t.$$

One can remark that the conditions of Theorem 10.12 are satisfied hence, the uniqueness of solution holds. The analytical solution is given by

$$a(t) = 1 + t, \quad h(t) = 1 + 2t, \quad u(x, t) = (1 + x)^2 + 8t. \tag{10.95}$$

Also, the analytical solution of the problem (10.88)–(10.91) is

$$a(t) = 1 + t, \quad h(t) = 1 + 2t, \quad v(y,t) = u(yh(t),t) = (1 + y(1 + 2t))^2 + 8t.$$

The initial guess is taken as $\underline{a}^{(0)} = \underline{h}^{(0)} = \underline{1}$. In case there is no noise in the input data (10.91), the unregularized ($\beta = 0$) objective function (10.93) decreases rapidly and takes a stationary value of $O(10^{-8})$ in about 7 iterations, and the numerical results for the corresponding unknowns $a(t)$ and $h(t)$ have been obtained in very good agreement with the exact values (10.95), [178]. Next, $p = 2\%$ noise is added to the measured data (10.94). Figure 10.4 presents the graphs of the recovered functions and it can be seen that there is not much difference between the numerical solution obtained with $\beta = 0$ or $\beta = 10^{-3}$, but there is some slight improvement in accuracy obtained for $\beta = 10^{-1}$.

Extensions to include the determination of unknown convection $b(t)u_x$ and reaction $c(t)u$ coefficients in the heat equation (10.84), in addition to the unknowns $a(t)$ and $h(t)$, are also possible [183, 370].

Chapter 11

Space-Dependent Conductivity

In the previous chapter, the unknown conductivity was time-dependent, whereas in this chapter the conductivity is considered to be mainly space-dependent charaterizing a functionally-graded or heterogeneous medium. First, in section 11.1, the determination of a spatially dependent permeability of a one-dimensional reservoir from measurements of the permeability itself at a few locations along with pressure data is considered [257, 387]. Second, in section 11.2, the recovery of a discontinuous anisotropic conductivity is analyzed. Third, in section 11.3 a special type of layered and time-dependent orthotropic conductivity is recovered from boundary data.

11.1 Reconstruction of a Permeability Function from Core Measurements and Pressure Data

Let $(0,1)$ represent a one-dimensional reservoir and let $K = K(x)$ be an unknown permeability function that is to be estimated. Suppose that pressure data $\{p_i\}_{i=\overline{1,N_0}}$ are available at N_0 locations $\{x_i\}_{i=\overline{1,N_0}}$ along with permeabillity core measurements $\{K_i\}_{i=\overline{1,N_1}}$ at N_1 locations $\{\tilde{x}_i\}_{i=\overline{1,N_1}}$. Assume that at x_{N_0}, the condition $p(x_{N_0}) = 0$ holds and that the reservoir boundary is insulated. If $f \in L^2(0,1)$ represents a hydraulic source and if $K \in \Lambda_{ad} := \{K \in H^1(0,1) \mid \infty > K_{max} \geq K \geq K_{min} > 0\}$, where K_{min} and K_{max} are (given) positive constants, then the pressure p is obtained as the unique weak solution in $H^1(0,1)$ of the problem,

$$-(K(x)p'(x))' = f(x), \quad x \in (0,1) \tag{11.1}$$

$$p'(0) = p'(1) = 0, \quad p(x_{N_0}) = 0. \tag{11.2}$$

The approach used to recover K from the measurements

$$p(x_i) := p_i, \quad i = \overline{1,N_0}, \quad K(\tilde{x}_i) := K_i, \quad i = \overline{1,N_1} \tag{11.3}$$

is the so-called regularized output least squares (ROLS) method [18], based on minimizing over Λ_{ad} the functional

$$L(K) := \sum_{i=1}^{N_0} (p(x_i;K) - p_i)^2 + \sum_{i=1}^{N_1} (K(\tilde{x}_i) - K_i)^2 + \lambda \int_0^1 K'^2(x)dx, \tag{11.4}$$

DOI: 10.1201/9780429400629-11

where $\lambda \geq 0$ is a regularization parameter to be prescribed and $p(\cdot; K)$ denotes the solution of the direct problem (11.1) and (11.2) for given $K \in \Lambda_{ad}$. One can easily recognize that the ROLS given by (11.4) is nothing else than the first-order nonlinear Tikhonov regularization method which imposes that $K \in H^1(0,1)$. It can be shown that a minimizer to (11.4) exists [388]. An approximate solution to it is sought in the form of the linear combinations

$$p(x) = \sum_{i=1}^{N} c_i \alpha_i(x), \tag{11.5}$$

$$K(x) = \sum_{i=1}^{M} a_i \psi_i(x), \tag{11.6}$$

where $\{\alpha_i\}_{i=\overline{1,N}}$ and $\{\psi_i\}_{i=\overline{1,M}}$ are sets of N and M linearly independent functions in $H^1(0,1)$, respectively. Applying the Ritz-Galerkin method, the problem (11.1)–(11.3) can be reduced to the system of equations

$$G(\underline{a})\underline{c} = \underline{\rho}, \tag{11.7}$$

where $\underline{c} = (c_i)_{i=\overline{1,N}}$, $\underline{a} = (a_k)_{k=\overline{1,M}}$, $G(\underline{a}) = (G_{ij}(\underline{a}))_{i,j=\overline{1,N}}$, $\underline{\rho} = (\rho_i)_{i=\overline{1,N}}$,

$$G_{ij}(\underline{a}) = \sum_{k=1}^{M} a_k G_{ij}^{(k)},\ G_{ij}^{(k)} = \int_0^1 \psi_k(x)\alpha_i'(x)\alpha_j'(x)dx,\ \rho_j = \int_0^1 \alpha_j(x)f(x)dx.$$

On choosing $\psi_i(x) = x^{i-1}$ for $i = \overline{1,M}$ and $\alpha_i(x) = (x - x_{N_0})x^{i-1}$ for $i = \overline{1,N}$,

$$G_{11}^{(k)} = 1, \quad G_{1j}^{(k)} = \frac{j}{j+k-1} - x_{N_0}\frac{(j-1)}{j+k-2},$$

$$G_{i1}^{(k)} = \frac{i}{i+k-1} - x_{N_0}\frac{(i-1)}{i+k-2},$$

$$G_{ij}^{(k)} = \frac{ij}{i+j+k-2} - x_{N_0}\frac{(2ij-i-j)}{i+j+k-3} + x_{N_0}^2\frac{(i-1)(j-1)}{i+j+k-4},$$

$$i,j = \overline{2,N},\ k = \overline{1,M}.$$

The functional (11.4) then takes the form

$$L(\underline{a}) = \{C_p - 2\underline{\zeta}^{\mathrm{T}}\underline{c} + \underline{c}^{\mathrm{T}}H\underline{c}\} + \{C_K - 2\underline{\kappa}^{\mathrm{T}}\underline{a} + \underline{a}^{\mathrm{T}}H_K\underline{a}\} + \lambda\underline{a}^{\mathrm{T}}G_0\underline{a}, \tag{11.8}$$

where $\underline{c} = \underline{c}(\underline{a})$ satisfies (11.7) and

$$C_p = \sum_{k=1}^{N_0} p_k^2, \quad C_K = \sum_{k=1}^{N_1} K_k^2,$$

$$\zeta_i = \sum_{k=1}^{N_0} \alpha_i(x_k)p_k = \sum_{k=1}^{N_0}(x_k - x_{N_0})x_k^{i-1}p_k,$$

$$\kappa_n = \sum_{k=1}^{N_1} \psi_n(\tilde{x}_k)K_k = \sum_{k=1}^{N_1} \tilde{x}_k^{n-1}K_k,$$

$$H_{ij} = \sum_{k=1}^{N_0} \alpha_i(x_k)\alpha_j(x_k) = \sum_{k=1}^{N_0} (x_k - x_{N_0})^2 x_k^{i+j-2},$$

$$(H_K)_{ln} = \sum_{k=1}^{N_1} \psi_l(\tilde{x}_k)\psi_n(\tilde{x}_k) = \sum_{k=1}^{N_1} \tilde{x}_k^{l+n-2},$$

$$(G_0)_{ln} = \int_0^1 \psi_l'(x)\psi_n'(x)dx = \begin{cases} 0, & \text{if } l+n=3, \\ \frac{(l-1)(n-1)}{l+n-3}, & \text{otherwise,} \end{cases}$$

for $i, j = \overline{1,N}$ and $l, n = \overline{1,M}$. This is minimized subject to the constraints

$$\infty > K_{max} \geq \sum_{k=1}^{M} a_k \psi_k(x) \geq K_{min} > 0, \quad x \in (0,1) \tag{11.9}$$

using the NAG routine E04UCF, which is based on a sequential programming method designed to minimize an arbitrary smooth function subject to simple bounds on the variables and linear and nonlinear constraints. In the computation, apart from the expression for $L(\underline{a})$ given by (11.8), one also needs to supply its gradient, which is given by [387],

$$\nabla L(\underline{a}) = 2\left[-\underline{\kappa} + \underline{\Theta} + (\lambda G_0 + H_K)\underline{a}\right], \tag{11.10}$$

where

$$\Theta_k = \underline{\phi}^{\mathrm{T}} G^{(k)} \underline{c}, \quad k = \overline{1,M}, \quad G\underline{\phi} = H\underline{c} - \underline{\zeta}. \tag{11.11}$$

If one chooses the vector $\underline{\rho}$ to be sufficiently small in magnitude, then there is at most one minimizer $K \in \mathring{\Lambda}_{ad}$ to (11.4), [386, 387]. Based on (11.7), (11.10) and (11.11), the optimality conditions satisfied by the (interior) minimum of (11.4) are

$$G(\underline{a})\underline{c} = \underline{\rho}, \quad G(\underline{a})\underline{\phi} = H\underline{c} - \underline{\zeta},$$

$$\Theta_k = \underline{\phi}^{\mathrm{T}} G^{(k)} \underline{c}, \quad k = \overline{1,M}, \quad (\lambda G_0 + H_K)\underline{a} - \underline{\kappa} + \underline{\Theta} = \underline{0},$$

or, in component form,

$$\sum_{j=1}^{N}\sum_{k=1}^{M} a_k G_{ij}^{(k)} c_j = \rho_i, \tag{11.12}$$

$$\sum_{j=1}^{N}\sum_{k=1}^{M} a_k G_{ij}^{(k)} \phi_j = \sum_{j=1}^{N} H_{ij} c_j - \zeta_i, \tag{11.13}$$

$$\sum_{l=1}^{M} \left[\lambda (G_0)_{kl} + (H_K)_{kl}\right] a_l - \kappa_k + \sum_{i=1}^{N} \phi_i \sum_{j=1}^{N} G_{ij}^{(k)} c_j = 0, \tag{11.14}$$

for $i = \overline{1,N}$ and $k = \overline{1,M}$. If the measurements (11.3) are noisy, namely,

$$p_i^{\epsilon} = p_i + \epsilon, \quad i = \overline{1,N_0}, \quad K_i^{\epsilon} = K_i + \epsilon, \quad i = \overline{1,N_1}, \tag{11.15}$$

where ϵ is a (small) amount of noise, then, the regularization parameter $\lambda > 0$ can be chosen according to the discrepancy principle criterion, such that

$$\sqrt{\sum_{i=1}^{N_0}(p(x_i;K) - p_i^\epsilon)^2 + \sum_{i=1}^{N_1}(K(\tilde{x}_i) - K_i^\epsilon)^2} \approx |\epsilon|\sqrt{N_0 + N_1 - 1}. \quad (11.16)$$

11.1.1 Results

Let us take $x_{N_0} = 1/2$ and consider the following test example [257]:

$$K(x) = 1 + x, \quad f(x) = 6 - 18x^2, \quad p(x) = (2x-1)(2x^2 - 2x - 1)/2.$$

Taking $N = 3$, $M = 2$, the previous analysis give

$$p(x) = (2x-1)(c_1 + c_2x + c_2x^2)/2, \quad K(x) = a_1 + a_2x,$$

$$G_0 = \begin{pmatrix} 0 & 0 \\ 0 & 1 \end{pmatrix}, \quad G = a_1G^{(1)} + a_2G^{(2)},$$

$$G^{(1)} = \begin{pmatrix} 1 & 1/2 & 1/2 \\ 1/2 & 7/12 & 7/12 \\ 1/2 & 7/12 & 19/30 \end{pmatrix}, \quad G^{(2)} = \begin{pmatrix} 1/2 & 5/12 & 5/12 \\ 5/12 & 11/24 & 59/120 \\ 5/12 & 59/120 & 11/20 \end{pmatrix},$$

$$\rho_1 = -\frac{3}{2}, \quad \rho_2 = -17/20, \quad \rho_3 = -7/10.$$

The constraint (11.9) recasts as the simple bounds on the variables

$$K_{max} \geq a_1 \geq K_{min}, \quad K_{max} - K_{min} \geq a_2 \geq K_{min} - K_{max} \quad (11.17)$$

and the linear constraint

$$K_{max} \geq a_1 + a_2 \geq K_{min}. \quad (11.18)$$

The following three cases are analyzed.

Case (a): $N_0 = 1$, $N_1 = 2$.
In this case, take $\tilde{x}_1 = 0$ and $\tilde{x}_2 = 1$, such that only the core boundary measurements of K are employed. Then,

$$p_1 = 0,\ K_1 = 1+\epsilon,\ K_2 = 2+\epsilon,\ C_p = 0,$$

$$C_K = (1+\epsilon)^2 + (2+\epsilon)^2,\ \underline{\zeta} = \underline{0},\ \kappa_1 = 3 + 2\epsilon,\ \kappa_2 = 2+\epsilon,\ H = 0_{3\times 3},$$

$$\underline{\phi} = \underline{0},\ \underline{\Theta} = \underline{0},\ H_K = \begin{pmatrix} 2 & 1 \\ 1 & 1 \end{pmatrix}.$$

Then, equation (11.8) gives

$$L(a_1, a_2) = \lambda a_2^2 + [(1+\epsilon)^2 + (2+\epsilon)^2 - 2(3+2\epsilon)a_1 - 2(2+\epsilon) \\ +2a_1^2 + a_2^2 + 2a_1a_2], \quad (11.19)$$

whilst equation (11.10) gives

$$\nabla L(a_1, a_2) = 2(2a_1 + a_2 - 3 - 2\epsilon, (1+\lambda)a_2 + a_1 - 2 - \epsilon) = (0,0),$$

yielding

$$a_1 = \epsilon + \frac{3\lambda+1}{2\lambda+1}, \quad a_2 = \frac{1}{2\lambda+1}. \tag{11.20}$$

Then, $K_\lambda^\epsilon(x) = \epsilon + \frac{x+3\lambda+1}{2\lambda+1}$, and according to (11.16) one can choose

$$\lambda = \frac{|\epsilon|}{1-2|\epsilon|} > 0, \tag{11.21}$$

where it is assumed that $|\epsilon| < 0.5$, i.e. the amount of noise is less than 50% error in the measurement of K_1 and less that 25% error in the measurement of K_2. This gives $a_1^\epsilon = 1 + \epsilon + |\epsilon|$, $a_2^\epsilon = 1 - 2|\epsilon|$, $K^\epsilon(x) := a_1^\epsilon + a_2^\epsilon x = 1 + \epsilon + |\epsilon| + x(1-2|\epsilon|)$. When $\epsilon = 0$, one obtains that $K^0(x) = 1 + x$, which is the exact solution. This shows that the retrieval of the permeability K is unique. One can also evaluate the L^2-norm $\|K^\epsilon - K\| = \frac{2}{\sqrt{3}}|\epsilon|$, which shows that the retrieval of permeability K is stable if λ is chosen as in (11.21). Finally, the pressure is obtained by solving the system of equations (11.7), i.e.

$$\begin{cases} (18+12\epsilon)c_1 + (11+2\epsilon)c_2 + (11+2\epsilon)c_3 = -18, \\ 10(11+2\epsilon)c_1 + 5(25+6\epsilon)c_2 + (129+22\epsilon)c_3 = -102, \\ 10(11+2\epsilon)c_1 + (129+22\epsilon)c_2 + 2(71+10\epsilon)c_3 = -84, \end{cases}$$

where, for simplicity, $\epsilon \geq 0$ was taken. The solution of this system is

$$c_1 = -(8\epsilon^2 + 86\epsilon + 63)/\Delta, \quad c_2 = -(16\epsilon^2 + 268\epsilon + 126)/\Delta,$$
$$c_3 = (-8\epsilon^2 + 232\epsilon + 126)/\Delta,$$

where $\Delta = (3+2\epsilon)(4\epsilon^2 + 36\epsilon + 21)$. Observe that when $\epsilon = 0$, one obtains $p^0(x) = (2x-1)(2x^2-2x-1)/2$, which is the exact solution. This shows that the retrieval of the pressure p is unique. One can also evaluate the L^2-norm

$$\|p^\epsilon - p\|^2 = 4\epsilon^2(68\epsilon^4 + 1160\epsilon^3 + 5761\epsilon^2 + 6547\epsilon + 2189)/(35\Delta^2),$$

which indicates linear stability.

Case (b): $N_0 = 3$, $N_1 = 0$.
In this case, take $x_1 = 0$ and $x_2 = 1$, such that only boundary data of p are employed. Then,

$$p_1 = 1/2 + \epsilon,\ p_2 = -1/2 + \epsilon,\ p_3 = 0, \quad C_p = 1/2 + 2\epsilon^2,\ C_K = 0,\ \underline{\kappa} = \underline{0},$$

$$\zeta_1 = -1/2,\ \zeta_2 = \zeta_3 = -(1/2+\epsilon)/2,\ H_K = 0_{2\times 2},\ H = \begin{pmatrix} 1/2 & 1/4 & 1/4 \\ 1/4 & 1/4 & 1/4 \\ 1/4 & 1/4 & 1/4 \end{pmatrix}.$$

Then (11.12)–(11.14) give rise to a system of 8 nonlinear equations with 8 unknowns a_1, a_2, c_1, c_2, c_3, ϕ_1, ϕ_2 and ϕ_3. Solving the system formed from the last 3 equations yields [257]

$$c_1 = -(11a_2^2 + 61a_1a_2 + 54a_1^2)/(2\Delta_1), \quad c_2 = (a_2^2 - 61a_1a_2 - 66a_1^2)/\Delta_1,$$
$$c_3 = 2(a_1^2 + 32a_1a_2 + 30a_1^2)/\Delta_1,$$

where $\Delta_1 = (2a_1+a_2)(a_2^2+10a_1a_2+10a_1^2)$. The remaining system of 6 nonlinear equations is solved using MAPLE and only the solution which ensures that a_1 and a_2 satisfy the constraints (11.17) and (11.18) with $K_{max} = 10^{10}$ and $K_{min} = 10^{-10}$, is accepted. According to the discrepancy principle (11.16):
$\lambda = 1/180$ for $\epsilon = 5\%$ noise, and $\|K^\epsilon - K\| = 0.017$, $\|p^\epsilon - p\| = 0.003$;
$\lambda = 1/90$ for $\epsilon = 10\%$ noise, and $\|K^\epsilon - K\| = 0.031$, $\|p^\epsilon - p\| = 0.007$;
$\lambda = 1/45$ for $\epsilon = 20\%$ noise, and $\|K^\epsilon - K\| = 0.065$, $\|p^\epsilon - p\| = 0.016$.

Case (c): $N_0 = 2$, $N_1 = 1$
In this case, take $\tilde{x}_1 = 0$ and $x_1 = 1$, such that K is measured at $x = 0$ and the pressure p at $x = 1$. Then,

$$p_1 = -1/2 + \epsilon,\ p_2 = 0,\ K_1 = 1 + \epsilon, \quad C_p = (-1/2 + \epsilon)^2,\ C_K = (1 + \epsilon)^2,$$
$$\zeta_1 = \zeta_2 = \zeta_3 = -(1/2 + \epsilon)/2, \quad \kappa_1 = 1 + \epsilon,\ \kappa_2 = 0,\ H_K = \begin{pmatrix} 1 & 0 \\ 0 & 0 \end{pmatrix},$$
$$H = \begin{pmatrix} 1/4 & 1/4 & 1/4 \\ 1/4 & 1/4 & 1/4 \\ 1/4 & 1/4 & 1/4 \end{pmatrix}.$$

Then equations (11.13) and (11.14) form a system of 5 nonlinear equations, which together with the 3 equations in (11.12) complete the system of 8 equations in the 8 unknowns a_1, a_2, c_1, c_2, c_3, ϕ_1, ϕ_2 and ϕ_3. According to the discrepancy principle (11.16), this yields $\lambda = 1/41$ for $\epsilon = 5\%$, and $\|K^\epsilon - K\| = 0.104$ and $\|p^\epsilon - p\| = 0.022$.

It was shown elsewhere [257] that Case (b) provides more information to the inverse problem compared to Cases (a) and (c).

11.2 Discontinuous Anisotropic Conductivity

Factors such as manufacturing and curing process may affect the material properties of a structure often introducing additional variations such as heterogeneity and anisotropy [176], which are difficult to measure directly, and for which inverse problem modelling is considered as the most important stage for monitoring the robustness of materials. Raw materials are anisotropic and heterogeneous in nature, and recovering their conductivity is of utmost importance to the oil, aerospace and medical sectors concerned with imaging soils, reinforced fiber composites and organs. However, reconstructing the conductivity of an inhomogeneous and anisotropic medium is ill-posed because it depends on indirect observable measurements which contain errors that lead to large changes in the solution. In addition, there is the notorious issue of non-identifiability in the sense that even the full knowledge of the Dirichlet-to-Neumann (D-to-N) boundary map is not sufficient to determine the anisotropic and inhomogeneous conductivity tensor of a material subject to steady-state heat conduction or electrostatic excitations [279]. Moreover, even if an isotropic inhomogeneous conductivity can be uniquely retrieved from the complete knowledge of the D-to-N map [238], this is still not practical because an infinite number of measurements has to be performed for providing all the necessary boundary temperature/potential and heat/current flux Cauchy data pairs. To deal with this non-identifiability, previous studies, e.g. [195, 236, 237, 347], proposed to measure the dependent variable (temperature, piezometric head or electric potential) everywhere inside the solution domain such that the original second-order elliptic equation is recast as a first-order hyperbolic equation. In [15], it was suggested to measure many internal current densities to guarantee a unique and explicit reconstruction of a fully anisotropic conductivity tensor entering a steady-state Darcy flow conductivity equation. Another approach is to assume that the conductivity is piecewise constant or linearly dependent on the space variables [147, 312], such that a finite dimensional problem is to be solved in a nonlinear least-squares sense.

At steady-state, the determination of the conductivity of a layered and orthotropic medium was initiated in [64] with the general case of an anisotropic space-dependent conductivity in the Laplace-Beltrami elliptic equation reviewed in [378]. Nevertheless, due to the ill-posedness of the anisotropic inverse conductivity problem certain simplifications are required, such as assuming that the conductivity is discontinuous with known piecewise values but with unknown geometrical support. Therefore, in this section, an inverse problem is considered to estimate the geometry of discontinuities in a conductive material $\Omega \subset \mathbb{R}^d$ with anisotropic space-dependent conductivity $(I + (K - I)\chi_D)$ from Cauchy data on $\partial\Omega$, where $D \subset \Omega$, K is a symmetric and positive definite tensor not equal to the identity I and χ_D is the characteristic function of the domain D. Then, the refraction (transmission, conjugate) problem for the potential ϕ is given by [197]

$$\begin{cases} \nabla \cdot ((I + (K - I)\chi_D)\nabla\phi) = 0 & \text{in } \Omega, \\ \phi = f & \text{on } \partial\Omega, \end{cases} \tag{11.22}$$

subject to refraction conditions related to the continuity of the boundary potential ϕ and its fluxes $(\partial\phi/\partial n^-)$ and $(K\nabla\phi) \cdot \underline{n}^+$ across the interface ∂D, where $\underline{n}$, $\underline{n}^-$ and $\underline{n}^+$ are the outward unit normals to the boundaries $\partial\Omega$, $\partial(\Omega\backslash D)\backslash\partial\Omega$ and ∂D, respectively. If $f \in H^{1/2}(\partial\Omega)$, the direct problem satisfying (11.22), when K and D are known, has a unique solution $\phi \in H^1(\Omega)$, [243]. Assuming that K is known, the inverse conductivity problem requires finding D from the knowledge of the D-to-N map $\Lambda_D(f) = \frac{\partial\phi}{\partial n} := g \in H^{-1/2}(\partial\Omega)$

given by

$$\langle \Lambda_D f, q \rangle = \int_{\partial\Omega} gq \, dS = \int_{\Omega} ((I + (K - I)\chi_D)\nabla\phi) \cdot \nabla w \, d\Omega \tag{11.23}$$

for any $q \in H^{1/2}(\partial\Omega)$ and $w \in H^1(\Omega)$ with $w|_{\partial\Omega} = q$, for finitely many Dirichlet data f on $\partial\Omega$. Uniqueness results can be found in [187, 188, 189, 190, 196, 197]. In the following sub-section, the size of the unknown inclusion D is estimated from a single boundary measurement only [187, 253].

11.2.1 Size estimation of the inclusion

Let $\phi_0 \in H^1(\Omega)$ be the unique solution of the direct problem in the isotropic case with $K = I$, i.e.

$$\begin{cases} \nabla^2\phi_0 = 0 & \text{in } \Omega, \\ \phi_0 = f & \text{on } \partial\Omega. \end{cases} \tag{11.24}$$

Its D-to-N map $\Lambda_0(f) = \frac{\partial\phi_0}{\partial n} := g_0 \in H^{-1/2}(\partial\Omega)$, is defined, via Green's formula, as

$$\langle \Lambda_0 f, q \rangle = \int_{\partial\Omega} g_0 q \, dS = \int_{\Omega} \nabla\phi_0 \cdot \nabla w \, d\Omega, \tag{11.25}$$

for any $q \in H^{1/2}(\partial\Omega)$ and $w \in H^1(\Omega)$ with $w|_{\partial\Omega} = q$. In particular,

$$\langle \Lambda_0 f, f \rangle = \int_{\partial\Omega} g_0 f \, dS = \int_{\Omega} |\nabla\phi_0|^2 d\Omega \tag{11.26}$$

and

$$\langle \Lambda_D f, f \rangle = \int_{\partial\Omega} gf \, dS = \int_{\Omega} ((I + (K - I)\chi_D)\nabla\phi) \cdot \nabla\phi \, d\Omega. \tag{11.27}$$

Since K is symmetric and positive definite there exists $\delta_0 > 0$ such that

$$K\underline{x} \cdot \underline{x} \geq \delta_0 |\underline{x}|^2, \quad \forall \underline{x} \in \mathbb{R}^d. \tag{11.28}$$

From equations (11.27) and (11.28),

$$\langle \Lambda_D f, f \rangle \geq \int_{\Omega\backslash D} |\nabla\phi|^2 d(\Omega\backslash D) + \delta_0 \int_D |\nabla\phi|^2 dD \geq 0. \tag{11.29}$$

Also, $\langle \Lambda_0 f, f \rangle = \int_\Omega |\nabla\phi_0|^2 dD \geq 0$ and thus $\langle \Lambda_0 f, f \rangle > 0$ if and only if $f \not\equiv$ constant, or equivalently if and only if $\langle \Lambda_D f, f \rangle > 0$. The following results and discussion [187, 253] characterize the size estimation of the inclusion D.

Lemma 11.1. ([187]) *Let C_1 and C_2 be symmetric and positive definite tensors and u_1, $u_2 \in H^1(\Omega)$ be the unique solutions of*

$$\begin{cases} \nabla \cdot (C_j \nabla u_j) = 0 & \text{in } \Omega, \\ u_j = \phi_j \in H^{1/2}(\partial\Omega) & \text{on } \partial\Omega, \end{cases} \tag{11.30}$$

for $j = 1, 2$. Then, the following identity holds:

$$\int_\Omega [(C_1\nabla(u_1 - u_2)) \cdot \nabla(u_1 - u_2) + ((C_2 - C_1)\nabla u_2) \cdot \nabla u_2]\, d\Omega$$
$$= \langle \Lambda_{C_1}(\phi_1) - \Lambda_{C_2}(\phi_2), \phi_1 - \phi_2\rangle + \langle(\Lambda_{C_2} - \Lambda_{C_1})(\phi_2), \phi_1\rangle, \quad (11.31)$$

where

$$\langle \Lambda_{C_j}(\phi_1), q\rangle := \int_\Omega (C_j\nabla u_i) \cdot \nabla w\, d\Omega, \quad i, j = 1, 2 \quad (11.32)$$

for any $q \in H^{1/2}(\partial\Omega)$ and $w \in H^1(\Omega)$ with $w|_{\partial\Omega} = q$.

Remark 11.1. The proof of (11.31) makes use of Alessandrini's identity,

$$\langle(\Lambda_{C_2} - \Lambda_{C_1})(\phi_2), \phi_1\rangle = \int_\Omega ((C_2 - C_1)\nabla u_2) \cdot \nabla u_1\, d\Omega \quad (11.33)$$

Lemma 11.2. ([187]) *Let C_j, u_j for $j = 1, 2$ be as in Lemma 11.1. Then,*

$$\int_\Omega ((I - C_2^{-1}C_1)\nabla u_1) \cdot (C_1\nabla u_1)\, d\Omega$$
$$\leq \langle \Lambda_{C_1}(\phi_1) - \Lambda_{C_2}(\phi_2), \phi_1 - \phi_2\rangle + \langle(\Lambda_{C_2} - \Lambda_{C_1})(\phi_2), \phi_1\rangle. \quad (11.34)$$

Proposition 11.1. ([187]) *Assume that f is not a constant function, which is equivalent to $\langle \Lambda_0 f, f\rangle > 0$. Then,*

$$\int_D ((I - K^{-1})\nabla\phi_0) \cdot \nabla\phi_0\, dD \leq \frac{\langle \Lambda_0 f, f\rangle}{\langle \Lambda_D f, f\rangle}\langle(\Lambda_D - \Lambda_0)f, f\rangle, \quad (11.35)$$

$$\langle(\Lambda_D - \Lambda_0)f, f\rangle \leq \int_D ((K - I)\nabla\phi_0) \cdot \nabla\phi_0\, dD. \quad (11.36)$$

Proof. Following [253], take in Lemma 11.1, $C_1 = I$, $\phi_1 = f$, $C_2 = I + (K - I)\chi_D$, $\phi_2 = (1 + t)f$, with t an arbitrary real number. Then, the solutions of (11.30) are $u_1 = \phi_0$ and $u_2 = (1 + t)\phi$. Using (11.34),

$$\int_D ((I - K^1)\nabla\phi_0) \cdot \nabla\phi_0\, dD \leq -t\langle \Lambda_0 f - (1 + t)\Lambda_D f, f\rangle$$
$$+(1 + t)\langle(\Lambda_D - \Lambda_0)f, f\rangle = \langle \Lambda_D f, f\rangle \left(t + \frac{\langle(\lambda_D - \Lambda_0)f, f\rangle}{\langle \Lambda_D f, f\rangle}\right)^2$$
$$+\frac{\langle \Lambda_0 f, f\rangle}{\langle \Lambda_D f, f\rangle}\langle(\Lambda_D - \Lambda_0)f, f\rangle. \quad (11.37)$$

Since f is not a constant, equation (11.29) implies that $\langle \Lambda_D f, f\rangle > 0$. Putting $t = -\frac{\langle(\lambda_D - \Lambda_0)f, f\rangle}{\langle \Lambda_D f, f\rangle}$ in (11.37) yields the inequality (11.35). To establish the second inequality, take in Lemma 11.1, $C_1 = I + (K - I)\chi_D$, $C_2 = I$, $\phi_1 = \phi_2 = f$ in which case $u_1 = \phi$ and $u_2 = \phi_0$. Using (11.26) and (11.27),

$$\langle(\Lambda_0 - \Lambda_D)f, f\rangle = \int_\Omega ((I + (K - I)\chi_D)\nabla(\phi - \phi_0)) \cdot \nabla(\phi - \phi_0)\, d\Omega$$
$$- \int_\Omega ((K - I)\nabla\phi_0) \cdot \nabla\phi_0\, d\Omega. \quad (11.38)$$

Since the first term in the right-hand side of (11.38) is non-negative, the inequality (11.36) is obtained.

Theorem 11.1. ([187]) *Let $f \in H^{1/2}(\partial\Omega)$ be such that $inf_\Omega|\nabla\phi_0| > 0$. Then, if $(K - I)\chi_D$ is positive definite, so is $I - K^{-1}\chi_D$ and*

$$min\{a^{-1}\}(sup_\Omega|\nabla\phi_0|)^{-2}\langle(\Lambda_D - \Lambda_0)f, f\rangle \le |D|$$
$$\le max\{b^{-1}\}(inf_\Omega|\nabla\phi_0|)^{-2}\frac{\langle\Lambda_0 f, f\rangle}{\langle\Lambda_D f, f\rangle}\langle(\Lambda_D - \Lambda_0)f, f\rangle, \quad (11.39)$$

where a is an eigenvalue of $(K - I)\chi_D$ and b is an eigenvalue of $I - K^{-1}\chi_D$. A similar size estimate holds if $(K - I)\chi_D$ is negative definite.

Proof: Following [253], $I - K^{-1} = (K - I)[(K - I)^{-1} - K^{-1}](K - I)$, so $(K - I)$ positive definite implies that $(I - K^{-1})$ is positive definite. Then,

$$\min\{b\}\ |D|\ \inf_\Omega|\nabla\phi_0|^2 \le \int_D ((I - K^{-1})\nabla\phi_0) \cdot \nabla\phi_0\, dD,$$
$$\int_D ((K - I)\nabla\phi_0) \cdot \nabla\phi_0\, dD \le \max\{a\}\ |D|\ \sup_\Omega|\nabla\phi_0|^2,$$

and the inequality (11.39) follows from (11.35) and (11.36).

Remark 11.2. In $d = 2$-dimensions, the condition $\inf_\Omega|\nabla\phi_0| > 0$ is ensured if one chooses the Dirichlet data f to be non-constant and monotonic on $\partial\Omega_1$ and on $\partial\Omega_2$, separately, where $\partial\Omega$ is decomposed into two disjoint arcs $\partial\Omega_1$ and $\partial\Omega_2$, [6]. If the Dirichlet boundary condition $u|_{\partial\Omega} = f$ is replaced by the Neumann condition $\partial_n\phi|_{\partial\Omega} = g$, then the monotonicity conditions should be replaced by the conditions $g \not\equiv 0$, $g \ge 0$ on $\partial\Omega_1$ and $g \le 0$ on $\partial\Omega_2$. Further, in any dimensions, the condition $\inf_\Omega|\nabla\phi_0| > 0$ is ensured if one chooses, for example, $g(\underline{x}) = \underline{\nu} \cdot \underline{n}(\underline{x})$, where $\underline{\nu}$ is a constant vector [225]. Also, boundary conditions of the mixed type can be considered [5].

Theorem 11.2. ([187]) *Let λ be a nonzero eigenvalue of $(K - I)\chi_D$ and $\underline{a}$ the corresponding normalized eigenvector. Set the Dirichlet boundary data $f(\underline{s}) = \underline{a} \cdot \underline{s}$ for $\underline{s} \in \partial\Omega$. Then,*

$$\frac{\lambda}{\lambda + 1}|D| \le \frac{\langle\Lambda_0 f, f >}{< \Lambda_D f, f\rangle}\langle(\Lambda_D - \Lambda_0)f, f\rangle,$$
$$\langle(\Lambda_D - \Lambda_0)f, f\rangle \le \lambda|D|. \quad (11.40)$$

Proof: Following [253], since $K = I + (K - I)$ is positive definite, $\lambda + 1 > 0$. Also, $(K - I)\underline{a} = \lambda \underline{a}$ and hence, $(I - K^{-1})\underline{a} = \frac{\lambda}{\lambda+1}\underline{a}$. Since ϕ_0 is the solution of problem (11.24) with $f(\underline{s}) = \underline{a} \cdot \underline{s}$ on $\partial\Omega$, it implies that $\phi_0(\underline{s}) = \underline{a} \cdot \underline{s}$ in Ω and thus $\nabla\phi_0 = \underline{a}$. Using that $|\underline{a}| = 1$, from (11.35) and (11.36), and $f \not\equiv$ constant,

$$\frac{\lambda}{\lambda+1}|D| = \int_D ((I - K^{-1})\underline{a}) \cdot \underline{a}\, dD \leq \frac{\langle \Lambda_0 f, f\rangle}{\langle \Lambda_D f, f\rangle}\langle(\Lambda_D - \Lambda_0)f, f\rangle, \quad (11.41)$$

$$\langle(\Lambda_D - \Lambda_0)f, f\rangle \leq \int_D ((K - I)\underline{a}) \cdot \underline{a}\, dD = \lambda|D|. \quad (11.42)$$

Remark 11.3. (i) From Theorem 11.2, one can estimate the size $|D|$ in terms of $\langle(\Lambda_D - \Lambda_0)f, f\rangle$.

(ii) $\langle \Lambda_0 f, f\rangle = \int_\Omega |\nabla\phi_0|^2 d\Omega = \int_\Omega |\underline{a}|^2 d\Omega = |\Omega|$ and $\langle \Lambda_D f, f\rangle = \int_{\partial\Omega}(\underline{a} \cdot \underline{s})g(\underline{s})\, dS =: \mathrm{T} > 0$.

(iii) $\langle \Lambda_D f, f\rangle = \langle \Lambda_0 f, f\rangle$ if and only if $|D| = 0$. So, if $\mathrm{T} = |\Omega|$, then there is no inclusion.

The estimates (11.40) give

$$\mathrm{T} - |\Omega| \leq \lambda|D| \leq \frac{(\lambda+1)|\Omega|}{\mathrm{T}}(\mathrm{T} - |\Omega|). \quad (11.43)$$

It is interesting to look at the case when $\mathrm{T} - |\Omega| = \frac{(\lambda+1)|\Omega|}{\mathrm{T}}(\mathrm{T} - |\Omega|)$. This can happen if and only if $\mathrm{T} = |\Omega|$ (in which case $|D| = 0$) or $\mathrm{T} = (\lambda+1)|\Omega|$ (in which case $|D| = |\Omega|$).

Example 11.1 Let Ω be the unit circle and consider an orthotropic material with conductivity $K = \begin{bmatrix} 1 & 0 \\ 0 & 2 \end{bmatrix}$. Then, $\lambda = 1$, $\underline{a} = \begin{bmatrix} 0 \\ 1 \end{bmatrix}$, $|\Omega| = \pi$, $f(\underline{s}) = \underline{a} \cdot \underline{s} = y$ and $\mathrm{T} = \int_{\partial\Omega} y g(x, y) dS = \int_0^{2\pi} \sin(\theta) g(\cos(\theta), \sin(\theta)) d\theta$. In this case, the estimate (11.43) gives

$$\mathrm{T} - \pi \leq |D| \leq \frac{2\pi}{\mathrm{T}}(\mathrm{T} - \pi), \quad (11.44)$$

so a necessary condition for the existence of a solution is $2\pi \geq \mathrm{T} \geq \pi$. Of course, the Neumann flux data g cannot be taken arbitrarily once the Dirichlet data f has already been given. However, assuming that a solution exists, one can then consider various choices of the flux g satisfying $\int_0^{2\pi} g(\cos(\theta), \sin(\theta)) d\theta = 0$:

(i) $g = \sin(\theta)$, then $\mathrm{T} = \int_0^{2\pi} \sin^2(\theta) d\theta = \pi = |\Omega|$ and hence $|D| = 0$, i.e. there is no inclusion, as expected since in this case $\phi = \phi_0 = y$.

(ii) $g = \cos(\theta)$, then $\mathrm{T} = \int_0^{2\pi} \sin(\theta)\cos(\theta) d\theta = 0$ and hence f would be a constant, which is a contradiction. Hence, no size estimation can be established.

(iii) $g = \frac{3}{2}\sin(\theta)$, then $\mathrm{T} = \int_0^{2\pi} \frac{3}{2}\sin^2(\theta) d\theta = \frac{3\pi}{2}$ and (11.44) gives

$\frac{\pi}{2} \le |D| \le \frac{2\pi}{3}$.

Let $K = \begin{bmatrix} k_{11} & k_{12} \\ k_{12} & k_{22} \end{bmatrix}$ be a positive definite matrix, i.e. $k_{11}k_{22} - k_{12}^2 > 0$. The eigenvalues of $(K - I)$ are given by $\lambda_{1,2} = \frac{(k_{11}+k_{22}-2)\pm\sqrt{(k_{11}-k_{22})^2+4k_{12}^2}}{2}$. Note that since $K \ne I$, it follows that $\lambda_1 \ne \lambda_2$ and that an eigenvalue can be zero if and only if $k_{11}k_{22} - k_{12}^2 = k_{11} + k_{22} - 1$. In particular, in the orthotropic case, i.e. $k_{12} = 0$, $\lambda_{1,2} = \frac{(k_{11}+k_{22}-2)\pm|k_{11}-k_{22}|}{2}$ can be zero if and only if $k_{11} + k_{22} = 1 + k_{11}k_{22}$, i.e. when $k_{11} = 1$ or $k_{22} = 1$. So, there is the possibility to choose better experiments for an orthotropic inclusion for which $k_{11} \ne 1$ and $k_{22} \ne 1$. Let us therefore consider $K = \begin{bmatrix} 2 & 0 \\ 0 & 3 \end{bmatrix}$. Then $\lambda_1 = 1$ and $\lambda_2 = 2$ with the corresponding normalized eigenvectors for $(K-I)$ given by $\underline{a}_1 = \begin{bmatrix} 1 \\ 0 \end{bmatrix}$ and $\underline{a}_2 = \begin{bmatrix} 0 \\ 1 \end{bmatrix}$. Then, one can experiment with $f_1(\underline{s}) = \underline{a}_1 \cdot \underline{s} = x$ or $f_2(\underline{s}) = \underline{a}_2 \cdot \underline{s} = y$, which gives $\mathrm{T}_1 = \int_0^{2\pi} \cos(\theta) g(\cos(\theta), \sin(\theta)) d\theta$ or $\mathrm{T}_2 = \int_0^{2\pi} \sin(\theta) g(\cos(\theta), \sin(\theta)) d\theta$, respectively, and from (11.44) decide which experiment is better, i.e., which estimates are sharper.

11.2.2 Boundary integral formulation

The discontinuous anisotropic model under investigation given by (11.22), and the corresponding transmission conditions, can be recast in a more convenient form by defining $\phi = \phi_1$ in $\Omega \backslash D$ and $\phi = \phi_2$ in D, where

$$\nabla^2 \phi_1 = 0 \quad \text{in } \Omega \backslash D, \tag{11.45}$$

$$\phi_1 = f \quad \text{on } \partial\Omega, \tag{11.46}$$

$$\nabla \cdot (K \nabla \phi_2) = 0 \quad \text{in } D, \tag{11.47}$$

$$\phi_1 = \phi_2, \quad \frac{\partial \phi_1}{\partial n^-} = -(K \nabla \phi_2) \cdot \underline{n}^+ = -\frac{\partial \phi_2}{\partial n^*} \quad \text{on } \partial D, \tag{11.48}$$

where $\underline{n}^* = K\underline{n}^+$. Assume that Ω and D are simply connected and have C^2 boundary. Then, $\phi_1 \in H^2(\Omega \backslash D) \cap C(\overline{\Omega})$ and $\phi_2 \in H^2(D)$, [243], and thus, Green's formula is applicable. Let G and G_K be fundamental solutions of equations (11.45) and (11.47) in $\mathbb{R}^d$, respectively,

$$G(\underline{x}, \underline{\xi}) = \begin{cases} -\frac{1}{2\pi}\ln(r), & d = 2, \\ \frac{1}{4\pi r}, & d = 3, \end{cases} \qquad G_K(\underline{x}, \underline{\xi}) = \begin{cases} -\frac{|K^{-1}|^{1/2}}{2\pi}\ln(R), & d = 2, \\ \frac{|K^{-1}|^{1/2}}{4\pi R}, & d = 3, \end{cases}$$

where $r = |\underline{x} - \underline{\xi}|$, $|K^{-1}|$ is the determinant of the inverse matrix K^{-1} and the geodesic distance R is defined by $R^2 = K^{-1}(\underline{x} - \underline{\xi}) \cdot (\underline{x} - \underline{\xi})$. Applying the Dirichlet boundary condition (11.46) and the interface transmission conditions

(11.48) result in the following integral representation formulae:

$$c(\underline{x})\phi_1(\underline{x}) = \int_{\partial\Omega} \left[G(\underline{x},\underline{\xi})\frac{\partial \phi_1}{\partial n}(\underline{\xi}) - f(\underline{\xi})\frac{\partial G}{\partial n(\underline{\xi})}(\underline{x},\underline{\xi}) \right] dS$$

$$- \int_{\partial D} \left[G(\underline{x},\underline{\xi})(K\nabla\phi_2(\underline{\xi}))\cdot \underline{n}^{+}(\underline{\xi}) + \phi_2(\underline{\xi})\frac{\partial G}{\partial n^{-}(\underline{\xi})}(\underline{x},\underline{\xi}) \right] dS,$$

$$x \in \overline{\Omega\backslash D}, \qquad (11.49)$$

$$c(\underline{x})\phi_2(\underline{x}) = \int_{\partial D} \left[G_K(\underline{x},\underline{\xi})\frac{\partial \phi_2}{\partial n^*}(\underline{\xi}) - \phi_2(\underline{\xi})\frac{\partial G_K}{\partial n^*(\underline{\xi})}(\underline{x},\underline{x}') \right] dS, \quad x \in \overline{D}, \qquad (11.50)$$

where $c(\underline{x}) = 1$ if $\underline{x} \in \Omega\backslash\partial D$ and $c(\underline{x}) = 0.5$ if $\underline{x} \in \partial\Omega \cup \partial D$. In two-dimensions, the boundary $\partial\Omega$ is discretized uniformly into M constant boundary elements in a anticlockwise sense, and the boundary ∂D into M constant boundary elements in both a anticlockwise and clockwise sense. Then, applying (11.49) at the nodes on $\partial\Omega \cup \partial D$ and (11.50) at the nodes on ∂D result in a system of $3M$ nonlinear equations with $5M$ unknowns, say $A(\underline{x})\underline{u} = \underline{b}$, where $\underline{u}$ contains the unspecified values of $\phi_1 = \phi_2$ on ∂D, $\partial\phi_1/\partial n$ on $\partial\Omega$ and $\partial\phi_2/\partial n^*$ on ∂D, and $\underline{x} = ((x_j, y_j))_{j=\overline{1,M}}$ are the coordinates of the boundary element nodes on ∂D. For a given initial guess $\underline{x}^0$, this system of equations becomes linear and it can be solved to determine the (calculated) current flux data $\partial\phi_1/\partial n$ on $\partial\Omega$. One can then minimize

$$\text{Objf}(\underline{x}) := \|\, (\partial_n\phi_1)_{calculated} - (\partial_n\phi_1)_{measured} \,\|^2. \qquad (11.51)$$

However, even by minimizing the functional (11.51), the above system of equations is still under-determined having $3M$ equations with $4M$ unknowns. Additional information is therefore necessary, which may include:

(i) $\underline{x} \in \Omega$, such that the unknown object D is always contained in Ω.

(ii) The inclusion in (11.51) of penalty regularizing terms such as $\lambda_1\|\underline{x}\|^2$, $\lambda_2\|\underline{x}'\|^2$ or $\lambda_3\|\underline{x}''\|^2$, where λ_1, λ_2, $\lambda_3 > 0$ are regularization parameters which may allow for C^0, C^1 or C^2-boundaries ∂D.

(iii) The domain D is star-shaped [226] or the boundary ∂D is the union of two disjoint graphs of functions, say $y^1 = y^1(x)$, $y^2 = y^2(x)$ for $x \in [a, b]$, such that the number of unknowns is reduced by M, with only the components y_j for $j = \overline{1, M}$ needed to be recovered.

(iv) The domain D is a circle [299] or an ellipse [227], such that only a finite small set of parameters need to be retrieved.

11.3 Reconstruction of an Orthotropic Conductivity

The previous sections have addressed stationary problems of elliptic type. In the transient parabolic case, studies on conductivity reconstruction include the nonlinear identification of a temperature-dependent orthotropic material [356], the space-dependent anisotropic case [237] and the recovery of leading coefficients of a heterogeneous orthotropic medium [20, 67, 177, 181, 287]. As before, due to the ill-posedness of the anisotropic inverse conductivity problem certain simplifications are required, and in this section such a model reduction in which the conductivity tensor is orthotropic with the main diagonal components independent of one space variable is considered [182]. Then, the conductivity components can be taken outside the divergence operator and the inverse problem requires reconstructing one or two components of the orthotropic conductivity tensor of a two-dimensional rectangular conductor using initial and Dirichlet boundary conditions, as well as non-local heat flux over-specifications on two adjacent sides of the boundary.

11.3.1 Mathematical formulation and analysis

The identification of coefficients in two-dimensional transient heat conduction was investigated in [89, 90, 208, 404]. In particular, the recovery of thermal conductivity coefficients $a(y,t) > 0$ and $b(x,t) > 0$ of an orthotropic rectangular conductor along with the temperature $u(x,y,t)$ in a two-dimensional model given by the parabolic heat equation

$$\frac{\partial u}{\partial t}(x,y,t) = a(y,t)\frac{\partial^2 u}{\partial x^2}(x,y,t) + b(x,t)\frac{\partial^2 u}{\partial y^2}(x,y,t) + f(x,y,t),$$
$$(x,y,t) \in Q_T := (0,h) \times (0,\ell) \times (0,T), \quad (11.52)$$

where h, ℓ, T are given positive quantities and $f(x,y,t)$ is a given heat source, subject to the initial condition

$$u(x,y,0) = u_0(x,y), \quad (x,y) \in \Omega := (0,h) \times (0,\ell), \quad (11.53)$$

the Dirichlet boundary conditions

$$u(0,y,t) = \mu_{11}(y,t), \quad u(h,y,t) = \mu_{12}(y,t), \quad (y,t) \in [0,\ell] \times [0,T], \quad (11.54)$$

$$u(x,0,t) = \mu_{21}(x,t), \quad u(x,\ell,t) = \mu_{22}(x,t), \quad (x,t) \in [0,h] \times [0,T], \quad (11.55)$$

and the heat flux over-specifications

$$a(y,t)\frac{\partial u}{\partial x}(0,y,t) = \kappa_1(y,t), \quad (y,t) \in [0,\ell] \times [0,T], \quad (11.56)$$

$$b(x,t)\frac{\partial u}{\partial y}(x,0,t) = \kappa_2(x,t), \quad (x,t) \in [0,h] \times [0,T], \quad (11.57)$$

where φ, μ_{1i}, μ_{2i} for $i = 1,2$ are given functions satisfying compatibility conditions, and κ_1 and κ_2 are given heat flux measured data, was investigated in [181]. In this section, the local heat flux measurements (11.56) and (11.57) are generalized to include the non-local over-specifications

$$a(y,t)\left[\nu_{11}(y,t)\frac{\partial u}{\partial x}(0,y,t)+\nu_{12}(y,t)\frac{\partial u}{\partial x}(h,y,t)\right]=\varkappa_1(y,t),$$
$$(y,t)\in[0,\ell]\times[0,T], \qquad (11.58)$$

$$b(x,t)\left[\nu_{21}(x,t)\frac{\partial u}{\partial y}(x,0,t)+\nu_{22}(x,t)\frac{\partial u}{\partial y}(x,\ell,t)\right]=\varkappa_2(x,t),$$
$$(x,t)\in[0,h]\times[0,T], \qquad (11.59)$$

where $\varkappa_1$ and $\varkappa_2$ are given functions and $(\nu_{ij})_{i,j=1,2}$ are given coefficients. Of course, when $\nu_{11}=\nu_{21}=1$ and $\nu_{12}=\nu_{22}=0$, expressions (11.58) and (11.59) become (11.56) and (11.57), respectively. Expressions (11.58) and (11.59) are linear combinations of heat fluxes across the opposite sides of the rectangular heat conductor $\Omega=(0,h)\times(0,\ell)$.

The above problem combines the features of multi-dimensions, multiple coefficient identification [183], as well as non-local overdetermination [203]. All these studies are unified by the approach utilized in [182] to prove the existence of solution: the inverse problem is reformulated as a fixed point problem for a certain nonlinear compact operator, so that the Schauder theorem can be applied to it. Afterwards, uniqueness of solution follows from the theory of Volterra integral equations of the second kind. This unique solvability of the inverse problem is summarized in the Theorems 11.1 and 11.2 below.

Consider the following assumptions:

$$\begin{aligned}
(A1)\ &\varphi\in H^{2+\varepsilon}(\overline{\Omega}),\ \mu_{1i}\in H^{2+\varepsilon,1+\varepsilon/2}([0,\ell]\times[0,T]),\\
&\mu_{2i}\in H^{2+\varepsilon,1+\varepsilon/2}([0,h]\times[0,T]),\ \nu_{1i}\in H^{\varepsilon,\varepsilon/2}([0,\ell]\times[0,T]),\\
&\nu_{2i}\in H^{\varepsilon,\varepsilon/2}([0,h]\times[0,T]), i\in\{1,2\},\ \varkappa_1\in H^{\varepsilon,\varepsilon/2}([0,\ell]\times[0,T]),\\
&\varkappa_2\in H^{\varepsilon,\varepsilon/2}([0,h]\times[0,T]),\ f\in H^{\varepsilon,\varepsilon/2}(\overline{Q}_T)\text{ for some }\varepsilon\in(0,1);\\
(A2)\ &\partial_x u_0(x,y)>0,\ \partial_y u_0(x,y)>0,\ (x,y)\in\overline{\Omega};\ \varkappa_1(y,t)>0,\\
&\nu_{11}(y,t)+\nu_{12}(y,t)>0,\ (y,t)\in[0,\ell]\times[0,T],\ \varkappa_2(x,t)>0,\\
&\nu_{21}(x,t)+\nu_{22}(x,t)>0,\ (x,t)\in[0,h]\times[0,T];\\
(A3)\ &\textit{compatibility conditions of zeroth and first orders hold.}
\end{aligned}$$

In $(A1)$, $H^{k+\varepsilon,(k+\varepsilon)/2}$ for $k\in\{0,2\}$ and $\varepsilon\in(0,1)$, denotes the space of functions which are k-times continuously differentiable in space and $k/2$-times continuously differentiable in time, with the space partial derivatives of order k being Hölder continuous with exponent ε and the time partial derivative of order $k/2$ being Hölder continuous with exponent $\varepsilon/2$.

Theorem 11.3. ([182]) (Local existence) *Suppose that (A1)–(A3) hold. Then, for some $T_0\in(0,T]$ there exists a solution $(a(y,t),b(x,t),u(x,y,t))$ of the* *problem (11.52)–(11.55), (11.58) and (11.59) such that* $0 < a \in H^{\varepsilon,\varepsilon/2}([0,\ell] \times [0,T_0])$, $0 < b \in H^{\varepsilon,\varepsilon/2}([0,h] \times [0,T_0])$ *and* $u \in H^{2+\varepsilon,1+\varepsilon/2}(\overline{Q}_{T_0})$.

Theorem 11.4. ([182]) (Uniqueness) *Suppose that the assumption*
(A4) $\varkappa_1(y,t) \neq 0$, $(y,t) \in [0,\ell] \times [0,T]$, $\varkappa_2(x,t) \neq 0$, $(x,t) \in [0,h] \times [0,T]$,
is satisfied. Then, the solution $(a(y,t), b(x,t), u(x,y,t))$ *of the problem (11.52)–(11.55), (11.58) and (11.59) is unique in the space* $H^{\varepsilon,\varepsilon/2}([0,\ell] \times [0,T]) \times H^{\varepsilon,\varepsilon/2}([0,h] \times [0,T]) \times H^{2+\varepsilon,1+\varepsilon/2}(\overline{Q}_T)$ *with* $a(y,t) > 0$, $(y,t) \in [0,\ell] \times [0,T]$ *and* $b(x,t) > 0$, $(x,t) \in [0,h] \times [0,T]$.

11.3.2 Formulation of a simplified inverse problem

A simplified inverse problem obtained when b is known and taken, for simplicity, to be unity, is briefly discussed. Then, (11.52) simplifies to

$$\frac{\partial u}{\partial t}(x,y,t) = a(y,t)\frac{\partial^2 u}{\partial x^2}(x,y,t) + \frac{\partial^2 u}{\partial y^2}(x,y,t) + f(x,y,t), \quad (x,y,t) \in Q_T. \tag{11.60}$$

The local existence and uniqueness of solution of the inverse problem (11.53)–(11.56), (11.60) were established in [208] and read as follows.

Theorem 11.5. ([182]) *Suppose that the following assumptions are satisfied:*

$(B1)$ $u_0 \in C^2(\overline{D}), \mu_{1i} \in C^{2,1}([0,\ell] \times [0,T]), \mu_{2i} \in C^{2,1}([0,h] \times [0,T]), i = 1,2,$
$\kappa_1 \in H^{\varepsilon,0}([0,\ell] \times [0,T]),\ f \in H^{1+\varepsilon,\varepsilon,0}(\overline{Q_T})$ *for some* $\varepsilon \in (0,1)$;

$(B2)$ $\partial_x u_0(x,y) > 0,\ (x,y) \in \overline{\Omega},\ \mu_{11_t}(y,t) - \mu_{11_{yy}}(y,t) - f(0,y,t) \leq 0,$
$\mu_{12_t}(y,t) - \mu_{12_{yy}}(y,t) - f(h,y,t) \geq 0,\ \kappa_1(y,t) > 0,\ (y,t) \in [0,\ell] \times [0,T],$
$\mu_{2i_x}(x,t) > 0,\ i = 1,2,\ (x,t) \in [0,h] \times [0,T],\ f_x(x,y,t) \geq 0,\ (x,y,t) \in \overline{Q}_T;$

$(B3)$ *compatibility conditions of order zero [244] between the initial condition* (11.53) *and the Dirichlet boundary conditions* (11.54) *and* (11.55) *hold.*

Then, there exists $T_0 \in (0,T]$, *which is determined by the input data, such that the problem (11.53)–(11.56), (11.60) has a solution* $(a(y,t), u(x,y,t)) \in H^{\varepsilon,0}([0,\ell] \times [0,T_0]) \times C^{2,1}(\overline{Q}_{T_0})$ *with* $a(y,t) > 0$ *for* $(y,t) \in [0,\ell] \times [0,T_0]$.

Theorem 11.6. ([182]) *Suppose that the condition* $H^{\varepsilon,0}([0,\ell] \times [0,T]) \ni \kappa_1(y,t) \neq 0$ *for* $(y,t) \in [0,\ell] \times [0,T]$, *is satisfied. Then, the inverse problem (11.53)–(11.56), (11.60) cannot have more than one solution in the class* $(a(y,t), u(x,y,t)) \in H^{\varepsilon,0}([0,\ell] \times [0,T]) \times C^{2,1}(\overline{Q_T})$ *with* $a(y,t) > 0$ *for* $(y,t) \in [0,\ell] \times [0,T]$.

11.3.3 Numerical solution of the direct problem

The numerical solution of the direct problem (11.52)–(11.55), where $a(y,t)$, $b(x,t)$, $f(x,y,t)$, $\varphi(x,y)$ and μ_{ij}, $i,j=1,2$, are known and $u(x,y,t)$ is to be determined is approached using the conditionally stable forward time central space (FTCS) finite-difference scheme [182]. For this, the domain Q_T is subdivided into M_1, M_2 and N subintervals of equal step lengths Δx and Δy, and uniform time step Δt, where $\Delta x = h/M_1$, $\Delta y = \ell/M_2$ and $\Delta t = T/N$, for space and time, respectively. At the node (i,j,n), denote $u_{i,j}^n := u(x_i,y_j,t_n)$, where $x_i = i\Delta x$, $y_j = j\Delta y$, $t_n = n\Delta t$, $a_{j,n} := a(y_j,t_n)$, $b_{i,n} := b(x_i,t_n)$ and $f_{i,j}^n := f(x_i,y_j,t_n)$ for $i=\overline{0,M_1}$, $j=\overline{0,M_2}$ and $n=\overline{0,N}$. The explicit difference scheme for equations (11.52)–(11.55) is

$$\begin{aligned}\frac{u_{i,j}^{n+1}-u_{i,j}^n}{\Delta t} &= a_{j,n}\left(\frac{u_{i+1,j}^n-2u_{i,j}^n+u_{i-1,j}^n}{(\Delta x)^2}\right)\\ &\quad +b_{i,n}\left(\frac{u_{i,j+1}^n-2u_{i,j}^n+u_{i,j-1}^n}{(\Delta y)^2}\right)+f_{i,j}^n,\\ &\quad i=\overline{1,M_1-1},\ j=\overline{1,M_2-1},\ n=\overline{0,N}.\\ u_{i,j}^0 &= u_0(x_i,y_j), \quad i=\overline{0,M_1}, \quad j=\overline{0,M_2},\\ u_{0,j}^n = \mu_{11}(y_j,t_n), \quad u_{M_1,j}^n &= \mu_{12}(y_j,t_n), \quad j=\overline{0,M_2}, \quad n=\overline{1,N},\\ u_{i,0}^n = \mu_{21}(x_i,t_n), \quad u_{i,M_2}^n &= \mu_{22}(x_i,t_n), \quad i=\overline{0,M_1}, \quad n=\overline{1,N}.\end{aligned}$$

Let $\tilde{a}$ and $\tilde{b}$ be the maximum values of $a(y,t)$ and $b(x,t)$, respectively, then, the stability condition for the above explicit FDM scheme is [306],

$$\frac{\tilde{a}\Delta t}{(\Delta x)^2}+\frac{\tilde{b}\Delta t}{(\Delta y)^2} \le \frac{1}{2}. \tag{11.61}$$

The combination of the heat fluxes (11.58) and (11.59) can be calculated using the second-order FDM approximations:

$$\begin{aligned}\varkappa_1(y_j,t_n) &= a_{j,n}\Big(\nu_{11}(y_j,t_n)u_x(0,y_j,t_n)+\nu_{12}(y_j,t_n)u_x(h,y_j,t_n)\Big),\\ &\qquad j=\overline{1,M_2-1}, \quad n=\overline{1,N},\\ \varkappa_2(x_i,t_n) &= b_{i,n}\Big(\nu_{21}(x_i,t_n)u_y(x_i,0,t_n)+\nu_{22}(x_i,t_n)u_y(x_i,\ell,t_n)\Big),\\ &\qquad i=\overline{1,M_1-1}, \quad n=\overline{1,N},\end{aligned}$$

where

$$\begin{aligned}u_x(0,y_j,t_n) &= \frac{4u(x_1,y_j,t_n)-u(x_2,y_j,t_n)-3\mu_{11}(y_j,t_n)}{2\Delta x},\\ &\qquad j=\overline{1,M_2-1}, \quad n=\overline{1,N},\\ u_x(h,y_j,t_n) &= \frac{4u(x_{M_1-1},y_j,t_n)-u(x_{M_1-2},y_j,t_n)-3\mu_{12}(y_j,t_n)}{-2\Delta x},\\ &\qquad j=\overline{1,M_2-1}, \quad n=\overline{1,N},\end{aligned}$$

and similar expressions for $u_y(x_i, 0, t_n)$, $u_y(x_i, \ell, t_n)$, $i = \overline{1, M_1 - 1}$, $n = \overline{1, N}$.

11.3.4 Numerical solution of the inverse problem

The inverse problem requires obtaining the principal direction components $a(y,t) > 0$ and $b(x,t) > 0$ of the two-dimensional orthotropic rectangular medium together with the temperature $u(x,y,t)$ satisfying (11.52)–(11.55), (11.58) and (11.59). At the initial time $t = 0$, the values $a(y,0)$ and $b(x,0)$ can be obtained from (11.58) and (11.59) as

$$a(y,0) = \frac{\varkappa_1(y,0)}{\nu_{11}(y,0)\partial_x u_0(0,y) + \nu_{12}(y,0)\partial_x u_0(h,y)}, \tag{11.62}$$

$$b(x,0) = \frac{\varkappa_2(x,0)}{\nu_{21}(x,0)\partial_y u_0(x,0) + \nu_{22}(x,0)\partial_y u_0(x,\ell)}. \tag{11.63}$$

The inverse problem is solved based on the nonlinear minimization of the least-squares objective function

$$F(a,b) := \left\| a(y,t)\Big(\nu_{11}(y,t)u_x(0,y,t) + \nu_{12}(y,t)u_x(h,y,t)\Big) - \varkappa_1(y,t)\right\|^2 + \left\| b(x,t)\Big(\nu_{21}(x,t)u_y(x,0,t) + \nu_{22}(x,t)u_y(x,\ell,t)\Big) - \varkappa_2(x,t)\right\|^2,$$

or, in discretized form,

$$\begin{aligned} F(\underline{a},\underline{b}) = \sum_{n=1}^{N}\sum_{j=0}^{M_2} \Big[a_{j,n}\Big(\nu_{11}(y_j,t_n)u_x(0,y_j,t_n) \\ +\nu_{12}(y_j,t_n)u_x(h,y_j,t_n)\Big) - \varkappa_1(y_j,t_n)\Big]^2 \\ +\sum_{n=1}^{N}\sum_{i=0}^{M_1} \Big[b_{i,n}\Big(\nu_{21}(x_i,t_n)u_y(x_i,0,t_n) \\ +\nu_{22}(x_i,t_n)u_y(x_i,\ell,t_n)\Big) - \varkappa_2(x_i,t_n)\Big]^2, \end{aligned} \tag{11.64}$$

where $u(x,y,t)$ solves (11.52)–(11.55) for given $\underline{a}$ and $\underline{b}$. The minimization of (11.64), subject to the simple bounds $\underline{a} > \underline{0}$ and $\underline{b} > \underline{0}$ is accomplished using the MATLAB optimization toolbox routine *lsqnonlin* with the Trust Region Reflective (TRR) algorithm [92] based on the interior-reflective Newton method [87, 88]. Each iteration involves a large linear system of equations whose solution, based on a preconditioned conjugate gradient method, allows a regular and sufficiently smooth decrease of the objective functional (11.64).

The inverse problem (11.52)–(11.55) is solved subject to both exact and noisy data (11.58) and (11.59). The noisy data is simulated as

$$\varkappa_1^{\epsilon 1}(y_j, t_n) = \varkappa_1(y_j, t_n) + \epsilon 1_{j,n}, \quad j = \overline{0, M_2}, \quad n = \overline{1, N} \tag{11.65}$$

$$\varkappa_2^{\epsilon 2}(x_i, t_n) = \varkappa_2(x_i, t_n) + \epsilon 2_{i,n}, \quad i = \overline{0, M_1}, \quad n = \overline{1, N}, \tag{11.66}$$

where $\epsilon 1_{j,n}$ and $\epsilon 2_{i,n}$ are random variables generated from a Gaussian normal distribution with mean zero and standard deviations

$$\sigma 1 = p \times \max_{(y,t)\in[0,\ell]\times[0,T]} |\varkappa_1(y_j, t_n)|, \quad \sigma 2 = p \times \max_{(y,t)\in[0,h]\times[0,T]} |\varkappa_2(x_i, t_n)|,$$

where p represents the percentage of noise. In the case of noisy data (11.65) and (11.66), $\varkappa_1(y_j, t_n)$ and $\varkappa_2(x_i, t_n)$ are replaced by $\varkappa_1^{\epsilon 1}(y_j, t_n)$ and $\varkappa_2^{\epsilon 2}(x_i, t_n)$, respectively, in (11.64).

11.3.5 Numerical results and discussion

To assess the accuracy of the numerical solution, we define:

$$\text{RMSE}(a) = \left[\frac{1}{N(M_2+1)} \sum_{n=1}^{N} \sum_{j=0}^{M_2} \left(a^{numerical}(y_j, t_n) - a^{exact}(y_j, t_n)\right)^2 \right]^{1/2},$$

$$\text{RMSE}(b) = \left[\frac{1}{N(M_1+1)} \sum_{n=1}^{N} \sum_{i=0}^{M_1} \left(b^{numerical}(x_i, t_n) - b^{exact}(x_i, t_n)\right)^2 \right]^{1/2}.$$

Consider the input data φ, μ_{ij}, ν_{ij} and $\varkappa_i$, $i, j = \overline{1, 2}$:

$$h = \ell = T = 1, \quad u_0(x, y) = u(x, y, 0) = -(-2 + x)^2 - (-2 + y)^2,$$
$$f(x, y, t) = 2 + \frac{3 + 2t + x + y}{100},$$
$$\mu_{11}(y, t) = u(0, y, t) = -4 + 2t - (-2 + y)^2,$$
$$\mu_{12}(y, t) = u(1, y, t) = -1 + 2t - (-2 + y)^2,$$
$$\mu_{21}(x, t) = u(x, 0, t) = -4 + 2t - (-2 + x)^2,$$
$$\mu_{22}(x, t) = u(x, 1, t) = -1 + 2t - (-2 + x)^2,$$
$$\nu_{11}(y, t) = \nu_{12}(y, t) = \nu_{21}(x, t) = \nu_{22}(x, t) = 1,$$
$$\varkappa_1(t) = \frac{3(y + t + 1)}{100}, \ \varkappa_2(t) = \frac{3(2 + x + t)}{100}.$$

Conditions of Theorem 11.4 are satisfied and therefore, the uniqueness of the solution is guaranteed. In fact, the analytical solution is given by

$$a(y, t) = \frac{y + t + 1}{200}, \quad (y, t) \in [0, 1] \times [0, 1],$$
$$b(x, t) = \frac{2 + x + t}{200}, \quad (x, t) \in [0, 1] \times [0, 1],$$
$$u(x, y, t) = -(x - 2)^2 - (y - 2)^2 + 2t, \quad (x, y, t) \in \overline{Q}_T.$$

TABLE 11.1: Numerical results for various noise levels $p \in \{0, 1, 10\}\%$.

	$p = 0$	$p = 1\%$	$p = 10\%$
No. of iterations	8	8	9
Value of (11.64) at final iteration	2.7E−19	3.9E−19	5.6E−19
RMSE(a)	4.7E−12	1.6E−4	1.6E−3
RMSE(b)	6.6E−12	1.8E−4	1.9E−3
Computational time (in min)	21	23	25

Take $M_1 = M_2 = N = 10$ which together with the upper bound $1/40$ for a and b ensure that the stability condition (11.61) is always satisfied at each iteration of the minimization process. Take the lower bound 10^{-9} to impose the physical constraint that the thermal conductivity coefficients must be positive. The initial guesses for $a(y, t)$ and $b(x, t)$ are taken as

$$a^0(y,t) = a(y,0) = \frac{y+1}{200}, \ y \in [0,1], \ b^0(x,t) = b(x,0) = \frac{2+x}{200}, \ x \in [0,1],$$

while noting that the values of $a(y, 0)$ and $b(x, 0)$ are available from (11.62) and (11.63), respectively. Table 11.1 shows details of the computation including the results for $p \in \{0, 1, 10\}\%$ noise. As expected, the numerical results become more stable and accurate as the percentage of noise p decreases.

Chapter 12

Nonlinear Conductivity

In the previous two chapters, the unknown conductivity was assumed time- and/or space-dependent. In this chapter, the inverse determination of the thermal properties of heat conductors which are temperature dependent is considered in section 12.1. Since there is too much to hope for an analytical solution to the nonlinear heat equation when the thermal coefficients are general functions depending on the temperature, the thermal properties are sought in the special class of functions of simple metals [372]. Then, not only can the nonlinear heat equation be exactly linearized but also the thermal properties and the temperature can be uniquely determined from a time-dependent temperature measurement function at a single sensor location within the conductor, provided that its continuous derivative never vanishes. The analytical method developed in [252] and described in this section yields exact solutions, generally in implicit form, which can then be solved numerically. Solutions are obtained for infinite, semi-infinite and finite slabs with particular physical applications related to the cases of an instantaneous source, metals subjected to a constant heat flux and the quencing of a flat plate in a tank filled with fluid. In section 12.2, some uniqueness results are revisited and analyzed for the more general case when the conductivity depends not only on the temperature but also on the space variable in a separable way.

12.1 Determination of Nonlinear Thermal Properties

The unsteady heat conduction in a material is nonlinear when either, or both, the thermal conductivity $K(u) > 0$ and the heat capacity $C(u) > 0$ are temperature dependent, and in one-dimension it is given by

$$\frac{\partial}{\partial x}\left(K(u)\frac{\partial u}{\partial x}\right) = C(u)\frac{\partial u}{\partial t}, \quad t > 0, \quad x \in (0, l), \tag{12.1}$$

where l is the length of the heat conductor which may be finite or infinite. An initial temperature is prescribed at $t = 0$, namely,

$$u(x, 0) = u_0(x), \quad x \in [0, l] \tag{12.2}$$

DOI: 10.1201/9780429400629-12

and the face $x = 0$ is subjected to a prescribed heat flux

$$q_0(t) = -K(u(0,t))\frac{\partial u}{\partial x}(0,t), \quad t > 0. \tag{12.3}$$

If the slab is finite, i.e. $l < \infty$, the face $x = l$ can also be subjected to a prescribed heat flux

$$q_l(t) = K(u(l,t))\frac{\partial u}{\partial x}(l,t), \quad t > 0. \tag{12.4}$$

If the slab is semi-infinite, i.e. $l \to \infty$, instead of the boundary condition (12.4), an infinity condition is prescribed, namely,

$$u(x,t) \to u_\infty(t), \quad \text{as } x \to \infty. \tag{12.5}$$

If the slab is infinite, i.e. the interval $(0,l)$ is replaced by $(-\infty,\infty)$, only the initial condition (12.2) is imposed.

12.1.1 Linearization

A sufficient condition for the nonlinear equation (12.1) to be linearized is [252, 372]

$$\frac{1}{\sqrt{KC}}\frac{d}{du}\left(\ln\sqrt{C/K}\right) = A = \text{constant}. \tag{12.6}$$

Under this assumption, the transformations

$$\overline{Q}(u) = \int_{u^0}^{u} \sqrt{K(\lambda)C(\lambda)}d\lambda, \tag{12.7}$$

where u^0 is an arbitrary constant reference temperature,

$$Q(x,t) = \overline{Q}(u(x,t)) = Q^*(\mathrm{X}(x,t),t), \tag{12.8}$$

where

$$\mathrm{X}(x,t) = \int_{x^0}^{x} \sqrt{C(u(x',t))/K(u(x',t))}dx' \tag{12.9}$$

with x^0 an arbitrary fixed spatial reference position, which can be taken as the position where a flux boundary condition is specified, say $x^0 = 0$, and finally, if $A \neq 0$,

$$Q^*(\mathrm{X},t) = -(1/A)\ln(\zeta(\mathrm{X},t)) \tag{12.10}$$

recast the nonlinear heat conduction equation (12.1) into the linear form [372],

$$\frac{\partial^2\zeta}{\partial \mathrm{X}^2} = \frac{\partial\zeta}{\partial t} + Aq_0(t)\frac{\partial\zeta}{\partial \mathrm{X}}. \tag{12.11}$$

Setting

$$\zeta(\mathrm{X},t) = U(\xi(\mathrm{X},t),t), \tag{12.12}$$

where

$$\xi(X,t) = X - A\int_0^t q_0(\tau)d\tau, \tag{12.13}$$

equation (12.11) reduces to the classical linear heat equation

$$\frac{\partial^2 U}{\partial \xi^2} = \frac{\partial U}{\partial t}. \tag{12.14}$$

The most general forms for $K(u)$ and $C(u)$ which satisfy (12.6) are

$$K(u) = K_0 g(u) \exp\left[-A\sqrt{K_0 C_0}\int_{u^0}^{u} g(\lambda)d\lambda\right], \tag{12.15}$$

$$C(u) = C_0 g(u) \exp\left[A\sqrt{K_0 C_0}\int_{u^0}^{u} g(\lambda)d\lambda\right], \tag{12.16}$$

where the subscript zero means that the functions involved are to be evaluated at $u = u^0$, and the otherwise arbitrary positive function $g(u)$ satisfies the normalization condition $g(u^0) = 1$.

Remark 12.1. (i) Many materials, in addition to having the property that the thermal conductivity $K(u)$ and the heat capacity $C(u)$ are temperature dependent, also have direct proportionality between these two quantities, i.e., their thermal diffusivity $\alpha(u) = K(u)/C(u)$ is constant [167, 392]. Then, the constant A in (12.6) is zero, and only the so-called Kirchhoff transformation (12.7) needs to be employed to recast (12.1) into a linear form. For inverse determinations of $K(u)$ and $C(u)$ in this quasi-linear case, see [57, 262].

(ii) For a variety of simple metals, Storm [372] found that the product $K(u)C(u)$ is essentially constant for a wide range of temperatures and therefore $g(u) \equiv 1$, which, on using (12.15) and (12.16), give

$$K(u) = K_0 \exp\left[-A\sqrt{K_0 C_0}(u - u^0)\right] \approx K_0[1 - b(u - u^0)], \tag{12.17}$$

$$C(u) = C_0 \exp\left[A\sqrt{K_0 C_0}(u - u^0)\right] \approx C_0[1 + b(u - u^0)], \tag{12.18}$$

where, for simple metals, the quantity $b = A\sqrt{K_0 C_0}$ is typically $O(10^{-4})$. The linear assumptions (12.17) and (12.18) were adopted in [173] for simultaneously estimating $K(u)$ and $C(u)$ from interior temperature measurements.

(iii) By taking $C(u) \equiv 1$, it was remarked [234] that $g(u) = [1 + A\sqrt{K_0}(u - u^0)]^{-1}$, and then the thermal conductivity (diffusivity) satisfying (12.6) is given by $K(u) = K_0[1 + A\sqrt{K_0}(u - u^0)]^{-2}$. Inverse determinations of $K(u)$ when $C(u) \equiv 1$ can be found in [58, 102].

12.1.2 Inverse formulation

In an inverse formulation, one may require finding the triplet $(K(u) > 0, C(u) > 0, T)$ satisfying some of the equations (12.1)–(12.5) and the additional temperature measurement at an arbitrary location $x = X_0 \in [0, l]$,

$$u(X_0, t) = \chi(t), \quad t > 0. \tag{12.19}$$

At this point it is worth emphasizing the distinction between what is referred to as intial and boundary data, such as (12.2)–(12.4), and data that is referred to as overspecified (additional), such as (12.19). In any experiment for which equation (12.1) is a mathematical model, one can force the conditions (12.2)–(12.4) to be satisfied by appropriately arranging the experiment. At the same time, one can measure $u(X_0, t)$ and record the fact that it is equal to $\chi(t)$. One cannot, however, arbitrarily require $u(X_0, t)$ to attain specific values independent of (12.2)–(12.4). In other words, the specified data is that over which one exercises control and thus force to be satisfied, whilst the overspecified (additional) data is simply observed and recorded [60]. Clearly, any hypothesis on the overspecified data $\chi(t)$ cannot be completely independent of whatever hypotheses are placed on the specified data $u_0(x)$, $q_0(t)$ and $q_l(t)$.

There are two main approaches to solving the type of inverse problem aforementioned. These can be classified into either parameterization schemes, e.g. linear variations of the thermal properties, [173], or quadratic, [261], or coefficient identification schemes, [53, 57, 58, 102, 262]. The parameterization schemes reduce the inverse problem to finding a finite set of unknowns (usually less than six) and then seeking to adjust these parameters such that the gap between the computed output generated by an associated mathematical model and the experimentally measured data is minimized in a least-squares sense. These methods have the advantage that the inverse problem can be recast as a (constrained) minimization problem for which efficient computational routines are available. However, the solution obtained using parameterization based on a minimization procedure is not necessarily the mathematical solution of the inverse problem. Therefore, it would be useful to develop a reconstructional approach in which the unknown thermal properties can be expressed directly in terms of the overspecified data. For this purpose, a special space of functions to which the coefficients $K(u)$ and $C(u)$ should belong, see (12.6), in order to linearise equation (12.1), is analyzed. In this case, $K(u)$ and $C(u)$ have the special form given by (12.15) and (12.16) with K_0, C_0 and A being specified; a situation in which only the pair $(g(u), u)$ needs to be identified. This provides insight into what data is required for the more general inverse problem when $K(u)$ and $C(u)$ are not connected by any relationship and need to be determined simultaneously.

12.1.3 Inverse analysis

Using (12.15) and (12.16), transformations (12.7)–(12.10) become

$$\begin{aligned} -\frac{1}{A}\ln(\zeta(\mathrm{X}(x,t),t)) &= Q^*(\mathrm{X}(x,t),t) \\ &= Q(x,t) = \int_{u^0}^{u(x,t)} \sqrt{K_0 C_0}\, g(\lambda)\, d\lambda, \end{aligned} \tag{12.20}$$

where

$$\mathrm{X}(x,t) = \int_0^x \sqrt{C_0/K_0}\exp\left(A\sqrt{K_0C_0}\int_{u^0}^{u(x',t)} g(\lambda)d\lambda\right)dx'$$
$$= \int_0^x \frac{\sqrt{C_0/K_0}}{\zeta(X(x',t),t)}dx'. \quad (12.21)$$

Inverting equation (12.21) yields

$$x = \sqrt{K_0/C_0}\int_0^{\mathrm{X}} \zeta(X',t)dX'. \quad (12.22)$$

In particular, $x = 0$ if and only if $\mathrm{X}(0,t) \equiv 0$. Applying (12.20) at $x = X_0$ and using (12.19) result in

$$Q(X_0,t) = \sqrt{K_0C_0}\int_{u^0}^{\chi(t)} g(\lambda)d\lambda. \quad (12.23)$$

By differentiating this expression with respect to t, it follows that

$$\sqrt{K_0C_0}\, g(\chi(t))\chi'(t) = \frac{\partial Q}{\partial t}(X_0,t), \quad t > 0. \quad (12.24)$$

Now using equations (12.8)–(12.10), it follows that

$$\frac{\partial Q}{\partial t}(X_0,t) = \frac{\partial^2 Q^*}{\partial \mathrm{X}^2}(\mathrm{X}(X_0,t),t) = \left(\frac{\zeta_{\mathrm{X}}^2 - \zeta\zeta_{\mathrm{XX}}}{A\zeta^2}\right)(\mathrm{X}(X_0,t),t). \quad (12.25)$$

If $\chi'(t) \neq 0$ for all $t > 0$, then χ is invertible and using (12.12), (12.13), (12.24) and (12.25),

$$g(t) = \frac{(\chi^{-1})'(t)}{A\sqrt{K_0C_0}}\left(\frac{\zeta_{\mathrm{X}}^2 - \zeta\zeta_{\mathrm{XX}}}{\zeta^2}\right)(\mathrm{X}(X_0,\chi^{-1}(t)),\chi^{-1}(t))$$
$$= \frac{(\chi^{-1})'(t)}{A\sqrt{K_0C_0}}\left(\frac{U_\xi^2 - UU_{\xi\xi}}{U^2}\right)(\xi(\mathrm{X}(X_0,\chi^{-1}(t)),\chi^{-1}(t)),\chi^{-1}(t)). \quad (12.26)$$

Once the function g, and thus K and C, as given by (12.15) and (12.16), are obtained, using equations (12.7), (12.8), (12.20) and (12.26) yield

$$\overline{Q}(u) = \int_{u^0}^{u} \frac{(\chi^{-1})'(\lambda)}{A}\left(\frac{\zeta_{\mathrm{X}}^2 - \zeta\zeta_{\mathrm{XX}}}{\zeta^2}\right)(\mathrm{X}(X_0,\chi^{-1}(\lambda)),\chi^{-1}(\lambda))d\lambda. \quad (12.27)$$

Finally, inverting equation (12.8) yields

$$u(x,t) = \overline{Q}^{-1}(Q(x,t)). \quad (12.28)$$

12.1.4 Results and discussion

In what follows, only the case $A \neq 0$ is considered, with the case $A = 0$ being referred to [57, 262]. The constant reference temperature u^0 is taken to be zero and the sensor recording the additional temperature measurement (12.19) is located at the (boundary) $x = X_0 = 0$ where a flux condition is prescribed. Then, from (12.21) it follows that $\mathrm{X}(0,t) \equiv 0$, and assuming that equation (12.25) holds at the boundary $x = 0$, then equations (12.24) and (12.25) give

$$g(\chi(t))\chi'(t) = \frac{1}{A\sqrt{K_0 C_0}}\left(\frac{\zeta_{\mathrm{X}}^2 - \zeta\zeta_{\mathrm{XX}}}{\zeta^2}\right)(0,t), \quad t > 0. \tag{12.29}$$

All the examples presented have physical meaning in engineering applications and relate to the cases of an instantaneous source, metals subjected to a constant heat flux and quenching of a flat plate in a tank filled with fluid.

12.1.4.1 The infinite slab

In this case the interval $(0, l)$ is replaced by the real-axis interval $(-\infty, \infty)$, and then equation (12.1) has to be solved subject to the initial condition (12.2) and to the additional measurement (12.19). Consider a physical example similar to that studied in [234] describing the diffusion which follows the instantaneous release of permeable material in the region $(-1, 1)$ at the instant $t = 0$. In this case, the initial temperature $u_0(x)$ is given by

$$u_0(x) = \begin{cases} 0, & |x| \leq 1, \\ 1, & |x| > 1. \end{cases} \tag{12.30}$$

Then, the transformations (12.20) and (12.21) give

$$Q(x,0) = \begin{cases} 0, & |x| \leq 1, \\ \ell, & |x| > 1, \end{cases} \tag{12.31}$$

$$\mathrm{X}(x,0) = \begin{cases} \mathrm{a}x, & |x| \leq 1, \\ \mathrm{a}[-1 + (x+1)e^{A\ell}], & x < -1, \\ \mathrm{a}[1 + (x-1)e^{A\ell}], & x > 1, \end{cases} \tag{12.32}$$

$$\zeta(\mathrm{X},0) = \begin{cases} 1, & |\mathrm{X}| \leq \mathrm{a}, \\ e^{-A\ell}, & |\mathrm{X}| > \mathrm{a}, \end{cases} \tag{12.33}$$

where

$$\ell = \sqrt{K_0 C_0}\int_0^1 g(\lambda)d\lambda, \quad \mathrm{a} = \sqrt{C_0/K_0}. \tag{12.34}$$

Because of symmetry $\partial_x u(0,t) = 0$ and then (12.11) reduces to the heat equation. The solution of (12.11) with $q_0 \equiv 0$, subject to (12.33) is [28],

$$\zeta(\mathrm{X},t) = e^{-A\ell} + \frac{1 - e^{-A\ell}}{2}\left\{\mathrm{erfc}\left(\frac{\mathrm{X}-\mathrm{a}}{2\sqrt{t}}\right) - \mathrm{erfc}\left(\frac{\mathrm{X}+\mathrm{a}}{2\sqrt{t}}\right)\right\} \tag{12.35}$$

and equation (12.29) gives

$$g(\chi(t))\chi'(t) = \frac{(1-e^{-A\ell})\mathrm{a}e^{-\mathrm{a}^2/4t}}{2A\sqrt{\pi t^3 K_0 C_0}[1-(1-e^{-A\ell})\mathrm{erfc}(\mathrm{a}/(2\sqrt{t}))]}, \quad t>0. \quad (12.36)$$

The function $\chi(t)$ represents the temperature measurement at $x = X_0 = 0$ and it should be continuously differentiable with non-zero derivative to enable the inversion in equation (12.26).

Example 12.1 For simplicity, let us take $K_0 = C_0 = 1$ and hence from (12.34), $\mathrm{a} = 1$. Consider the test example

$$\chi(t) = -\ln\left[1-(1-e^{-1})\mathrm{erfc}(1/(2\sqrt{t}))\right], \quad (12.37)$$

which is a continuously differentiable function with $\chi'(t) \neq 0$. Also, $\lim_{t\to 0}\chi(t) = 0$ and $\lim_{t\to\infty}\chi(t) = 1$. On differentiating χ with respect to t and using (12.36) give

$$g(\chi(t)) = \frac{(1-e^{-A\ell})}{A(1-e^{-1})}\frac{[1-(1-e^{-1})\mathrm{erfc}(1/(2\sqrt{t}))]}{[1-(1-e^{-A\ell})\mathrm{erfc}(1/(2\sqrt{t}))]}. \quad (12.38)$$

Letting $t \to 0$ and imposing that $g(0) = 1$ give

$$\ell = -\frac{\ln[1+(e^{-1}-1)A]}{A}. \quad (12.39)$$

For example, if $A = 1$, then $\ell = 1$ and from (12.15), (12.16) and (12.38),

$$g(u) = 1, \quad K(u) = \exp(-u), \quad C(u) = \exp(u), \quad (12.40)$$

which characterises a wide range of simple metals [372]. Using these expressions in equations (12.7)–(12.10) and (12.28) result in

$$u(x,t) = -\ln[\zeta(\mathrm{X}(x,t),t)], \quad (12.41)$$

where $\zeta(\mathrm{X}(x,t),t)$ is given by (12.35) and $\mathrm{X}(x,t)$ is found from (12.22), as

$$x = \int_0^{\mathrm{X}} \zeta(X',t)dX' = \frac{\mathrm{X}-1}{e} + 1$$
$$-\sqrt{t}(1-e^{-1})\left\{\mathrm{ierfc}\left(\frac{\mathrm{X}-1}{2\sqrt{t}}\right) - \mathrm{ierfc}\left(\frac{\mathrm{X}+1}{2\sqrt{t}}\right)\right\}, \quad (12.42)$$

where

$$\mathrm{ierfc}(z) = \int_z^{\infty} \mathrm{erfc}(z')dz' = \pi^{-1/2}\exp(-z^2) - z\,\mathrm{erfc}(z).$$

The numerical results for the temperature $u(x,t)$ are plotted in Figure 12.1. This shows that, as t decreases to zero, the temperature approaches the step profile (12.30). At the other end, as t increases, the temperature approaches unity, see also equations (12.35), (12.41) and (12.42). This behavior is similar to that obtained in [234] for the same problem in diffusion, but with constant C and singular diffusion coefficient $K(u) = (1-u)^{-2}$. However, the current analysis avoids such unphysical situations and only continuous and strictly positive physical thermal properties are retrieved.

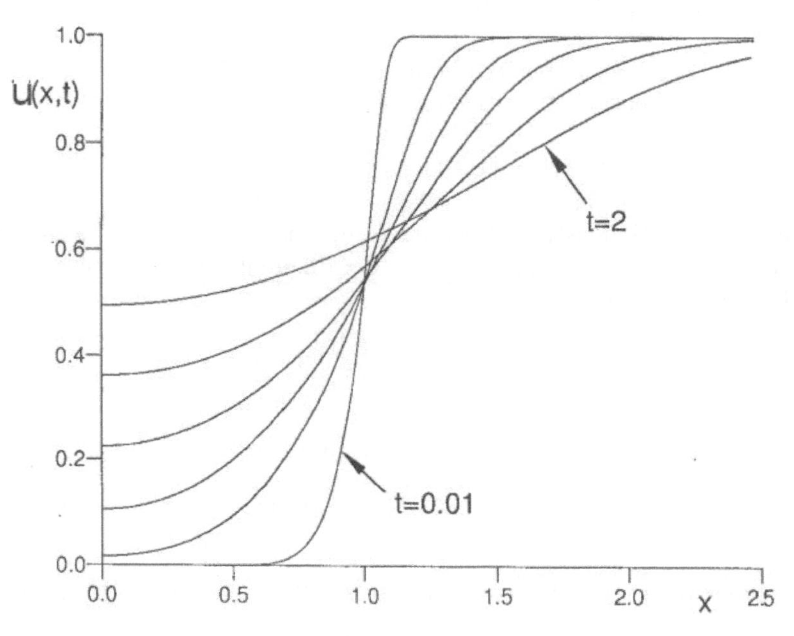

FIGURE 12.1: The results for the temperature $u(x,t)$, as a function of x for various values of time $t \in \{0.01, 0.1, 0.25, 0.5, 1, 2\}$, for Example 12.1, [252].

Example 12.2 For simplicity, let us take $K_0 = C_0 = 1$ and hence a $= 1$. Consider the test example

$$\chi(t) = \mathrm{erfc}(1/(2\sqrt{t})), \tag{12.43}$$

which is a continuously differentiable function with $\chi'(t) \neq 0$. Also, $\lim_{t\to 0}\chi(t) = 0$ and $\lim_{t\to\infty}\chi(t) = 1$. Then, from (12.36) one obtains

$$g(u) = \frac{A^{-1}(1-e^{-A\ell})}{1-(1-e^{-A\ell})u}. \tag{12.44}$$

Imposing that $g(0) = 1$ and $l = \int_0^1 g(\lambda)d\lambda$ yield

$$\ell = -\frac{\ln(1-A)}{A}. \tag{12.45}$$

For example, if $A = -1$, then $\ell = \ln(2)$ and from (12.16), (12.17) and (12.44),

$$g(u) = \frac{1}{1+u}, \quad K(u) = 1, \quad C(u) = \frac{1}{(1+u)^2}. \tag{12.46}$$

Using these expressions in equations (12.7)–(12.10) and (12.35) result in

$$\begin{aligned} u(x,t) &= -1 + \zeta(\mathrm{X}(x,t),t) \\ &= 1 - \frac{1}{2}\left\{\mathrm{erfc}\left(\frac{\mathrm{X}(x,t)-1}{2\sqrt{t}}\right) - \mathrm{erfc}\left(\frac{\mathrm{X}(x,t)+1}{2\sqrt{t}}\right)\right\}, \end{aligned} \tag{12.47}$$

where from equation (12.22)

$$\begin{aligned} x &= 2\mathrm{X} - \frac{1}{2}\int_0^{\mathrm{X}} \left\{\mathrm{erfc}\left(\frac{X'-1}{2\sqrt{t}}\right) - \mathrm{erfc}\left(\frac{X'+1}{2\sqrt{t}}\right)\right\} dX' \\ &= 2\mathrm{X} - 1 + \sqrt{t}\left\{\mathrm{ierfc}\left(\frac{\mathrm{X}-1}{2\sqrt{t}}\right) - \mathrm{ierfc}\left(\frac{\mathrm{X}+1}{2\sqrt{t}}\right)\right). \end{aligned} \tag{12.48}$$

For a given x and t, equation (12.48) can be inverted and the obtained $\mathrm{X}(x,t)$ substituted into (12.47) to find $u(x,t)$. Similarly to the behaviour in Figure 12.1, as t decreases towards zero the temperature approaches the step profile given by (12.30); however, this rate of decrease is slower because $\chi(t)$ given by (12.37) is less than that given by (12.43).

12.1.4.2 The semi-infinite slab

In this case, $(0,l) = (0,\infty)$ and (12.1) has to be solved subject to (12.2), (12.3), (12.5) and (12.19).

Example 12.3 The initial and infinity conditions (12.2) and (12.5) are taken to be constant and equal to zero, i.e. $u_0 = u_\infty = 0$. Then, based on (12.7)–(12.10) and (12.20), in terms of ζ, conditions (12.2), (12.3) and (12.5) become

$$\zeta(\mathrm{X},0) = 1, \tag{12.49}$$

$$\frac{\partial\zeta}{\partial\mathrm{X}}(0,t) = Aq_0(t)\zeta(0,t), \tag{12.50}$$

$$\zeta(\mathrm{X},t) \to 1 \quad \text{as } \mathrm{X} \to \infty. \tag{12.51}$$

In the case of a metal subjected to a constant heat flux at $x = 0$, i.e. $q_0(t) = q_0 = \text{constant} \neq 0$, [372], then the problem given by equation (12.11) subject to the conditions (12.49)–(12.51) has the analytical solution [252],

$$\begin{aligned} \zeta(\mathrm{X},t) = 1 &+ \frac{e^\eta}{2}(\tau^2 + 1 + \eta)\mathrm{erfc}\left(\frac{\eta}{2\tau} + \frac{\tau}{2}\right) - \frac{1}{2}\mathrm{erfc}\left(\frac{\eta}{2\tau} - \frac{\tau}{2}\right) \\ &- \frac{\tau}{\sqrt{\pi}}\exp\left[-\left(\frac{\eta}{2\tau} - \frac{\tau}{2}\right)^2\right], \end{aligned} \tag{12.52}$$

where

$$\eta = Aq_0\mathrm{X}, \quad \tau = Aq_0\sqrt{t}. \tag{12.53}$$

Further, on using (12.52), expression (12.22) yields

$$\tilde{x}(\mathrm{X},t) = \frac{e^{\eta}}{2}(\tau^2+\eta)\mathrm{erfc}\left(\frac{\eta}{2\tau}+\frac{\tau}{2}\right)+(\eta-\tau^2)\left[1-\frac{1}{2}\mathrm{erfc}\left(\frac{\eta}{2\tau}-\frac{\tau}{2}\right)\right], \tag{12.54}$$

where

$$\tilde{x} = Aq_0 x\sqrt{C_0/K_0}. \tag{12.55}$$

Then (12.29) gives

$$g(\chi(t))\chi'(t) = (K_0C_0)^{-1/2}Aq_0^2\frac{\mathrm{ierfc}(\tau/2)}{\tau[\mathrm{erfc}(\tau/2)-\tau\ \mathrm{ierfc}(\tau/2)]}. \tag{12.56}$$

For simplicity, let us take $K_0 = C_0 = q_0 = A = 1$ and consider

$$\chi(t) = -\ln\left(\mathrm{erfc}(\sqrt{t}/2) - \sqrt{t}\ \mathrm{ierfc}(\sqrt{t}/2)\right), \tag{12.57}$$

which is a continuously differentiable function with $\chi'(t) \neq 0$, $\chi(0) = 0$ and $\lim_{t\to\infty}\chi(t) = \infty$. On differentiating χ yields that $\chi'(t)$ is equal to the right-hand side of (12.56), hence $g(\chi(t)) \equiv 1$. From the properties of the function χ it results that $g(u) \equiv 1$, as in Example 12.1. Finally, from (12.15) and (12.16) one obtains the thermal properties given by (12.40). Furthermore, $u(x,t)$ can be obtained as being given by (12.41), where $\zeta(\mathrm{X}(x,t),t)$ is given by (12.52) and $\mathrm{X}(x,t)$ is found by solving (12.54), which for $K_0 = C_0 = q_0 = A = 1$ is

$$x = \frac{e^{\mathrm{X}}}{2}(t+\mathrm{X})\mathrm{erfc}\left(\frac{\mathrm{X}}{2\sqrt{t}}+\frac{\sqrt{t}}{2}\right)+(\mathrm{X}-t)\left[1-\frac{1}{2}\mathrm{erfc}\left(\frac{\mathrm{X}}{2\sqrt{t}}-\frac{\sqrt{t}}{2}\right)\right]. \tag{12.58}$$

For a given x and t, equation (12.58) can be inverted and the obtained value of $\mathrm{X}(x,t)$ substituted into (12.41) to find $u(x,t)$.

Example 12.4 Since as $t \to 0$, the right-hand side of (12.56) has a singularity of the type $t^{-1/2}$, consider an example in which the data (12.19) removes this singularity, e.g. $\chi(t) = 2\sqrt{t/\pi}$, which is also a continuously differentiable with $\chi'(t) \neq 0$. Taking for simplicity, $q_0 = K_0 = C_0 = 1$ and $A = 2/\sqrt{\pi}$, then from equation (12.56) one obtains

$$g(u) = \frac{\pi^{1/2}\ \mathrm{ierfc}(u/2)}{\mathrm{erfc}(u/2) - u\ \mathrm{ierfc}(u/2)}, \tag{12.59}$$

which also ensures that $g(0) = 1$. Further, using equations (12.15) and (12.16),

$$K(u) = \sqrt{\pi}\ \mathrm{ierfc}(u/2), \quad C(u) = \frac{\sqrt{\pi}\ \mathrm{ierfc}(u/2)}{[\mathrm{erfc}(u/2) - u\ \mathrm{ierfc}(u/2)]^2}. \tag{12.60}$$

These functions are plotted in Figure 12.2 along with their departures from the exponential profiles (12.40) of Example 12.1. It can also be remarked that $C(u)$ and $K(u)$ are monotonically increasing and decreasing functions of u, respectively. Finally, using (12.20) one obtains

$$\zeta(\mathrm{X}(x,t),t) = \mathrm{erfc}(u(x,t)/2) - u(x,t)\ \mathrm{ierfc}(u(x,t)/2), \tag{12.61}$$

where $\zeta(\mathrm{X}(x,t),t)$ is given by (12.52) and $\mathrm{X}(x,t)$ is found by solving equation (12.54). For a given x and t, equation (12.54) is inverted and the obtained value of $\mathrm{X}(x,t)$ substituted into (12.61) to obtain a nonlinear equation for $u(x,t)$, which can be solved using the NAG routine C05NCF.

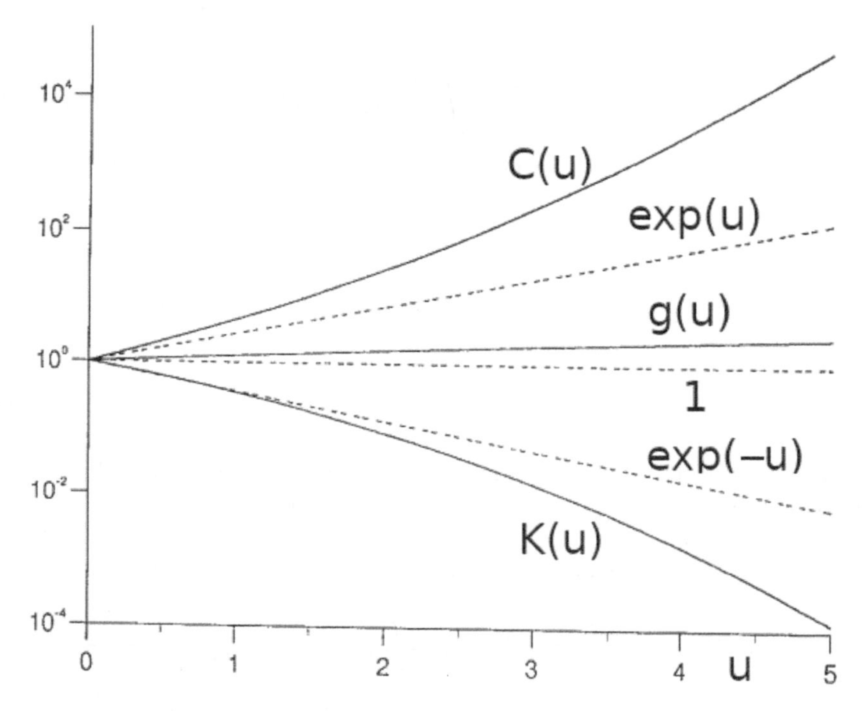

FIGURE 12.2: The results for $g(u)$, $K(u)$ and $C(u)$ as functions of u, as given by equations (12.59) and (12.60), for Example 12.4 (the dashed lines represent the functions given by (12.40)), [252].

12.1.4.3 The finite slab

In this case $l < \infty$, and (12.1) has to be solved subject to the initial boundary conditions (12.2)–(12.4) and to the additional measurement (12.19).

Example 12.5 Consider the nonlinear heat conduction in a finite slab $(0, l)$ in which the side $x = 0$ is kept insulated, i.e., $q_0 \equiv 0$, whilst the side $x = l$ is subjected to a prescribed heat flux $q_l \not\equiv 0$. Also, at $t = 0$ the initial temperature is uniform and is taken to be zero, i.e. $u_0 \equiv 0$. This mathematical formulation models the quencing of a flat plate in a tank filled with fluid [28].

Based on the transformations (12.20) and (12.21), in terms of ζ, the problem given by the (12.11) subject to the conditions (12.2)–(12.4) is

$$\frac{\partial \zeta}{\partial t}(\mathrm{X}, t) = \frac{\partial^2 \zeta}{\partial \mathrm{X}^2}(\mathrm{X}, t), \quad t > 0, \quad 0 < \mathrm{X} < a(t), \tag{12.62}$$

$$\zeta(\mathrm{X}, 0) = 1, \quad 0 \leq \mathrm{X} \leq a(0), \tag{12.63}$$

$$\frac{\partial \zeta}{\partial \mathrm{X}}(0, t) = 0, \quad \frac{\partial \zeta}{\partial \mathrm{X}}(a(t), t) = -Aq_l(t)\zeta(a(t), t), \quad t > 0, \tag{12.64}$$

where $a(t) = \mathrm{X}(l, t)$. From equation (12.21), $\mathrm{X}(0, t) = 0$, $a(0) = \mathrm{X}(l, 0) = l\sqrt{C_0/K_0}$ and from (12.22),

$$l = \sqrt{K_0/C_0} \int_0^{a(t)} \zeta(X', t)dX'. \tag{12.65}$$

Differentiating (12.65) with respect to t and using (12.62) and (12.64),

$$\begin{aligned} 0 &= \zeta(a(t), t)a'(t) + \int_0^{a(t)} \zeta_t(X', t)dX' \\ &= \zeta(a(t), t)a'(t) + \zeta_X(a(t), t) = \zeta(a(t), t)[a'(t) - Aq_l(t)]. \end{aligned} \tag{12.66}$$

Since $\zeta \neq 0$, the moving boundary $a(t)$ is given by

$$a(t) = a(0) + A \int_0^t q_l(t')dt'. \tag{12.67}$$

The problem (12.62)–(12.64) is an inverse Stefan-type moving boundary problem which can be transformed into a boundary integral form [391],

$$\begin{aligned} \eta(\mathrm{X})\zeta(\mathrm{X}, t) &= \int_0^t G(\mathrm{X}, t; a(\tau), \tau)\frac{\partial \zeta}{\partial \mathrm{X}}(a(\tau), \tau)d\tau \\ &+ \int_0^t G(\mathrm{X}, t; 0, \tau)\frac{\partial \zeta}{\partial \mathrm{X}}(0, \tau)d\tau - \int_0^t \frac{\partial G}{\partial \mathrm{X}}(\mathrm{X}, t; a(\tau), \tau)\zeta(a(\tau), \tau)d\tau \\ &- \int_0^t \frac{\partial G}{\partial \mathrm{X}}(\mathrm{X}, t; 0, \tau)\zeta(0, \tau)d\tau + \frac{1}{2}\left[\mathrm{erf}\left(\frac{\mathrm{X}}{2\sqrt{t}}\right) - \mathrm{erf}\left(\frac{\mathrm{X} - a(0)}{2\sqrt{t}}\right)\right] \\ &+ \int_{a(0)}^{a(t)} G(\mathrm{X}, t; y, \tau(y))\zeta(y, \tau(y))dy, \quad t > 0, \quad 0 \leq \mathrm{X} \leq a(t), \end{aligned} \tag{12.68}$$

where $\eta(\mathrm{X}) = 1$ if $\mathrm{X} \in (0, a(t))$, $\eta(0) = \eta(a(t)) = 1/2$, $\tau(y) = a^{-1}(y)$ which exists if $q_1 \neq 0$. Also, G is the fundamental solution (3.23). Further, using (12.64) and the change of variables $y = a(\tau)$, equation (12.68) becomes

$$\begin{aligned} \eta(\mathrm{X})\zeta(\mathrm{X}, t) &= -\int_0^t \frac{\partial G}{\partial \mathrm{X}}(\mathrm{X}, t; a(\tau), \tau)\zeta(a(\tau), \tau)d\tau \\ &- \int_0^t \frac{\partial G}{\partial \mathrm{X}}(\mathrm{X}, t; 0, \tau)\zeta(0, \tau)d\tau + \frac{1}{2}\left[\mathrm{erf}\left(\frac{\mathrm{X}}{2\sqrt{t}}\right) - \mathrm{erf}\left(\frac{\mathrm{X} - a(0)}{2\sqrt{t}}\right)\right], \end{aligned} \tag{12.69}$$

for $t > 0$, $0 \leq \mathrm{X} \leq a(t)$, since the first and the last terms of (12.68) are equal and of opposite signs. Equation (12.69) can be solved numerically using the BEM, [391]. Assuming that $\zeta_{\mathrm{XX}}(0, t) = \zeta_t(0, t)$ and using (12.64), equation (12.29) gives

$$g(\chi(t))\chi'(t) = -\frac{\zeta_t(0, t)}{A\sqrt{K_0 C_0}\zeta(0, t)}, \tag{12.70}$$

where $\zeta_t(0, t)$ can be evaluated using finite differences. As additional temperature measurement (12.19), consider the simplest continuously differentiable function with $\chi'(t) \neq 0$, namely, $\chi(t) = t - \gamma$, and then (12.70) gives

$$g(u) = -\frac{\zeta_t(0, u + \gamma)}{A\sqrt{K_0 C_0}\zeta(0, u + \gamma)}, \quad u > 0, \tag{12.71}$$

where $\gamma > 0$ is chosen such that $g(0) = 1$. Imposing this condition, it is obtained that $\gamma = 0.3537$ for $K_0 = C_0 = 0.6748$, [252]. The thermal properties $K(T)$ and $C(T)$ can also be found from equations (12.15) and (12.16), as

$$K(u) = -\frac{\sqrt{K_0/C_0}\ \zeta_t(0, u + \gamma)}{A},$$
$$C(u) = -\frac{\sqrt{C_0/K_0}\ \zeta_t(0, u + \gamma)}{A\zeta^2(0, u + \gamma)}, \quad u \geq 0. \tag{12.72}$$

The results in Figure 12.3 show that whilst $g(u)$ and $K(u)$ are monotonically decreasing functions of u, the function $C(u)$ possesses a maximum point.

12.2 Nonlinear and Heterogeneous Conductivity

The governing equation for u in the steady-state diffusion within a sample $\Omega \subset \mathbb{R}^d$ of isotropic conductivity $k(\underline{x}, u) > 0$, in the absence of sources, is

$$\nabla \cdot (K(\underline{x}, u)\nabla u(\underline{x})) = 0, \quad \underline{x} \in \Omega. \tag{12.73}$$

The corresponding Cauchy problem associated to this non-linear elliptic equation was investigated in [35]. In this section, the identification of the conductivity $K(\underline{x}, u)$ is considered. If the conductivity is a separable function, i.e.,

$$K(\underline{x}, u) = F_1(u)F_2(\underline{x}), \tag{12.74}$$

then the Kirchhoff transformation

$$\psi = \int^u F_1(u)du \tag{12.75}$$

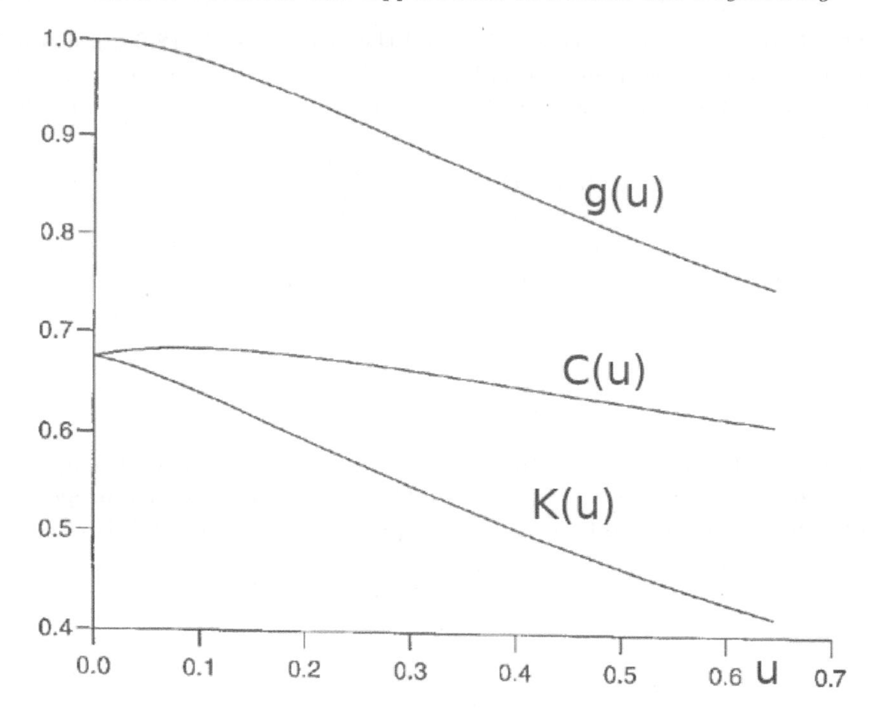

FIGURE 12.3: The results for $g(u)$, $K(u)$ and $C(u)$, as functions of u, as given by equations (12.71) and (12.72), for Example 12.5, [252].

reduces (12.73) to the linear equation

$$\nabla \cdot (F_2(\underline{x})\nabla\psi(\underline{x})) = 0, \quad \underline{x} \in \Omega. \tag{12.76}$$

If the conductivity $K(\underline{x}, u)$ is known and is separable as in (12.74), then, for a wide range of spatial variations of $F_2(\underline{x})$, one can compute the unique solution u of (12.73) subject to the Dirichlet (or Neumann, mixed) boundary condition

$$u(\underline{x}) = f(\underline{x}), \quad \underline{x} \in \partial\Omega \tag{12.77}$$

using the BEM, [359]. However, if $K(\underline{x}, u)$ is unknown, then additional information is necessary to uniquely determine the pair solution $(K(\underline{x}, u), u)$ satisfying (12.73) and (12.77). Such additional information may include seeking for a separable $K(\underline{x}, u)$, as given by (12.74), in the admissible set

$$\Lambda_{ad} := \{k = F_1(u)F_2(\underline{x}) \in L^\infty(\Omega) \mid \infty > c_1 \geq k \geq c_0 > 0\}.$$

The remaining additional information may include the knowledge of the Dirichlet-to-Neumann (D-to-N) map [188], which is defined as

$$\Lambda_D : H^{1/2}(\partial\Omega) \to H^{-1/2}(\partial\Omega), \quad \Lambda_D(f) := q := K\frac{\partial u}{\partial n}. \tag{12.78}$$

In practical terms, in electrical impedance tomography, $\Lambda_D(f)$ is the electric current distribution applied on the boundary $\partial\Omega$ which produces the voltage f. These quantities can be measured by means of a combination of a voltmeter and amperemeter at the boundary. Under the transformation (12.75), the Dirichlet condition (12.77) and its corresponding Neumann data reduce to

$$\psi := h(\underline{x}) := \int^{f(\underline{x})} F_1(u)du, \quad \underline{x} \in \partial\Omega, \tag{12.79}$$

$$F_2 \frac{\partial \psi}{\partial n} = q(\underline{x}), \quad \underline{x} \in \partial\Omega. \tag{12.80}$$

From (12.79), by differentiation with respect to the arc length s along $\partial\Omega$,

$$h'(s) = f'(s) F_1(f(s)), \quad s \in \partial\Omega. \tag{12.81}$$

Remark 12.2. (i) If $F_2(\underline{x})$ is known then $F_1(u)$ can uniquely be retrieved on $\partial\Omega$ from (12.81), from a single Cauchy data boundary measurement $(f,q)|_{\partial\Omega}$ with $f'(s) \neq 0$ for $s \in \partial\Omega$, [53].

(ii) If $F_1(u)$ is known then $h|_{\partial\Omega}$ is known, in which case Λ_D determines uniquely $F_2(\underline{x})$, [43, 238, 313].

(iii) If $F_1(u)$ is unknown, then to determine both $F_1(u)$ and $F_2(\underline{x})$, the knowledge of the Neumann to Dirichlet (N-to-D) map is also needed [187], corresponding to the problem given by equations (12.76) and (12.80), namely,

$$\Lambda_N : H_*^{-1/2} := \left\{ q \in H^{-1/2}(\partial\Omega) \,\middle|\, \int_{\partial\Omega} q \, dS = 0 \right\} \to H^{1/2}(\partial\Omega),$$
$$\Lambda_N(q) := h \,|_{\partial\Omega} = \psi \,|_{\partial\Omega} \,. \tag{12.82}$$

(iv) For an anisotropic separable conductivity $K_{ij}(\underline{x}, u) = F_1(u) F_2^{ij}(\underline{x})$, $i.j = 1, 2$, when $F_1(u)$ is known, then Λ_D determines uniquely the equivalent class of $F_2^{ij}(\underline{x})$, see [373] and the references therein.

12.2.1 An example

The variation with both $\underline{x}$ and u of the conductivity $K(\underline{x}, u)$ combines heat conduction in metals which are mainly homogeneous but temperature dependent, and flows in porous media which are mainly independent of the pressure head but they are inhomogeneous. Such a situation arises in connection with an unsaturated flow in a heterogeneous porous medium [349], in which the hydraulic conductivity in a one-dimensional sample $0 \leq x \leq 1$ is both function and space dependent as in equation (12.74) with

$$F_1(u) = (\alpha + \beta u)^{-2}, \quad F_2(x) = 1 + cx, \tag{12.83}$$

where α, β are constants satisfying $\alpha + \beta u \neq 0$ and $c > -1$ is a constant.

The case $\beta = 0$
In this case, $F_1(u) = \alpha^{-2}$ and $\alpha \neq 0$. Consider the problem of determining the triplet solution $(u(x), \alpha, c)$ with $\alpha \neq 0$, $c > -1$ of

$$\frac{d}{dx}\left((1+cx)\frac{du}{dx}\right) = 0, \quad x \in (0,1), \tag{12.84}$$

$$u(0) = \mu_0, \quad u(1) = \mu_1, \tag{12.85}$$

$$u'(0) = -\alpha^2 q_0, \tag{12.86}$$

where μ_0, μ_1 and q_0 are given numbers. This has the solution

$$u(x) = \begin{cases} \mu_0 + (\mu_1 - \mu_0)x, & \text{if } c = 0, \\ \mu_0 - \frac{\alpha^2 q_0}{c}\ln(1+cx), & \text{if } c \neq 0. \end{cases} \tag{12.87}$$

Proposition 12.1. *If $c = 0$ and $\frac{\mu_0 - \mu_1}{q_0} > 0$, then $\alpha \neq 0$ can be found as*

$$\alpha = \pm\sqrt{\frac{\mu_0 - \mu_1}{q_0}}. \tag{12.88}$$

Consider now the case $c \neq 0$. Imposing $u(1) = \mu_1$ results in

$$\frac{\alpha^2}{c}\ln(1+c) = \frac{\mu_0 - \mu_1}{q_0}, \tag{12.89}$$

so, one extra internal measurement, say at $x = 0.5$, $u(0.5) = \tilde{\mu}$ yields

$$\frac{\alpha^2}{c}\ln(1+c/2) = \frac{\mu_0 - \tilde{\mu}}{q_0}. \tag{12.90}$$

Dividing equations (12.89) and (12.90) gives

$$1 + c/2 = (1+c)^p, \tag{12.91}$$

where $p := \frac{\mu_0 - \tilde{\mu}}{\mu_0 - \mu_1}$. Consider now the function

$$\theta : (-1, \infty)\backslash\{0\} \to \mathbb{R}, \quad \theta(c) = (1+c)^p - c/2 - 1. \tag{12.92}$$

Several cases are distinguised:

(i) $p < 0$. Then, $\theta'(c) < 0$ and since $\lim_{c\to -1}\theta(c) = \infty$, $\lim_{c\to 0}\theta(c) = 0$, there is no solution of (12.91) with $c \neq 0$.

(ii) $p = 0$. Then, $\theta(c) = -c/2$, so there is no solution of (12.91) with $c \neq 0$.

(iii) $0 < p < 1$. Then, $\theta''(c) < 0$, and $\theta'(c) = 0$ if and only if $c := c_0 = \left(\frac{1}{2p}\right)^{1/(p-1)} - 1$ is the unique maximum point of θ. Now, since $\lim_{c\to -1}\theta(c) = -1/2$, $\lim_{c\to 0}\theta(c) = 0$, $\lim_{c\to\infty}\theta(c) = -\infty$, there exists a unique solution $0 \neq c_1 > -1$ of (12.91) if $p \neq 1/2$. If $p = 1/2$ then, $c_0 = c_1 = 0$.

(iv) $p = 1$. Then, $\theta(c) = c/2$ and there is no solution for (12.91) with $c \neq 0$.

(v) $p > 1$. Then, $\theta''(c) > 0$ and c_0 is the unique minimum point for θ. Since $\lim_{c\to-1}\theta(c) = -1/2$, $\lim_{c\to 0}\theta(c) = 0$, $\lim_{c\to\infty}\theta(c) = \infty$ and $\theta(c_0) < -1/2$, there is no solution to (12.91) with $c \neq 0$.

Summarizing, the following proposition is obtained.

Proposition 12.2. *There is a unique solution $-1 < c_1 \neq 0$ of (12.91) if and only if $0 < p < 1$ and $p \neq 1/2$. Furthermore, if $\frac{c_1}{\ln(1+c_1)}\left(\frac{\mu_0-\mu_1}{q_0}\right) > 0$, then $\alpha \neq 0$ can be found from (12.88).*

The case $\beta \neq 0$

In this case, transformation (12.75) gives $\psi(u) = -\frac{1}{\beta(\alpha+\beta u)}$, and one seeks for the solution $(u(x), \alpha, \beta, c)$ with $\alpha + \beta u \neq 0$ and $c > -1$ of

$$\begin{cases} \frac{d}{dx}\left((1+cx)\frac{d\psi}{dx}\right) = 0, & x \in (0,1), \\ \psi(0) = -\frac{1}{\beta(\alpha+\beta\mu_0)} =: h_0, & \psi(1) = -\frac{1}{\beta(\alpha+\beta\mu_1)} =: h_1, \\ -\psi'(0) = q_0 & . \end{cases}$$

Integrating it yields

$$\psi(x) = \begin{cases} h_0 + (h_1 - h_0)x, & \text{if } c = 0 \\ h_0 - \frac{q_0}{c}\ln(1+cx), & \text{if } c \neq 0. \end{cases}$$

If $c = 0$, then imposing $\psi'(0) = -q_0$ gives

$$(\alpha + \beta\mu_1)(\alpha + \beta\mu_0)q_0 = \mu_0 - \mu_1. \tag{12.93}$$

In order to determine both α and β, one extra internal measurement, say at $x = 0.5$, $u(0.5) = \tilde{\mu}$ is needed to yield

$$[2\alpha + \beta(\mu_0 + \mu_1)](\alpha + \beta\tilde{\mu})q_0 = 2(\mu_0 - \mu_1). \tag{12.94}$$

Solving equations (12.93) and (12.94) yield

$$\beta = \pm\left(\frac{\delta}{\alpha_0 + \beta_0\epsilon + \epsilon^2}\right)^{1/2}, \quad \alpha = \epsilon\beta, \tag{12.95}$$

where $\alpha_0 = \mu_0\mu_1$, $\beta_0 = \mu_0 + \mu_1$, $\delta = \frac{\mu_0-\mu_1}{q_0}$, $\epsilon = \frac{2\alpha_0-\beta_0\tilde{\mu}}{2\tilde{\mu}-\beta_0}$. As in Proposition 12.1, the condition for the existence of solution (12.95) requires $\frac{\delta}{\alpha_0+\beta_0\epsilon+\epsilon^2} > 0$.

Let us now consider the case $c \neq 0$. Imposing $\psi(1) = h_1$ gives

$$c(\mu_0 - \mu_1) = (\alpha + \beta\mu_1)(\alpha + \beta\mu_0)q_0\ln(1+c). \tag{12.96}$$

If one now takes another two Dirichlet boundary measurements μ_0^1, μ_1^1 and μ_0^2, μ_1^2 which give rise, through the D-to-N map (12.78), to the flux data q_0^1 and q_0^2, and imposing $\psi_1(1) = h_1^1$ and $\psi_2(1) = h_1^2$, one obtains $\alpha = z\beta$, where

$$z = \frac{(\delta_1\alpha - \delta\alpha_1)(\delta_2 - \delta) - (\delta_2\alpha - \delta\alpha_2)(\delta_1 - \delta)}{(\delta_2\beta - \delta\beta_2)(\delta_1 - \delta) - (\delta_1\beta - \delta\beta_1)(\delta_2 - \delta)} \tag{12.97}$$

and $\alpha_1 = \mu_0^1\mu_1^1$, $\alpha_2 = \mu_0^2\mu_1^2$, $\beta_1 = \mu_0^1 + \mu_1^1$, $\beta_2 = \mu_0^2 + \mu_1^2$, $\delta_1 = \frac{\mu_0^1 - \mu_1^1}{q_0^1}$, $\delta_2 = \frac{\mu_0^2 - \mu_1^2}{q_0^2}$. Then, equation (12.96) yields

$$c(\mu_0 - \mu_1) = b^2(z + \mu_1)(z + \mu_0)q_0 \ln(1 + c). \tag{12.98}$$

Taking one extra internal measurement, say at $x = 0.5$, $u(0.5) = \tilde{\mu}$, yields

$$c(\mu_0 - \tilde{\mu}) = \beta^2(z + \tilde{\mu})(z + \mu_0)q_0 \ln(1 + c/2). \tag{12.99}$$

Dividing equations (12.98) and (12.99), it yields

$$1 + \frac{c}{2} = (1 + c)^P, \tag{12.100}$$

where $P := \frac{p(z+\mu_1)}{z+\tilde{\mu}}$. Then, a similar result to Proposition 12.2 is obtained.

Proposition 12.3. *There is a unique solution* $-1 < c_1 \neq 0$ *to* (12.99) *if and only if* $0 < P < 1$ *and* $P \neq 1/2$. *Furthermore, if* $\frac{c_1\delta}{(z+\mu_1)(z+\mu_0)\ln(1+c_1)} > 0$, *then* β *can be found from* (12.95), *and finally* $\alpha = z\beta$, *provided that* z *in equation* (12.97) *is well-defined.*

12.2.2 Extention to transient problems

The governing diffusion equation in the unsteady-case is given by

$$\nabla \cdot (K(\underline{x}, u)\nabla u(\underline{x}, t)) = C(\underline{x}, u)\frac{\partial u}{\partial t}(\underline{x}, t), \quad (\underline{x}, t) \in \Omega \times (0, \infty), \tag{12.101}$$

where $C(\underline{x}, t) > 0$ is a property of the sample representing the heat capacity of a heat conductor or the specific storage of a porous medium. In time-dependent processes, an initial condition is usually prescribed, namely,

$$u(\underline{x}, 0) = u_0(\underline{x}), \quad \underline{x} \in \overline{\Omega}. \tag{12.102}$$

If K is separable as in (12.74) and also C is separable as

$$C(\underline{x}, u) = F_1(u)F_3(\underline{x}),$$

then the transformation (12.75) applied to (12.101) and (12.102) yields

$$\nabla \cdot (F_2(\underline{x})\nabla\psi(\underline{x}, t)) = F_3(\underline{x})\frac{\partial\psi}{\partial t}(\underline{x}, t)), \quad (\underline{x}, t) \in \Omega \times (0, \infty),$$

$$\psi(\underline{x}, 0) = \int^{u_0(\underline{x})} F_1(u)du, \quad \underline{x} \in \overline{\Omega}.$$

In this case, the transient analysis requires less additional information to obtain a unique solution [127].

Chapter 13

Anti-Reflection Coatings

Anti-reflection coatings (ARC's) are a key and vital feature for high-efficiency silicon solar cell design [76]. They are also widely used to increase transmission and reduce glare resulting from window coatings in a diverse range of industries such as photovoltaics, buildings, displays, and ophthalmics. ARC's currently in use enhance the transparency of certain surfaces by the introduction of a smooth and gradual change in effective refractive index between two media, see Figure 13.1, which results in improved efficiency of some commercial architectural glazing and solar collectors.

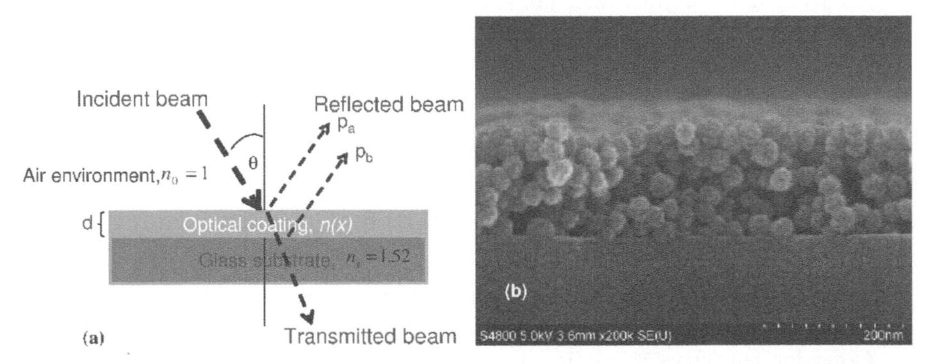

FIGURE 13.1: (a) Schematic diagram of an anti-reflection coating of thickness d. When p_a and p_b are an angle of π out of phase and have equal amplitude, the magnitude of the reflected wave is zero. (b) An electron micrograph of a cross section of an anti-reflection coating on glass [276].

The reflection from any given interface at normal incidence is related to the ratio of refractive indices of the materials forming the interface and is characterized by the % reflectance given by $100(n_0 - n_s)^2/(n_0 + n_s)^2$, where n_0 is the refractive index of the first layer (air) and n_s is the refractive index of the second layer (window) [407]. Thus, for a crown glass window, $n_0 = 1$ and $n_s = 1.52$ giving a reflectance at normal incidence of 4.3% per surface, i.e. a total reflectance of 8.6% from the window. To minimize or remove this reflectance completely a further layer of refractive index n_1 is coated onto the window such that reflections from the air/coating and coating/window interfaces undergo destructive interference. In this case, for a certain wavelength λ related to the film thickness d, namely if $\lambda = 4dn_1$ and if $n_1 = \sqrt{n_0 n_s}$, see [407] and

DOI: 10.1201/9780429400629-13

later on Example 13.1, the reflection may become zero. Thus, for example, at a normally incident wavelength of 550 nm (green light) a perfect ARC on a crown glass window will have a thickness of 112 nm and a refractive index of 1.23. However, this assumes that the refractive index is constant throughout the thickness of the coating. If the experimental reflectivity curves deviate from those predicted by this simplified homogeneous single-layer model, this is indicative of some refractive index gradient within the film. However, there is no way to measure this gradient directly. The gradient is important as it provides useful information regarding the relationship between the optical and mechanical properties of the ARC.

In the optical coating synthesis problem, the unknown refractive index is varied in space, either continuously or piecewisely, so as to approximate the spectral characteristics, such as the transmission or reflection coefficient. The solution of this problem in turn can then be used to optimize the design of optical coatings. The difficulty in solving this inverse problem comes from the fact that it is ill-posed; for example, the solution may be non-unique such that widely different refractive indices may have close spectral characteristics, or the solution may be unstable, i.e. small errors in the spectral characteristics measurements can cause large errors in the retrieved refractive index.

13.1 Mathematical Model and Analysis

In this section, a class of inverse scattering problems in one dimension is formulated for the determination of the index of refraction of a scatterer, i.e. the optical coating, from scattering data, i.e. the reflection coefficient.

Consider the wave refraction propagation initiated by a transverse wave in which the electric and magnetic fields are at right angles to each other, and to the direction of propagation which is taken as the $x-$axis, incident from $x = -\infty$ of the form $u_{inc}(x) = e^{ikn_0 x}$, where $k = \omega/c$ is the free-space wavenumber, ω is the frequency, $c = 1/\sqrt{\epsilon_0 \mu_0}$ is the speed of light, ϵ_0 and μ_0 are the permittivity and magnetic permeability of the air free-space, respectively, and $n_0 = 1$ is the refractive index of the outside air environment. The wave impinges on a dielectric graded-index obstacle of known thickness $d > 0$ and unknown refractive index $n(x)$, in contact with an infinitely thick glass substrate absorbing material of uniform index of refraction $n_s = 1.52$, imagine Figure 13.1(a) for $\theta = 0$ angle of incidence.

Upon imposing the continuity at the interfaces $x = 0$ and $x = d$, from the time-harmonic Maxwell's equations one obtains that the y-component $u(x)$ of the electric field satisfies the following non-dimensionalised problem [105]:

$$u''(x) + \beta^2 \epsilon(x) u(x) = 0, \quad x \in (0,1), \tag{13.1}$$

$$u'(0) + in_0 \beta u(0) = 2in_0\beta, \tag{13.2}$$

$$u'(1) - in_s \beta u(1) = 0, \tag{13.3}$$

where $\epsilon(x) = n^2(x)$ is the dielectric permittivity of the optical coating, $\beta = kd = 2\pi d/\lambda$ is the non-dimensionalised wavenumber and λ is the wavelength. The inverse problem consists of determining $\epsilon(x)$ (and hence $n(x) = \sqrt{\epsilon(x)}$) from the reflection coefficient measurements

$$R(\beta) := u(0) - 1 \quad \text{for } \beta \in [\beta_{min}, \beta_{max}]. \tag{13.4}$$

One could also attempt to measure the transmission coefficient

$$T(\beta) := u(1)e^{-in_s\beta} \quad \text{for } \beta \in [\beta_{min}, \beta_{max}], \tag{13.5}$$

in addition to (13.4), but this additional information will not be considered, as it is not available yet (to us) from practical measurements. Remark that from (13.2) and (13.4),

$$u(0) = 1 + R(\beta), \quad u'(0) = in_0\beta(1 - R(\beta)), \tag{13.6}$$

thus, the boundary value problem (13.1)–(13.4) can be recast as the initial value problem (13.1), (13.3) and (13.6), which is a nonlinear and ill-posed inverse coefficient identification problem [98].

First, concerning the existence of a solution, a coating that fits the given reflection coefficient $R(\beta)$ for all values of $\beta \in [\beta_{min}, \beta_{max}]$ does not necessarily exist. Hence, the noisy measurement (13.4) should be understood in a least-squares sense which minimizes, with respect to $n \in N_{ad} := \{n \in L_\infty(0,1) \mid n_0 \le n(x) \le n_s, \forall x \in [0,1]\}$, the functional $F : N_{ad} \to \mathbb{R}_+$ defined by

$$F(n) := \|R(\beta) + 1 - u(0; \beta, n)\|^2_{L^2[\beta_{min}, \beta_{max}]}. \tag{13.7}$$

Second, concerning uniqueness, different coatings may produce the same given reflection coefficient $R(\beta)$ for $\beta \in [\beta_{min}, \beta_{max}]$. Questions relate to whether the finite range $[\beta_{min}, \beta_{max}]$ is sufficient for uniqueness, i.e., should the reflection coefficient $R(\beta)$ be known for all positive wavenumbers $\beta \in (0, \infty)$ or, whether uniqueness still holds if only the reflectance $|R(\beta)|$, i.e. the magnitude of the complex reflection coefficient $R(\beta)$, is measured as

$$|R(\beta)| = |u(0) - 1| \quad \text{for } \beta \in [\beta_{min}, \beta_{max}]. \tag{13.8}$$

The uniqueness will be considered in more detail in subsection 13.1.1.

Third, concerning stability, small errors in the practical measurements (13.4) or (13.8), may cause large errors in the desired index of refraction. To prevent unphysical unstable solutions, regularization has to be applied, using some prior information about $n(x)$. This may include physical bounds such as $n \in N_{ad}$ and the continuity (or smoothness) of the function $n(x)$ which is imposed by adding to the least-squares function (13.7) a penalty regularization term. Hence, one minimizes the Tikhonov regularization functional

$$F_\Lambda(n) := \|R(\beta) + 1 - u(0;\beta,n)\|^2_{L^2[\beta_{min},\beta_{max}]} + \Lambda\|n - n^*\|^2_{L^2(0,1)}, \quad (13.9)$$

where $n^*(x) \in [n_0, n_s]$ is an *a priori* guess on the solution $n(x)$ and $\Lambda \geq 0$ is a regularization parameter prescribed, e.g., using the L-curve criterion, which choses Λ at the corner of the L-curve generated by plotting the solution norm $\|n-n^*\|_{L^2(0,1)}$ against the residual norm $\|R(\beta)+1-u(0;\beta,n)\|_{L^2[\beta_{min},\beta_{max}]}$ for a range of values of Λ. Despite some limitations [133], the L-curve has gained attention for the convenient selection of the regularization parameter [135]. It will be employed to choose Λ in Example 13.2. A more rigorous choice of the regularization parameter Λ, based on the *a priori* knowledge of the amount of noise with which the input data is contaminated, is discussed in more detail in Example 13.1.

The case of multilayer dielectric systems in which the reflection at the main contact surface is decreased, due to the interference of the reflected light from each interface so that the refractive index $n(x)$ is an unknown piecewise constant function, was investigated in [407] (single and double layer), [320] (triple layer) and [98] (N layers). Although multilayer systems combined with surface texturization can reduce reflection losses over the useful solar spectrum and thus improving the device performance and efficiency, the cost rapidly becomes higher. Another practical manner in which to minimize reflection is to deposit an inhomogeneous dielectric with a gradual decrease of refractive index from the glass substrate to the ambient air [286]. Then, the spacewise variation of $n(x)$ through the depth of the rugate film is continuous, in contrast to the classical multilayer stacks [209]. In the literature, it is reported that coatings with a smooth continuous variation of the refractive index, compared with classical multilayer stacks show higher laser induced damage threshold [348], less scattering losses and better optical properties [209].

The identification of a continuous refractive index was investigated by various methods, e.g. the Gelfand-Levitan-Marchenko spectral method [71], the integral equation method [128], the spline approximation projection [105], techniques based on trace formulae [77], low-coherence interferometric imaging [75] and coupled-mode Zakharov-Shabat equations [29].

13.1.1 Uniqueness of solution

The inverse medium diffraction problem (13.1)–(13.4), or equivalently (13.1), (13.3) and (13.6), under the key assumption that the reflection coefficient $R(\beta)$ is known for all values of $\beta \in (0, \infty)$, is analyzed in more detail as an inverse scattering problem [276]. Employing the well-known ansatz [29],

$$u(x) = \frac{1}{\sqrt{\beta n(x)}}\left(A(x)e^{iS(x)} + B(x)e^{-iS(x)}\right),$$
$$u'(x) = i\sqrt{\beta n(x)}\left(A(x)e^{iS(x)} - B(x)e^{-iS(x)}\right), \tag{13.10}$$

where $S(x) = \beta\int_0^x n(\zeta)d\zeta$, for this to hold one must have

$$i\sqrt{\beta n(x)}\left(A(x)e^{iS(x)} - B(x)e^{-iS(x)}\right) = u'(x)$$
$$= -\frac{n'(x)}{2\sqrt{\beta n^3(x)}}\left(A(x)e^{iS(x)} + B(x)e^{-iS(x)}\right)$$
$$+\frac{1}{\sqrt{\beta n(x)}}\left[(A'(x) + i\beta n(x)A(x))e^{iS(x)} + (B'(x) - i\beta n(x)B(x))e^{-iS(x)}\right].$$

This is ensured if

$$A'(x) = \frac{n'(x)}{2n(x)}B(x)e^{-2iS(x)}, \quad B'(x) = \frac{n'(x)}{2n(x)}A(x)e^{2iS(x)}. \tag{13.11}$$

Assuming that $n \in C[0,1]$, in particular $n(0) = n_0$ and $n(1) = n_s$, and then (13.2) and (13.3) yield $A(0) = \sqrt{\beta n_0}$ and $B(1) = 0$. Also, in terms of the reflection coefficient, condition (13.6) gives $B(0) = R(\beta)\sqrt{\beta n_0}$. Now define

$$P(x) = A(x)e^{iS(x)}, \quad Q(x) = B(x)e^{-iS(x)}. \tag{13.12}$$

Differentiating (13.12) and using (13.11) give

$$\begin{cases} P'(x) = i\beta n(x)P(x) + \frac{n'(x)}{2n(x)}Q(x), \\ Q'(x) = -i\beta n(x)Q(x) + \frac{n'(x)}{2n(x)}P(x), \end{cases} \tag{13.13}$$

which has to be solved subject to the boundary conditions

$$P(0) = \sqrt{\beta n_0}, \quad Q(0) = R(\beta)\sqrt{\beta n_0}, \quad Q(1) = 0. \tag{13.14}$$

Writing the optical path length as

$$\xi(x) = \frac{S(x)}{\beta} = \int_0^x n(\zeta)d\zeta, \tag{13.15}$$

equations (13.13) become the Zakharov-Shabat system

$$\dot{P} = qQ + i\beta P, \quad \dot{Q} = qP - i\beta Q, \tag{13.16}$$

where the dot denotes the differentiation with respect to ξ, and

$$q(\xi) = \frac{\dot{n}(\xi)}{2n(\xi)} = \frac{d}{d\xi}\left(\ln(\sqrt{n(\xi)})\right). \tag{13.17}$$

From (13.14), the boundary conditions associated to (13.16) are

$$P(0) = \sqrt{\beta n_0}, \quad Q(0) = R(\beta)\sqrt{\beta n_0}, \quad Q(\overline{n}) = 0, \tag{13.18}$$

where the average quantity $\overline{n} = \xi(1) = \int_0^1 n(\zeta)d\zeta$ is called the full optical path length. Note that since $n > 0$, the function $\xi(x)$ is strictly increasing and thus $0 \leq \xi(x) \leq \overline{n}$ for all $x \in [0,1]$. Upon dividing by $\sqrt{\beta n_0}$ in (13.18), the resulting problem given by equations (13.16) and (13.18) is precisely the problem P_2 discussed in [352]. It is well-known [71] that retrieving the real potential $V(\xi) = \dot{q}(\xi) + q^2(\xi)$ from the (left hand) reflection coefficient $\{R(\beta)|\beta \in \mathbb{R}\}$ is unique. Further, it can be shown that the real $q(\xi)$ is uniquely determined from standard results of the inverse scattering method [95]. Finally, if the function $q(\xi)$ is known, one can find the refractive index from (13.17) and the original coordinate x from (13.15), using the inversion formulae

$$n(\xi) = n_0 \exp\left[2\int_0^\xi q(\zeta)d\zeta\right], \quad x = \int_0^\xi \frac{d\zeta}{n(\zeta)}. \tag{13.19}$$

13.2 Numerical Implementation

When substituted into equations (13.1)–(13.3), the real and imaginary parts of the complex function $u = A + iB$ satisfy

$$A''(x) + \beta^2\epsilon(x)A(x) = 0, \quad B''(x) + \beta^2\epsilon(x)A(x) = 0, \quad x \in (0,1), \tag{13.20}$$

$$A'(0) - n_0\beta B(0) = 0, \quad B'(0) + n_0\beta A(0) = 2n_0\beta, \tag{13.21}$$

$$A'(1) + n_s\beta B(1) = 0, \quad B'(1) - n_s\beta A(1) = 0. \tag{13.22}$$

Note that from (13.4),

$$\begin{aligned} R(\beta) = u(0) - 1 = A(0) - 1 + iB(0) = |R(\beta)|\exp(i\ \arg(R(\beta))), \\ |R(\beta)| = \sqrt{(A(0)-1)^2 + B^2(0)}, \\ \arg(R(\beta)) = \tan^{-1}\left(\frac{B(0)}{A(0)-1}\right). \end{aligned} \tag{13.23}$$

A central FDM numerical discretization of the equations (13.20)–(13.22) is employed [276], as follows. Let $x_j = j/N$ for $j = \overline{0,N}$ be a uniform discretization of the interval $[0,1]$ with the mesh size $h = 1/N$. Denote $a_j = A(x_j)$, $b_j = B(x_j)$, $\epsilon_j = \epsilon(x_j)$ for $j = \overline{0,N}$. Then, equations (13.20)–(13.22), in discretized form, are given by

$$\frac{a_1 - a_{-1}}{2h} - n_0\beta b_0 = 0, \quad \frac{a_{j+1} - 2a_j + a_{j-1}}{h^2} + \beta^2\epsilon_j a_j = 0, \ j = \overline{0, N},$$

$$\frac{a_{N+1} - a_{N-1}}{2h} + n_s\beta b_N = 0,$$

$$\frac{b_1 - b_{-1}}{2h} + n_0\beta b_0 = 2n_0\beta, \quad \frac{b_{j+1} - 2b_j + b_{j-1}}{h^2} + \beta^2\epsilon_j b_j = 0, \ j = \overline{0, N},$$

$$\frac{b_{N+1} - b_{N-1}}{2h} - n_s\beta a_N = 0,$$

where a_{-1}, a_{N+1}, b_{-1} and b_{N+1} are fictitious values introduced in order to approximate using central finite differences the Robin boundary conditions (13.21) and (13.22). The above equations form a system of $(2N+6)$ equations with $(2N+6)$ unknowns $(a_{-1}, a_0, a_1, ..., a_N, a_{N+1}, b_{-1}, b_0, b_1, ..., b_N, b_{N+1})$, if $\underline{\epsilon} = (\epsilon_0, \epsilon_1, ..., \epsilon_N)$ were known. However, in the inverse problem under investigation $\underline{\epsilon}$ is unknown, and it has to be inferred from some additional information. In practice, the reflection coefficient $R(\beta)$ is usually measured for $M \geq N+1$ wavenumber (or wavelength) values

$$\beta_l = \frac{2\pi d}{\lambda_l} = \beta_{min} + \frac{(l-1)}{(M-1)}(\beta_{max} - \beta_{min}), \quad l = \overline{1, M}. \tag{13.24}$$

Let us denote these practical measurements by $R_l := R(\beta_l)$ for $l = \overline{1, M}$. From a given initial guess $\epsilon^*(x)$ for $\epsilon(x)$, the direct problem (13.20)–(13.22) is solved M times to provide the complex values of the reflection coefficients $u(0; \beta_l, \underline{\epsilon}) - 1$ for $l = \overline{1, M}$. The gap between these computed reflection coefficients and the measured ones R_l is minimized. Corresponding to (13.9), the discretized functional $F_\Lambda : [n_0^2, n_s^2]^{N+1} \to \mathbb{R}_+$ defined by

$$F_\Lambda(\underline{\epsilon}) = \sum_{l=1}^{M} |u(0; \beta_l, \underline{\epsilon}) - 1 - R_l|^2 + \Lambda \sum_{j=0}^{N} (\epsilon_j - \epsilon_j^*)^2, \tag{13.25}$$

where $\underline{\epsilon}^* = (\epsilon_j)_{j=\overline{0,N}} \in [n_0^2, n_s^2]^{N+1}$ is an *a priori* guess of the solution $\epsilon(x)$, is minimized. If measurements are limited to the modulus of the reflection coefficient, i.e. the amplitude, also called the energy reflection coefficient, then instead of (13.25), the functional

$$\tilde{F}_\Lambda(\underline{\epsilon}) := \sum_{l=1}^{M} \left(|u(0; \beta_l, \underline{\epsilon}) - 1| - |R_l|\right)^2 + \Lambda \sum_{j=0}^{N} (\epsilon_j - \epsilon_j^*)^2 \tag{13.26}$$

is minimized. The minimization of (13.26) contains obviously less information than the minimization of (13.25). The minimization of the functional F_Λ (or $\tilde{F}_\Lambda$) defined by (13.25) (or (13.26)), subject to the simple physical bounds on the variables $1 = n_0^2 \leq \epsilon_j \leq n_s^2 = 2.3104$ for $j = \overline{0, N}$, is performed using the

NAG Fortran routine E04JYF. This routine consists of an easy-to-use quasi-Newton algorithm for finding a minimum of a function subject to fixed upper and lower bounds on the independent variables, using function values only. Other NAG routines such as E04KYF or E04KZF, which requires supplying the gradient of the objective function, did not significantly improve the numerical results.

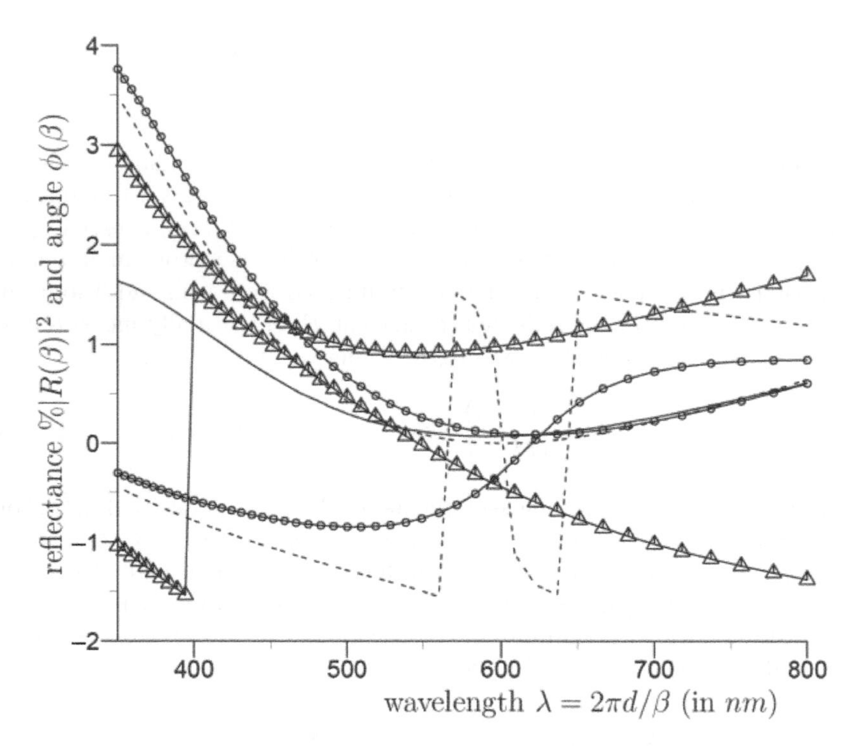

FIGURE 13.2: The experimentally (—) measured percentance of reflectance data $\%|R(\beta)|^2$ and the exact solutions (13.28) and (13.29), as functions of the wavelength $\lambda = 2\pi d/\beta$ (in nm), for homogeneous single layer coatings of various constant indices of refraction $n(x) \equiv n_1 = 1.12(-\Delta-)$, $1.23(- - -)$ and $1.27(- \circ -)$. Reading down the intercepts with the vertical axis, the upper group of 3 marked curves corresponds to $\%|R(\beta)|^2$, whilst the lower group of 3 marked curves corresponds to $\phi(\beta)$, [276].

Example 13.1 To test the accuracy, convergence and stability of the numerical solution of the inverse problem, consider first a test example for which an analytical solution for a single uniform layer is available, as given by

$$n(x) \equiv n_1,$$

$$u(x) = \frac{2n_0 n_1 \cosh(i\beta n_1(1-x)) - 2n_0 n_s \sinh(i\beta n_1(1-x))}{n_1(n_0+n_s)\cosh(i\beta n_1) - (n_1^2 + n_0 n_s)\sinh(i\beta n_1)}, \tag{13.27}$$

for $x \in [0,1]$. From (13.23) and (13.27), the reflectance is given by

$$|R(\beta)|^2 = |u(0) - 1|^2$$
$$= \frac{n_1^2(n_0 - n_s)^2 \cos^2(\beta n_1) + (n_1^2 - n_0 n_s)^2 \sin^2(\beta n_1)}{n_1^2(n_0 + n_s)^2 \cos^2(\beta n_1) + (n_1^2 + n_0 n_s)^2 \sin^2(\beta n_1)}. \quad (13.28)$$

The phase angle is given by

$$\phi(\beta) := \arg(R(\beta))$$
$$= \tan^{-1}\left[\frac{n_0 n_1 (n_1^2 - n_s^2)\sin(2\beta n_1)}{n_1^2(n_0^2 - n_s^2)\cos^2(\beta n_1) - (n_1^4 - n_0^2 n_s^2)\sin^2(\beta n_1)}\right]. \quad (13.29)$$

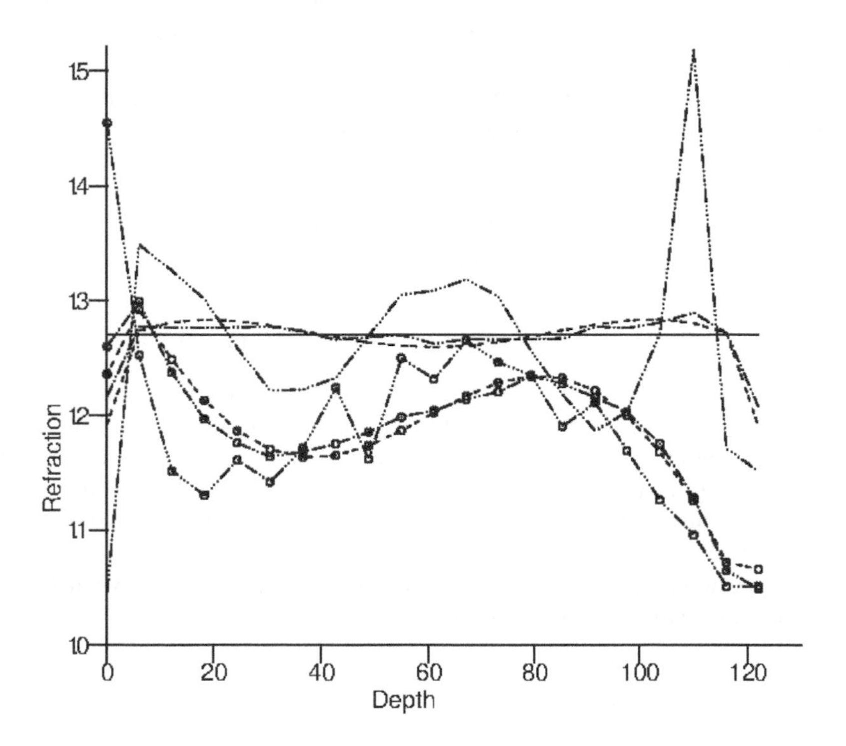

FIGURE 13.3: Numerical refraction index profiles $n(x)$, as functions of the depth $x \in [0, d = 122]$ nm, in comparison with the exact value $n_1 = 1.27$ shown by continuous horizontal line (—), obtained with various regularization parameters $\Lambda = 10^{-6}$(–....–), $\Lambda = 10^{-5}$(–..–) and $\Lambda = 10^{-4}$(– – –). The additional data is the full complex reflection coefficient $R(\beta)$ given by (13.30) for the curves without markers, and the square root of reflectance limited data $|R(\beta)|$ given by (13.28) for the curves with markers on [276].

Both quantities (13.28) and (13.29) are shown in Figure 13.2 for various values of the constant $n_1 \in \{1.12, 1.23, 1.27\}$. The thickness of the slab is $d = 122$ nm and $M = 46$ values of the wavelength λ_l are chosen in the interval [350,800] nm by taking $\beta_{min} = 2\pi d/800$ and $\beta_{max} = 2\pi d/350$ in (13.24). Note that $R(\beta) = |R(\beta)|\cos(\phi(\beta)) + i|R(\beta)|\sin(\phi(\beta))$. When $n_1 = \sqrt{n_0 n_s} \approx 1.23$, the reflection coefficient (13.28) can become zero for $\lambda = 2\pi d/\beta = 4dn_1$. For $n_1 \in \{n_0, n_s\}$, $|R(\beta)| = (n_s - n_0)/(n_s + n_0)$ is independent of β. Also, when $n_1 = n_0$, $\phi(\beta) = 2\beta$, whilst when $n_1 = n_s$, $\phi(\beta) = 0$. In Figure 13.2, experimentally measured data for the reflectance supplied by the company Oxford Advanced Surfaces Group Plc are also included. From this figure it can be seen that this experimental data cannot in fact correspond to a simple uniform coating with constant index of refraction. Trials with piecewise constant functions, i.e. layered coatings, did not produce significantly better agreement with the experiment. Instead, as shall be seen in Example 13.2, a significantly better fit is obtained by considering continuously varying spacewise-dependent indices of refraction.

In all the inversion results that follow, $N = 20$, i.e. $h = 0.05$, is used. From (13.23) and (13.27), the (complex) reflection coefficient is

$$\begin{aligned} R(\beta) &= u(0) - 1 \\ &= \frac{n_1^2(n_0 - n_s)^2\cos^2(\beta n_1) - (n_1^4 - n_0^2 n_s^2)\sin^2(\beta n_1)}{n_1^2(n_0 + n_s)^2\cos^2(\beta n_1) + (n_1^2 + n_0 n_s)^2\sin^2(\beta n_1)} \\ &+ \frac{i n_0 n_1 (n_1^2 - n_s^2)\sin(2\beta n_1)}{n_1^2(n_0 + n_s)^2\cos^2(\beta n_1) + (n_1^2 + n_0 n_s)^2\sin^2(\beta n_1)}. \end{aligned} \tag{13.30}$$

Numerical results for the index of refraction $n(x)$ obtained by minimizing the functional (13.25), based on the additional data containing the full complex reflection coefficient $R(\beta)$ given by equation (13.30), are shown in Figure 13.3 for various regularization parameters $\Lambda \in \{10^{-6}, 10^{-5}, 10^{-4}\}$. The initial guess is taken as the prior estimate $\underline{\epsilon}^* = \underline{1.12^2}$. The exact solution is the constant function $n(x) \equiv n_1 = 1.27$. There exists already numerical noise given by the residual between the numerical value of the FDM direct solver with $h = 0.05$ and the exact value (13.30) given by

$$\eta := \sum_{l=1}^{M} |u(0; \beta_l, \underline{n_1}) - 1 - R_l|^2 = 3.6E - 5. \tag{13.31}$$

To determine the regularization parameter Λ in (13.25), one can apply the discrepancy principle and choose the value of Λ for which

$$\sum_{l=1}^{M} |u(0; \beta_l, \underline{\epsilon}) - 1 - R_l|^2 \approx \eta. \tag{13.32}$$

For the synthetic Example 13.1, the value of η is available and the discrepancy principle can be used. However, for Example 13.2 the value of the noise level η is not available and then, more heuristic choices of the regularization parameter Λ, such as the L-curve criterion [135], need to be employed. From

Figure 13.3 it can be seen that regularization is necessary, otherwise for Λ too small, see results for $\Lambda = 10^{-6}$, an unstable least-squares solution is produced. A value of Λ between 10^{-5} and 10^{-4} is appropriate to ensure that a stable and reasonably accurate solution is obtained.

Next, Example 13.1 is investigated for limited additional data, as given by just the amplitude $|R_l|$ for $l = \overline{1, M}$. Then, the functional (13.26) is minimized and the numerical results obtained are also included with markers on in Figure 13.2. From these results it can be clearly seen that, for Example 13.1, the limited amplitude data is not enough to produce a unique retrieval of the index of refraction, and additional information is required. On the other hand, the use of the complete information provided by the full complex reflection coefficient enables a unique retrieval of the index of refraction, as was expected from the uniqueness analysis described in subsection 13.1.1. Once the numerical method and solution have been validated in terms of accuracy and stability for the benchmark test Example 13.1, the technique is next applied in Example 13.2 to a case study concerning inverting real reflectance data.

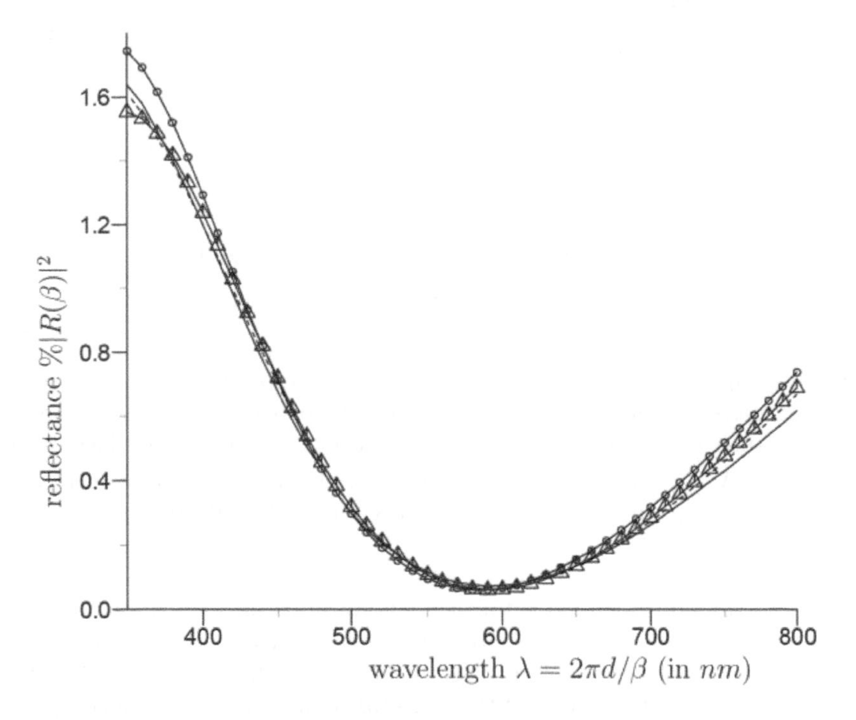

FIGURE 13.4: The numerical results for the regularized fit of the percentance of reflectance data $\%|R(\beta)|^2$, as functions of the wavelength $\lambda = 2\pi d/\beta$ (in nm), obtained with various regularization parameters $\Lambda = 10^{-5}(--\,-)$, $\Lambda = 10^{-4}(-\Delta-)$ and $\Lambda = 10^{-3}(-\circ-)$, in comparison with the measurement data shown with continuous line (—), for Example 13.2, [276].

Example 13.2 (Inversion of real data)
The real data is shown by continuous line (—) in Figure 13.2. Moreover, from the practical experiment an additional phyical measurement of the full integrated refraction index

$$1.27 = \overline{n} = \int_0^1 n(x)dx = \int_0^1 \sqrt{\epsilon(x)}dx \tag{13.33}$$

is available. Assuming also that $n \in C[0,1]$, the additional constraints corresponding to an unbounded layer are

$$1 = n_0^2 = \epsilon(0) = \epsilon_0, \quad 1.6129 = n_s^2 = \epsilon(1) = \epsilon_N. \tag{13.34}$$

Although achieving a value of unity for $\epsilon(0)$ at the film-air interface is not possible since there are no low index solid materials that equal the index of air, it is still possible in practice to approach unity by reducing the packing density [111]. However, the coating becomes less dense and rugged as its index of refraction approaches unity.

With the *a priori* physical information (13.33) and (13.34), the modified functional $\tilde{F}_\Lambda : [n_0^2, n_s^2]^{N-1} \to \mathbb{R}_+$ defined by

$$\tilde{F}_\Lambda(\underline{\epsilon}') := \sum_{l=1}^{M} \left(|u(0;\beta_l,\underline{\epsilon}') - 1| - |R_l|\right)^2$$
$$+ \left[\frac{h}{2}\left(\sqrt{\epsilon_0} + \sqrt{\epsilon_N} + 2\sum_{j=1}^{N-1}\sqrt{\epsilon_j}\right) - 1.27\right]^2 + \Lambda \sum_{j=1}^{N-1}(\epsilon_j - \epsilon_j^*)^2 \tag{13.35}$$

is minimized, where $h = 1/N = 1/20 = 0.05$, $\underline{\epsilon}' = (\epsilon_j)_{j=\overline{1,(N-1)}} \in [n_0^2, n_s^2]^{N-1}$, and the trapezoidal rule has been employed to approximate the integral in (13.33); smoother constraints can be employed if one assumes higher regularity for $n(x)$ such as $n \in C^1[0,1]$, and then one has to replace the last term in (13.35) by $\Lambda \sum_{j=1}^{N}(\epsilon_j - \epsilon_{j-1})^2$. The measured reflectance data is recorded at $M = 46$ uniform stations taken in the interval $[350, 800]$ nm, i.e. $\lambda_l = 800 - 450(l-1)/(M-1)$, $\beta_l = 2\pi d/\lambda_l$ for $l = \overline{1,M}$. The thickness of the optical coating measured by cross-sectional scanning electron microscopy is $d = 122$ nm. The initial guess is taken as the prior estimate $\underline{\epsilon}'^* = \underline{1.27^2}$.

Numerical results obtained for various regularization parameters $\Lambda \in \{10^{-5}, 10^{-4}, 10^{-3}\}$ are shown in Figures 13.4 and 13.5. In Figure 13.4, there are the results for the best fit of the reflectance $|u(0;\beta_l,\underline{\epsilon}') - 1|$ in comparison with the measurement data $|R_l|$ for $l = \overline{1,M}$. In Figure 13.5, there are the numerical results for the index of refraction $n(x)$, as a function of the depth x (in nm). By comparing Figures 13.2 and 13.4 it can be seen that a better fit of the experimentally measured reflection data is obtained when a continuously varying index of refraction shown in Figure 13.5 is sought than when this coefficient is sought as a piecewise constant function. Although the fit seems to improve as Λ becomes smaller, the instability in the numerical solution increases. This is to be expected since the well-known least-squares solution obtained by taking $\Lambda = 0$ is unstable and physically meaningless although the fit to the input data is excellent. In addition, there may be issues related to the non-uniqueness of solution when using the reflectance data only as input, as discussed in Example 13.1. From Figure 13.5, it seems that any of the solutions obtained with the regularization parameters $\Lambda \in \{10^{-5}, 10^{-4}, 10^{-3}\}$ could be taken as a good candidate for the true unknown value of the index of refraction. On displaying the norm of the solution $||\underline{\epsilon}' - \underline{\epsilon}'^*||$ versus the residual $\sqrt{\tilde{F}_\Lambda(\underline{\epsilon}') - \Lambda||\underline{\epsilon}' - \underline{\epsilon}'^*||^2}$ for various values of Λ, the resulting L-curve has been found to have a corner located near $\Lambda \approx 10^{-4}$, [276]. Thus, the curve marked with triangles $(-\Delta-)$ in Figure 13.5, corresponding to this value of $\Lambda = 10^{-4}$, is recommended as a good stable approximation of the index of refraction for the physical Example 13.2.

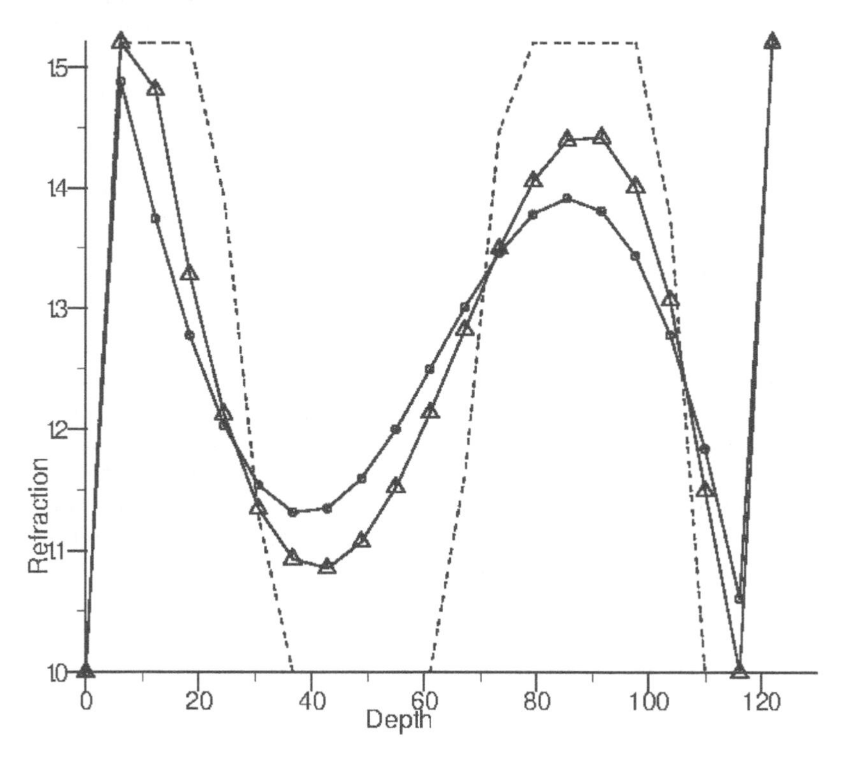

FIGURE 13.5: Refraction index profiles $n(x)$, as functions of the depth $x \in [0, d = 122]$ nm, obtained with various regularization parameters $\Lambda = 10^{-5}(- - -)$, $\Lambda = 10^{-4}(-\Delta-)$ and $\Lambda = 10^{-3}(-\circ-)$, for Example 13.2, [276].

13.3 Conclusions

In this chapter, the real-world application of inverse problems has concerned the determination of a space-dependent index of refraction of an optical anti-reflection coating structure. The mathematical model was based on solving an inverse coefficient identification problem for the one-dimensional Helmholtz equation. The additional data necessary for the inversion can be the full complex reflection coefficient or its absolute value only, measured for many wavenumbers. The numerical method was based on a finite-difference direct solver combined with a nonlinear Tikhonov regularization procedure. The choice of the regularization parameter was based on the discrepancy principle or the L-curve criterion. The need to use both the real and imaginary parts of the complex reflection coefficient in order to render a unique solution has been investigated. It was shown that, in general, the knowledge of the full complex reflection coefficient is necessary to determine uniquely a spacewise continuous index of refraction. When only the absolute value of the reflection coefficient is used as input data, constraints need to be imposed, for example, the knowledge of the full integrated refraction index, additional smoothness assumptions on the index of refraction, or more reflectance data measured for many wavelengths. Apart from this insight into the uniqueness of solution of the inverse problem, the principal conclusion is that, as shown in Figure 13.4, a better fit of the reflectance measured data is obtained by using a continuously varying index of refraction (see Figure 13.5 for $\Lambda = 10^{-4}$) than when this coefficient is sought as piecewise constant function. The use of additional reflectance data at non-normal incidence [360], the finite-dimensional parameterization of the unknown coefficient using cubic splines, or the piecewise constant layered material assumption could be useful ideas to reduce the non-uniqueness of solution of the inverse problem. Furthermore, the practical measurement of the phase shift $\phi_l := \arg(R_l)$ for $l = \overline{1, M}$, can be realized by using a combination of modulated reflectance spectroscopy, X-ray reflectivity and modulated interferometry [332].

Chapter 14

Flexural Rigidity of a Beam

Many physical phenomena can be described by equations of the type

$$\frac{\partial^2}{\partial x^2}\left(a\frac{\partial^2 u}{\partial x^2}\right)+b\frac{\partial^2 u}{\partial t^2}+c\frac{\partial u}{\partial t}-\frac{\partial}{\partial x}\left(\kappa\frac{\partial u}{\partial x}\right)+gu=f \tag{14.1}$$

Among such phenomena, both second-order equations, i.e. $a = 0$, governing heat conduction and wave propagation in solids, fluid flow through porous media, pollutant diffusion in the absence of convection, and fourth-order equations, i.e. $a > 0$, governing the Euler-Bernoulli beam theory in elasticity, are included. Equation (1) connects the physical parameters a, b, c, κ and g with the distribution of the dependent variable u and the load f.

In this chapter, the identification and reconstruction of the elastic properties of a beam, such as its heterogeneous flexural rigidity and/or mass per unit length, from deflection measurements along the beam or limited boundary measurements are considered.

14.1 Distributed Parameters in Beam-Type Systems

In practical applications, it is sometimes almost impossible to get a precise knowledge of the physical parameters of the system by experimental methods, whereas it is usually easier to take measurements of the 'potential' u. From such considerations it arises the well-known inverse problem of 'parameter identification', i.e. evaluating the physical parameter values by 'potential' measurements. Such a problem is ill-posed, since a solution, whenever it exists, need not be unique and generally does not depend continuously on the input data. The problem of uniqueness is linked to that of identifiability. A parameter is identifiable if different choices of parameters lead to different 'potential' distributions. In this setup, the identifiability is equivalent to the uniqueness of solution of the inverse problem in its direct formulation.

In (14.1), the coefficients may depend on t, x and/or u. For second-order equations, i.e. $a = 0$, results on identifiability for the space-dependent coefficients $\kappa = \kappa(x)$, $g = g(x)$ in a parabolic equation, i.e. $c = 1$ and $b = 0$, were obtained in [231]. A generalization of these results for $\kappa = \kappa(x)$, $c = c(x)$,

DOI: 10.1201/9780429400629-14

when $b = g = 0$ with applications were given in [125]. For elliptic equations, i.e. $c = b = 0$, in many dimensions, analogous results can be found in [81, 97, 347]. The identifiability of the coefficient $\kappa = \kappa(u)$ in a one-dimensional parabolic equation was also considered in [104]. Identifiability results for coefficients $\kappa = \kappa(x)$, $g = g(x)$ or $\kappa = \kappa(u)$, $g = g(u)$ in the class of problems introduced above were obtained in [129]. Generalizations to identifying coefficients of the form $\kappa = \kappa(x, u)$, $g = g(x, u)$ were considered in [130].

For fourth-order equations, i.e. $a \neq 0$, the identification of $a = a(x)$, when $c = \kappa = g = 0$, $b = 1$, was studied in [17] for constant coefficients and in [202] for a space-dependent coefficient. In this section, the (simultaneous) identifiability of the parameters $a(x)$ and/or $b(x)$ for the Euler-Bernoulli equation, i.e. $c = \kappa = g = 0$, namely,

$$\frac{\partial^2}{\partial x^2}\left(a(x)\frac{\partial^2 u}{\partial x^2}\right) + b(x)\frac{\partial^2 u}{\partial t^2} = f(x,t), \quad (x,t) \in \Omega_T := (0,l) \times (0,T) \tag{14.2}$$

is considered [251]. In (14.2), $T > 0$ is a final time of interest, l is the length of the beam, $u(x,t)$ is the deflection of the beam, $f(x,t)$ is the load to which the beam is subjected and $a(x)$ and $b(x)$ are the flexural rigidity and the mass per unit length of the beam, respectively. Since it is usually difficult to measure the physical parameters of the system (a and b), compared to measuring the deflection u and the load f, a and b are deduced from u and f by solving an ill-posed inverse problem.

In what follows, necessary and/or sufficient conditions for the identifiability, or non-identifiability, of the coefficients a and/or b in equation (14.2) are provided by extending and generalizing the approaches of [124, 125, 231] and particularising some of the results of [269].

It is natural to assume that the physical properties a and b are strictly positive and sufficiently differentiable, i.e. the space of admissible functions Λ_{ad} in which the pair (a,b) is sought is given by

$$\Lambda_{ad} = \big\{\zeta = (a,b) \in \Lambda := C^2[0,l] \times C[0,l] \mid \infty > a_{max} \geq a(x) \geq a_{min} > 0, \\ \infty > b_{max} \geq b(x) \geq b_{min} > 0, \forall x \in [0,l]\big\}$$

where a_{min}, a_{max}, b_{min} and b_{max} are some given strictly positive numbers. Let us also assume that the deflection u belongs to the admissible space of functions

$$U_{ad} = \big\{u \in U := C^{4,2}(\Omega_T) \cap C(\overline{\Omega}_T) \mid u \text{ satisfies boundary} \\ \text{and initial conditions}\big\},$$

where the boundary conditions at $x \in \{0,l\}$ may involve the specification of the primary variables u (deflection), $\frac{\partial u}{\partial x}$ (slope) and/or secondary variables $a\frac{\partial^2 u}{\partial x^2}$ (bending moment), $\frac{\partial}{\partial x}\left(a\frac{\partial^2 u}{\partial x^2}\right)$ (shear force), whilst the initial conditions involve the specification of $u(x,0)$ and $\frac{\partial u}{\partial t}(x,0)$. Define the operator

$$\Psi : \Lambda_{ad} \times U_{ad} \longrightarrow C(\Omega_T), \quad \Psi(\zeta, u) := \frac{\partial^2}{\partial x^2}\left(a\frac{\partial^2 u}{\partial x^2}\right) + b\frac{\partial^2 u}{\partial t^2}, \tag{14.3}$$

where $\zeta = (a, b)$. If the direct problem has a unique solution, i.e. given $\zeta \in \Lambda_{ad}$ and $f \in C(\Omega_T)$ there is a unique $u \in U_{ad}$ satisfying $\Psi(\zeta, u) = f$, then for each $f \in C(\Omega_T)$, the surjective operator

$$\Phi_f : \Lambda_{ad} \longrightarrow \Phi_f(\Lambda_{ad}) \subset U_{ad}, \quad \Phi_f(\zeta) = u \tag{14.4}$$

can be defined. Then, the unique solvability of the inverse problem, i.e. given $f \in C(\Omega_T)$ and $u \in \Phi_f(\Lambda_{ad})$ there is a unique solution $\zeta \in \Lambda_{ad}$ satisfying (14.4), is equivalent to the operator Φ_f having a unique inverse. Of course, if $f \equiv 0$ and $u \in \Phi_0(\Lambda_{ad})$ is a linear function of x and t, then from (14.2), this will not be the case. Therefore, one should look instead for a subset $U_0 \subset \Phi_f(\Lambda_{ad})$ on which the inverse operator

$$\Phi_f^{-1} : U_0 \longrightarrow \Phi_f^{-1}(U_0) \subset \Lambda_{ad}, \quad \Phi_f^{-1}(u) = \zeta \tag{14.5}$$

can be defined. This formulation leads to the folowing definition of (conditional) identifiability [125].

Definition 14.1 *The parameter $\zeta_1 = (a_1, b_1) \in \Lambda_{ad}$ is said to be (conditionally) identifiable on a set U_0 if given $\Phi_f(\zeta_1) = u_1 \in U_0$ and $\Phi_f(\zeta_2) = u_2 \in U_0$ for a $\zeta_2 = (a_2, b_2) \in \Lambda_{ad}$, from $u_1 = u_2$ it follows that $a_1 = a_2$ and $b_1 = b_2$.*

Based on this definition and using equation (14.3), the complete characterization of the set U_0 for identifiability is given by

$$U_0(\zeta) = \{u \in U_{ad} \mid \Psi(\zeta, u) = 0 \;\Rightarrow\; \zeta = 0\}. \tag{14.6}$$

Individual (single) identifiability of a or b on the sets $U_0(a)$ or $U_0(b)$ follows naturally from (14.6) as

$$\begin{aligned} U_0(a) &= \{u \in U_{ad} \mid \Psi((a, 0), u) = 0 \;\Rightarrow\; a = 0\}, \\ U_0(b) &= \{u \in U_{ad} \mid \Psi((0, b), u) = 0 \;\Rightarrow\; b = 0\}. \end{aligned} \tag{14.7}$$

Lemma 14.1. ([251]) *If u_1, $u_2 \in U_{ad}$, the equality $u_1(x, t) = u_2(x, t)$ for all $(x, t) \in \overline{\Omega}_T$ holds if and only if*

$$\frac{\partial^2}{\partial x^2}\left(q(x)\frac{\partial^2 u_1}{\partial x^2}\right) + p(x)\frac{\partial^2 u_1}{\partial t^2} = 0, \quad \forall (x, t) \in \Omega_T, \tag{14.8}$$

where

$$q(x) = a_1(x) - a_2(x), \quad p(x) = b_1(x) - b_2(x). \tag{14.9}$$

Proof. Let $u_1(x, t) = u_2(x, t)$ for all $(x, t) \in \overline{\Omega}_T$. Then, from (14.1) it follows that $\Psi(\zeta_i, u_i) = f$ for $i = 1, 2$. Subtracting $\Psi(\zeta_1, u_1)$ from $\Psi(\zeta_2, u_2)$ and setting $u_1 = u_2$, equation (14.8) is obtained. Conversely, let (14.8) hold. The function $u_2 \in U_{ad}$ satisfies (14.1) with the coefficients a_2 and b_2 and $u_1 \in U_{ad}$ satisfies the same initial and boundary conditions as u_2, i.e. $u_1 - u_2 \in U_{ad}$ satisfying homogeneous initial and boundary conditions. From

(14.8) and since $u_1 \in U_{ad}$ satisfies (14.1) with the coefficients a_1 and b_1 it follows that $\Psi(\zeta_2, u_1) = \Psi(\zeta_1, u_1) = f$. However, $\Psi(\zeta_2, u_2) = f$ and hence, $\Psi(\zeta_2, u_1) = f = \Psi(\zeta_2, u_2)$. Finally, from the uniqueness of the direct problem for the coefficient ζ_2 it yields that $u_1 = u_2$.

Let us now introduce the following sets:

$$A_i(t) := \left\{ x \in [0, l] \middle| \frac{\partial^i u}{\partial x^i}(x, t) = 0 \right\}, \quad i = 2, 3, 4, \tag{14.10}$$

$$B(t) := \left\{ x \in [0, l] \middle| \frac{\partial^2 u}{\partial t^2}(x, t) = 0 \right\}. \tag{14.11}$$

Throughout what follows let $u_1(x, t) = u_2(x, t) := u(x, t)$ for all $(x, t) \in \overline{\Omega}_T$ and the superscript ^{c} denote the complementary of a set, e.g. $A_i^c(t) = [0, l] \backslash A_i(t)$, etc. Also, when only the coefficient a is identified the coefficient b is taken to be zero and vice-versa.

14.1.1 Identifiability results for the constant case

Theorem 14.1. ([251]) *(a) The single constant parameter a or b is identifiable if and only if there exists $(x_0, t_0) \in \Omega_T$ such that $\frac{\partial^4 u}{\partial x^4}(x_0, t_0) \neq 0$ or $\frac{\partial^2 u}{\partial t^2}(x_0, t_0) \neq 0$, respectively.*

(b) The constant parameters a and b are simultaneously identifiable if and only if there exist $(x_i, t_i) \in \Omega_T$ for $i = 1, 2$ such that

$$\frac{\partial^4 u}{\partial x^4}(x_1, t_1) \frac{\partial^2 u}{\partial t^2}(x_2, t_2) \neq \frac{\partial^4 u}{\partial x^4}(x_2, t_2) \frac{\partial^2 u}{\partial t^2}(x_1, t_1). \tag{14.12}$$

Proof. (a) For the identifiability of the the single constant parameter a, from (14.8) it follows that $q \frac{\partial^4 u}{\partial x^4}(x, t) = 0$. Thus, $q = 0$, i.e. a is identifiable, if and only if there exists $(x_0, t_0) \in \Omega_T$ such that $\frac{\partial^4 u}{\partial x^4}(x_0, t_0) \neq 0$. Similarly, for the identifiability of the single constant parameter b.

(b) Applying (14.8) at two points $(x_i, t_i) \in \Omega_T$ for $i = 1, 2$, yield

$$q \frac{\partial^4 u}{\partial x^4}(x_i, t_i) + p \frac{\partial^2 u}{\partial t^2}(x_i, t_i) = 0, \quad i = 1, 2. \tag{14.13}$$

This homogeneous system of equations has only the trivial solution $q = p = 0$, i.e. a and b are simultaneously identifiable, if and only if (14.12) holds.

14.1.2 Identifiability results for the space-dependent case

Theorem 14.2. ([251]) (single identifiability of $a(x)$)
(a) The single parameter $a(x)$ is identifiable if one of the following conditions is satisfied:

(i) There exists $t_0 \in (0,T)$ *such that* $A_2(t_0) \cap A_3(t_0) \neq \emptyset$ *and* $\overline{A_2^c(t_0)} = [0,l]$.
(ii) $A_2(t) \cap A_3(t) \neq \emptyset$ *for all* $t \in (0,T)$ *and* $\overline{\cup_{t\in(0,T)} A_2^c(t)} = [0,l]$.
(iii) The functions $\frac{\partial^i u}{\partial x^i}(x,t)$ *for* $i = \overline{2,4}$ *are linearly independent as functions of* t *for all* x *in a dense subset in* $[0,l]$.

(b) If there exists $t_0 \in (0,T)$ *such that* $A_2(t_0) = \emptyset$, *then* $a(x)$ *is identifiable if and only if there exists* $x_0 \in [0,l]$ *such that* $q(x_0) = q'(x_0) = 0$.

Proof. If $b = 0$ then $p = 0$ and integrating (14.8) with respect to x yields that there is a function $F(t)$ such that

$$F(t) = \frac{\partial}{\partial x}\left(q(x)\frac{\partial^2 u}{\partial x^2}(x,t)\right) = q'(x)\frac{\partial^2 u}{\partial x^2}(x,t) + q(x)\frac{\partial^3 u}{\partial x^3}(x,t), \quad \forall (x,t) \in \Omega_T.$$

(a) (i) Since $A_2(t_0) \cap A_3(t_0) \neq \emptyset$, it follows that $F(t_0) = 0$ and

$$\frac{\partial}{\partial x}\left(q(x)\frac{\partial^2 u}{\partial x^2}(x,t_0)\right) = 0, \quad \forall x \in (0,l). \tag{14.14}$$

Integrating equation (14.14) and using that $A_2(t_0) \neq \emptyset$ yield

$$q(x)\frac{\partial^2 u}{\partial x^2}(x,t_0) = 0, \quad \forall x \in (0,l). \tag{14.15}$$

From (14.15) and $\overline{A_2^c(t_0)} = [0,l]$, the set $\{x \in [0,l] \mid q(x) = 0\}$ is dense in $[0,l]$ and, since q is continuous, $q \equiv 0$, i.e. $a(x)$ is identifiable.

(ii) Since $A_2(t_0) \cap A_3(t_0) \neq \emptyset$ for all $t > 0$, similarly as in (i),

$$q(x)\frac{\partial^2 u}{\partial x^2}(x,t) = 0, \quad \forall (x,t) \in \Omega_T. \tag{14.16}$$

Further if we let $\mathcal{M} := \cup_{t\in(0,T)} A_2^c(t)$, then, for any $x \in \mathcal{M}$, there is $t_x \in (0,T)$ such that $x \in A_2^c(t_x)$, i.e. $u_{xx}(x,t_x) \neq 0$ and thus, from (14.16), $q(x) = 0$ for all $x \in \mathcal{M}$. Finally, from the continuity of q and since $\overline{\mathcal{M}} = [0,l]$, it follows that $q \equiv 0$, i.e. $a(x)$ is identifiable.

(iii) From Lemma 14.1,

$$q''(x)\frac{\partial^2 u}{\partial x^2}(x,t) + 2q'(x)\frac{\partial^3 u}{\partial x^3}(x,t) + q(x)\frac{\partial^4 u}{\partial x^4}(x,t) = 0. \tag{14.17}$$

From the linear independence of the functions $\frac{\partial^i u}{\partial x^i}(x,t)$ for $i = \overline{2,4}$, it follows that $q(x) = q'(x) = q''(x) = 0$ on a dense set in $[0,l]$, and by continuity, $q(x) \equiv 0$ for all $x \in [0,l]$, i.e. $a(x)$ is identifiable.

(b) If there exists $t_0 \in (0,T)$ such that $A_2(t_0) = \emptyset$, then for all $x_0 \in [0,l]$,

$$q(x) = \frac{u_{xx}(x_0,t_0)(q'(x_0)x - x_0 q'(x_0) + q(x_0)) + u_{xxx}(x_0,t_0)(x-x_0)q(x_0)}{u_{xx}(x,t_0)}$$

is a solution of (14.8). From this, $q(x) \equiv 0$, i.e. $a(x)$ is identifiable, if and only if there exists $x_0 \in [0, l]$ such that $q(x_0) = q'(x_0) = 0$.

Theorem 14.3. ([251]) (single non-identifiability of $a(x)$)
The single parameter $a(x)$ is not identifiable if one of the following conditions is satisfied:
(i) The set $\mathcal{M} := \cup_{t\in(0,T)} A_2^c(t)$ is not dense in $[0, l]$.
(ii) There exists $t_0 \in (0, T)$ such that $A_2(t_0) = \emptyset$ and the derivative $u_{xx}(x, t)$ can be represented as $v(x)(w(t)x + z(t))$ or, in particular, $u(x, t)$ is separable.

Proof. (i) Since $\overline{\mathcal{M}} \neq [0, l]$ there is an open interval centered at a point, say $x_0 \in (0, l)$, such that $J := (x_0 - \epsilon, x_0 + \epsilon) \subset [0, l]\backslash\overline{\mathcal{M}}$. Let s be a twice continuously differentiable positive function such that its support $\text{supp}(s) := \{x \in [0, l] \mid s(x) \neq 0\} = J$ and let us construct $a_1(x) := a_2(x) + s(x)$. Then,

$$u_{xx}(x, t) = 0, \quad \forall x \in J \times (0, T) \tag{14.18}$$

since $J \subset [0, l]\backslash\overline{\mathcal{M}} \subset [0, l]\backslash\mathcal{M} = \cap_{t\in(0,T)} A_2(t)$. Also, since $\text{supp}(s) = J$,

$$s(x)u_{xx}(x, t) = 0, \quad \forall x \in J^c \times (0, T). \tag{14.19}$$

Thus, by Lemma 14.1, $u(x, t) = 0$ for all $(x, t) \in \Omega_T$. However, $a_1(x_0) \neq a_2(x_0)$ since $s(x_0) \neq 0$ and thus, $a(x)$ is not identifiable.

(ii) If $A_2(t_0) = \emptyset$ then $v(x) \neq 0$, $\forall x \in [0, l]$. Let $a_1(x) := a_2(x) + \frac{\alpha}{v(x)}$, where $\alpha = 1$ if $v(x) > 0$ and $\alpha = -1$ if $v(x) < 0$. Then, $a_1(x) > 0$ and $a_1(x) \neq a_2(x)$, $\forall x \in [0, l]$. However, $\frac{\partial^2}{\partial x^2}(q(x)v(x)(w(t)x + z(t))) = 0$ and from Lemma 14.1, $u \equiv 0$ and $a(x)$ is not identifiable. The separable case $u(x, t) = V(x)Z(t)$ is a particular form of representing $u_{xx}(x, t)$ as $V''(x)Z(t)$.

From Theorems 14.2(ii) and 14.3(i), the following corollary is obtained.

Corollary 14.1. *If $A_2(t) \cap A_3(t) \neq \emptyset$ for all $t \in (0, T)$, then $a(x)$ is identifiable if and only if $\overline{\cup_{t\in(0,T)} A_2^c(t)} = [0, l]$.*

Theorem 14.4. ([251]) (single identifiability of $b(x)$)
The single parameter $b(x)$ is identifiable if and only if $\overline{\cup_{t\in(0,T)} B^c(t)} = [0, l]$.

Proof. If $a = 0$ then $q \equiv 0$, and from (14.8), $p(x)u_{xx}(x, t) = 0$. On letting $\mathcal{N} := \cup_{t\in(0,T)} B^c(t)$, then for any $x \in \mathcal{N}$ there is $t_x \in (0, T)$ such that $x \in B^c(t_x)$, i.e. $u_{xx}(x, t_x) \neq 0$ and thus, $p(x) = 0$, $\forall x \in \mathcal{N}$. For necessity, from the continuity of p and the density of $\mathcal{N}$ in $[0, l]$, $p \equiv 0$, i.e. $b(x)$ is identifiable. The proof of sufficiency is similar to that of Theorem 14.3(i).

Theorem 14.5. (simultaneous identifiability of $a(x)$ and $b(x)$)
The parameters $a(x)$ and $b(x)$ are simultaneously identifiable if the functions $\frac{\partial^i u}{\partial x^i}(x, t)$ for $i = \overline{2, 4}$ and $\frac{\partial^2 u}{\partial t^2}(x, t)$ are linearly independent as functions of t

for all x in a dense subset in $[0, l]$.

Proof. The proof is similar to that of Theorem 14.2(iii).

Results for the simultaneous identifiability of both $a(x)$ and $b(x)$ which are stronger than that of Theorem 14.5 are more difficult to establish hence they are analyzed next through one example and linked with the single identifiability of $a(x)$ and $b(x)$ separately. Based on the identifiability sets defined by equations (14.6) and (14.7), the following remark can be made [125].

Remark 14.1. *(i) If $\zeta = (a, b)$ is identifiable on $U_0(\zeta)$, then $U_0(\zeta) \subset U_0(a) \cap U_0(b)$.*
(ii) If a is identifiable on $U_0(a)$ and b is identifiable on $U_0(b)$, then $\zeta = (a, b)$ is not identifiable on $U_0^c(a) \cap U_0^c(b)$.

One could now think that from the identifiability of a on $U_0(a)$ and of b on $U_0(b)$, there follows the identifiability of ζ on $U_0(a) \cap U_0(b)$. However, this is not the case, as it can be shown from the following example.

Example 14.1 Let us take $l = T = 1$ and $u(x, t) = e^t(x^4 + 1)$. Then, $a(x)$ is single identifiable since $A_2(t) \cap A_3(t) = \{0\}$ and $A_2^c(t) = (0, 1]$, $\forall t \in (0, T)$, see Theorem 14.2, and $b(x)$ is identifiable since $B(t) = \emptyset$, $\forall t \in (0, T)$, see Theorem 14.4. However, $q(x) = x^4 + 15$ and $p(x) = -360$ is a non-zero solution of equation (14.8) and thus, $a(x)$ and $b(x)$ are not simultaneously identifiable.

So the problem of simultaneous identifiability has to be examined independently from that of the single identifiability of $a(x)$ or $b(x)$, as described next by extending the study of [125]. Let us denote

$$\Lambda_{ad}^1 = \left\{a \in C^2[0, l] \mid \infty > a_{max} \geq a(x) \geq a_{min} > 0, \forall x \in [0, l]\right\},$$
$$\Lambda_{ad}^2 = \{b \in C[0, l] \mid \infty > b_{max} \geq b(x) \geq b_{min} > 0, \forall x \in [0, l]\},$$

in which case $\Lambda_{ad} = \Lambda_{ad}^1 \times \Lambda_{ad}^2$, and let us define the operators

$$\psi_x : \Lambda_{ad}^1 \times U_{ad} \longrightarrow C(\Omega_T), \quad \psi_x(a, u) := \frac{\partial^2}{\partial x^2}\left(a(x)\frac{\partial^2 u}{\partial x^2}\right),$$
$$\psi_t : \Lambda_{ad}^2 \times U_{ad} \longrightarrow C(\Omega_T), \quad \psi_t(b, u) := -b(x)\frac{\partial^2 u}{\partial t^2},$$

in which case $\Psi(\zeta, u) = \psi_x(a, u) - \psi_t(b, u)$.

Remark 14.2. *Let us assume that there exists an operator $\Theta : C(\Omega_T) \longrightarrow C(\Omega_T)$ such that for any $u \in \hat{U} \subset U_0(a) \cap U_0(b)$ the following properties hold:*
(i) $\Theta\psi_t(p, u) = 0$;
(ii) $\Theta u \in U_{ad}$ and $\Theta\psi_x(q, u) = \psi_x(q, \Theta u)$;
(iii) $\psi_x(q, \Theta u) = 0$ implies $q = 0$.

Then, $a(x)$ and $b(x)$ are simultaneously identifiable on $\hat{U}$.

Indeed, let us take $u \in \hat{U} \subset U_0(a) \cap U_0(b)$ such that $\psi_t(p,u) = \psi_x(q,u)$. Applying to this relation the operator Θ and using (i) and (ii) give $0 = \Theta\psi_t(p,u) = \psi_x(q,\Theta u)$. Using (iii) implies $q = 0$, i.e. $a(x)$ is identifiable. Also, $\psi_t(p,u) = 0$ and since $u \in U_0(b)$, it follows that $p = 0$, i.e. $b(x)$ is identifiable. Therefore, $a(x)$ and $b(x)$ are simultaneously identifiable on $\hat{U}$.

14.1.3 Steady-state analysis

Stronger results can be obtained in the steady case of the Euler-Bernoulli fourth-order ordinary differential equation

$$\frac{d^2}{dx^2}\left(a(x)\frac{d^2u}{dx^2}\right) = f(x), \quad x \in (0,l). \tag{14.20}$$

When the load $f(x)$ is strictly positive and bounded, identifiability holds [268], when $a \in \{a \in L^\infty(0,l) \mid \infty > a_{max} \geq a(x) \geq a_{min} > 0, \|a'\| \leq E < \infty\}$. Otherwise, without these hypotheses, the following theorem holds. Its proof follows a similar method to that in Theorems 14.2 and 14.3.

Theorem 14.6. ([251]) (identifiability and non-identifiability of $a(x)$)
(i) The parameter $a(x)$ is identifiable if $A_2 \cap A_3 \neq \emptyset$ and $\overline{A_2^c} = [0,l]$. Therefore, a is identifiable over the set $U_{0s} := \{u \in U_{ad} \mid A_2 \cap A_3 \neq \emptyset,\ \overline{A_2^c} = [0,l]\}$.
(ii) Alternatively, if $A_2 = \emptyset$, then $a(x)$ is identifiable if and only if there exists $x_0 \in [0,l]$ such that $q(x_0) = q'(x_0) = 0$.
(iii) The parameter $a(x)$ is not identifiable if $A_2 = \emptyset$.

Remark 14.3. Let $\hat{a}(x)$ be the flexural rigidity coefficient that is desired to determine over the set U_{0s}. Then, the unique solution, which can be obtained by direct integration of equation (14.20) between an $x_0 \in A_2 \cap A_3 \neq \emptyset$ and x in a neighbourhood of x_0, reads as

$$\hat{a}(x) = \frac{G_2(x) - G_2(x_0) - (x - x_0)G_1(x_0)}{u''(x)}, \quad x \in (0,l), \tag{14.21}$$

where

$$G_1(x) = \int_0^x f(\xi)d\xi, \quad G_2(x) = \int_0^x G_1(\xi)d\xi. \tag{14.22}$$

Example 14.2 Let $u(x) = (x - l/2)^4$ and $f(x) = 24(3x + l)$. Then, $A_2 = A_3 = \{l/2\}$ and the conditions of Theorem 14.6(i) for the identifiability of $a(x)$ are satisfied. Using (14.21) and (14.22) with $x_0 = l/2$ yield

$$G_1(x) = 12(3x^2 + 2xl), \quad G_2(x) = 12(x^3 + x^2 l), \quad \hat{a}(x) = 2l + x. \tag{14.23}$$

When $A_2 \cap A_3 = \emptyset$ or $\overline{A_2^c} \subset [0,l]$ strictly, one needs more information to restore identifiability. In particular, this could be done by using two independent sets of data, i.e. measuring the deflections $u_1(x)$ and $u_2(x)$ for the same flexural rigidity coefficient $a(x)$, but for two linearly independent load functions $f_1(x)$ and $f_2(x)$, respectively. Without giving a new rigorous definition, one can say that two sets of data (u_1, f_1) and (u_2, f_2), i.e. formally $\Phi_{f_1}(a) = u_1$ and $\Phi_{f_2}(a) = u_2$, lead to the identifiability of $a(x)$ if and only if from

$$\frac{d^2}{dx^2}\left(q(x)\frac{d^2u_1}{dx^2}\right) = \frac{d^2}{dx^2}\left(q(x)\frac{d^2u_2}{dx^2}\right) = 0. \tag{14.24}$$

it follows that $q \equiv 0$.

Theorem 14.7. ([251]) *If f_1 and f_2 are linearly independent loads over any open sub-interval of $(0,l)$ such that f_i, xf_i and $F_i := \int f_i dx$ for $i = 1,2$ are linearly independent on $(0,l)$, then the flexural rigidity $a(x)$ is identifiable.*

Proof. From (14.24), there are some constants c_1, c_2, d_1 and d_2 such that

$$\frac{d}{dx}\left(q(x)\frac{d^2u_i}{dx^2}(x)\right) = \mathrm{d}_i, \quad q(x)\frac{d^2u_i}{dx^2}(x) = \mathrm{c}_i + \mathrm{d}_i x, \quad i = 1,2, \tag{14.25}$$

and thus

$$\begin{cases} \mathrm{d}_2\frac{d}{dx}\left(q(x)\frac{d^2u_1}{dx^2}(x)\right) - \mathrm{d}_1\frac{d}{dx}\left(q(x)\frac{d^2u_2}{dx^2}(x)\right) = 0, \\ q(x)\left[(\mathrm{c}_2 + \mathrm{d}_2 x)\frac{d^2u_1}{dx^2}(x) - (\mathrm{c}_1 + \mathrm{d}_1 x)\frac{d^2u_2}{dx^2}(x)\right] = 0. \end{cases} \tag{14.26}$$

On integration, it follows that there is a constant c such that

$$\begin{cases} q(x)\left[\mathrm{d}_2\frac{d^2u_1}{dx^2}(x) - \mathrm{d}_1\frac{d^2u_2}{dx^2}(x)\right] = \mathrm{c}, \\ q(x)\left[\mathrm{c}_2\frac{d^2u_1}{dx^2}(x) - \mathrm{c}_1\frac{d^2u_2}{dx^2}(x)\right] = -\mathrm{c}x. \end{cases} \tag{14.27}$$

If $q(x) \neq 0$ for all $x \in (0,l)$, then,

$$(\mathrm{c}_2 + \mathrm{d}_2 x)\frac{d^2u_1}{dx^2}(x) - (\mathrm{c}_1 + \mathrm{d}_1 x)\frac{d^2u_2}{dx^2}(x) = 0, \quad \forall x \in (0,l).$$

Multiplying this equation with $a(x)$ and differentiating give

$$\mathrm{d}_2 a(x)\frac{d^2u_1}{dx^2}(x) + (\mathrm{c}_2 + \mathrm{d}_2 x)F_1(x) - \mathrm{d}_1 a(x)\frac{d^2u_2}{dx^2}(x) - (\mathrm{c}_1 + \mathrm{d}_1 x)F_2(x) = 0.$$

Further, on differentiating this equation gives

$$2\mathrm{d}_2 F_1(x) - 2\mathrm{d}_1 F_2(x) + (\mathrm{c}_2 + \mathrm{d}_2 x)f_1(x) - (\mathrm{c}_1 + \mathrm{d}_1 x)f_2(x) = 0, \quad \forall x \in (0,l).$$

Since f_i, xf_i and F_i for $i = 1,2$ are linearly independent it follows that $c_i = d_i = 0$, $i = 1,2$. Then, since $q(x) \neq 0$ for all $x \in (0,l)$, from (14.25) it follows that $\frac{d^2u_i}{dx^2}(x) = 0$, $i = 1,2$, i.e. $f_1 = f_2 = 0$, which contradicts the linear independence of f_1 and f_2. Thus, there exists an $x_0 \in (0,l)$ such that $q(x_0) = 0$ and then applying (14.27) at $x = x_0$, it implies that c $= 0$. Let us assume, by contradiction, that $q(x) \not\equiv 0$ on $(0,l)$. Then, using (14.27), there is an open interval $x_0 \notin I \subset (0,l)$ on which $q(x) \neq 0$, and hence

$$\mathrm{d}_2 \frac{d^2 u_1}{dx^2}(x) - \mathrm{d}_1 \frac{d^2 u_2}{dx^2}(x) = 0 = \mathrm{c}_2 \frac{d^2 u_1}{dx^2}(x) - \mathrm{c}_1 \frac{d^2 u_2}{dx^2}(x), \quad \forall x \in I. \tag{14.28}$$

From the linear independence of f_1 and f_2 over I it follows that $\mathrm{c}_i = \mathrm{d}_i = 0$, $i = 1, 2$. Then, since $q(x) \neq 0$ for all $x \in I$, from (14.25), it follows that $\frac{d^2 u_i}{dx^2}(x) = 0$, $i = 1, 2$, i.e. $f_1 = f_2 = 0$ over I, which contradicts the linear independence of f_1 and f_2 over I. Hence, $q \equiv 0$, i.e. $a(x)$ is identifiable.

Next, the analysis is extended by considering slightly different hypotheses which allow us to define more general loading conditions and wider variability of the flexural rigidity function. We enlarge the class of admissible functions Λ_{ad} for identification, namely, let us seek $a \in \Lambda_{ad}^{ext}$, where

$$\Lambda_{ad}^{ext} := \{a \in L^\infty(0,l) \mid \infty > a_{max} \geq a(x) \geq a_{min} > 0 \text{ a.e. } (0,l)\},$$

so that discontinuous flexural rigidity functions can also be identified. Furthermore, in the next two subsections, boundary conditions which correspond to physical cases of clamped-clamped and clamped-free beams are considered.

14.1.3.1 Clamped-clamped beams

In this configuration, the beam is subjected to the boundary conditions

$$u(0) = u_0, \quad \frac{du}{dx}(0) = u_0', \quad u(l) = u_l, \quad \frac{du}{dx}(l) = u_l', \tag{14.29}$$

where u_0, u_0', u_l and u_l' are given constants. For clamped beams at both ends, $u_0 = u_0' = u_l = u_l' = 0$, so the boundary conditions (14.29) are more general. In order to define a weak solution of equation (14.20) subject to the boundary conditions (14.29), the space of functions

$$H_0^2(0,l) := \left\{v \in H^2(0,l) \mid v(0) = v'(0) = v(l) = v'(l) = 0\right\}$$

is introduced, and it is assumed that $f \in H_0^2(0,l)'$ (the dual space of $H_0^2(0,l)$) so that point loads are allowed.

Definition 14.2. (Statement of the forward problem (FP)).
Given $a \in \Lambda_{ad}^{ext}$ and $f \in H_0^2(0,l)'$, find $u \in H^2(0,l)$ such that equation (14.20) *is satisfied in a weak sense, i.e.*

$$\int_0^L a(x) u''(x) v''(x) dx = \int_0^l f(x) v(x) dx, \quad \forall v \in H_0^2(0,l), \tag{14.30}$$

subject to the clamped (Dirichlet) beam boundary conditions (14.29).

The FP requires a to be known. However, sometimes it is easier to measure the deflection u, whereas the flexural rigidity is generally unknown. Therefore, the identification inverse problem, i.e. finding a from u and f, arises as follows.

Definition 14.3. (Statement of the inverse problem (IP)).
Given $f \in H_0^2(0,l)'$ and $u \in H^2(0,l)$, find $a \in \Lambda_{ad}^{ext}$ such that u is the solution of the corresponding FP.

Let $u \in C^1[0,l]$ have a piecewise continuous second-order derivative. Then there exists on $(0,l)$ a lattice of $N+1$ points $\{0 = \alpha_0 < \alpha_1 < \dots < \alpha_N = l\}$ such that $u \in C^2(I_n)$ for $n = \overline{0,(N-1)}$, where $I_n = (\alpha_n, \alpha_{n+1})$. Define

$$D_n := \left\{x \in I_n \mid \lim_{\xi \in I_n, \xi \to x} u''(\xi) = 0\right\}, \quad n = \overline{0,(N-1)}$$

and $D := \cup_{n=0}^{N-1} D_n$, and let μ denote the Lebesgue measure.

Theorem 14.8. *Let $u \in H^2(0,l)$ have a piecewise continuous second-order derivative in $(0,l)$. Then $a \in \Lambda_{ad}^{ext}$ is not identifiable if $D = \emptyset$ or $\mu(D) > 0$.*

Proof. The proof is very similar to the analysis of [124] for diffusive-like systems. Finally, remark that $D \neq \emptyset$ if $u_0' = u_l'$ in (14.29).

Example 14.3 Take for simplicity $l = 1$ and let

$$a(x) = \begin{cases} 1, & x \in (0,1/2], \\ 2, & x \in (1/2,1), \end{cases} \tag{14.31}$$

which belongs to Λ_{ad}^{ext}, and $f(x) = \alpha\chi_{[1/2,1]}(x) + \beta\delta(x - 1/2)$, where $\chi_{[1/2,1]}$ is the characteristic function of the interval $[1/2,1]$, δ is the Dirac delta function and α and β are given numbers. Then, equation (14.30) gives

$$\begin{aligned} &\int_0^{1/2} u''(x)v''(x)dx + 2\int_{1/2}^1 u''(x)v''(x)dx \\ &= \alpha \int_{1/2}^1 v(x)dx + \beta v(1/2), \quad \forall v \in H_0^2(0,1) \end{aligned} \tag{14.32}$$

and the solution u of the FP can be found by seeking

$$u(x) = \begin{cases} a_0 + b_0 x + c_0 x^2, & x \in (0,1/2], \\ a_1 + b_1 x + c_1 x^2 + d_1 x^3, & x \in (1/2,1), \end{cases} \tag{14.33}$$

as a solution of (14.32) satisfying the boundary conditions (14.29) with, say

$$u_0 = u_0' = u_1 = 0, \quad u_1' = 1. \tag{14.34}$$

This yields

$$a_0 = b_0 = a_1 + b_1 + c_1 + d_1 = 0, \quad b_1 + 2c_1 + 3d_1 = 1 \tag{14.35}$$

and on imposing that $u \in C^1(0,1)$ give

$$a_1 + b_1/2 + c_1/4 + d_1/8 = c_0/4, \quad b_1 + c_1 + 3d_1/4 = c_0. \tag{14.36}$$

Taking $v(x) = x^2(x-1)^2 \in H_0^2(0,1)$ in equation (14.32) gives

$$d_1 = \alpha/45 + \beta/12. \tag{14.37}$$

Solving the system (14.35)–(14.37), the solution (14.33) is obtained as

$$u(x) = \begin{cases} (-1/2 + d_1/8)x^2, & x \in (0,1/2], \\ (1/2 - 3d_1/8) + (-2 + 7d_1/4)x & \\ \quad +(3/2 - 19d_1/8)x^2 + d_1x^3, & x \in (1/2,1). \end{cases} \tag{14.38}$$

In this case, $I_0 = (0,1/2)$, $I_1 = (1/2,1)$ and $u \in C^1(I_k)$, $k = 0,1$. If $d_1 = 4$ then $\mu(D) = 1/2 > 0$, or if $d_1 \in [-12/5, 12/7]$ then $D = \emptyset$, and in these cases from Theorem 14.8 it follows that $a(x)$ is not identifiable. In the other cases, i.e. $d_1 \in (-\infty, -12/5) \cup (12/7, 4) \cup (4, \infty)$, $\mu(D) = 0$ and $D = \{(19d_1 - 12)/(24d_1)\} \neq \emptyset$, but it is not clear from Theorem 14.8 whether a is identifiable or not. However, a closer look at the weak form (14.30) with u given by (14.38), which requires seeking other solutions $\overline{a}(x)$ of the IP such that

$$(\overline{a}(x) - a(x))u''(x) = d_3x + c_3, \tag{14.39}$$

where d_3 and c_3 are arbitrary constants and $a(x)$ is given by (14.31), yields

$$\overline{a}(x) = \begin{cases} 1 + (a_3 - 2)(12 - 19d_1 + 24d_1x)/(d_1 - 4), & x \in (0,1/2], \\ a_3, & x \in (1/2,1), \end{cases} \tag{14.40}$$

where a_3 is to be chosen such that $\overline{a} \in \Lambda_{ad}^{ext}$. This recasts as the condition $\overline{a} > 0$, which based on (14.40) results in

$$a_3 \in \begin{cases} (0, (28 - 39d_1)/(12 - 19d_1)], & d_1 \in (-\infty, -12/5) \cup (4, \infty), \\ ((28 - 39d_1)/(12 - 19d_1), \infty), & d_1 \in (12/7, 4). \end{cases} \tag{14.41}$$

This means that any $\overline{a}$ as given by equations (14.40) and (14.41) will satisfy equation (14.30) and hence, a is not identifiable for all values of d_1. Finally, remark that $\overline{a} = a$ if and only if $a_3 = 2$, which is also included in (14.40).

14.1.3.2 Clamped-free beams

In this configuration, the beam is subjected to the mixed boundary conditions

$$u(0) = u_0, \quad \frac{du}{dx}(0) = u_0', \quad a(l)\frac{d^2u}{dx^2}(l) = 0, \quad \frac{d}{dx}\left(a(x)\frac{d^2u}{dx^2}(x)\right)\Big|_{x=l} = 0. \tag{14.42}$$

When $u_0 = u_0' = 0$ such an example was considered in [202]. Similarly to the previous case, introduce the space of functions

$$V := \{v \in H^2(0,l) \mid v(0) = v'(0) = 0\} \tag{14.43}$$

and, based on the Definitions 14.2 and 14.3 with $H_0^2(0,l)$ replaced by V, state the following theorem.

Theorem 14.9. ([251]) *Let $u \in H^2(0,l)$ have a piecewise continuous second-order derivative in $(0,l)$. Then, $a \in \Lambda_{ad}^{ext}$ is identifiable if and only if $\mu(D)=0$.*

Proof. First, if $\mu(D)>0$ then

$$\overline{a} = \begin{cases} a & \text{in } (0,l)\backslash D, \\ \tilde{a} & \text{in } D, \end{cases} \tag{14.44}$$

is a solution of the IP, for every $\tilde{a} \in \Lambda_{ad}^{ext}$. Thus, a would not be identifiable. Let us now assume that $\mu(D) - 0$, and remark that based on the bending moment zero boundary condition in (14.42), it follows that $D \neq \emptyset$. Then,

$$\forall \epsilon > 0, \ \exists \delta_0 > 0 \text{ such that } |u''(x)| < \epsilon, \ \forall x \in (l-\delta_0, l). \tag{14.45}$$

Let $\overline{a} \in \Lambda_{ad}^{ext}$ be another solution of the IP. Then, using (14.30),

$$\int_0^l a(x)u''(x)v''(x)dx = \int_0^l \overline{a}(x)u''(x)v''(x)dx = \int_0^l f(x)v(x)dx, \ \forall v \in V.$$

Since u satisfies the boundary conditions (14.42) and since $v \in V$ is arbitrary,

$$\frac{d}{dx}\left((a(x)-\overline{a}(x))u''(x)\right) = \mathrm{d} = \text{constant} \Rightarrow (a(x)-\overline{a}(x))u''(x) = \mathrm{d}x + \mathrm{c}.$$

The shear force boundary condition in (14.42) implies that $\mathrm{d}=0$ and thus

$$(a(x)-\overline{a}(x))u''(x) = \mathrm{c}. \tag{14.46}$$

Then, using equations (14.45) and (14.46), it implies that $|\mathrm{c}| \leq \epsilon \|a-\overline{a}\|_{L^\infty(0,l)}$, and since $\epsilon > 0$ is arbitrary, $\mathrm{c}=0$. Then, (14.46) gives

$$(a(x)-\overline{a}(x))u''(x) = 0, \quad \forall x \in (0,l).$$

Now since $\mu(D)=0$, this equation leads to $a=\overline{a}$, and hence a is identifiable.

Example 14.4 Consider now an example which is similar to Example 14.3, but instead of (14.29), the boundary conditions (14.42) with $u_0 = u_0' = 0$ are imposed. Following a similar analysis to that used for Example 14.3, with $v(x)=x^2$, the solution of the FP is given by

$$u(x) = \begin{cases} \mathrm{c}_0 x^2, & 0 \leq x \in (0,1/2], \\ \mathrm{c}_0(x-1/4), & x \in (1/2,1), \end{cases}$$

where $\mathrm{c}_0 = (7\alpha+6\beta)/48$. It is easy to remark that $\mu(D)=1/2>0$, hence from Theorem 14.9, a is not identifiable.

Example 14.5 In the physical case of clamped-free beams, the boundary conditions (14.42) apply with $u_0 = u_0' = 0$. Taking $l=1$, $f=1$ and $u(x) = x^4/24 - x^3/6 + x^2/4$ it clear that $\mu(D)=0$ and hence, from Theorem 14.9, $a(x) \equiv 1$ is identifiable.

14.1.4 Conclusions

In this section, results concerning the identifiability of physical parameters for linear beam-type systems have been established. The identification of the flexural rigidity $a(x)$ depends on the simultaneous annihilation of the second and third-order partial derivatives $u_{xx}(x,t)$ and $u_{xxx}(x,t)$, whilst the identification of the mass per unit length $b(x)$ depends on the annihilation of $u_{tt}(x,t)$. To provide identifiability, the sets of points at which the partial derivatives vanish should be of zero measure, but, in general non-empty. The simultaneous identifiability of a and b cannot be, in general, deduced from the identifiability of a and b separately, and should be studied independently. In the steady-state, conditions for the identification of a are given in terms of whether the set of inflexion points of u has zero measure, and physical cases of clamped-clamped and clamped-free beams have been investigated. Point load terms and discontinuous flexural rigidity functions could also be considered.

In the infinite-dimensional setting of the present analysis, identifiability followed from regularity properties of the functions involved. However, these properties disappear in the discrete case of pointwise measurements of deflection. Moreover, in applications, the hypotheses required for the operator Φ_f in (14.4) to be invertible are generally not satisfied since the actual noisy measurement of the deflection u may not belong to a regular space of functions such as U_{ad}. Also, even if Φ_f is invertible, the solution $\zeta = \Phi_f^{-1}(u)$ is unstable since U_{ad} is an infinitely dimensional space. Therefore, to overcome the ill-posedness, a regularized least-squares formulation can be considered, namely, find $\zeta \in \Lambda_{ad}$ which minimizes the functional

$$R_\lambda(\zeta) = \|\Phi_f(\zeta) - u\|^2 + \lambda\|\zeta\|^2,$$

where the norms are in $L^2(0,l)$ and $\lambda > 0$ is a regularization parameter.

14.2 The Comparison Model Method

Let us consider again the physical situation described in subsection 14.1.3.1 consisting of the steady-state deflection $u(x)$ of a beam of length l governed by the Euler-Bernoulli fourth-order ordinary differential beam equation (14.20), subject to Dirichlet (clamped beams) boundary conditions (14.29). As before, the identification of the spacewise dependent flexural rigidity $a(x)$ of the beam, given the load $f(x)$ and the measured deflection $u(x)$ is considered. For related studies on this inverse problem, see [251, 258, 268, 273, 292, 351].

Denote by a a general solution, which may or may not be in the admissible set Λ_{ad}^{ext}; however, $\hat{a}$ will stand for the reference solution which belongs to Λ_{ad}^{ext}.

Theorem 14.10. ([259]) *Assume that $u \in C^4(0,l)$ is given such that*

$$u''''(x) \geq \gamma, \quad \forall x \in (0,l) \tag{14.47}$$

for some positive constant γ, and that there exists an $x_0 \in (0,l)$ such that

$$u''(x_0) = u'''(x_0) = 0. \tag{14.48}$$

Then, for all $0 < f \in L^1(0,l)$, there exists a unique $0 < a \in L^\infty(0,l)$, such that equation (14.20) is satisfied, and

$$\|a\|_\infty \leq \frac{1}{\gamma}\|f\|_\infty. \tag{14.49}$$

Proof. Integrating twice equation (14.20) and imposing (14.48) result in

$$a(x)u''(x) = \int_{x_0}^{x} dx' \int_{x_0}^{x'} f(x'')dx''. \tag{14.50}$$

On the other hand, from (14.47) and (14.48) it follows that $x_0 \in (0,l)$ is the unique (global) minimimum point for $u''(x)$. Thus, $u''(x) > 0$ for all $x \in [0,l]\backslash\{x_0\}$ and $u''(x_0) = 0$. Based on this, from (14.50), upon applying L'Hospital's rule twice, $a(x)$ is uniquely given by

$$a(x) = \begin{cases} \frac{\int_{x_0}^{x} dx' \int_{x_0}^{x'} f(x'')dx''}{u''(x)}, & \text{if } x \neq x_0, \\ \\ \frac{f(x_0)}{u''''(x_0)}, & \text{if } x = x_0. \end{cases} \tag{14.51}$$

From this expression, if $f > 0$ then $a > 0$, and the estimate

$$|a(x)| \leq \begin{cases} \frac{\|f\|_\infty (x-x_0)^2}{2u''(x)}, & \text{if } x \neq x_0, \\ \\ \frac{|f(x_0)|}{\gamma}, & \text{if } x = x_0, \end{cases} \tag{14.52}$$

holds. Since $u \in C^4(0,l)$, from Taylor's series expansion, for each $x \in (0,l)$ there is ξ between x and x_0 such that

$$u''(x) = u''(x_0) + (x - x_0)u'''(x_0) + \frac{(x-x_0)^2}{2}u''''(\xi) = \frac{(x-x_0)^2}{2}u''''(\xi).$$

So,

$$\frac{(x-x_0)^2}{2u''(x)} = \frac{1}{u''''(\xi)} \leq \frac{1}{\gamma},$$

and since also $|f(x_0)| \leq \|f\|_\infty$, the estimate (14.49) follows from (14.52).

Remark 14.4. (i) Theorem 14.10 remains true if assumption (14.48) is replaced by $u''''(x) \leq -\gamma$, $\forall x \in (0,l)$ for some positive constant γ.

(ii) If $u''(x) \neq 0$, $\forall x \in (0,l)$, then a is not unique as $a_1(x) := a(x) + (\mathrm{c}x + \mathrm{d})/u''(x)$ with c and d arbitrary real constants is still a solution. Further, if $a > 0$, then c and d can be chosen such that $a_1 > 0$. Thus, a necessary condition for uniqueness is that there exists $x_0 \in (0,l)$ such that $u''(x_0) = 0$.

(iii) If $u'''(x) \neq 0$, $\forall x \in (0,l)$, then u'' is strictly increasing or decreasing function and thus, injective. From this and (ii) above, for uniqueness, x_0 must be the unique zero of u''. Then, from (14.50),

$$a(x_0) = \lim_{x\to x_0} \frac{\int_{x_0}^{x} dx' \int_{x_0}^{x'} f(x'')dx''}{u''(x)} = \lim_{x\to x_0} \frac{\int_{x_0}^{x} f(x')dx'}{u'''(x)} = 0,$$

and equation (14.50) yields

$$a(x) = \begin{cases} \dfrac{\int_{x_0}^{x} dx' \int_{x_0}^{x'} f(x'')dx''}{u''(x)}, & \text{if } x \neq x_0, \\[2ex] 0, & \text{if } x = x_0. \end{cases}$$

Thus, for $a \in \Lambda_{ad}^{ext}$ there should $\exists x_1 \in (0,l)$ such that $u'''(x_1) = 0$.

(iv) From equation (14.51) it follows that if condition (14.48) is satisfied, but $u''''(x_0) = 0$ and $f(x_0) \neq 0$, then a does not exist.

From points (ii) and (iii) of the above remark, the following corollary holds.

Corollary 14.2. *If $u''(x) \neq 0$ for all $x \in (0,l)$, then a is not unique. If $u'''(x) \neq 0$ for all $x \in (0,l)$, then a is not unique or it does not exist in Λ_{ad}^{ext}.*

The uniqueness of a reference solution $\hat{a} \in \Lambda_{ad}^{ext}$ does not rule out the existence of other general solutions (distributions) which are, of course, not in $L^\infty(0,l)$ and yet satisfy (14.20), as shown in the next example.

Example 14.6 Let

$$l = 1, \quad f(x) = 1, \quad u(x) = \frac{1}{24}\left(x - \frac{1}{2}\right)^4, \quad u'(x) = \frac{1}{6}\left(x - \frac{1}{2}\right)^3,$$
$$u''(x) = \frac{1}{2}\left(x - \frac{1}{2}\right)^2, \quad u'''(x) = x - \frac{1}{2}, \quad u''''(x) = 1. \quad (14.53)$$

Hence, $x_0 = 1/2$ and $\gamma = 1$ can be chosen in Theorem 14.10. Then, the unique $\hat{a} \in \Lambda_{ad}^{ext}$ is $\hat{a} \equiv 1$. However, the distributions $a_1(x) := 1 + c/(x - 1/2)$ with c arbitrary non-zero constant are not in $L^\infty(0,1)$, but they are still generalized solutions of (14.20).

The method of solution is the *comparison model method* [259]—a direct method which has been known and applied for sometime in diffusion processes of transport in porous media [333, 335, 357].

Assume that a reference solution, which is consistent with the data u and f exists, i.e., there exists

$$\hat{a} \in \Lambda_{ad} := \{\tilde{a} \in C^2[0,l] \mid \infty > a_{max} \geq \tilde{a}(x) \geq a_{min} > 0, \ \forall x \in [0,l]\} \subset \Lambda_{ad}^{ext},$$

such that equation (14.20) holds with given $f \in C(0,l)$ and $u \in C^4(0,l)$. If the positivity constraint is not taken into account then, as pointed out before, in the distribution sense, $a_1(x) := \hat{a}(x) + (\mathrm{c}x + \mathrm{d})/u''(x)$ with c and d arbitrary real constants is still a solution of (14.20). In general, the constants c and d cannot be determined because, from the direct method point of view, one is solving a second-order ordinary differential equation in a without any initial condition. There are however circumstances [251], where the existence of $\hat{a}$ implies its uniqueness.

First, a well-posed direct problem, for an auxiliary function p, called the comparison model (CM), is introduced. In the CM, the (four-point) boundary conditions (14.29) and the load f are the same as in the original problem, and the flexural rigidity is a given constant auxiliary coefficient b, namely,

$$\mathrm{b}p''''(x) = f(x), \quad x \in (0,l), \tag{14.54}$$

$$p(0) = u_0, \quad p(l) = u_l, \quad p'(0) = u_0', \quad p'(l) = u_l'. \tag{14.55}$$

Then, the CMM solution is defined by the substitution rule

$$a(x) = \frac{\mathrm{b}p''(x)}{u''(x)}, \quad x \in (0,l). \tag{14.56}$$

This solution is well-defined because, if (14.54) is subtracted from (14.20),

$$(a(x)u''(x) - \mathrm{b}p''(x))'' = 0 \Rightarrow a(x) = \frac{\mathrm{c}x + \mathrm{d} + \mathrm{b}p''(x)}{u''(x)}, \quad x \in (0,l), \tag{14.57}$$

where c and d are arbitrary constants of integration. From (14.54)–(14.56), one obtains the following stability result with respect to the second-order derivative of the data $u''(x)$.

Theorem 14.11 ([259]) *Let l, u_0, u_l, u_0', u_l' and $f(x)$ be given. Let a_1 and a_2 be the unique solutions obtained by the CMM from the deflection data $u^{(1)}$ and $u^{(2)}$, respectively. Then, there exists $C > 0$ such that*

$$\|a_1 - a_2\|_{L^\infty(0,l)} \leq \frac{C\|u^{(1)} - u^{(2)}\|_{L^\infty(0,l)}}{\inf_{x\in(0,l)}|u_{xx}^{(1)}(x)| \cdot \inf_{x\in(0,l)}|u_{xx}^{(2)}(x)|}.$$

By CMM, a unique a is not determined, but rather a class of solutions

$$a(x;\mathrm{b}) = \hat{a}(x) + w(x;\mathrm{b}), \tag{14.58}$$

where b determines $w(x;\mathrm{b})$, namely, from (14.56) and (14.58),

$$w(x;\mathrm{b}) = \frac{\mathrm{b}p''(x) - \hat{a}(x)u''(x)}{u''(x)}, \quad x \in (0,l). \tag{14.59}$$

This equation completely characterises the dependence on b of the solution $a(\cdot;\mathrm{b})$ identified by the CMM. One can study the dependence of flexural rigidity solution $a(x;\mathrm{b})$ on the input deflection inverse problem data $u(x)$. Integrating (14.54) twice and imposing (14.55), equation (14.56) yields formally the solution $a(x;\mathrm{b})$ of the CM, as a function of $u(x)$ and $f(x)$, as:

$$a(x;\mathrm{b}) = \frac{1}{u''(x)}\left\{F(x) + \frac{6\mathrm{b}(u_l - u_0)}{l^2} - \frac{2\mathrm{b}(2u_0' + u_l')}{l} + 2F_1(l)\right.$$
$$\left. -6F_2(l) + \frac{6x}{l}\left[\frac{2\mathrm{b}(u_0 - u_l)}{l^2} + \frac{\mathrm{b}(u_0' + u_l')}{l} - F_1(l) + 2F_2(l)\right]\right\}, \tag{14.60}$$

where

$$F(x) = \int_0^x \int_0^{x'} f(x'')dx''dx', \quad F_1(x) = \frac{1}{l}\int_0^x F(x')dx',$$
$$F_2(x) = \frac{1}{l}\int_0^x F_1(x')dx'. \tag{14.61}$$

Remark that if $u_l - u_0 = lu_0' = lu_l'$, then $a(\cdot)$ from (14.60) and thus $w(\cdot)$ from (14.59) are independent of b.

Let us now define the reference auxiliary (constant) parameter $\hat{b}$ as the one satisfying $w(\cdot;\hat{b}) = 0$. Then, from (14.58)–(14.60),

$$\hat{\mathrm{b}}p''(x) = \hat{a}(x)u''(x) = a(x;\hat{b})u''(x) = F(x) + \frac{6\hat{\mathrm{b}}(u_l - u_0)}{l^2}$$
$$-\frac{2\hat{\mathrm{b}}(2v_0 + v_l)}{l} + 2F_1(l) - 6F_2(l)$$
$$+\frac{6x}{l}\left[\frac{2\hat{\mathrm{b}}(u_0 - u_l)}{l^2} + \frac{\hat{\mathrm{b}}(v_0 + v_l)}{l} - F_1(l) + 2F_2(l)\right]. \tag{14.62}$$

Integrating twice equation $(\hat{a}(x)u''(x))'' = f(x)$ from x_0 to x, where $x_0 \in [0,l]$,

$$\hat{a}(x)u''(x) = F(x) - F(x_0) - (x - x_0)\int_0^{x_0} f(\xi)d\xi + \hat{a}(x_0)u''(x_0)$$
$$+(x - x_0)\left(\hat{a}(x)u''(x)\right)'\big|_{x=x_0}. \tag{14.63}$$

Then from (14.62) and (14.63),

$$\begin{cases} 0 = 2F_1(l) - 6F_2(l) - \hat{a}(x_0)u''(x_0) + x_0\left(\hat{a}(x)u''(x)\right)'\big|_{x=x_0} \\ \quad +F(x_0) - x_0\int_0^{x_0} f(\xi)d\xi + \frac{2\hat{\mathrm{b}}}{l}\left(\frac{3(u_l - u_0)}{l} - (2v_0 + v_l)\right), \\ \\ 0 = \int_0^{x_0} f(\xi)d\xi - \left(\hat{a}(x)u''(x)\right)'\big|_{x=x_0} \\ \quad +\frac{6}{l}\left[-F_1(l) + 2F_2(l) + \frac{\hat{\mathrm{b}}}{l}\left(\frac{2(u_0 - u_l)}{l} + v_0 + v_l\right)\right] \end{cases} \tag{14.64}$$

This system of two linear equations contains in general 3 unknowns, namely $\hat{\mathrm{b}}$, $\hat{a}(x_0)$ and $(\hat{a}(x)u''(x))'\big|_{x=x_0}$. Thus, to uniquely determine $\hat{\mathrm{b}}$, some additional specification of $\hat{a}(x_0)$ and/or $(\hat{a}(x)u''(x))'\big|_{x=x_0}$ is needed.

The case of both $\hat{a}(x_0)$ and $\hat{a}'(x_0)$ specified

Assuming that both $\hat{a}(x_0)$ and $\hat{a}'(x_0)$ are given as additional data, expression (14.64) represents a system of two linear equations with one unknown. To have a solution, the following compatibility condition is needed:

$$\begin{aligned}\Big(\hat{a}(x_0)u''(x_0) - 2F_1(l) + 6F_2(l) - x_0\,(\hat{a}(x)u''(x))'\,|_{x=x_0} - F(x_0) \\ + x_0 \int_0^{x_0} f(\xi)d\xi\Big)\left(\frac{2(u_0-u_l)}{l} + v_0 + v_l\right) \\ = \left[\left((\hat{a}(x)u''(x))'\,|_{x=x_0} - \int_0^{x_0} f(\xi)d\xi\right)\frac{l}{3} + 2F_1(l)\right. \\ \left. -4F_2(l)\right]\left(\frac{3(u_l-u_0)}{l} - (2v_0+v_l)\right). \qquad (14.65)\end{aligned}$$

Case I. If $\frac{2(u_0-u_l)}{l} + u_0' + u_l' \neq 0$, then from the second equation of (14.64),

$$\hat{\mathrm{b}} = \frac{\left\{\left[(\hat{a}(x)u''(x))'\,|_{x=x_0} - \int_0^{x_0} f(\xi)d\xi\right]\frac{l}{6} + F_1(l) - 2F_2(l)\right\}l}{\frac{2(u_0-u_l)}{l} + u_0' + u_l'}. \qquad (14.66)$$

Substituting this into (14.60) and using (14.65) yield the unique solution

$$\begin{aligned}a(x;\hat{\mathrm{b}}) = \hat{a}(x) = \frac{F(x) - F(x_0) + \hat{a}(x_0)u''(x_0)}{u''(x)} \\ + \frac{(x-x_0)\left[(\hat{a}(x)u''(x))'\big|_{x=x_0} - \int_0^{x_0} f(\xi)d\xi\right]}{u''(x)}. \qquad (14.67)\end{aligned}$$

Case II. If $\frac{2(u_0-u_l)}{l} + u_0' + u_l' = 0$, then from the second equation of (14.64), it is necessary for the numerator in (14.66) to be zero for $\hat{\mathrm{b}}$ to exist. Still in Case II, consider two subcases, as follows.

Subcase II.1. If $\frac{3(u_l-u_0)}{l} - (2u_0' + u_l') \neq 0$, then from the first equation of (14.64),

$$\begin{aligned}\hat{\mathrm{b}} = \frac{[\hat{a}(x_0)u''(x_0) - 2F_1(l) + 6F_2(l) - x_0(\hat{a}(x)u''(x))'\big|_{x=x_0}]l}{2\left(\frac{3(u_l-u_0)}{l} - (2u_0'+u_l')\right)} \\ + \frac{[-F(x_0) + x_0\int_0^{x_0} f(\xi)d\xi]l}{2\left(\frac{3(u_l-u_0)}{l} - (2u_0'+u_l')\right)}. \qquad (14.68)\end{aligned}$$

Substituting this into (14.60), and using that both the numerator and denominator of the fraction in (14.66) are zero, yield again the unique solution (14.67).

Subcase II.2. If $\frac{3(u_l-u_0)}{l}-(2u_0'+u_l')=0$, then together with $\frac{2(u_0-u_l)}{l}+u_0'+u_l'=0$ yield $u_l-u_0=lu_0'=lu_l'$. Also, from the first equation of (14.64), see also (14.68), the numerator in (14.67) has to be zero. In this case, $\hat{b}$ satisfying the system (14.64) is undetermined, see also the previous remark below (14.61). From $u'(0)=u_0'=u_l'=u'(l)$, it follows that there exists $x_0 \in (0,l)$ such that $u''(x_0)=0$. Then the CMM equation (14.67) simplifies as

$$\hat{a}(x)=\frac{F(x)-F(x_0)+(x-x_0)\left[(\hat{a}(x)u''(x))'\,|_{x=x_0}-\int_0^{x_0} f(\xi)d\xi\right]}{u''(x)},$$

$$x \neq x_0, \quad (14.69)$$

and only the value of $(\hat{a}(x)u''(x))'\,|_{x=x_0}$ needs to be specified.

Remark 14.5. Another particular case is obtained when there is $x_0 \in (0,l)$ such that $\hat{a}'(x_0)=\hat{a}''(x_0)=0$ and $u''''(x_0)\neq 0$. Then, (14.20) applied at $x=x_0$ gives $\hat{a}(x_0)=f(x_0)/u''''(x_0)$ and (14.67) simplifies as

$$\hat{a}(x)$$

$$=\frac{F(x)-F(x_0)-(x-x_0)\int_0^{x_0} f(\xi)d\xi+\frac{f(x_0)[u''(x_0)+(x-x_0)u'''(x_0)]}{u''''(x_0)}}{u''(x)}. \quad (14.70)$$

Remark 14.6. Under the assumptions (14.48), equation (14.67) simplifies as

$$\hat{a}(x)=\frac{F(x)-F(x_0)-(x-x_0)\int_0^{x_0} f(\xi)d\xi}{u''(x)}.$$

Next, some examples on the application of the CMM are illustrated.

Consider first again the Example 14.6 given by equation (14.53) with $u_0=u_1=1/384$, $u_0'=-1/48$ and $u_1'=1/48$. The assumptions of Theorem 14.11 are satisfied, hence uniqueness of the solution $\hat{a}\in\Lambda_{ad}^{ext}$ is expected. From (14.54) and (14.55),

$$p(x)=\frac{x^4}{24b}-\frac{x^3}{12b}+\frac{(b+2)x^2}{48b}-\frac{x}{48}+\frac{1}{384}.$$

Then, (14.56) yields the CMM solution

$$a(x)=\frac{x^2-x+\frac{b}{12}+\frac{1}{6}}{\left(x-\frac{1}{2}\right)^2},$$

which is understood in the sense of (14.57). Since $f(x) = 1$, from (14.61), $F(x) = x^2/2$, $F_1(1) = 1/6$ and $F_2(1) = 1/24$. Then, the compatibility relation (14.65) requires that

$$\frac{1}{2}(\hat{a}(x)(x-1/2)^2)'\big|_{x=x_0} = x_0 - \frac{1}{2},$$

which implies $x_0 = 1/2$ or

$$\frac{\hat{a}'(x_0)}{2}\left(x_0 - \frac{1}{2}\right) + \hat{a}(x_0) = 1. \tag{14.71}$$

Applying (14.68) yields

$$\hat{\mathrm{b}} = 12\left[\hat{a}(x_0)\left(x_0 - \frac{1}{2}\right)^2 - \frac{1}{6} - x_0^2 + x_0\right].$$

Substituting this into (14.62) results in

$$a(x;\hat{\mathrm{b}}) = \hat{a}(x) = \frac{(x-x_0)(x+x_0-1) + \hat{a}(x_0)\left(x_0 - \frac{1}{2}\right)^2}{\left(x - \frac{1}{2}\right)^2}. \tag{14.72}$$

Taking $x_0 = 1/2$, the unique solution $\hat{a}(x) = 1$ is obtained, as requested. If $x_0 \neq 1/2$, then for $\hat{a} \in L^\infty(0,1)$, it is required that $\lim_{x\to 1/2} \hat{a}(x) = \lim_{x\to 1/2} \frac{(\hat{a}(x_0)-1)\left(x_0-\frac{1}{2}\right)^2}{\left(x-\frac{1}{2}\right)^2}$ be finite, implying $\hat{a}(x_0) = 1$. Then, (14.71) gives $\hat{a}'(x_0) = 0$ and (14.72) yields $\hat{a}(x) = 1$ again.

The dependence of $w(\cdot;\mathrm{b})$ on b is expressed by (14.59), which gives $w(x;\mathrm{b}) = \frac{\mathrm{b}-1}{12(x-1/2)^2}$ and the value of $\hat{\mathrm{b}}$ which yields $w(\cdot;\hat{\mathrm{b}}) = 0$ is $\hat{\mathrm{b}} = 1$, see also (14.68) .

Example 14.7 Let $l = 1$, $u(x) = x^3$, $f(x) = 1$, $u_0 = 0$, $u_1 = 1$, $u_0' = 0$ and $u_1' = 3$. From (14.54) and (14.55),

$$p(x) = \frac{x^2}{24\mathrm{b}}(x^2 + 2(12\mathrm{b}-1)x + 1).$$

Then, (14.56) yields the CMM solution

$$a(x) = \frac{x + 12\mathrm{b} - 1}{12} + \frac{1}{72x}, \tag{14.73}$$

which is understood in the sense of (14.57). Since $f(x) = 1$, from (14.61) it is obtained that $F(x) = x^2/2$, $F_1(1) = 1/6$ and $F_2(1) = 1/24$. Then, the compatibility relation (14.65) requires that

$$\hat{a}'(x_0) = \frac{1}{12} - \frac{1}{72x_0^2}. \tag{14.74}$$

Note that from (14.73), the derivative $a'(x) = \frac{1}{12} - \frac{1}{72x^2}$ is consistent with (14.74). Applying (14.66), it yields

$$\hat{\mathrm{b}} = \hat{a}(x_0) + \frac{1 - x_0}{12} - \frac{1}{72x_0}.$$

Substituting this into (14.62), or (14.73), results in

$$a(x; \hat{\mathrm{b}}) = \hat{a}(x) = \hat{a}(x_0) + \frac{x - x_0}{12} + \frac{1}{72}\left(\frac{1}{x} - \frac{1}{x_0}\right).$$

In this case the CMM solution satisfies (14.20), but it is not in Λ_{ad}^{ext}. Also, note that the assumption (14.47) of Theorem 14.10 is not satisfied, whilst the assumptions of Corollary 14.2 are satisfied, hence one expects a not to be unique.

The case of $\hat{a}(x_0)$ specified

Multiplying the second equation in (14.64) with x_0 and adding to the first one, $\hat{\mathrm{b}}$ is obtained from the following equation:

$$2F_1(l) - 6F_2(l) - \hat{a}(x_0)u''(x_0) + F(x_0) + \frac{6x_0}{l}[-F_1(l) + 2F_2(l)]$$
$$+\frac{\hat{\mathrm{b}}}{l}\left[\frac{6(u_l - u_0)}{l} - 2(2u_0' + u_l') + \frac{6x_0}{l}\left(\frac{2(u_0 - u_l)}{l} + u_0' + u_l'\right)\right] = 0. \quad (14.75)$$

Example 14.8 Let $l = \pi$, $u(x) = \cos(x)$, $u_0 = 1$, $u_\pi = -1$, $u_0' = u_\pi' = 0$ and $f(x) = 2\sin(2x)$. From (14.61),

$$F(x) = x - \frac{\sin(2x)}{2}, \quad F_1(\pi) = \frac{\pi}{2}, \quad F_2(\pi) = \frac{\pi^2}{6} - \frac{1}{4}.$$

Taking $x_0 = \pi/2$ and $\hat{a}(x_0) = 1$, from (14.75) it can be seen that $\hat{\mathrm{b}}$ is undetermined. However, if one takes $x_0 \neq \pi/2$, say $x_0 = \pi/4$, and $\hat{a}(x_0) = 1/\sqrt{2}$ specified, then (14.75) gives

$$\hat{\mathrm{b}} = \frac{\pi^2}{12}\left(\frac{\pi}{2} - \pi^2 + \frac{3}{2}\right). \quad (14.76)$$

Substituting this into (14.62) yields the unique solution $\hat{a}(x) = \sin(x) \in \Lambda_{ad}^{ext}$.

The case of $(\hat{a}(x)u''(x))'\,|_{x=x_0}$ specified

In this case the second equation in (14.64) provides a separate expression to determine $\hat{\mathrm{b}}$. Considering the same Example 13.8 with $x_0 = \pi/2$ and $(\hat{a}(x)u''(x))'\,|_{x=x_0} = 1$ specified, this gives $\hat{\mathrm{b}}$ as in (14.76), and substituting into (14.62) yields again the unique solution $\hat{a}(x) = \sin(x) \in \Lambda_{ad}^{ext}$.

In conclusion, the CMM is a simple and easy-to-use algebraic direct method, based on a substitution rule, which avoids numerical discretization

[333]. Furthermore, the method can easily be implemented in the discrete case [333, 334]. Of course, the stability of flexural rigidity with respect to noise in the deflection values still suffers from the typical drawbacks of direct methods and regularization needs to be incorporated.

14.3 Determination of the Flexural Rigidity of a Beam from Limited Boundary Measurements

For fourth-order ODEs, most of the studied inverse problems for the reconstruction of the flexural rigidity have been formulated as classical Sturm-Liouville problems, [21, 22, 120, 169, 284, 294, 324, 325]. However, spectral data is not always available. In the case of steady-state vibrations of a beam (of length l) on an elastic foundation, where the governing equation is the steady-state Euler-Bernoulli beam equation,

$$(a(x)u''(x))'' + g(x)u(x) = f(x), \quad x \in [0, l], \tag{14.77}$$

where $g(x)$ is the foundation modulus, one can also attempt to determine $a(x)$ when $g(x) \equiv 0$ from the full knowledge of $u(x)$ and $f(x)$, as described in the previous sections. However, this data, although it guarantees the uniquess of the solution implies that unlimited measurements of the deflection $u(x)$ along the beam have to be performed, which again may not be practically feasible. Therefore, in this section the aim of to study what it can be retrieved about $a(x)$, when the additional data is limited and is confined to boundary measurements only; hence non-destructive material testing techniques may be employed. More concretely, the following inverse coefficient problems (ICPs) are considered [258]: Find the flexural rigidity $a \in \Lambda_{ad}^{ext}$ and the deflection $u(x)$ from the class of functions (clamped essential conditions at the end $x = 0$ of the beam)

$$\tilde{H}^2[0, l] := \{u \in H^2[0, l] \mid u(0) = u'(0) = 0\} \tag{14.78}$$

such that equation (14.77) is satisfied in the form

$$M''(x) + g(x)u(x) = f(x), \quad M(x) = a(x)u''(x), \quad x \in [0, l], \tag{14.79}$$

subject to the natural boundary conditions on the bending moment $M(x)$ and the shear force $M'(x)$ at the end $x = l$ of the beam, namely,

$$M(l) = \gamma_l, \quad M'(l) = \gamma_l' \tag{14.80}$$

and the additional data:

ICP1

$$u(l) = u_l, \tag{14.81}$$

ICP2

$$u'(l) = u'_l, \tag{14.82}$$

ICP3

$$M(0) = \gamma_0, \tag{14.83}$$

ICP4

$$M'(a) = \gamma'_0. \tag{14.84}$$

Assuming that $0 \leq g \in L^2[0,l]$ and $f \in L^2[0,l]$, then the direct problem (DP) given by equations (14.79) and (14.80) when $a \in \Lambda_{ad}^{ext}$ is given, has a unique solution $u \in \tilde{H}^2[0,l]$ satisfying

$$B(u,v) := \int_0^l [a(x)u''(x)v''(x) + g(x)u(x)v(x)]dx$$

$$= \int_0^l f(x)v(x)dx + \gamma_l v'(b) - \gamma'_l v(b) =: \ell(v), \quad \forall v \in \tilde{H}^2[0,l]. \tag{14.85}$$

Alternatively, instead of solving the variational problem (14.85), since B is a bilinear and symmetric form and ℓ is linear, one can minimize over $\tilde{H}^2[0,l]$ the total energy $I(u) = \frac{1}{2}B(u,u) - \ell(u)$.

Taking $v = u \in \tilde{H}^2[0,l]$ in (14.85) gives

$$\int_0^l [a(x)u''(x)^2 + q(x)u^2(x)]dx = \int_0^l f(x)u(x)dx + \gamma_l u'(l) - \gamma'_l u(l). \tag{14.86}$$

Also, integrating equation (14.77) and applying (14.80) result in

$$\gamma'_l(l-x) + k(a)u''(x) - \gamma_l = \int_x^l \int_{x'}^b [f(x'') - g(x'')u(x'')]dx''dx' =: r(x). \tag{14.87}$$

Using $u \in \tilde{H}^2[0,l]$ and integrating twice (14.87) yield

$$u(x) = \int_0^x \int_0^{x'} \frac{r(x'') + \gamma_l + \gamma'_l(x''-l)}{a(x'')} dx''dx'. \tag{14.88}$$

The following remarks are obtained by extending the analysis of [148] for a second-order ODE.

Remark 14.7. (i) In a well-conditioned situation, i.e. if $u''(x) \neq 0$, $\forall x \in [0,l]$, identity (14.87) gives

$$a(x) = \frac{1}{u''(x)}(\gamma_l - \gamma'_l(l-x) + r(x)), \quad \forall x \in (0,l). \tag{14.89}$$

(ii) In a moderately (mildy) ill-conditioned situation, i.e. if there exists $x_0 \in (0,l)$ such that $u''(x_0) = 0$, $u''(x) \neq 0$, $\forall x \in [0,l] \backslash \{x_0\}$, $u'''(x) \neq 0$,

$\forall x \in [0, l]$, then applying the l'Hospital rule in (14.89) gives $a(x_0) = \frac{\gamma_l' + r'(x_0)}{u'''(x_0)}$, provided that $\gamma_l - \gamma_l'(l - x_0) + r(x_0) = 0$.

(iii) In a severely ill-conditioned situation, i.e. if there exists $x_0 \in (0, l)$ such that $u''(x_0) = u'''(x_0) = 0$, $u''(x) \neq 0$, $u'''(x) \neq 0$, $\forall x \in [0, l] \setminus \{x_0\}$, $u''''(x) \neq 0$, $\forall x \in [0, l]$, then a further application of the l'Hospital rule in (ii) above results in $a(x_0) = \frac{f(x_0) - g(x_0)u(x_0)}{u''''(x_0)}$, provided that $\gamma_l' + r'(x_0) = 0$.

(iv) If the ill-conditioned situations arise at several points x_1, x_2,..., i.e. $u''(x_i) = 0$ and/or $u'''(x_i) = 0$, $i = 1, 2, ...$, then one can extend the analysis of [150] for a second-order ODE.

14.3.1 Quasi-solutions

Consider for e.g., the ICP1, which from (14.81), it consists in solving

$$u(l; a) = u_l \tag{14.90}$$

for $a(x)$. In practice, the measured value of u_l is not exact and, thus, it is preferable to find a *quasi-solution* of the ICP1 as a solution of

$$I_1 := \min_{\tilde{a} \in \Lambda_{ad}^{ext}} I_1(\tilde{a}) \tag{14.91}$$

for the auxiliary functional

$$I_1(\tilde{a}) = |u(l; \tilde{a}) - u_l|. \tag{14.92}$$

Similarly, based on (14.82)–(14.84) for ICP2, ICP3 and ICP4, one can introduce the auxiliary functionals and search for quasisolutions as solutions of the corresponding minimization problems, respectively, [258],

$$I_2(\tilde{a}) = |u'(l; \tilde{a}) - \beta_b|, \quad I_2 := \min_{\tilde{a} \in \Lambda_{ad}^{ext}} I_2(\tilde{a}), \tag{14.93}$$

$$I_3(\tilde{a}) = |M(0; \tilde{a}) - \gamma_a|, \quad I_3 := \min_{\tilde{a} \in \Lambda_{ad}^{ext}} I_3(\tilde{a}), \tag{14.94}$$

$$I_4(\tilde{a}) = |M'(0; \tilde{a}) - \delta_a|, \quad I_4 := \min_{\tilde{a} \in \Lambda_{ad}^{ext}} I_4(\tilde{a}). \tag{14.95}$$

The quasi-solution exists if the minimization in (14.91) is performed on a compact subset of Λ_{ad}^{ext}, e.g. the class of polynomial functions [160],

$$K_{\mathrm{p}} = \left\{ a(x) = a_0 + (a_1 - a_0) \left(\frac{x}{l} \right)^{\mathrm{p}} \right\}, \tag{14.96}$$

where $\mathrm{p} \geq 1$ is the unknown to be determined in each of the ICP1-ICP4, and $a(0) = a_0$, $a(l) = a_1$ are assumed to be known.

Next, additional information obtained by combining versions of the ICPs previously defined in equations (14.81)–(14.84) are considered such that a wider class of flexural rigidity coefficients can be identified.

14.3.2 Combined inverse problem formulations

Let us now formulate new ICPs as combinations of ICP1–ICP4.

Integrating equation (14.77) on (x, l) and using (14.80) result in

$$-(a(x)u''(x))' + \gamma_l' = \int_x^l [f(x') - g(x')u(x')]dx' = r'(x). \tag{14.97}$$

Using (14.84), then (14.97) applied at $x = 0$ yields

$$\gamma_l' - \gamma_0' = \int_0^l [f(x') - g(x')u(x')]dx' = r'(0). \tag{14.98}$$

Using (14.83), then (14.87) applied at $x = 0$ yields

$$\gamma_l' l + \gamma_0 - \gamma_l = r(0). \tag{14.99}$$

Conditions (14.98) and (14.99) can be treated as compatibility conditions for the problem (14.79)–(14.82) formed by combining the ICP1 and ICP2.

Lemma 14.2. ([258]) *Let $f \in L^2[0,l]$ and $0 < g \in L^2[0,l]$. Let u_1, $u_2 \in \tilde{H}^2[0,l]$ be the solution of the direct problems* (14.79) *and* (14.80) *corresponding to the coefficients a_1, $a_2 \in \Lambda_{ad}ext$.*
(i) If u_1 and u_2 satisfy the boundary condition (14.83) *or* (14.84), *then there exists $x_0 \in (0,l)$ such that $u_1(x_0) = u_2(x_0)$.*
(ii) If u_1 and u_2 satisfy the boundary conditions (14.83) *and* (14.84), *then there exists $x_0 \neq x_0' \in (0,l)$ such that $u_1(x_0) = u_2(x_0)$ and $u_1(x_0') = u_2(x_0')$.*

Proof. Let $M_i = a_i u_i''(x)$ for $i = 1, 2$. Then

$$M_i'(x) = \gamma_l' + \int_x^l [g(x')u_i(x') - f(x')]dx',$$

$$M_i(x) = \gamma_l - \gamma_l'(l - x) + \int_x^l \int_{x'}^l [f(x'') - g(x'')u_i(x'')]dx''dx'. \tag{14.100}$$

Let $G(x) = \int_x^l g(x')w(x')dx'$, where $w(x) = u_1(x) - u_2(x)$.

(i) If (14.84) holds, then $M_1'(0) = \gamma_0' = M_2'(0)$ and (14.100) gives $G(0) = 0$, i.e. $w(x)$ is orthogonal on $g(x)$ in $L^2[0,l]$. Also, $G(l) = 0$ and from Rolle's theorem there exists $x_0 \in (0,l)$ such that $0 = G'(x_0) = -g(x_0)w(x_0)$. Since $g > 0$, it follows that $w(x_0) = 0$, i.e. $u_1(x_0) = u_2(x_0)$.

If (14.83) holds, then $M_1(0) = \gamma_0 = M_2(0)$ and (14.100) gives $H(0) = 0$, where $H(x) = \int_x^l G(x')dx'$. Also, $H(l) = 0$ and from Rolle's theorem there exists $x_1 \in (0,l)$ such that $0 = H'(x_1) = -G(x_1)$. Since $G(l) = 0$ from Rolle's theorem there exists $x_0 \in (x_1, l) \subset (0,l)$ such that $0 = G'(x_0) = -g(x_0)w(x_0)$ and then $w(x_0) = 0$, as before.

(ii) If both (14.83) and (14.84) hold, then from (i), $G(0) = G(x_1) = G(l) = 0$, and from Rolle's theorem there exists $x_0 \neq x_0' \in (0,l)$ such that $G'(x_0) = G'(x_0') = 0$ and then $w(x_0) = w(x_0') = 0$, as before.

Based on Lemma 14.2, unicity conditions for the inverse problems can be formulated, as follows. For an integer $P \geq 2$, let

$$Pol(P) = \left\{ \sum_{i=2}^{P} c_i x^i \mid c_i \in \mathbb{R},\ i = \overline{2,P},\ x \in [0,l] \right\} \tag{14.101}$$

be the class of real polynomials of degree $\leq P$ belonging to $\tilde{H}^2[0,l]$.

Theorem 14.12. ([258]) *Let $f \in L^2[0,l]$ and $0 < g \in L^2[0,l]$. For $P \in \{2,3,4,5\}$, the inverse problem (IP) satisfying any $(P-1)$ of the boundary conditions* (14.81)–(14.84) *has at most one solution $u \in Pol(P)$.*

Proof. Let $(a_i(x), u_i) \in \Lambda_{ad}^{ext} \times Pol(P)$ be two solutions for $i = 1,2$ of the IP and let $w(x) = u_1(x) - u_2(x)$. Based on Lemma 14.2, any $(P-1)$ of the boundary conditions (14.81)–(14.84) give $w(x_0) = 0$, $w(x_0') = 0$, $w(l) = 0$ and/or $w'(l) = 0$. Since $0 < x_0 \neq x_0' < l$, $(P-1)$ linearly independent homogeneous equations are obtained to determine the $(P-1)$ coefficients in $w \in Pol(P)$. Hence, $w(x) = 0$ and thus $u_1 = u_2$.

Finally, one can formulate the inverse coefficient problem (ICP) as the whole combination between ICP1–ICP4, namely:

ICP. Find $a(x) \in \Lambda_{ad}^{ext}$ and $u = u(x) \in \tilde{H}^2[0,l]$ satisfying (14.79)–(14.84).

An analogous inverse problem for second-order Sturm-Liouville operator was investigated in [148, 149, 159, 160].

Introducing the auxiliary functional

$$\begin{aligned} I(a) = {} & |(u[x;a])_{x=l} - u_l| + |(u'[x;a])_{x=l} - u_l'| \\ & + |(M[x;a])_{x=0} - \gamma_0| + |(M'[x;a])_{x=0} - \gamma_0'|, \end{aligned} \tag{14.102}$$

a quasi-solution to the ICP can be defined from the following minimization problem: Find $a \in K_0$ such that $I(a) = \min_{\tilde{a} \in K_0} I(\tilde{a})$, where K_0 is a compact set of admissible coefficients. Recovering the unknown function $a = a(x)$ by using only 4 boundary measurements evidently is a non-unique solution problem. For this reason, in the next section, the discussion is restricted to a particular class of approximations for u, similar to the so-called engineering approach employed in [159] for the conductivity problem of a second-order ODE.

14.3.3 An engineering approach

In this section, by using Theorem 14.12, a simple analytical formula for the solution of the ICP is presented, in the case when the solution $u(x)$ of the (DP) is in the class of polynomial functions $Pol(5) \subset \tilde{H}^2[0,l]$, namely,

$$u(x) = u_5(x) = c_2x^2 + c_3x^3 + c_4x^4 + c_5x^5 \tag{14.103}$$

with unknown coefficients to be determined from the boundary measurements (14.80)–(14.83).

Let us write explicitly (14.77) as

$$a''u'' + 2a'u''' + au'''' + gu = f \tag{14.104}$$

which, on using (14.103), yields

$$\begin{aligned} &2a''(x)(c_2 + 3c_3x + 6c_4x^2 + 10c_5x^3) + 12a'(x)(c_3 + 4c_4x + 10c_5x^2) \\ &\quad +24a(x)(c_4 + 5c_5x) + g(x)(c_2x^2 + c_3x^3 + c_4x^4 + c_5x^5) = f(x). \end{aligned}$$

Equations (14.98) and (14.99) give

$$\gamma_l' - \gamma_0' = r'(0) = \int_0^l [f(x) - g(x)u(x)]dx, \tag{14.105}$$

$$l\gamma_l' + \gamma_0 - \gamma_l = r(0) = \int_0^l \int_x^l [f(x') - g(x')u(x')]dx'dx. \tag{14.106}$$

Equations (14.81), (14.82), (14.105) and (14.106) form a system of 4 equations with 4 unknowns c_2, c_3, c_4 and c_5 from which $u(x)$ can be determined according to (14.103). Then $a(x)$ can be determined from (14.89) whenever

$$u''(x) = 2(a_2 + 3a_3x + 6a_4x^2 + 10a_5x^3) \neq 0, \quad x \in (0,l). \tag{14.107}$$

14.3.3.1 Numerical examples

Taking $g(x) \equiv 1$, $l = 1$, and assuming $u \in Pol(5)$, as given by (14.103), equations (14.81), (14.82), (14.105) and (14.106) give the following system of 4 linear equations with 4 unknowns

$$\left\{ \begin{array}{c} c_2 + c_3 + c_4 + c_5 = u_1, \\ 2c_2 + 3c_3 + 4c_4 + 5c_5 = u_1', \\ -\frac{c_2}{3} - \frac{c_3}{4} - \frac{c_4}{5} - \frac{c_5}{6} = \gamma_1' - \gamma_0' - \int_0^1 f(x)dx =: \rho_1, \\ -\frac{c_2}{4} - \frac{c_3}{5} - \frac{c_4}{6} - \frac{c_5}{7} = \gamma_0 - \gamma_l + \gamma_1' - \int_0^1 dx \int_x^1 f(x')dx' =: \rho_2. \end{array} \right. \tag{14.108}$$

This system of equations can be re-written as

$$A\underline{c} = \underline{b}, \tag{14.109}$$

where

$$A = \begin{bmatrix} 1 & 1 & 1 & 1 \\ 2 & 3 & 4 & 5 \\ -1/3 & -1/4 & -1/5 & -1/6 \\ -1/4 & -1/5 & -1/6 & -1/7 \end{bmatrix}, \quad \underline{c} = \begin{bmatrix} c_2 \\ c_3 \\ c_4 \\ c_5 \end{bmatrix}, \quad \underline{b} = \begin{bmatrix} u_1 \\ u_1' \\ \rho_1 \\ \rho_2 \end{bmatrix}.$$

The system of linear equations (14.109) has a unique solution since the determinant of its matrix is equal to 1/25200 which is non-zero. However, the system is ill-conditioned since the condition number of the matrix A is $O(10^6)$.

Example 14.9 Choose a simple polynomial deflection $u \in Pol(5)$, namely, $u(x) = x^2/2 + x^5/20$ which satisfies equation (14.77) with $g(x) = 1$, $a(x) = 1 + x$ and $f(x) = 6x + 25x^2/2 + x^5/20$. Then, $u_1 = 11/20$, $u_1' = 5/4$, $\gamma_0 = 1$, $\gamma_1 = 4$, $\gamma_0' = 1$, $\gamma_1' = 8$, $\int_0^1 f(x)dx = 287/40$ and $\int_0^1 dx \int_x^1 f(x')dx' = 1437/280$. This gives $\underline{b} = [11/20, 5/4, -7/40, -37/280]^{\mathrm{T}}$, and solving the the system of equations (14.109) gives $c_2 = 1/2$, $c_3 = c_4 = 0$, $c_5 = 1/20$ which yields the exact solution $u(x) = x^2/2 + x^5/20$. Finally, using (14.89) in which $p(x) = 5 - 7x + x^3 + x^4$, yields $a(x) = 1 + x$, as required.

Example 14.10 In this example the function

$$u(x) = e^x - 1 - x = \frac{x^2}{2!} + \frac{x^3}{3!} + \frac{x^4}{4!} + \frac{x^5}{5!} + ... \tag{14.110}$$

which does not belong to $Pol(5)$ is tested. Also, take $g(x) = a(x) = 1$ and $f(x) = 2e^x - x - 1$. Then, $u_1 = e - 2$, $u_1' = e - 1$, $\gamma_0 = 1$, $\gamma_1 = e$, $\gamma_0' = 1$, $\gamma_1' = e$, $\int_0^1 f(x)dx = 2e - 7/2$ and $\int_0^1 dx \int_x^1 f(x')dx' = 7/6$, and then (14.108) gives the system of equations

$$A\underline{c} = \underline{b}_\epsilon, \tag{14.111}$$

where

$$\underline{b}_\epsilon = [(e-2)(1+\epsilon), (e-1)(1+\epsilon), 5/2 - e - \epsilon, -1/6 + \epsilon]^{\mathrm{T}} \tag{14.112}$$

and ϵ represents the percentage of noise included in the exact data $[u_1, u_1', \gamma_0, \gamma_0'] = [e-2, e-1, 1, 1]^{\mathrm{T}}$. Once the system of equations (14.111) is solved, the approximate solution for the deflection u is obtained from (14.103), whilst the approximate solution for a can be obtained from (14.89) as

$$\begin{aligned} a(x) = \Bigg[& 2e^x - \frac{5}{6} - \frac{c_2}{4} - \frac{c_3}{5} - \frac{c_4}{6} - \frac{c_5}{7} \\ & + x\left(\frac{3}{2} - e + \frac{c_2}{3} + \frac{c_3}{4} + \frac{c_4}{5} + \frac{c_5}{6}\right) \\ & - \frac{x^2}{2} - \frac{x^3}{6} - \frac{c_2 x^4}{12} - \frac{c_3 x^5}{20} - \frac{c_4 x^6}{30} \\ & - \frac{c_5 x^7}{42} \Bigg] \Big/ \left(2c_2 + 6c_3 x + 12 c_4 x^2 + 20 c_5 x^3\right). \end{aligned} \tag{14.113}$$

Since the matrix A is ill-conditioned, the Tikhonov regularization method is employed for solving (14.111), to obtain

$$\underline{c}_\lambda = (A^{\mathrm{T}}A + \lambda I)^{-1}A^{\mathrm{T}}\underline{b}_\epsilon, \tag{14.114}$$

where $\lambda > 0$ is the regularization parameter, which can be chosen according to the discrepancy principle, i.e., choose $\lambda > 0$ for which

$$(A\underline{c}_\lambda - \underline{b}_\epsilon)^{\mathrm{T}}(A\underline{c}_\lambda - \underline{b}_\epsilon) \approx \epsilon^2[(e-2)^2+(e-1)^2+(-1)^2+1^2) = \epsilon^2(2e^2-6e+7) =: \theta.$$

Introducing (14.114) into this expression it yields that the value of $\lambda > 0$ should be chosen such that

$$\|((I + \lambda A^{-\mathrm{T}}A^{-1})^{-1} - I)\underline{b}_\epsilon\| \approx \sqrt{\theta}. \tag{14.115}$$

This gives the values of $\lambda = \{0.972, 0.218, 0, 0.189, 0.593\}$ for $\epsilon = \{-0.1, -0.05, 0, 0.05, 0.1\}$, respectively. Once the values of λ have been chosen, equation (14.114) is used to find the solution $\underline{c}$.

Figure 14.1 shows the numerical solutions for u and a given by equations (14.103) and (14.113), respectively, in comparison with the exact solution $(u(x), a(x)) = (e^x - 1 - x, 1)$. From this figure it can be seen that stable solutions are obtained which moreover, are consistent with the amount of noise ϵ included in the input data.

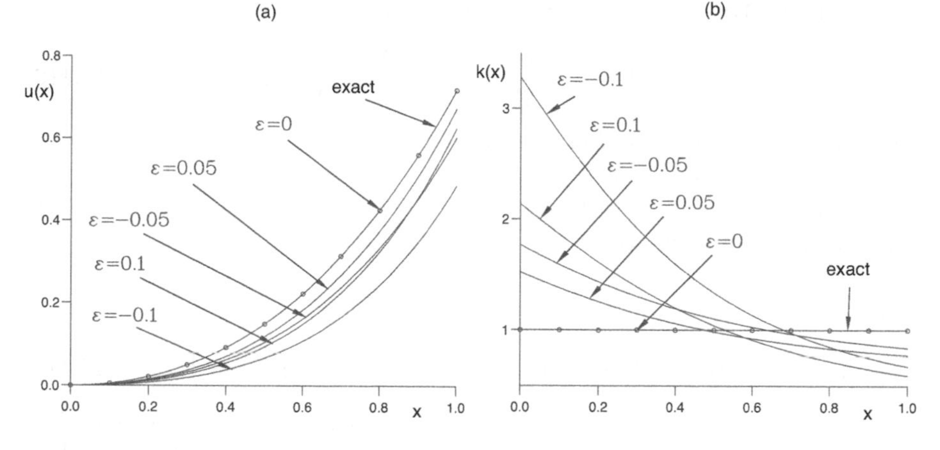

FIGURE 14.1: The numerical solution for the deflection $u(x)$ and the flexural rigidity $a(x)$, for various values of $\epsilon = \{\pm 0.1, \pm 0.05, 0\}$, in comparison with the exact solution $(u(x), a(x)) = (e^x - 1 - x, 1)$, for Example 14.10, [258].

More general approximations for the deflection $u(x)$ than the simple engineering approach based on the polynomial assumption (14.103) can be based on finite-differences, as described in the next subsection.

14.3.4 Finite-difference approximation

The numerical algorithm for solving the ICP can be based on the minimization of the function (14.102) over the solutions of the direct problem (DP). This means that one needs to solve a finite number of DPs and thus, the accuracy of the ICP depends also on the accuracy of the numerical algorithm that one would choose for solving the DP. Since the approach presented is based on the weak solution theory, for the numerical solution of the DP, a variational finite difference scheme (with a mesh size $h = l/N$) [354] is appropriate [273].

To describe a finite-dimensional analog of the ICP, the function $a(x)$ needs to be interpolated to approximate the DP and to parameterize the functional $I(k)$, as well. Since there are only 4 measured data, it is reasonable to interpolate the function $\nu(x) = 1/a(x)$ by the cubic Lagrange's polynomials

$$L_3(x) = \sum_{i=0}^{3} \nu(\xi_i) \left(\prod_{j \neq i}^{3} \frac{x - \xi_j}{\xi_i - \xi_j} \right) \tag{14.116}$$

where $\xi_0 = 0$, $\xi_1 = l/3$, $\xi_2 = 2l/3$ and $\xi_3 = l$. In this case, the ICP can be reformulated as a parameterised inverse problem of recovering the quadruplet $\nu_i := \nu(\xi_i) = L_3(x_i)$ for $i = \overline{0,3}$, from the measured data (14.81)–(14.84). Based on this approximation, the following objective function

$$\begin{aligned} \text{OBJF}(\underline{\nu}) &:= (u(b;\underline{\nu}) - \alpha_b)^2 + (u'(b;\underline{\nu}) - \beta_b)^2 \\ &+ (M(a;\underline{\nu}) - \gamma_a)^2 + (M'(a;\underline{\nu}) - \delta_a)^2 \end{aligned} \tag{14.117}$$

is minimized subject to $\underline{10}^{-10} \leq \underline{\nu} \leq \underline{10}^{10}$, using the NAG routine E04UCF. The gradient of the function (14.117) has been numerically calculated using a forward finite-difference method with a step of 10^{-5} such that any further decrease in this value did not affect significantly the numerical results; in addition, the NAG routine E04UCF employed indicates whether the directional derivative of the objective function and the finite-difference approximation are in closed agreement.

Example 14.11 Take $l = 1$, $g(x) = a(x) = 1$, $f(x) = 2e^x - x - 1$ and

$$u(x) = e^x - 1 - x, \quad x \in [0,1]. \tag{14.118}$$

Then, $\gamma_1 = e$ and $\gamma_1' = e$ in (14.80). Let the input data (14.81)–(14.84) be contaminated with ϵ percentage of noise as $u_1^\epsilon = (e-2)(1+\epsilon)$, $u_1'^\epsilon = (e-1)(1+\epsilon)$, $\gamma_0^\epsilon = 1+\epsilon$ and $\gamma_0'^\epsilon = 1+\epsilon$. In the ICP, one seeks to recover the deflection function (14.118) together with the flexural rigidity $\nu(x) = 1$ in the set (14.117) given by

$$\begin{aligned} L_3(x) = \frac{9[3(\nu_1 - \nu_2) + \nu_3 - \nu_0]x^3}{2} &+ \frac{9(2\nu_0 - \nu_3 - 5\nu_1 + 4\nu_2)x^2}{2} \\ &+ \frac{[-11\nu_0 + 2\nu_3 + 9(2\nu_1 - \nu_2)]x}{2} + \nu_0, \end{aligned} \tag{14.119}$$

i.e., seeks to retrieve $\nu_i = 1$ for $i = \overline{0,3}$. The initial guess to start the iterative nonlinear constrained minimization procedure is taken as $\nu_i^0 = 0.5$ for $i = \overline{0,3}$, and the iterations are stopped according to the discrepancy principle [304].

First, in order to investigate the convergence of the numerical solution as $h = 1/N$ decreases to zero, consider the case $\epsilon = 0$. Although for $\epsilon = 0$ the data is considered as exact, there still exists a numerical noise resulted from the error between the exact and the numerical values obtained by solving the direct problem DP with a strictly positive mesh size $h = 1/N$. Table 14.1 shows the retrieved values of $(\nu_i)_{i=\overline{0,3}}$ for various values of $N \in \{20, 40, 80\}$. From this table it can be seen that as N increases, the objective function (14.117) also decreases and the numerical solution converges to the exact solution $\nu_i = 1$ for $i = \overline{0, 3}$.

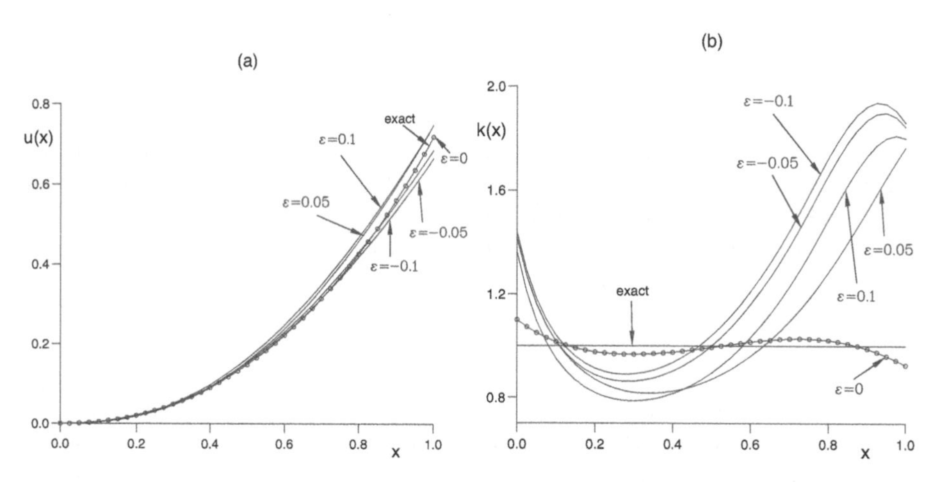

FIGURE 14.2: The numerical solution for the deflection $u(x)$ and the flexural rigidity $a(x)$, for various values of $\epsilon = \{\pm 0.1, \pm 0.05, 0\}$, in comparison with the exact solution $(u(x), a(x)) = (e^x - 1 - x, 1)$, for Example 14.11, [273].

TABLE 14.1: The numerical values of $(\nu_i)_{i=\overline{0,3}}$ obtained for various values of $N \in \{20, 40, 80\}$, when $\epsilon = 0$, for Example 14.11. The stopping iteration numbers k and the values of the objective function (14.117) are also included.

N	ν_0	ν_1	ν_2	ν_3	k	OBJF
20	0.88	1.04	0.97	1.05	23	2×10^{-7}
40	0.90	1.03	0.97	1.08	25	3×10^{-8}
80	0.97	1.00	0.99	1.04	41	7×10^{-9}

To investigate the stability of the numerical solution with respect to noise, fix $N = 40$ and vary the amount of noise ϵ. Table 14.2 shows reasonable stable retrieved values of $(\nu_i)_{i=\overline{0,3}}$ for various values of ϵ. Figure 14.2 shows the numerical solutions for u and a in comparison with the exact solution $(u(x), a(x)) = (e^x - 1 - x, 1)$. Reasonable accurate and stable solutions are obtained, especially for the deflection $u(x)$, which are consistent with the amount of noise ϵ included in the input data. Moreover, when compared with Figure 14.1, the results of Figure 14.2 are more accurate.

TABLE 14.2: The numerical values of $(\nu_i)_{i=\overline{0,3}}$ obtained for various values of $\epsilon \in \{\pm 0.1, \pm 0.05, 0\}$, when $N = 40$, for Example 14.11. The stopping iteration numbers k are also included.

ϵ	ν_0	ν_1	ν_2	ν_3	k
−0.1	0.69	1.11	0.73	0.53	2
−0.05	0.70	1.15	0.76	0.54	2
0	0.90	1.03	0.97	1.08	25
0.05	0.70	1.22	0.95	0.56	5
0.1	0.73	1.26	0.86	0.55	2

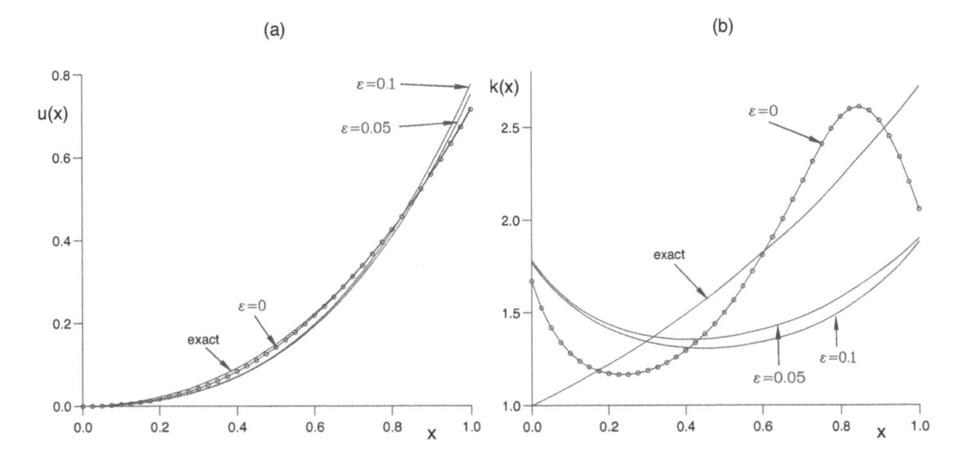

FIGURE 14.3: The numerical solution for the deflection $u(x)$ and the flexural rigidity $a(x)$, for various values of $\epsilon = \{0, 0.05, 0.1\}$, in comparison with the exact solution $(u(x), a(x)) = (e^x - 1 - x, e^x)$, for Example 14.12, [273].

Example 14.12 Consider the same deflection function as that given by (14.118), but now take a space-dependent reciprocal flexural rigidity $\nu(x) = 1/a(x) = e^{-x} = 1 - x + x^2/2! - x^3/3! + ...$, which does not belong to the set of Lagrange polynomials (14.116). Further, taking $g(x) = 1$ gives $f(x) = 4e^{2x} + e^x - x - 1$. Also, equations (14.81)–(14.84) perturbed by noise are $u_1^\epsilon = (e-2)(1+\epsilon)$, $u_1^{'\epsilon} = (e-1)(1+\epsilon)$, $\gamma_0^\epsilon = 1+\epsilon$ and $\gamma_0^{'\epsilon} = 2(1+\epsilon)$. One also has $\gamma_1 = e^2$ and $\gamma_1' = 2e^2$. From (14.119) and the Maclaurin expansion of the function e^{-x}, one expects $\nu_0 \approx 1$, $\nu_1 \approx 58/81$, $\nu_2 = 41/81$ and $\nu_3 = 1/3$, and these values have been used to calculate an estimate of the noise added in the input data, which is required to be known for the discrepancy principle to be applicable.

Figure 14.3 shows the numerical solutions for u and a in comparison with the exact solution $(u(x), a(x)) = (e^x - 1 - x, e^x)$. It can be seen that reasonably good and stable numerical approximations of the exact solution are obtained, especially for the deflection $u(x)$, although $\nu(x) = e^{-x} \notin L_3$ defined in (14.116). This latter fact is reflected in a quite poor estimation of the flexural rigidity.

Future directions of research may be concerned with extending the inverse formulation and analysis to the nonlinear equation [151],

$$(\kappa(u''^2)u'')'' + gu = f$$

and to systems of the Euler-Bernoulli-von Karman nonlinear theory of beams [341],

$$\begin{cases} \frac{d^2}{dx^2}\left(\kappa_1(x)\frac{d^2u_1}{dx^2}\right) - \frac{d}{dx}\left\{\kappa_2(x)\frac{du_1}{dx}\left[\frac{du_2}{dx} + \frac{1}{2}\left(\frac{du_1}{dx}\right)^2\right]\right\} = f_1(x), \\ -\frac{d}{dx}\left\{\kappa_2(x)\left[\frac{du_2}{dx} + \frac{1}{2}\left(\frac{du_1}{dx}\right)^2\right]\right\} = f_2(x). \end{cases}$$

Bibliography

[1] B. Abdelaziz, A. El Badia, and A. El Hajj. Direct algorithms for solving some inverse source problems in 2D elliptic equations. *Inverse Problems*, 31:105002 (26pp), 2015.

[2] G. Adomian. A new approach to the heat equation - an application of the decomposition method. *Journal of Mathematical Analysis and Applications*, 113:202–209, 1986.

[3] M.N. Ahmadabadi, M. Arab, and F.M. Maalek-Ghaini. The method of fundamental solutions for the inverse space-dependent heat source problem. *Engineering Analysis with Boundary Elements*, 33:1231–1235, 2009.

[4] T.S. Aleroev, M. Kirane, and S.A. Malik. Determination of a source term for a time fractional diffusion equation with an integral type over-determination condition. *Electronic Journal of Differential Equations*, 2013:No.270, (16pp), 2013.

[5] G. Alessandrini and V. Isakov. Analyticity and uniqueness for the inverse conductivity problem. *Rendiconti dell'Istituto di Matematica dell'Universita di Trieste*, 28:351–369, 1996.

[6] G. Alessandrini and R. Magnanini. Elliptic equations in divergence form, geometrical critical points of solutions, and Stekloff eigenfunctions. *SIAM Journal on Mathematical Analysis*, 25:1259–1268, 1994.

[7] O.M. Alifanov. *Inverse Heat Transfer Problems*. Springer-Verlag, Berlin, 1994.

[8] K.A. Ames, G.W. Clark, J.F. Epperson, and S.F. Oppenheimer. A comparison of regularizations for an ill-posed problem. *Mathematics of Computation*, 67:1451–1471, 1998.

[9] A. Kh. Amirov. On the question of solvability of inverse problems. *Soviet Mathematics Doklady*, 34:258–259, 1987.

[10] M.A. Anastasio, J. Zhang, D. Modgil, and P.J. Riviere. Application of inverse source concepts to photoacoustic tomography. *Inverse Problems*, 23:21–35, 2007.

[11] M. Andrle, F. Ben Belgacem, and A. El Badia. Identification of moving pointwise sources in an advection-dispersion-reaction equation. *Inverse Problems*, 27:025007 (21pp), 2011.

[12] G. Anger. *Inverse Problems in Differential Equations.* Plenum Press, New York, 1990.

[13] M.A. Atakhodjaev. *Ill-Posed Problems for the Biharmonic Equation.* Fan, Tashkent, 1986.

[14] H. Azari, W. Allegreto, Y. Lin, and S. Zhang. Numerical procedures for recovering a time dependent coefficient in a parabolic differential equation. *Dynamics of Continuous, Discrete and Impulsive Systems, Series B: Applications & Algorithms*, 11:181–199, 2004.

[15] G. Bal, C. Guo, and F. Monard. Inverse anisotropic conductivity from internal current densities. *Inverse Problems*, 30:025001 (21pp), 2014.

[16] H.T. Banks, S. Dediu, and S.L. Ernstberger. Sensitivity functions and their uses in inverse problems. *Journal of Inverse and Ill-Posed Problems*, 15:683–708, 2007.

[17] H.T. Banks, S.S. Gates, I.G. Rosen, and Y. Wang. The identification of a distributed parameter model for a flexible structure. *SIAM Journal on Control and Optimization*, 26:743–762, 1988.

[18] H.T. Banks and K. Kunisch. *Estimation Techniques for Distributed Parameter Systems.* Birkhauser, Boston, 1989.

[19] T.N. Baranger, S. Andrieux, and R. Rischette. Combined energy method and regularization to solve the Cauchy problems for the heat equation. *Inverse Problems in Science and Engineering*, 22:199–212, 2014.

[20] I.E. Barans'ka. The inverse problem in a domain with free bound for an anisotropic equation of parabolic type. *Naykovy Visnyk Chernivetskogo Universytetu Matematyka*, 374:13–28, 2008.

[21] V. Barcilon. Inverse problem for a vibration beam in a free-clamped configuration. *Philosophical Transactions of the Royal Society*, 304:211–251, 1982.

[22] V. Barcilon. Sufficient conditions for the solution of the inverse problem for a vibrating beam. *Inverse Problems*, 3:181–193, 1987.

[23] G. Bastay, T. Johansson, V.A. Kozlov, and D. Lesnic. An alternating method for the stationary Stokes system. *ZAMM*, 86:268–280, 2006.

[24] G. Bastay, V.A. Kozlov, and B.O. Turesson. Iterative methods for an inverse heat conduction problem. *Journal of Inverse and Ill-Posed Problems*, 9:375–388, 2001.

[25] F.S.V. Bazan. CGLS-GCV: a hybrid algorithm for low-rank-deficient problems. *Applied Numerical Mathematics*, 47:91–108, 2003.

[26] J.V. Beck. Surface heat flux determination using an integral method. *Nuclear Engineering and Design*, 7:170–178, 1968.

[27] J.V. Beck, B. Blackwell, and C.R. St. Clair Jr. *Inverse Heat Conduction: Ill-Posed Problems.* Wiley-Interscience, New York, 1985.

[28] J.V. Beck, K.D. Cole, A. Haji-Sheikh, and B. Litkouhi. *Heat Conduction Using Green's Functions.* Hemisphere Publ., London, 1992.

[29] O.V. Belai, L.L. Frumin, E.V. Podivilov, and D.A. Shapiro. Inverse scattering for the one-dimensional Helmholtz equation: fast numerical method. *Optics Letters*, 33:2101–2103, 2008.

[30] M. Belge, M. Kilmer, and E.L. Miller. Efficient determination of multiple regularization parameters in a generalized L-curve framework. *Inverse Problems*, 18:1161–1183, 2002.

[31] F. Ben Belgacem. Why is the Cauchy problem severely ill-posed? *Inverse Problems*, 23:823–836, 2007.

[32] N. Benmansour. *Boundary Element Solution to Stratified Shallow Water Wave Equations.* PhD Thesis, Wessex Institute of Technology, University of Portsmouth, 1993.

[33] J.R. Berger, P.A. Martin, V. Mantic, and L.J. Gray. The method of fundamental solutions for steady-state heat transfer in an exponentially graded material. *ZAMP*, 56:293–303, 2005.

[34] F. Berntsson. Sequential solution of the sideways heat equation by windowing of the data. *Inverse Problems in Engineering*, 11:91–103, 2003.

[35] F. Berntsson, V. Kozlov, and D. Wokiyi. Solvability of a non-linear Cauchy problem for an elliptic equation. *International Journal of Computer Mathematics*, 96:2317–2333, 2019.

[36] F. Berntsson, V.A. Kozlov, L. Mpinganzima, and B.O. Turesson. An alternating iterative procedure for the Cauchy problem for the Helmholtz equation. *Inverse Problems in Science and Engineering*, 22:45–62, 2014.

[37] R. Bialecki, E. Divo, and A. Kassab. Reconstruction of time-dependent boundary heat flux by a BEM-based inverse algorithm. *Engineering Analysis with Boundary Elements*, 30:767–773, 2006.

[38] P.A. Binding, P.J. Brown, and K. Seddeghi. Sturm-Liouville problems with eigenparameter dependent boundary conditions. *Proceedings of the Edinburgh Mathematical Society*, 37:57–72, 1993.

[39] K.B. Bischoff and O. Levenspiel. Fluid dispersion - generalization and comparison of mathematical models - I Generalisation of models. *Chemical and Engineering Science*, 17:245–255, 1962.

[40] I. Borachok, R. Chapko, and B.T. Johansson. Numerical solution of an elliptic 3-dimensional Cauchy problem by the alternating method and boundary integral equations. *Journal of Inverse and Ill-Posed Problems*, 24:711–725, 2016.

[41] N. Boussetila and F. Rebbani. Optimal regularization method for ill-posed Cauchy problems. *Electronic Journal of Differential Equations*, 147:1–15, 2006.

[42] B. Brown and D. Lesnic. Inverse problems in diffusion. In *Inverse Problems and Computational Mechanics*, pages 33–52. Editura Academiei, Bucharest, Romania, 2011.

[43] R.M. Brown and G. Uhlmann. Uniqueness in the inverse conductivity problem for nonsmooth conductivities in two dimensions. *Communications in Partial Differential Equations*, 22:1009–1027, 1997.

[44] O.R. Burggraf. An exact solution of the inverse problem in heat conduction theory and applications. *Journal of Heat Transfer*, 86C:373–382, 1964.

[45] E.M. Burghold, Y. Frekers, and R. Kneer. Determination of time-dependent thermal contact conductance through IR-thermography. *International Journal of Thermal Sciences*, 98:148–155, 2015.

[46] I. Bushuyev. Global uniqueness for inverse parabolic problems with final observation. *Inverse Problems*, 11:L11–L16, 1995.

[47] F. Cakoni and R. Kress. Integral equation method for the inverse obstacle problem with generalized impedance boundary condition. *Inverse Problems*, 29:015005 (19pp), 2013.

[48] D. Calvetti, S. Morigi, L. Reichel, and F. Sgallari. Tikhonov regularization and the L-curve for large discrete ill-posed problems. *Journal of Computational and Applied Mathematics*, 123:423–446, 2000.

[49] J.R. Cannon. The solution of the heat equation subject to the specification of energy. *Quarterly of Applied Mathematics*, 21:155–160, 1963.

[50] J.R. Cannon. A Cauchy problem for the heat equation. *Annali di Matematica Pura ed Applicata*, 66:155–166, 1964.

[51] J.R. Cannon. Determination of an unknown coefficient in a parabolic equation. *Duke Mathematical Journal*, 30:313–323, 1964.

[52] J.R. Cannon. Determination of certain parameters in heat conduction. *Journal of Mathematical Analysis and Applications*, 8:188–201, 1964.

[53] J.R. Cannon. Determination of the unknown coefficient $k(u)$ in the equation $\nabla \cdot k(u)\nabla u = 0$ from overspecified boundary data. *Journal of Mathematical Analysis and Applications*, 18:112–114, 1967.

[54] J.R. Cannon. Determination of an unknown heat source from overspecified boundary data. *SIAM Journal on Numerical Analysis*, 5:275–286, 1968.

[55] J.R. Cannon. *The One-Dimensional Heat Equation.* Addison-Wesley Publishing Company, Menlo Park, California, 1983.

[56] J.R. Cannon and P. DuChateau. Determination of unknown physical properties in heat conduction problems. *International Journal of Engineering Science*, 11:783–794, 1973.

[57] J.R. Cannon and P. DuChateau. Determining unknown coefficients in a nonlinear heat conduction problem. *SIAM Journal on Applied Mathematics*, 24:298–314, 1973.

[58] J.R. Cannon and P. DuChateau. An inverse problem for a nonlinear diffusion equation. *SIAM Journal on Applied Mathematics*, 39:272–289, 1980.

[59] J.R. Cannon, P. DuChateau, and D.L. Filmer. A method of determining unknown rate constants in a chemical reaction in the presence of diffusion. *Mathematical Biosciences*, 9:61–70, 1970.

[60] J.R. Cannon, P. DuChateau, and K. Steube. Trace type functional differential equations and the identification of hydraulic properties of porous media. *Transport in Porous Media*, 6:745–758, 1991.

[61] J.R. Cannon and D.R. Dunninger. Determination of an unknown forcing function in a hyperbolic equation from overspecified data. *Annali di Matematica Pura ed Applicata*, 1:49–62, 1970.

[62] J.R. Cannon and C.D. Hill. Existence, uniqueness, stability, and monotone dependence in a Stefan problem for the heat equation. *Journal of Mathematics and Mechanics*, 17:1–19, 1967.

[63] J.R. Cannon and J. Hoek. Diffusion subject to the specification of mass. *Journal of Mathematical Analysis and Applications*, 115:517–529, 1986.

[64] J.R. Cannon and F. Jones Jr. Determination of the diffusivity of an anisotropic medium. *International Journal of Engineering Science*, 1:457–460, 1963.

[65] J.R. Cannon, Y. Lin, and S. Wang. Determination of a control parameter in a parabolic partial differential equation. *Journal of the Australian Mathematical Society, Series B*, 33:149–163, 1991.

[66] J.R. Cannon and W. Rundell. Recovering a time-dependent coefficient in a parabolic differential equation. *Journal of Mathematical Anaysis and Applications*, 160:572–582, 1991.

[67] K. Cao and D. Lesnic. Determination of space-dependent coefficients from temperature measurements using the conjugate gradient method. *Numerical Methods for Partial Differential Equations*, 34:1370–1400, 2018.

[68] K. Cao and D. Lesnic. Simultaneous reconstruction of the spatially-distributed reaction coefficient, initial temperature and heat source from temperature measurements at different times. *Computers & Mathematics with Applications*, 78:3237–3249, 2019.

[69] K. Cao, D. Lesnic, and J.J. Liu. Simultaneous reconstruction of space-dependent heat transfer coefficients and initial temperature. *Journal of Computational and Applied Mathematics*, 375:112800 (18pp), 2020.

[70] E.J.F.R. Caron, K.J. Daun, and M.A. Wells. Experimental heat transfer coefficient measurements during hot forming die quenching of boron steel at high temperatures. *International Journal of Heat and Mass Transfer*, 71:396–404, 2014.

[71] K. Chadan and P.C. Sabatier. *Inverse Problems in Quantum Scattering Theory.* 2nd edn., Springer-Verlag, Berlin, 1989.

[72] Y.P. Chang, C.S. Kang, and D.J. Chen. The use of fundamental Green's functions for the solution of heat conduction in anisotropic media. *International Journal of Heat and Mass Transfer*, 16:1905–1918, 1973.

[73] B.T. Chao. A note on conduction of heat in anisotropic media. *Applied Scientific Research*, 12:134–138, 1963.

[74] P.C. Chatwin and C.M. Allen. Mathematical models of dispersion in rivers and estuaries. *Annual Reviews of Fluid Mechanics*, 17:119–149, 1985.

[75] M.J. Chaubell. *Low-coherence Interferometric Imaging: Solution of the One-dimensional Inverse Scattering Problem.* PhD Thesis, California Institute of Technology, 2004.

[76] D. Chen. Anti-reflection (AR) coatings made by sol-gel process: A review. *Solar Energy Materials & Solar Cells*, 69:313–336, 2001.

[77] Y. Chen and V. Rokhlin. On the inverse scattering problem for the Helmholtz equation in one dimension. *Inverse Problems*, 8:365–391, 1992.

[78] Z. Chen, Y. Lu, Y. Xu, and H. Yang. Multi-parameter Tikhonov regularization for linear ill-posed operator equations. *Journal of Computational Mathematics*, 26:37–55, 2008.

[79] A.H. D. Cheng. Darcy's flow with variable permeability: A boundary integral solution. *Water Resources Research*, 20:980–984, 1984.

[80] J. Cheng, S. Lu, and M. Yamamoto. Reconstruction of the Stefan-Boltzmann coefficients in a heat-transfer process. *Inverse Problems*, 28:045007, 2012.

[81] C. Chicone and J. Gerlach. A note on the identifiability of distributed parameters in elliptic equations. *SIAM Journal on Mathematical Analysis*, 18:1378–1384, 1987.

[82] M. Choulli and M. Yamamoto. Uniqueness and stability in determining the heat radiative coefficient, the initila temperature and a boundary coefficient in a parabolic equation. *Nonlinear Analysis: Theory, Methods and Applications*, 69:3983–3998, 2008.

[83] S.-N. Chow, K. Yin, H.-M. Zhou, and A. Behrooz. Solving inverse source problems by the orthogonal solution and kernel correction algorithm (OSKCA) with applications in fluorescence tomography. *Inverse Problems and Imaging*, 8:79–102, 2014.

[84] Y.T. Chow and J. Zou. A numerical method for reconstructing the coefficient in a wave equation. *Numerical Methods for Partial Differential Equations*, 31:289–307, 2015.

[85] M.J. Colaco and C.J.S. Alves. A fast non-intrusive method for estimating spatial thermal contact conductance by means of the reciprocity functional approach and the method of fundamental solutions. *International Journal of Heat and Mass Transfer*, 60:653–663, 2013.

[86] M.J. Colaco, C.J.S. Alves, and H.R.B. Orlande. Transient non-intrusive method for estimating spatial thermal contact conductance by means of the reciprocity functional approach and the method of fundamental solutions. *Inverse Problems in Science and Engineering*, 23:688–717, 2014.

[87] T.F. Coleman and Y. Li. On the convergence of interior-reflective Newton methods for nonlinear minimization subject to bounds. *Mathematical Programming*, 67:189–224, 1994.

[88] T.F. Coleman and Y. Li. An interior trust, region approach for nonlinear minimization subject to bounds. *SIAM Journal on Optimization*, 6:418–445, 1996.

[89] C. Coles and D.A. Murio. Identification of parameters in the 2-D IHCP. *Computers and Mathematics with Applications*, 40:939–956, 2000.

[90] C. Coles and D.A. Murio. Simultaneous space diffusivity and source term reconstruction in 2D IHCP. *Computers and Mathematics with Applications*, 42:1549–1564, 2001.

[91] D. Colton and R. Kress. *Inverse Acoustic and Electromagnetic Scattering Theory*. Springer, New York, 3rd edn., 2013.

[92] A. Conn, N. Gould, and P. Toint. *Trust Region Methods*. SIAM, Philadelphia, 1987.

[93] G. de Mey, S. de Smet, and M. Driscart. Forced convection cooling of flat electronic substrates. In *Boundary Elements Technology VI*, pages 39–50. Computational Mechanics Publications, Southampton, 1991.

[94] A.S. Deif. *Advanced Matrix Theory for Scientists and Engineers*. Halsted Press, New York, 1982.

[95] P. Deift and E. Trubowitz. Inverse scattering on the line. *Communications on Pure and Applied Mathematics*, 32:121–151, 1979.

[96] A.M. Denisov. *Elements of the Theory of Inverse Problems*. VSP, Utrecht, 1999.

[97] C.R. Dietrich and G.N. Newsam. Sufficient conditions for identifying transmissivity in a confined aquifer. *Inverse Problems*, 6:L21–L28, 1990.

[98] V.I. Dmitriev and A.S. Chernyavskii. Integral characteristic method in the inverse problem of optical coating design. *Computational Mathematics and Modelling*, 12:128–136, 2001.

[99] E.N. Domanskii. On the equivalence of convergence of a regularizing algorithm to the existence of a solution to an ill-posed problem. *Russian Mathematical Surveys*, 42:123–144, 1987.

[100] A. Doubova, E. Fernandez-Cara, and J.H. Ortega. On the identification of a single body immersed in a Navier-Stokes fluid. *European Journal of Applied Mathematics*, 18:57–80, 2007.

[101] S.S. Dragomir. *Some Gronwall Type Inequalities and Applications*. RGMIA Monographs, Victoria University, Australia, 2002.

[102] P. DuChateau. Monotonicity and uniqueness results in identifying an unknown coefficient in a nonlinear diffusion equation. *SIAM Journal on Applied Mathematics*, 41:310–323, 1981.

[103] P. DuChateau and A. Badran. Identification of spatially dependent coefficients in partial differential equations. *Dynamics of Continuous, Discrete and Impulsive Systems*, 10:827–850, 2003.

[104] S. Dummel and M. Pfaffe. Identification of a coefficient in the one dimensional heat equation. *Wiss.Zeitschrift der TU Chemnitz*, 34:45–51, 1992.

[105] M. Dunn and S.I. Hariharan. Numerical computations on one-dimensional inverse scattering problems. *Journal of Computational Physics*, 55:157–165, 1984.

[106] A. El Badia and A. El Hajj. Identification of dislocations in materials from boundary measurements. *SIAM Journal on Applied Mathematics*, 73:84–103, 2013.

[107] A. El Badia, T. Ha-Duong, and A. Hamdi. Identification of a point source in a linear advection-dispersion-reaction equation: application to a pollution source problem. *Inverse Problems*, 21:1121–1136, 2005.

[108] L. Elden. The numerical solution of a non-characteristic Cauchy problem for a parabolic equation. In *Numerical Treatment of Inverse Problems in Differential Equations and Integral Equations*, pages 246–268. Birkhauser, 1983.

[109] L. Elden, F. Berntsson, and T. Reginska. Wavelet and Fourier methods for solving the sideways heat equation. *SIAM Journal on Scientific Computing*, 21:2187–2205, 2000.

[110] T. Elfving and T. Nikazad. Stopping rules for Landweber-type iteration. *Inverse Problems*, 23:1417–1432, 2007.

[111] F.H. Elmer and F.W. Martin. Antireflection films and alkali-borosilicate glasses produced by chemical treatment. *American Ceramic Society Bulletin*, 58:1092–1097, 1979.

[112] H.W. Engl. Inverse problems and their regularization. In *Computational Mathematics Driven by Industrial Problems*, pages 127–150. Springer Verlag, Berlin, 2000.

[113] H.W. Engl, M. Hanke, and A. Neubauer. *Regularization of Inverse Problems.* Kluwer Academic Publishers, Dordrecht, 2000.

[114] H.W. Engl, O. Scherzer, and M. Yamamoto. Uniqueness and stable determination of forcing terms in linear partial differential equations with overspecified boundary data. *Inverse Problems*, 10:1253–1276, 1994.

[115] A. Erdelyi. *Higher Transcedental Functions.* McGraw-Hill, New York, 1953.

[116] A. Erdem, D. Lesnic, and A. Hasanov. Identification of a spacewise dependent heat source. *Applied Mathematical Modelling*, 37:10231–10244, 2013.

[117] A. Farcas and D. Lesnic. The boundary element method for the determination of a heat source dependent on one variable. *Journal of Engineering Mathematics*, 54:375–388, 2006.

[118] L. Garifo, V.E. Schrock, and E. Spedicato. On the solution of the inverse heat conduction problem by finite differences. *Energia Nucleare*, 22:452–464, 1975.

[119] J. Gaspar, F. Rigollet, J.-L. Gardarein, C. Le Niliot, and Y. Corre. Identification of space and time varying thermal resistance: Numerical feasibility for plasma facing materials. *Inverse Problems in Science and Engineering*, 22:213–231, 2014.

[120] I.M. Gelfand and B.M. Levitan. Identification of space and time varying thermal resistance: Numerical feasibility for plasma facing materials. *American Mathematical Society Transactions Series 2*, 1:253–304, 1951.

[121] K.F. Gerdes and R.L. Bell. A two-dimensional dispersion problem-model verification and parameter estimation. *Mathematical and Computer Modelling*, 8:66–71, 1987.

[122] J. Gill, E. Divo, and A.J. Kassab. Estimating thermal contact resistance using sensitivity analysis and regularization. *Engineering Analysis with Boundary Elements*, 33:54–62, 2009.

[123] F. Ginsberg. On the Cauchy problem for the one-dimensional heat equation. *Mathematics of Computation*, 17:257–269, 1963.

[124] M. Giudici. A result concerning identifiability of the inverse problem of groundwater hydrology. *Inverse Problems*, 5:L31–L36, 1989.

[125] M. Giudici. Identifiability of distributed physical parameters in diffusive-like systems. *Inverse Problems*, 7:231–245, 1991.

[126] G.H. Golub, M. Heath, and G. Wahba. Generalized cross-validation as a method for choosing a good ridge parameter. *Technometrics*, 21:215–223, 1979.

[127] J. Gottlieb. Reconstruction algorithms for hydraulic tomography. In *Inverse Problems in Engineering Mechanics*, pages 119–122. (Eds. H.D. Bui, M. Tanaka, M. Bonnet, H. Maigre, E. Luzzato and M. Reynier), A.A. Balkema, Rotterdam, 1994.

[128] F. Hagin. Some numerical approaches to solving one-dimensional inverse problems. *Journal of Computational Physics*, 43:16–30, 1981.

[129] S. Handrock-Meyer. The identifiability of the unknown parameters in quasilinear and linear differential equations. In *Inverse Problems: Principles and Applications in Geophysics, Technology and Medicine*, pages 167–173. Akademic Verlag, Berlin, 1993.

[130] S. Handrock-Meyer. Identifiability of distributed parameters for a class of quasilinear differential equations. *Journal of Inverse and Ill-Posed Problems*, 5:19–28, 1997.

[131] M. Hanke. *Conjugate Gradient-Type Methods for Ill-Posed Problems.* Longman Scientific and Technical, Harlow, Essex, 1995.

[132] M. Hanke. The minmal error conjugate gradient method is a regularization method. *Proceedings of the American Mathematical Society*, 123:3487–3497, 1995.

[133] M. Hanke. Limitations of the L-curve method in ill-posed problems. *BIT Numerical Mathematics*, 36:287–301, 1996.

[134] M. Hanke and P.C. Hansen. Regularization methods for large-scale problems. *Surveys on Mathematics for Industry*, 3:253–315, 1993.

[135] P.C. Hansen. Analysis of discrete ill-posed problems by means of the L-curve. *SIAM Review*, 34:561–580, 1992.

[136] P.C. Hansen. *Rank Deficient and Discrete Ill-Posed Problems: Numerical Aspects of Linear Inversion.* SIAM, Philadelphia, 1998.

[137] P.C. Hansen. The L-curve and its use in the numerical treatment of inverse problems. In *Computational Inverse Problems in Electrocardiology*, pages 119–142. (Ed. P. Johnston), WIT Press, Southampton, 2001.

[138] P.C. Hansen and D.P. O'Leary. The use of the L-curve in the regulatization of discrete ill-posed problems. *SIAM Journal on Scientific Computing*, 14:1487–1503, 1993.

[139] Dinh Nho Hao. Regularization of a non-characteristic Cauchy problem for the heat equation. *Mathematical Methods in the Applied Sciences*, 15:537–545, 1992.

[140] Dinh Nho Hao. *Methods for Inverse Heat Conduction Problems.* Peter Lang, Frankfurt am Main, 1998.

[141] Dinh Nho Hao, Nguyen Van Duc, and D. Lesnic. A non-local boundary value problem method for the Cauchy problem for elliptic equations. *Inverse Problems*, 25:055002 (27pp), 2009.

[142] Dinh Nho Hao, Nguyen Van Duc, and D. Lesnic. Regularization of parabolic equations backward in time by a non-local boundary value problem method. *IMA Journal of Applied Mathematics*, 75:291–315, 2010.

[143] Dinh Nho Hao, Nguyen Van Duc, and H. Sahli. A non-local boundary value problem method for parabolic equations backward in time. *Journal of Mathematical Analysis and Applications*, 345:805–815, 2008.

[144] Dinh Nho Hao and R. Gorenflo. A non-characteristic Cauchy problem for the heat equation. *Acta Applicandae Mathematicae*, 24:1–27, 1991.

[145] Dinh Nho Hao, Bui Viet Huong, Nguyen Thi Ngoc Oanh, and Phan Xuan Thanh. Determination of a term in the right-hand side of parabolic equations. *Journal of Computational and Applied Mathematics*, 309:28–43, 2017.

[146] Dinh Nho Hao, Phan Xuan Thanh, D. Lesnic, and M. Ivanchov. Determination of a source in the heat equation from integral observations. *Journal of Computational and Applied Mathematics*, 264:82–98, 2013.

[147] S.D. Harris, R. Mustata, L. Elliott, D.B. Ingham, and D. Lesnic. Numerical identification of the hydraulic conductivity of composite anisotropic materials. *Computer Modeling in Engineering and Sciences*, 25:69–79, 2008.

[148] A. Hasanov. An inverse polynomial method for the identification of the leading coefficient in the Sturm-Liouville operator from boundary measurements. *Applied Mathematics and Computation*, 140:501–515, 2003.

[149] A. Hasanov. The determination of the leading coefficient in the monotone potential Sturm-Liouville operator from boundary measurements. *Applied Mathematics and Computation*, 152:141–162, 2004.

[150] A. Hasanov. Error analysis of a multi-singular inverse coefficient problem for the Sturm-Liouville operator from boundary measurements. *Applied Mathematics and Computation*, 150:493–524, 2004.

[151] A. Hasanov. On existence, uniqueness and convergence of approximate solution to boundary value problems related to the nonlinear operator $Au := -(k((u')^2)u')' + g(u)$. *Applied Mathematics and Computation*, 153:659–672, 2004.

[152] A. Hasanov. Simultaneous determination of source terms in a linear parabolic problem from the final overdetermination: Weak solution approach. *Journal of Mathematical Analysis and Applications*, 330:766–779, 2007.

[153] A. Hasanov. Identification of an unknown source term in a vibrating cantilevered beam from final overdetermination. *Inverse Problems*, 25:115015 (19pp), 2009.

[154] A. Hasanov. Simultaneous determination of source terms in a linear hyperbolic problem from the final overdetermination: Weak solution approach. *IMA Journal of Applied Mathematics*, 74:1–19, 2009.

[155] A. Hasanov. Identification of spacewise and time dependent source terms in 1D heat conduction equation from temperature measurement at a final time. *International Journal of Heat and Mass Transfer*, 55:2069–2080, 2012.

[156] A. Hasanov and O. Baysal. Identification of an unknown spatial load distribution in a vibrating cantilevered beam from final overdetermination. *Journal of Inverse and Ill-Posed Problems*, 23:85–102, 2015.

[157] A. Hasanov and B. Pektas. Identification of an unknown time-dependent heat source term from overspecified Dirichlet boundary data by conjugate gradient method. *Computers and Mathematics with Applications*, 65:42–57, 2013.

[158] A. Hasanov and V.G. Romanov. *Introduction to Inverse Problems for Differential Equations*, Springer International Publishing AG, 2017.

[159] A. Hasanov and Z. Seyidmamedov. Determination of leading coefficients in Sturm-Liouville operator from boundary measurements, ii. Unicity and an engineering approach. *Applied Mathematics and Computation*, 125:23–34, 2002.

[160] A. Hasanov and T.S. Shores. Solution of an inverse coefficient problem for an ordinary differential equation. *Applicable Analysis*, 67:11–20, 1997.

[161] A. Hazanee, M.I. Ismailov, D. Lesnic, and N.B. Kerimov. An inverse time-dependent source problem for the heat equation. *Applied Numerical Mathematics*, 69:13–33, 2013.

[162] A. Hazanee and D. Lesnic. Determination of a time-dependent heat source from nonlocal boundary conditions. *Engineering Analysis with Boundary Elements*, 37:936–956, 2013.

[163] A. Hazanee and D. Lesnic. Reconstruction of an additive space- and time-dependent heat source. *European Journal of Computational Mechanics*, 22:304–329, 2013.

[164] A. Hazanee and D. Lesnic. Reconstruction of multiplicative space- and time-dependent heat sources. *Inverse Problems in Science and Engineering*, 24:1528–1549, 2016.

[165] A. Hazanee, D. Lesnic, M.I. Ismailov, and N.B. Kerimov. An inverse time-dependent source problem for the heat equation with a non-classical boundary condition. *Applied Mathematical Modelling*, 39:6258–6272, 2015.

[166] A. Hazanee, D. Lesnic, M.I. Ismailov, and N.B. Kerimov. Inverse time-dependent source problems for the heat equation with nonlocal boundary conditions. *Applied Mathematics and Computation*, 346:800–815, 2019.

[167] R.G. Hills and E. Hensel. One-dimensional nonlinear inverse heat conduction technique. *Numerical Heat Transfer*, 10:369–393, 1986.

[168] D.H.G. Hinestroza, G. Peralta, and E.O. Luis. Regularization algorithm within two parameters for the identification of the heat conduction coefficient in the parabolic equation. *Mathematical and Computer Modelling*, 57:1990–1998, 2013.

[169] H. Hochstedt. The inverse Sturm-Liouville problem. *Communications on Pure and Applied Mathematics*, 26:715–729, 1973.

[170] B. Hofmann and U. Tautenhahn. On ill-posedness measures and space change in Sobolev scales. *Journal of Analysis and its Applications*, 16:979–1000, 1997.

[171] R.V. Homsey. *Two-dimensional Transient Dispersion and Absorption in Porous Media.* UCRL-81970 Rev.1, Lawrence Livermore Laboratory, 1979.

[172] X. Hu, X. Xu, and W. Chen. Numerical method for the inverse heat transfer problem in composite materials with Stefan-Boltzmann conditions. *Advances in Computational Mathematics*, 33:471–489, 2010.

[173] C.H. Huang and M.N. Ozisik. Direct integration approach for simultaneously estimating temperature dependent thermal conductivity and heat capacity. *Numerical Heat Transfer*, 20:95–110, 1991.

[174] C.H. Huang, M.N. Ozisik, and B. Sawaf. Conjugate gradient method for determining unknown contact conductance during metal casting. *International Journal of Heat and Mass Transfer*, 35:1779–1786, 1992.

[175] C.H. Huang, D.M. Wang, and H.M. Chen. Prediction of local thermal contact conductance in plate thinned-tube heat exchangers. *Inverse Problems in Engineering*, 7:119–141, 1999.

[176] L. Huang, X. Sun, Y. Liu, and Z. Cen. Parameter identification for two-dimensional orthotropic material bodies by the boundary element method. *Engineering Analysis with Boundary Elements*, 28:109–121, 2004.

[177] M.S. Hussein, N. Kinash, D. Lesnic, and M. Ivanchov. Retrieving the time-dependent thermal conductivity of an orthotropic rectangular conductor. *Applicable Analysis*, 96:1–15, 2016.

[178] M.S. Hussein and D. Lesnic. Determination of a time-dependent thermal diffusivity and free boundary in heat conduction. *International Communications in Heat and Mass Transfer*, 53:154–163, 2014.

[179] M.S. Hussein and D. Lesnic. Identification of the time-dependent conductivity of an inhomogeneous diffusive material. *Applied Mathematics and Computation*, 269:35–58, 2015.

[180] M.S. Hussein, D. Lesnic, and M.I. Ismailov. An inverse problem of finding the time-dependent diffusion coefficient from an integral condition. *Mathematical Methods in the Applied Sciences*, 39:963–980, 2016.

[181] M.S. Hussein, D. Lesnic, and M. Ivanchov. Identification of a heterogeneous orthotropic conductivity in a rectangular domain. *International Journal of Novel Ideas*, 1:1–11, 2017.

[182] M.S. Hussein, D. Lesnic, M. Ivanchov, and N. Kinash. Reconstruction of an orthotropic thermal conductivity from nonlocal heat flux measurements. *International Journal of Mathematical Modelling and Numerical Optimisation*, 10:102–122, 2020.

[183] M.S. Hussein, D. Lesnic, M.I. Ivanchov, and H.A. Snitko. Multiple time-dependent coefficient identification thermal problems with a free boundary. *Applied Numerical Mathematics*, 99:24–50, 2016.

[184] S.O. Hussein and D. Lesnic. Determination of a space-dependent force function in the one-dimensional wave equation. *Electronic Journal of Boundary Elements*, 12:1–26, 2014.

[185] S.O. Hussein and D. Lesnic. Determination of forcing functions in the wave equation. Part I: the space-dependent case. *Journal of Engineering Mathematics*, 96:115–133, 2016.

[186] S.O. Hussein and D. Lesnic. Determination of forcing functions in the wave equation. Part II: the time-dependent case. *Journal of Engineering Mathematics*, 96:135–153, 2016.

[187] M. Ikehata. Size estimation of inclusion. *Journal of Inverse and Ill-Posed Problems*, 6:127–140, 1998.

[188] M. Ikehata. How to draw a picture of an unknown inclusion from boundary measurements. Two mathematical inversion algorithms. *Journal of Inverse and Ill-Posed Problems*, 7:255–271, 1999.

[189] M. Ikehata. On reconstruction in the inverse conductivity problem with one measurement. *Inverse Problems*, 16:785–793, 2000.

[190] M. Ikehata. Reconstruction of inclusion from boundary measurements. *Journal of Inverse and Ill-Posed Problems*, 10:37–65, 2002.

[191] V.A. Il'in. How to express basis conditions and conditions for the equiconvergence with trigonometric series of expansions related to non-self-adjoint differential operators. *Computers & Mathematics with Applications*, 34:641–647, 1997.

[192] D.B. Ingham and M.A. Kelmanson. *Boundary Integral Analysis of Singular, Potential and Biharmonic Problems.* Springer-Verlag, Berlin, 1984.

[193] D.B. Ingham and Y. Yuan. *The Boundary Element Method for Solving Improperly-Posed Problems.* Computational Mechanics Publications, Southampton, 1994.

[194] N.I. Ionkin. Solution of a boundary-value problem in heat conduction with a non-classical boundary condition. *Differential Equations*, 13:204–211, 1977.

[195] S. Irmay. Piezometric determination of inhomogenoeus hydraulic conductivity. *Water Resources Research*, 16:691–694, 1980.

[196] V. Isakov. On uniqueness of recovery of discontinuous conductivity coefficient. *Communications on Pure and Applied Mathematics*, 41:865–877, 1988.

[197] V. Isakov. *Inverse Problems for Partial Differential Equations.* 3rd edn., Springer International Publishing, New York, 2017.

[198] M.I. Ismailov and F. Kanca. An inverse coefficient problem for a parabolic equation in the case of nonlocal boundary and overdetermination conditions. *Mathematical Methods in the Applied Sciences*, 34:692–702, 2011.

[199] M.I. Ismailov and F. Kanca. The inverse problem of finding the time-dependent diffusion coefficient of the heat equation from integral overdetermination data. *Inverse Problems in Science and Engineering*, 20:463–476, 2012.

[200] M.I. Ismailov, F. Kanca, and D. Lesnic. Determination of a time-dependent heat source under nonlocal boundary and integral overdetermination conditions. *Applied Mathematics and Computation*, 218:4138–4146, 2011.

[201] K. Ito and J.-C. Liu. Recovery of inclusions in 2D and 3D domains for Poisson's equation. *Inverse Problems*, 29:075005 (20pp), 2013.

[202] K. Ito and S. Nakagiri. Identifiability of stiffness and damping coefficients in Euler-Bernoulli beam equations with Kelvin-Voigt damping. *Numerical Functional Analysis and Optimization*, 18:107–129, 1997.

[203] M.I. Ivanchov. Some inverse problems for the heat equation with nonlocal boundary conditions. *Ukrainian Mathematics Journal*, 45:1186–1192, 1994.

[204] M.I. Ivanchov. On the determination of unknown source in the heat equation with nonlocal boundary conditions. *Ukrainian Mathematics Journal*, 47:1647–1652, 1995.

[205] M.I. Ivanchov. Inverse problem for a multidimensional heat equation with an unknown source function. *Matematychni Studii*, 16:93–98, 2001.

[206] M.I. Ivanchov. Inverse problem with free boundary for heat equation. *Ukrainian Mathematical Journal*, 55:1086–1098, 2003.

[207] M.I. Ivanchov. *Inverse Problems for Equations of Parabolic Type*. VNTL Publishers, Lviv, 2003.

[208] M.I. Ivanchov and N.E. Kinash. Inverse problem for the heat-conduction equation in a rectangular domain. *Ukrainian Mathematical Journal*, 69:1865–1876, 2018.

[209] V. Janicki, J. Sancho-Parramon, and H. Zorc. Refractive index profile modelling of dielectric inhomogeneous coatings using effective medium theories. *Thin Solid Films*, 516:3368–3373, 2008.

[210] M.A. Jaswon and G.T. Symm. *Integral Equation Methods in Potential Theory and Elastostatics*. Academic Press, London, 1977.

[211] B. Jin and X. Lu. Numerical identification of a Robin coefficient in parabolic problems. *Mathematics of Computation*, 81:1369–1398, 2006.

[212] B. Jin and J. Zou. Numerical estimation of the Robin coefficient in a stationary diffusion equation. *IMA Journal of Numerical Analysis*, 30:677–701, 2010.

[213] B.T. Johansson, D. Lesnic, and T. Reeve. A method of fundamental solutions for the one-dimensional inverse Stefan problem. *Applied Mathematical Modelling*, 35:4367–4378, 2011.

[214] T. Johansson. An iterative procedure for solving a Cauchy problem for second order elliptic equations. *Mathematische Nachrichten*, 272:46–54, 2004.

[215] T. Johansson and D. Lesnic. Reconstruction of a stationary flow from incomplete boundary data using iterative methods. *European Journal of Applied Mathematics*, 17:651–663, 2006.

[216] T. Johansson and D. Lesnic. Determination of a spacewise dependent heat source. *Journal of Computational and Applied Mathematics*, 209:66–80, 2007.

[217] T. Johansson and D. Lesnic. An iterative method for the reconstruction of a stationary flow. *Numerical Methods for Partial Differential Equations*, 23:998–1017, 2007.

[218] T. Johansson and D. Lesnic. A variational method for identifying a spacewise-dependent heat source. *IMA Journal of Applied Mathematics*, 72:748–760, 2007.

[219] T. Johansson and D. Lesnic. A procedure for determining a spacewise dependent heat source and the initial temperature. *Applicable Analysis*, 87:265–276, 2008.

[220] B.F. Jones Jr. The determination of a coefficient in a parabolic differential equation, Part I. Existence and uniqueness. *Journal of Mathematics and Mechanics*, 11:907–918, 1962.

[221] B.F. Jones Jr. Various methods for finding unknown coefficients in parabolic differential equations. *Communications on Pure and Applied Mathematics*, 16:33–44, 1963.

[222] M. Jourhmane, D. Lesnic, and N.S. Mera. Relaxation procedures for an iterative algorithm for solving the Cauchy problem for the Laplace equation. *Engineering Analysis with Boundary Elements*, 28:655–665, 2004.

[223] M. Jourhmane and A. Nachaoui. An alternating method for an inverse Cauchy problem. *Numerical Algorithms*, 21:247–260, 1999.

[224] S.I. Kabanikhin. *Inverse and Ill-posed Problems.* De Gruyter, Berlin, 2011.

[225] H. Kang, J.K. Seo, and D. Sheen. The inverse conductivity problem with one measurement: stability and estimation of size. *SIAM Journal on Mathematical Analysis*, 28:1389–1405, 1997.

[226] A. Karageorghis and D. Lesnic. The method of fundamental solutions for the inverse conductivity problem. *Inverse Problems in Science and Engineering*, 18:567–583, 2010.

[227] A. Karageorghis and D. Lesnic. Reconstruction of an elliptical inclusion in the inverse conductivity problem. *International Journal of Mechanical Sciences*, 142-143:603–609, 2018.

[228] N.B. Kerimov and V.S. Mirzoev. On the basis properties of one spectral problem with a spectral parameter in a boundary condition. *Siberian Mathematical Journal*, 44:813–816, 2003.

[229] J.T. King. The minimal error conjugate gradient method for ill-posed problems. *Journal of Optimization Theory and Applications*, 60:297–304, 1989.

[230] A. Kirsch. *An Introduction to the Mathematical Theory of Inverse Problems*, 3rd edition, Springer Nature AG, 2021.

[231] A. Kitamura and S. Nakagiri. Identifiability of spatially-varying and constant parameters in distributed systems of parabolic type. *SIAM Journal on Control Optimization*, 15:785–802, 1977.

[232] M. Klibanov. Inverse problems and Carleman estimates. *Inverse Problems*, 8:575–596, 1992.

[233] M. Klibanov. Numerical solution of a time-like Cauchy problem for the wave equation. *Mathematical Methods in the Applied Sciences*, 15:559–570, 1992.

[234] J.H. Knight and J.R. Philip. Exact solutions in nonlinear diffusion. *Journal of Engineering Mathematics*, 8:219–227, 1974.

[235] I. Knowles. A variational algorithm for electrical impedance tomography. *Inverse Problems*, 14:1513–1525, 1998.

[236] I. Knowles. Parameter identification for elliptic problems. *Journal of Computational and Applied Mathematics*, 31:175–194, 2001.

[237] I. Knowles and A. Yan. The recovery of an anisotropic conductivity in groundwater modelling. *Applicable Analysis*, 81:1347–1365, 2002.

[238] R. Kohn and M. Vogelius. Determining conductivity by boundary measurments. II Interior results. *Communications on Pure and Applied Mathematics*, 38:643–667, 1985.

[239] V.A. Kozlov and V.G. Maz'ya. On iterative procedures for solving ill-posed boundary value problems for elliptic equations. *Leningrad Mathematical Journal*, 1:1207–1228, 1991.

[240] V.A. Kozlov, V.G. Maz'ya, and D.V. Fomin. An iterative method for solving the Cauchy problem for elliptic equation. *USSR Computers in Mathematics and Mathematical Physics*, 31:45–52, 1991.

[241] V.A. Kozlov, V.G. Maz'ya, and D.V. Fomin. The inverse problem of coupled thermo-elasticity. *Inverse Problems*, 10:153–160, 1994.

[242] O.A. Ladyzhenskaya. *The Mathematical Theory of Viscous Incompressible Flow.* Gordon and Breach, New York, 1963.

[243] O.A. Ladyzhenskaya. *The Boundary Value Problems of Mathematical Physics.* Springer-Verlag, Berlin, 1985.

[244] O.A. Ladyzhenskaya, V.V. Solonnikov, and N.N. Ural'tseva. *Linear and Quasi-linear Equations of Parabolic Type.* American Mathematical Society, Providence, 1968.

[245] R.E. Langer. A problem in diffusion or in the flow of heat for a solid in contact with a fluid. *Tohoku Mathematical Journal*, 35:360–375, 1932.

[246] D. Langford. New analytic solutions of the one-dimensional heat equation for temperature and heat flow both prescribed at the same fixed boundary (with applications to the phase change problem). *Quarterly of Applied Mathematics*, 24:315–322, 1966.

[247] R. Lattes and J.-L. Lions. *Methode de Quasi-reversibilite et Applications.* Dunod, Paris, 1967.

[248] M.M. Lavrentiev, V.G. Romanov, and S.P. Shishatskii. *Ill-Posed Problems of Mathematical Physics and Analysis.* American Mathematical Society, Providence, 1986.

[249] G. Le Meur and B. Bourouga. Electrothermal static contact: Inverse analysis and experimental approach. *Inverse Problems in Science and Engineering*, 17:977–998, 2009.

[250] A. Leitao. An iterative method for solving elliptic Cauchy problems. *Numerical Functional Analysis and Optimization*, 21:715–742, 2000.

[251] D. Lesnic. Identifiability of distributed parameters in beam-type systems. *Journal of Inverse and Ill-Posed Problems*, 8:379–398, 2000.

[252] D. Lesnic. The determination of the thermal properties of a heat conductor in a nonlinear heat conduction problem. *ZAMP*, 53:175–196, 2002.

[253] D. Lesnic. Inversion of discontinuous anisotropic conductivities. In *Recent Developments in Theories and Numerics: International Conference on Inverse Problems*, pages 123–133. (eds. Y-C. Hon, M. Yamamoto, J. Cheng and J-Y. Lee), World Scientific, Singapore, 2003.

[254] D. Lesnic. A numerical investigation which proves the existence or non-existence of a solution for the Cauchy problem for elliptic equations. In *Fourth UK Conference on Boundary Integral Methods*, pages 137–146. (Ed. S. Amini), Salford University, UK, 2003.

[255] D. Lesnic. The determination of the unknown thermal properties of homogeneous heat conductors. *International Journal of Computational Methods*, 1:431–443, 2004.

[256] D. Lesnic. The decomposition method for Cauchy advection-diffusion problems. *Computers & Mathematics with Applications*, 49:525–537, 2005.

[257] D. Lesnic. The determination of a permeability function from core measurements and pressure data. In *Inverse Problems, Design and Optimization Symposium*, pages 143–150. (Eds. M.J. Colaco, H.R.B. Orlando and G.S. Dulikravich), E-papers, Rio de Janeiro, Brazil, 2005.

[258] D. Lesnic. Determination of the flexural rigidity of a beam from limited boundary measurements. *Journal of Applied Mathematics & Computing*, 20:17–34, 2006.

[259] D. Lesnic. The comparison model method for determining the flexural rigidity of a beam. *Journal of Inverse and Ill-Posed Problems*, 18:577–590, 2010.

[260] D. Lesnic and L. Elliott. The decomposition approach to inverse heat conduction. *Journal of Mathematical Analysis and Applications*, 232:82–98, 1999.

[261] D. Lesnic, L. Elliott, and D.B. Ingham. Identification of the thermal conductivity and heat capacity in unsteady nonlinear heat conduction problems using the boundary element method. *Journal of Computational Physics*, 126:410–420, 1995.

[262] D. Lesnic, L. Elliott, and D.B. Ingham. A note on the determination of the thermal properties of a material in a transient nonlinear heat conduction problem. *International Communications in Heat and Mass Transfer*, 22:475–482, 1995.

[263] D. Lesnic, L. Elliott, and D.B. Ingham. Application of the boundary element method to inverse heat conduction problems. *International Journal of Heat and Mass Transfer*, 5:1503–1517, 1996.

[264] D. Lesnic, L. Elliott, and D.B. Ingham. An alternating boundary element method for solving Cauchy problems for the biharmonic equation. *Inverse Problems in Engineering*, 5:145–168, 1997.

[265] D. Lesnic, L. Elliott, and D.B. Ingham. The boundary element solution of the Laplace and biharmonic equations subjected to noisy boundary data. *International Journal for Numerical Methods in Engineering*, 43:479–492, 1998.

[266] D. Lesnic, L. Elliott, and D.B. Ingham. An iterative boundary element method for solving the backward heat conduction problem using an elliptic approximation. *Inverse Problems in Engineering*, 6:255–279, 1998.

[267] D. Lesnic, L. Elliott, and D.B. Ingham. The solution of an inverse heat conduction problem subject to the specification of energies. *International Journal of Heat and Mass Transfer*, 74:25–32, 1998.

[268] D. Lesnic, L. Elliott, and D.B. Ingham. Analysis of coefficient identification problems associated to the Euler-Bernoulli beam theory. *IMA Journal of Applied Mathematics*, 62:101–116, 1999.

[269] D. Lesnic, L. Elliott, and D.B. Ingham. Identifiability of distributed parameters for high-order quasi-linear differential equations. *Journal of Inverse and Ill-Posed Problems*, 8:1–22, 2000.

[270] D. Lesnic, L. Elliott, D.B. Ingham, B. Clennell, and R.J. Knipe. A mathematical model and numerical investigation for determining the hydraulic conductivity of rocks. *International Journal of Rock Mechanics and Mining Sciences*, 34:741–759, 1997.

[271] D. Lesnic, L. Elliott, D.B. Ingham, and A. Zeb. The boundary element method regularized solution for the biharmonic equation with underspecified boundary conditions and interior measurements. In *Second UK Conference on Boundary Integral Methods*, pages 245–256. Brunel University Press, 1999.

[272] D. Lesnic, L. Elliott, D.B. Ingham, and A. Zeb. A numerical method for an inverse biharmonic problem. *Inverse Problems in Engineering*, 7:145–168, 1999.

[273] D. Lesnic and A. Hasanov. Determination of the leading coefficient in fourth-order Sturm-Liouville operator from boundary measurements. *Inverse Problems in Science and Engineering*, 16:413–424, 2008.

[274] D. Lesnic, S.O. Hussein, and B.T. Johansson. Inverse space-dependent force problems for the wave equation. *Journal of Computational and Applied Mathematics*, 306:10–39, 2016.

[275] D. Lesnic and G.C. Wake. A mollified method for the solution of the Cauchy problem for the convection-diffusion equation. *Inverse Problems in Science and Engineering*, 15:293–302, 2007.

[276] D. Lesnic, G. Wakefield, B.D. Sleeman, and J.R. Ockendon. Determination of the index of refraction of anti-reflection coatings. *Maths-in-Industry Case Studies Journal*, 2:155–173, 2010.

[277] D. Lesnic, S.A. Yousefi, and M. Ivanchov. Determination of a time-dependent diffusivity form nonlocal conditions. *Journal of Applied Mathematics and Computation*, 41:301–320, 2013.

[278] W. Liao, M. Deghan, and A. Mohebbi. Direct numerical method for an inverse problem of a parabolic partial differential equation. *Journal of Computational and Applied Mathematics*, 232:351–360, 2009.

[279] W.R.B. Lionheart. Conformal uniqueness results in anisotropic electrial impedance imaging. *Inverse Problems*, 13:125–134, 1997.

[280] J.-L. Lions and E. Magenes. *Non-homogeneous Boundary Value Problems and Applications*. Vol. 1, Springer, Berlin, 1972.

[281] C.-S. Liu, C.-L. Kuo, and F. Wang. Recovering both the space-dependent heat source and the initial temperature by using a fast convergent iterative method. *Numerical Heat Transfer, Part B: Fundamentals*, 72:233–249, 2017.

[282] J.E. Losch. *Tables of Higher Functions.* 6th edn., McGraw-Hill, New York, 1960.

[283] T. Loulou and E.P. Scott. An inverse heat conduction problem with heat flux measurements. *International Journal for Numerical Methods in Engineering*, 67:1587–1616, 2006.

[284] B. Lowe and W. Rundell. An inverse problem for a Sturm-Liouville operator. *Journal of Mathematical Analysis and Applications*, 181:188–199, 1994.

[285] J. Lundgren. *Recontruction of Stresses in Plates by Incomplete Cauchy Data.* Theses No. 960, Linkoping Studies in Science and Technology, Sweden, 2002.

[286] A. Mahdjoub and L. Zighed. New designs for graded refractive index antireflection coatings. *Thin Solid Films*, 478:299–304, 2005.

[287] M.S. Mahmood and D. Lesnic. Identification of thermal conductivity in inhomogeneous orthotropic media. *International Journal of Numerical Methods for Heat and Fluid Flow*, 29:165–183, 2019.

[288] P. Manselli and K. Miller. Calculation of the surface temperature and heat flux on one side of a wall from measurements on the opposite side. *Annali di Matematica Pura ed Applicata*, 123:161–183, 1980.

[289] L. Marin, L. Elliott, P.J. Heggs, D.B. Ingham, D. Lesnic, and X. Wen. An alternating iterative algorithm for the Cauchy problem associated to the Helmholtz equation. *Computer Methods in Applied Mechanics and Engineering*, 192:709–722, 2003.

[290] L. Marin, L. Elliott, D.B. Ingham, and D. Lesnic. Boundary element method for the Cauchy problem in linear elasticity. *Engineering Analysis with Boundary Elements*, 25:783–793, 2001.

[291] L. Marin, A. Karageorghis, and D. Lesnic. The method of fundamental solutions for problems in static thermo-elasticity with incomplete boundary data. *Inverse Problems in Science and Engineering*, 25:652–673, 2017.

[292] T.T. Marinov and A.S. Vatsala. Inverse problem for coefficient identification in the Euler-Bernoulli equation. *Computers & Mathematics with Applications*, 56:400–410, 2008.

[293] D. Maxwell. Kozlov-Maz'ya iteration as a form of Landweber iteration. *Inverse Problems and Imaging*, 8:537–560, 2014.

[294] J.R. McLaughlin. On constructing solutions to an inverse Euler-Bernoulli problem. In *Inverse Problems of Acoustic and Elastic Waves*, pages 341–347. SIAM, Philadelphia, 1984.

[295] I.V. Mel'nikova. Regularization of ill-posed differential problems. *Siberian Mathematical Journal*, 33:289–298, 1992.

[296] I.V. Mel'nikova and A. Filinkov. *Abstract Cauchy Problems: Three Approaches.* Chapman & Hall/CRC, Boca Raton, FL, 2001.

[297] N.S. Mera, L. Elliott, D.B. Ingham, and D. Lesnic. The boundary element solution of the Cauchy steady heat conduction problem in an anisotropic medium. *International Journal for Numerical Methods in Engineering*, 49:481–499, 2000.

[298] N.S. Mera, L. Elliott, D.B. Ingham, and D. Lesnic. An iterative boundary element method for solving the one-dimensional backward heat conduction problem. *International Journal of Heat and Mass Transfer*, 44:1937–1946, 2001.

[299] N.S. Mera and D. Lesnic. A boundary element method for the numerical inversion of discontinuous anisotropic conductivities. *Engineering Analysis with Boundary Elements*, 27:1937–1946, 2003.

[300] K. Miller. Least squares methods for ill-posed problems with a prescribed bound. *SIAM Journal on Mathematical Analysis*, 1:52–74, 1970.

[301] K. Miller. Stabilized quasi-reversibility and other nearly-best-possible methods for non-well-posed problems. In *Symposium on Non-well-posed Problems and Logarithmic Convexity*, pages 161–176. Springer Lecture Notes No.316, 1973.

[302] T. Miyauchi, H. Imai, and H. Kubo. Longitudinal dispersion of liquid in sieve trays. *Chemical Engineering Japan*, 5:20–24, 1962.

[303] B.B. Mkic. Thermal contact conductance: Theoretical considerations. *International Journal of Heat and Mass Transfer*, 17:205–214, 1974.

[304] V.A. Morozov. On the solution of functional equations by the method of regularization. *Soviet Mathematics Doklady*, 7:414–417, 1966.

[305] V.A. Morozov. *Methods for Solving Incorrectly Posed Problems.* Springer-Verlag, New York, 1984.

[306] K.W. Morton and D.F. Mayers. *Numerical Solution of Partial Differential Equations: An Introduction.* Cambridge University Press, Cambridge, 2005.

[307] L. Mpinganzima. *Iterative Methods for Solving the Cauchy Problem for the Helmholtz Equation.* Linkoping Studies in Science and Technolgy, No. 1593, 2014.

[308] L.A. Muravei and A.V. Filinovskii. On a problem with nonlocal boundary condition for a parabolic equation. *Math. USSR Sbornik*, 74:219–249, 1993.

[309] D.A. Murio. The mollification method and the numerical solution of an inverse heat conduction problem. *SIAM Journal on Scientific and Statistical Computing*, 2:17–34, 1981.

[310] D.A. Murio. Numerical identification of interface source functions in heat transfer problems under nonlinear boundary conditions. *Computers & Mathematics with Applications*, 24:65–76, 1992.

[311] D.A. Murio. *The Mollification Method and the Numerical Solution of Ill-Posed Problems*. John Wiley, New York, 1993.

[312] R. Mustata, S.D. Harris, L. Elliott, D. Lesnic, D.B. Ingham, R.A. Khachfe, and Y. Jarny. The determination of the properties of orthotropic heat conductors. In *Inverse Problems and Experiemental Design in Thermal and Mechanical Engineering*, pages 325–332. (Eds. D. Petit, D.B. Ingham, Y. Jarny and F. Plourde), Proceedings of the Eurotherm Seminar 68, March 5-7, 2001, Poitiers, France, 2001.

[313] A. Nachman. Global uniqueness for a two-dimensional inverse boundary value problem. *Annals of Mathematics*, 143:71–96, 1996.

[314] M.A. Naimark. *Linear Differential Operators: Elementary Theory of Linear Differential Operators.* Frederick Ungar Publishing Co., New York, 1967.

[315] G. Nakamura, S. Wang, and Y. Wang. Numerical differentiation for the second order derivatives of functions of two variables. *Journal of Computational and Applied Mathematics*, 212:341–358, 2008.

[316] A.S. Nemirovskii. The regularizing properties of the adjoint gradient method in ill-posed problems. *USSR Computational Mathematics and Mathematical Physics*, 2:7–16, 1986.

[317] J.W. Neuberger. *Sobolev Gradient and Differential Equations.* Springer-Verlag, Berlin, 1997.

[318] Y. Nievergelt. Solution to an inverse problem in diffusion. *SIAM Review*, 40:74–80, 1998.

[319] J. Nocedal and S.J. Wright. *Numerical Optimization.* Springer, New York, 2006.

[320] P. Nubile. Analytical design of antireflection coatings for silicon photovoltaic devices. *Thin Solid Films*, 342:257–261, 1999.

[321] A.O. Olatunji-Ojo, S.K.S. Boetcher, and T.R. Cundari. Thermal conduction analysis of layered functionally graded materials. *Computational Materials Science*, 54:329–335, 2012.

[322] M.N. Ozisik and H.R.B. Orlande. *Inverse Heat Transfer: Fundamentals and Applications.* Taylor & Francis, New York, 2000.

[323] R.S. Padilha, M.J. Colaco, H.R.B. Orlande, and L.A.S. Abreu. An analytical method to estimate spatially-varying thermal contact conductances using the reciprocity functional and the integral transform methods: Theory and experimental validation. *International Journal of Heat and Mass Transfer*, 100:599–607, 2016.

[324] V.G. Papanicolau. The spectral theory of the vibrating periodic beam. *Communications in Mathematical Physics*, 170:359–373, 1995.

[325] V.G. Papanicolau and D. Kravvaritis. An inverse spectral problem for the Euler-Bernoulli equation for the vibrating beam. *Inverse Problems*, 13:1083–1092, 1997.

[326] L. Payne. *Improperly Posed Problems in Partial Differential Equations.* SIAM, Philadelphia, 1975.

[327] D.W. Peaceman and H.H. Rachford Jr. The numerical solution of parabolic and elliptic differential equations. *Journal of the Society for Industrial and Applied Mathematics*, 3:28–41, 1955.

[328] Y. P. Petrov and V. S. Sizikov. *Well-posed, Ill-posed and Intermediate Problems.* VSP, The Netherlands, 2005.

[329] D.L. Philips. A technique for the numerical solution of certain integral equations of the first kind. *Journal of the Association for Computer Machinery*, 9:84–97, 1962.

[330] M.S. Pilant and W. Rundell. Undetermined coefficient problems for quasilinear parabolic equations. In *Inverse Problems in Partial Differential Equations*, pages 165–185. SIAM, Philadelphia, 1990.

[331] R. Plato. Optimal algorithms for linear ill-posed problems yield regularization methods. *Numerical Functional Analysis and Optimization*, 11:111–118, 1990.

[332] W.J. Plieth and H. Bruckner. Experimental determination of the complex fresnel reflection coefficient of a three-phase system by combination of modulated reflectance spectroscopy and modulated interferometry. *Surface Science*, 66:357–360, 1977.

[333] G. Ponzini and G. Crosta. The comparison model method: A new arithmetic approach to the discrete inverse problem of groundwater hydrology. *Transport in Porous Media*, 3:415–436, 1988.

[334] G. Ponzini, G. Crosta, and M. Giudici. Identification of thermal conductivities by temperature gradient profiles: One-dimensional steady flow. *Geophysics*, 54:643–653, 1989.

[335] G. Ponzini and A. Lozej. Identification of aquifer transmissivities: The comparison model method. *Water Resources Research*, 18:597–622, 1982.

[336] A.I. Prilepko, D.G. Orlovsky, and I.A. Vasin. *Methods for Solving Inverse Problems in Mathematical Physics.* M. Dekker, New York, 2000.

[337] A.I. Prilepko and V.V. Solov'ev. Solvability theorems and Rothe's method for inverse problems for a parabolic equation. I. *Differential Equations*, 23:1230–1237, 1988.

[338] A.I. Prilepko and V.V. Solov'ev. Solvability theorems and Rothe's method for inverse problems for a parabolic equation. II. *Differential Equations*, 23:1341–1349, 1988.

[339] A.I. Prilepko and D.S. Thachenko. Inverse problem for a parabolic equation with integral overdetermination. *Journal of Inverse and Ill-Posed Problems*, 11:191–218, 2003.

[340] Z. Ranjbar and L. Elden. Numerical analysis of an ill-posed Cauchy problem for a convection-diffusion equation. *Inverse Problems in Science and Engineering*, 15:191–211, 2007.

[341] J.N. Reddy. *An Introduction to the Finite Element Method.* McGraw-Hill, New York, 1993.

[342] R. Reemtsen and A. Kirsch. A method for the numerical solution of the one-dimensional inverse Stefan problem. *Numerische Mathematiche*, 45:253–273, 1984.

[343] T. Reginska and L. Elden. Solving the sideways heat equation by a wavelet-Galerkin method. *Inverse Problems*, 13:1093–1106, 1997.

[344] H.-J. Reinhardt and Dinh Nho Hao. A sequential conjugate gradient method for linear inverse heat conduction problems. *Inverse Problems in Engineering*, 2:263–272, 1996.

[345] R.J. Renka. Nonlinear least squares and Sobolev gradients. *Applied Numerical Mathematics*, 65:91–104, 2013.

[346] W.B. Richardson. Sobolev preconditioning for the Poisson-Boltzmann equation. *Computer Methods in Applied Mechanics and Engineering*, 181:425–436, 2000.

[347] G.R. Richter. An inverse problem for the steady state diffusion equation. *SIAM Journal on Applied Mathematics*, 41:210–221, 1981.

[348] D. Ristau, H. Schink, F. Mittendorf, S.M.J. Akhtar, J. Ebert, and H. Welling. Laser induced damage of dielectric systems with gradual interfaces at $1.064\mu m$. *NIST Special Publications*, 775:414–426, 1988.

[349] C. Rogers and P. Broadbridge. On a nonlinear moving boundary problem with heterogeneity: application of a reciprocal transformation. *ZAMP*, 39:122–128, 1988.

[350] W. Rundell. Determination of an unknown non-homogeneous term in a linear partial differential equation from overspecified boundary data. *Applicable Analysis*, 10:231–242, 1980.

[351] S. Rydstrom. *Regularization of Parameter Problems for Dynamic Beam Models.* PhD Thesis, Linnaeus University, Sweden, 2010.

[352] P. Sacks. An inverse problem in coupled mode theory. *Journal of Mathematical Physics*, 45:1699–1710, 2004.

[353] A.A. Samarskii. Some problems in differential equations theory. *Differential Equations*, 16:1221–1228, 1980.

[354] A.A. Samarskii. *The Theory of Difference Schemes.* Marcel Dekker, New York, 2001.

[355] E.G. Savateev. On problems of determining the source function in a parabolic equation. *Journal of Inverse and Ill-Posed Problems*, 3:83–102, 1995.

[356] B. Sawaf, M.N. Ozisik, and Y. Jarny. An inverse analysis to estimate linearly temperature dependent thermal conductivity components and heat capacity of an orthotropic medium. *International Journal of Heat and Mass Transfer*, 38:3005–3010, 1995.

[357] S. Scarascia and G. Ponzini. An approximate solution of the inverse problem in hydraulics. *L'Energia Electtrica*, 49:518–531, 1972.

[358] T.I. Seidman and L. Elden. An 'optimal filtering' method for the sideways heat equation. *Inverse Problems*, 6:681–696, 1990.

[359] R.P. Shaw. Steady-state heat conduction with a separable position and temperature dependent conductivity. *Boundary Element Communications*, 10:15–18, 1999.

[360] B. Sheldon, J.S. Haggerty, and A.G. Emslie. Exact computation of the reflectance of a surface layer of arbitrary refractive-index profile and an approximate solution of the inverse problem. *Journal of the Optical Society of America*, 72:1049–1055, 1982.

[361] C. Shi, C. Wang, and T. Wei. Numerical solution for an inverse heat source problem by an iterative method. *Applied Mathematics and Computation*, 244:577–597, 2014.

[362] V. Singhal, P.J. Litke, A.F. Black, and S.V. Garimella. An experimentally validated thermo-mechanical model for the prediction of thermal contact conductance. *International Journal of Heat and Mass Transfer*, 48:5446–5459, 2005.

[363] M. Slodicka. Some uniqueness theorems for inverse spacewise dependent source problems in nonlinear PDEs. *Inverse Problems in Science and Engineering*, 22:2–9, 2014.

[364] M. Slodicka. Determination of a solely time-dependent source in a semilinear parabolic problem by means of boundary measurements. *Journal of Computational and Applied Mathematics*, 289:433–440, 2015.

[365] M. Slodicka. A parabolic inverse source problem with a dynamical boundary condition. *Applied Mathematics and Computation*, 256:529–539, 2015.

[366] M. Slodicka and B.T. Johansson. Uniqueness and counterexamples in some inverse source problems. *Applied Mathematics Letters*, 58:56–61, 2016.

[367] W.R. Smith. Entry and exit conditions for flow reactors. *IMA Journal of Applied Mathematics*, 41:1–20, 1988.

[368] W.R. Smith and G.C. Wake. Mathematical analysis: an inverse problem arising in convective-diffusive flow. *IMA Journal of Applied Mathematics*, 45:225–231, 1990.

[369] W.R. Smith, G.C. Wake, J.E.A. McIntosh, R.P. McIntosh, M. Pettigrew, and R. Kao. Mathematical analysis of perifusion data: models predicting elution concentration. *American Journal of Physiology*, 261:R247–R256, 1991.

[370] H.A. Snitko. Inverse problem for determination of time-dependent coefficients of a parabolic equation in a free-boundary domain. *Journal of Mathematical Science*, 181:350–365, 2012.

[371] V. Solov'ev. Solvability of the inverse problem of finding a source, using overdetermination on the upper base for a parabolic equation. *Differential Equations*, 25:1114–1119, 1990.

[372] M.L. Storm. Heat conduction in simple metals. *Journal of Applied Physics*, 22:940–951, 1951.

[373] Z. Sun and G. Uhlmann. Anisotropic inverse problems in two dimensions. *Inverse Problems*, 19:1001–1010, 2003.

[374] D. Tiba. Sur l'approximation d'un probleme mal pose. *Comptes Rendus Academie des Sciences de Paris*, 320:619–624, 1995.

[375] A.N. Tikhonov and V.Y. Arsenin. *Solution of Ill-posed Problems.* John Wiley, New York, 1977.

[376] Nguyen Huy Tuan, Vo Van Au, D. Lesnic, and Tran Quoc Viet. Regularization of the semilinear sideways heat equation. *IMA Journal of Applied Mathematics*, 84:258–291, 2019.

[377] S. Twomey. On the numerical solution of Fredholm integral equations of the first kind by the inversion of the linear system produced by quadrature. *Journal of the Association for Computing Machinery*, 10:97–101, 1963.

[378] G. Uhlmann. Developments in inverse problems since Calderon's foundational paper. In *Harmonic Analysis and Partial Differential Equations*, pages 295–345. University of Chicago Press, Chicago, IL, USA, 1999.

[379] P.N. Vabishchevich. Nonlocal parabolic problems and the inverse heat-conduction problem. *Differential Equations*, 17:1193–1199, 1981.

[380] P.N. Vabishchevich and A.Yu. Denisenko. Regularization of nonstationary problems for elliptic equations. *Journal of Engineering Physics and Thermophysics*, 65:1195–1199, 1993.

[381] C.R. Vogel. Non-convergence of the L-curve regularization parameter selection method. *Inverse Problems*, 12:535–547, 1996.

[382] P. Wang and K. Zheng. Determination of an unknown coefficient in a nonlinear heat equation. *Journal of Mathematical Analysis and Applications*, 271:525–533, 2002.

[383] C. Weber. Analysis and solution of the ill-posed inverse heat conduction problems. *International Journal of Heat and Mass Transfer*, 24:1783–1792, 1981.

[384] T. Wei. Uniqueness of moving boundary for a heat conduction problem with nonlinear interface conditions. *Applied Mathematics Letters*, 23:600–604, 2010.

[385] T. Wei and J.C. Wang. Simultaneous determination of a space-dependent heat source and the initial data by the MFS. *Engineering Analysis with Boundary Elements*, 36:1848–1855, 2012.

[386] L.M. White. Resolution of regularized output least squares estimation procedures. *Applied Mathematics and Computation*, 81:139–172, 1997.

[387] L.M. White, Y.J. Jin, and D.J. O'Meara. Estimation of discontinuous elliptic coefficients. *Applied Mathematics and Computation*, 81:113–138, 1997.

[388] L.M. White and J. Zhou. Continuity and uniqueness of regularized output least squares optimal estimations. *Journal of Mathematical Analysis and Applications*, 196:53–83, 1995.

[389] J.C.F. Wong. Two-objective optimization strategies using the adjoint method and game theory in inverse natural convection problems. *International Journal for Numerical Methods in Fluids*, 70:1341–1366, 2012.

[390] S. Woodland, A.D. Crocombe, J.W. Chew, and S.J. Mills. A new method for measuring thermal contact conductance - experimental technique and results. *Journal of Engineering for Gas Turbines and Power*, 133:071601, 2011.

[391] L.C. Wrobel. A boundary element solution to Stefan's problem. In *Boundary Elements V*, pages 173–182. (Eds. C.A. Brebbia, T. Futagami and M. Tanaka), Springer-Verlag, Berlin, 1983.

[392] L.C. Wrobel and C.A. Brebbia. The dual reciprocity boundary element method for nonlinear diffusion problems. *Computer Methods in Applied Mechanics and Engineering*, 65:147–164, 1987.

[393] X. Xiong, Y. Yan, and J. Wang. A direct numerical method for solving inverse heat source problems. *Journal of Physics: Conference Series*, 290:012017 (10pp), 2011.

[394] X.-T. Xiong, C.-L. Fu, and H.-F. Li. Fourier regularization method of a sideways heat equation for determining surface heat flux. *Journal of Mathematical Analysis and Applications*, 317:331–348, 2006.

[395] M. Yamamoto. Stability, reconstruction formula and regularization for an inverse source hyperbolic problem by a control method. *Inverse Problems*, 11:481–496, 1995.

[396] M. Yamamoto and J. Zou. Simultaneous reconstruction of the initial temperature and heat radiative coefficient. *Inverse Problems*, 17:1181–1202, 2001.

[397] L. Yan, C.-L. Fu, and F.-F. Dou. A computational method for identifying a spacewise-dependent heat source. *International Journal for Numerical Methods in Biomedical Engineering*, 26:597–608, 2010.

[398] L. Yan, C. L. Fu, and F. L. Yang. The method of fundamental solutions for the inverse heat source problem. *Engineering Analysis with Boundary Elements*, 32:216–222, 2008.

[399] L. Yan, F.-L. Yang, and C.-L. Fu. A meshless method for solving an inverse spacewise-dependent heat source problem. *Journal of Computational Physics*, 228:123–136, 2009.

[400] G. Yang, M. Yamamoto, and J. Cheng. Heat transfer in composite materials with Stefan-Boltzmann interface conditions. *Mathematical Methods in the Applied Sciences*, 31:1297–1314, 2008.

[401] L. Yang, M. Dehghan, J.-N. Yu, and G.-W. Luo. Inverse problem of time-dependent heat sources numerical reconstruction. *Mathematics and Computers in Simulation*, 81:1656–1672, 2011.

[402] L. Yang, Z.-C. Deng, and Y.-C. Hon. Simultaneous identification of unknown initial temperature and heat source. *Dynamic Systems and Applications*, 25:583–602, 2016.

[403] L. Yang, J.-N. Yu, G.-W. Luo, and Z.-C. Deng. Reconstruction of a space and time dependent heat source from finite measurement data. *International Journal of Heat and Mass Transfer*, 55:6573–6581, 2012.

[404] Z. Yi and D. A. Murio. Source term identification in 2-D IHCP. *Computers & Mathematics with Applications*, 47:1517–1533, 2004.

[405] S.A. Yousefi, D. Lesnic, and Z. Barikbin. Satisfier function in Ritz-Galerkin method for the identification of a time-dependent diffusivity. *Journal of Inverse and Ill-Posed Problems*, 20:701–722, 2012.

[406] A. Zeb, L. Elliott, D.B. Ingham, and D. Lesnic. The boundary element method for the solution of Stokes equations in two-dimensional domains. *Engineering Analysis with Boundary Elements*, 22:317–326, 1998.

[407] J. Zhao and M.A. Green. Optimized antireflection coatings for high-efficiency silicon solar cells. *IEEE Transactions on Electron Devices*, 38:1925–1934, 1991.

[408] S. Zhou and M. Cui. New algorithm for determining the leading coefficient in a parabolic equation. *World Academy of Science, Engineering and Technology*, 55:512–516, 2009.

[409] L. Zhuo, D. Lesnic, and S. Meng. Reconstruction of the heat transfer coefficient at the interface of a bi-material. *Inverse Problems in Science and Engineering*, 28:374–401, 2020.

Index

adjoint problem, 20, 59, 67, 150, 177
alternating algorithm, 11
anisotropic conductivity, 230
anti-reflection coatings, 261

backward heat conduction problem (BHCP), 45
biharmonic equation, 26
boundary element method (BEM), 9, 17, 24, 29, 51, 76, 94, 101, 105, 112, 127, 135
boundary integral equation, 9, 17, 24, 28, 50, 64, 76, 94, 112, 127, 135, 192, 235, 254

Cauchy data, 7, 19, 32
Cauchy problem, 7, 15, 18, 32
clamped-clamped beams, 284
clamped-free beams, 286
comparison model method (CMM), 291
condition number, 4, 82, 141
conjugate gradient method (CGM), 22, 62, 166, 175

decomposition method, 35
degree of ill-posedness, 1
diffusion, 183
diffusion column, 34
direct problem, 1
Dirichlet-to-Neumann map, 230, 256
discrepancy principle, 6, 21, 48, 116, 117, 158, 180, 270, 304
dispersion, 188

elution chromatography, 34
Euler-Bernoulli equation, 276

finite-difference method (FDM), 159, 214
flexural rigidity, 276
foundation modulus, 297
Fredholm integral equation, 45, 123
free boundary, 219
fundamental solution, 9, 18, 24, 29, 51, 135, 192, 234

generalized cross validation (GCV), 6, 104, 116, 117
Green's formula, 9, 60, 150, 230
Green's function, 122, 205

Hölder space, 55, 72
heat equation, 32, 71, 73, 109, 173, 199, 211, 219, 236, 245
heat source, 109
heat transfer coefficient, 171
Heaviside function, 51, 135
Holmgren class two, 33, 39

ill-posedness, 1, 45, 174
index of refraction, 263
integral observation, 215
integrated refraction index, 272
inverse boundary-value problems, 7
inverse conductivity problem, 230
inverse heat conduction problem (IHCP), 32
inverse problem, 1
inverse scattering, 262

Kirchhoff transformation, 10, 245, 255

L-curve, 6, 30, 97, 116, 130
L-surface, 117

Lagrange polynomials, 305
Landweber-Fridman method (LFM), 20, 67, 158
Laplace's equation, 7
Laplace-Beltrami equation, 9
least-squares, 4, 22, 59, 115, 129, 142, 154, 156, 175, 217, 240
logarithmic convexity, 47
lsqnonlin, 220, 240

minimal error method (MEM), 23
mollification, 41

nonlinear conductivity, 243

orthotropic conductivity, 236

Poisson equation, 53
preconditioning, 178

quasi-reversibility, 46

reciprocity gap, 13
reflectance, 263
reflection coefficient, 263
regularization, 5, 16, 46, 130
regularization matrix, 5
regularization parameter, 16, 46, 48
regularized output least squares (ROLS), 223
relaxation factor, 11
Ritz-Galerkin method, 206, 224

satisfier function, 206
semi-group approach, 58
sensitivity problem, 176
silicon chips, 215
singular value decomposition (SVD), 5
space-dependent conductivity, 224
space-dependent heat source, 53
Stefan-Boltzman coefficient (SBC), 173
Stokes system, 18
Sturm-Liouville problem, 88, 100, 104, 216
system of linear algebraic equations, 4, 29, 30, 82

thermal contact conductance (TCC), 173
Tikhonov regularization, 5, 29, 72, 85, 97, 106, 116, 130, 140, 148, 264, 288, 304
time-dependent conductivity, 199
time-dependent heat source, 71
truncated SVD, 6, 30, 115
Trust Region Reflective (TRR), 214, 240

Volterra integral equation, 91, 123, 212

wave equation, 133, 144, 149